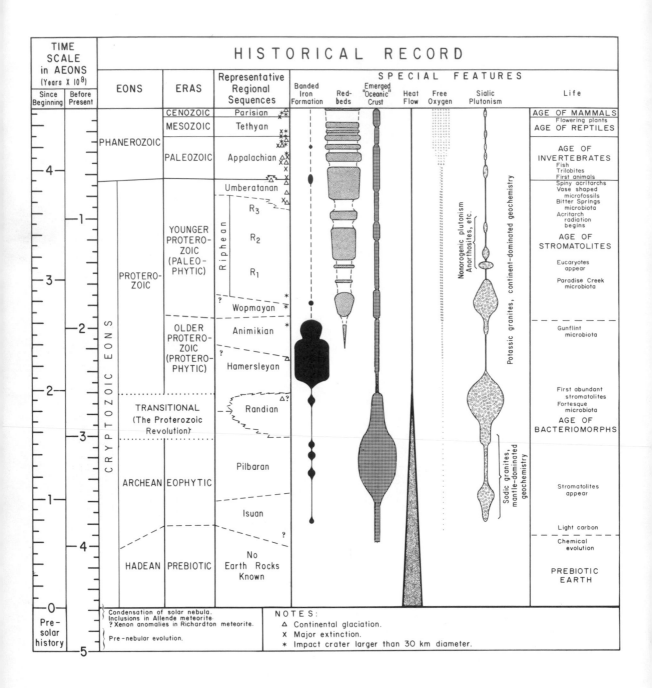

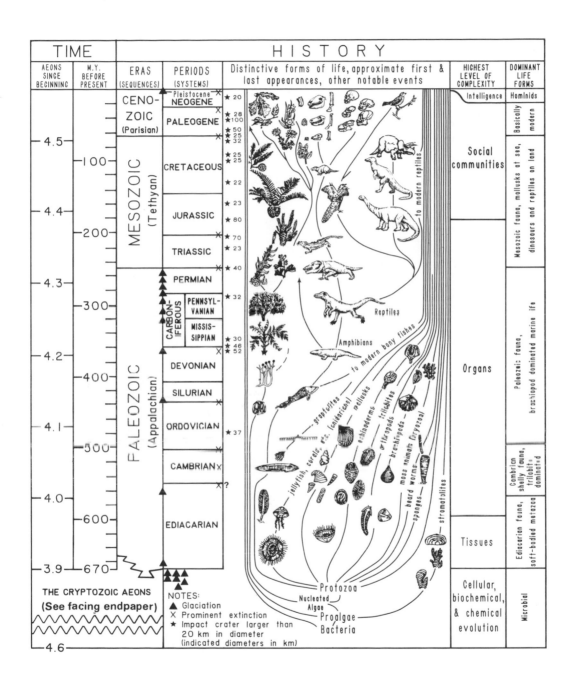

OASIS IN SPACE

EARTH HISTORY FROM THE BEGINNING

The Commonwealth Fund Book Program gratefully acknowledges the assistance of Memorial Sloan-Kettering Cancer Center in the administration of the Program.

OASIS IN SPACE

EARTH HISTORY

FROM THE BEGINNING

PRESTON CLOUD

A volume of
The Commonwealth Fund Book Program
under the editorship of
LEWIS THOMAS, M.D.

W · W · NORTON & COMPANY
NEW YORK · LONDON

Published simultaneously in Canada by Penguin Books Canada Ltd.,
2801 John Street, Markham, Ontario L3R 1B4.

Printed in the United States of America.

The text of this book is composed in Linotype Walbaum, with
display type set in Walbaum. Composition Vail-Ballou Press, Inc.
Manufacturing by Kingsport Press.
Book design by Jacques Chazaud.

First Edition

Library of Congress Cataloging-in-Publication Data

Cloud, Preston, 1912–
Oasis in space.

Bibliography: p.
Includes index.
1. Earth Sciences. 2. Biology. 3. The human habitat. I. Title.
QE26.2.C56 1987   550   87-5693
ISBN 0-393-01952-7

W. W. Norton & Company, Inc., 500 Fifth Avenue, New York, N.Y. 10110
W. W. Norton & Company Ltd., 37 Great Russell Street, London WC1B 3NU

2 3 4 5 6 7 8 9 0

For Jan

CONTENTS

CONTENTS

CONTENTS

16. The Penultimate Scene: The Cenozoic Era   415

17. The Human Habitat   447

FOREWORD

It is generally accepted these days that science, as explored in the last two centuries, has had a greater impact on the way we live out our lives than any other social force in the history of human culture. Arguments will always hang in the air about the relative value of the changes attributed to science, disputes over the usefulness or hazard of the numberless technologies resulting from science. These range from the positive value of indisputable improvements in medicine, agriculture, nutrition, sanitation, transportation (including space exploration), communication and the like, to the obvious negative ones: environmental pollution, depletion of Earth's resources, climatologic transformations, and, topping all lists, nuclear warfare. Even though it is true, in most instances, that the things being argued over are mostly items of technology, and these are developed by intellectual and political processes quite different from the processes involved in the undifferentiated, basic research from which the technologies take their origin, we could not have the one without the other. Science and technology are inextricably interconnected, like it or not.

It is therefore exceedingly important for the general public to have access to as much information as possible about what the scientists are up to; how the enterprise is carried out; and how each item of new information regularly leads, whether by intention or accident, to another, deeper item, usually unexpected. For this reason, but not this reason alone, the Commonwealth Fund has sponsored for several years a series of books written by scientists about the events in their own field. It is hoped that a well-informed, sophisticated grasp of the process of science will be useful for the making of important public decisions in the years ahead.

FOREWORD

But there is another reason for making science more accessible to the public, quite apart from its utilitarian or hazardous outcomes.

Science, especially in the latter years of the present century, has had important effects on human *thought.* Very deep matters have been explored, often with flabbergasting results, and these are affecting the ways in which we view ourselves as individuals, as societies, as a species, and as members of the larger family of life.

This book by Preston Cloud is an example of the sort of cultural illumination now being switched on by scientific research. Here, Cloud deals with the intricate details of the emergence and development of the planet which we make our home. If widely read, as we hope it will be, and used in the years ahead as a ranking source of information about Earth, it will surely influence the way we think about our place of residence.

The Advisory Committee for the Commonwealth Fund Book Program, which recommended the sponsorship of this volume, consists of the following members: Alexander G. Bearn, M.D.; Donald S. Fredrickson, M.D.; Lynn Margulis, Ph.D.; Maclyn McCarty, M.D.; Lady Jean Medawar; Berton Roueché; Frederick Seitz, Ph.D.; and Otto Westphal, M.D. The publisher is represented by Edwin Barber, senior vice president, W. W. Norton & Company. Antonina W. Bouis serves as managing editor of the series, and Stephanie Hemmert as secretary. Margaret Mahoney, president of the Commonwealth Fund, devised the Book Program in the first place, and has actively supported the work of the Advisory Committee at every turn.

Lewis Thomas, M.D.
Editor, Commonwealth Fund Book Program

PREFACE

This book tells the story of Earth's history from the time our Sun caught fire until now. It is intended for those having a non-trivial interest in how this clement and bounteous planet came to have the properties that made it an oasis in space and eventually a haven for mankind. It aims to put more history into the first 85 percent of geologic time, to bring that into more appropriate balance with the conventional last 15 percent, and to lighten the reading of all. In reaching toward more realistic views of our planet's evolution, the book also introduces some novel outlooks and challenges some current enthusiasms. It seeks, in fact, to simplify and integrate the whole 4.6 billion years of Earth history at an accessible level, with emphasis on historical processes, the succession of living systems, and the interactions of both with the physical environment.

Geology since mid-century has been experiencing a renaissance, a second flowering to rival the surge of interest and challenge that midwived its emergence as an historical science during the early nineteenth century. It has been transformed by advances in our perceptions of the planet, its history, and the processes that have governed its evolution.

Geochemistry laid the foundation for these advances with an expanded range of rigorous, self-correcting dating systems and methods for tracing Earth processes. Marine geology, paleomagnetics, and seismology joined with paleontology to give us the plate-tectonic revolution. And biogeology—reinforced by the convergence of inductions from paleomicrobiology, geochemistry, and sedimentology—is rapidly unveiling the early biospheric history of the planet, antecedent to familiar forms of life. From such new insights and data, supplemented by the outlook from space, we now see more clearly how the Earth works and what its history has been.

I deal here with evidence about the real world and the conceptual worlds we construct from it, selected to convey a sense of the grand flow of history in non-mathematical terms. Matter included is as timely and as accurate as I can make it in plain language. Interpretations, however, are matters of personal judgment. Those presented here are drawn from what I consider to be the best current opinion meeting the scientific requirements of consistency with natural law and available evidence, and having verifiable consequences.

Although based primarily on the works of others, and heedful of their opinions, the book is by no means a simple summary of prevailing judgment. Where differences exist they are acknowledged and left unresolved, or my own perceptions have taken precedence. I tell the story as I see it without dramatics or appeal to consensus. This wet, mobile, and ever-evolving Earth needs no hyperbole to be seen as a pretty amazing place. Its story needs no embellishment. But many an unresolved puzzle invites the attention of fertile minds.

History being the theme, particularly the history of Earth's surface features and conditions, subject matter is chosen to reflect that history. It focuses on the results of long-completed natural experiments, on things one can see, feel, move about on or in—on evidence one can experience directly. It deals with that in roughly historical order, but an order that takes up principles and current processes as they become significant for the interpretation of historical events. It delves into internal structure, mechanisms, and events known from indirect evidence and theory alone only as they bear on historical processes.

It is not intended for bedtime reading, but neither is it "science for the superman." I have aimed to make the book readable by anyone who owns a dictionary, has enough interest in science to think about it, and can hold a few brief definitions in mind. Words and expressions likely to be unfamiliar (or used in an unfamiliar sense) are introduced in a descriptive context or parenthetically explained on first appearance. Should recollection fail, reference to a definition can be found in the index.

Above all, the contents are still subject to the fundamental scientific requirement that nothing be accepted simply on faith or the word of authority. I try, therefore, to keep interpretations connected with reasons to the degree that readily communicable evidence allows.

It is both an advantage and an encumbrance that so much of Earth science is visual or best presented graphically. Photographs of geologic subjects used here are taken from my own files as well as those of helpful friends and organizations. Nearly all graphic matter was newly prepared for this book, mostly by adaptation (often extensive) from previously published sources, but in part new.

Thinking that others might read somewhat as I do—sampling to see how a book is structured, and sometimes reading sections out of order or omitting them—I try to help the reader with choices. All chapters begin with a brief italicized theme note, to clue one in on what's planned, and they end with a longer summary that condenses and reviews the highlights. Selected supplemental readings comprise the Appendix—they're exclusively in English and mostly recent. Although simply written and beautifully illustrated supplemental sources are available for a few chapters, most selections had to be chosen from professional sources. At least they will inform the interested reader of other sources or contrary opinion.

The story centers on fundamental issues—the solar system, the historical and geochronological foundations of Earth history, the nature of the primordial crust, the making and evolution of the continents, plate tectonism, the origin and early flowering of life and its interactions with the physical environment, and recurrent continental glaciation. It considers the beginnings of familiar animal and plant life, the rise and fall of the dinosaurs, their mammalian successors, the interactions of all with changing habitats, rates of evolution, extinction, the immediate premodern world, the emergence of humankind, and finally natural hazards. The critical reader will find some intentional repetition. Ideas that are introduced briefly in passing may be elaborated later as they come to play more central roles.

Now to some inadequate words of thanks for those who have helped me get this book into circulation. Over the decades during which I, from time to time, have considered what such a book might include, I have prepared for it by studying real rocks and geologic processes both in the laboratory and in the field worldwide, wherever possible with informed local geologists. I owe my education to literally scores of them, as well as to more than a few biologists and chemists—men and women who shared the fruits of their research and thoughts with me. Alas, there were so many over such a long time that I dare not try to list them because I would surely overlook some. Colleagues do such things for one another and gratitude is understood. I feel it deeply.

My field work, much of it costly, at times involving extensive surface travel on four-wheel-drive tracks or cross-country, as well as the use of boats and aircraft, was supported at different times and places by a score of different geological organizations and funding institutions worldwide. Here I mention only that superlative bureaucracy the U.S. Geological Survey, which, over a quarter century, provided the greater part of my research experience, support, and encouragement. I also deeply appreciate the support of those organizations whose funds enabled me to undertake and complete the actual writing of the book. They are the Australian Department of Science and their Queen Elizabeth II Senior Fellowship Program, the Guggenheim Foundation, the Canadian National Science and Engineering Research Council, and the Commonwealth Fund Book Program.

The drafting of all graphic works was done from my modifications of published figures or original rough sketches by a succession of students and other short-term drafts-people, among whom Wendy Borst, Morgan De Lucia, and Curtis Hopkins did the lion's share. I was also fortunate to have the help of talented geologic cartoonist John Holden in executing my ideas for a representation of Earth processes as a symphony orchestra (Fig. 5.1).

Sources of illustrations other than those original with me are indicated in individual figure legends. Most monochrome prints from my negatives were made by David Pierce. John Shelton, John Crowell, and NASA were generous in allowing me to use selections from their world-class geological imagery.

Where an illustration includes several parts, they are ordinarily arranged in order from the most distant in time (older or earlier) to more recent (younger or later)—older at the bottom.

Each chapter has been reviewed in draft by two or more colleagues, mainly from the University of California at Santa Barbara, and the U.S. Geological Survey, Menlo Park, California. They include Ivan Barnes, John A. Barron, James Boles, Arthur Boucot, Michael H. Carr, John C. Crowell, G. Brent Dalrymple, R. V. Fisher, Clifford Hopson, David G. Howell, David L. Jones, Paul Kieniewicz, James Mattinson, Ellen Moore, George W. Moore, Robert Norris, George Plafker, William V. Sliter, James Valentine, and Tracy L. Vallier. I thank all of these valued friends and colleagues, in particular Boucot, Ellen Moore, and Norris, who helped in ways above and beyond the call of collegiality. Boucot read the whole thing and tried the since-revised first twelve chapters on students in his historical geology course at Oregon State University. Norris also read all seventeen chapters, and Moore read all of Part III, as well as several other chapters. In addition, she arranged with all other USGS critics to help. I am similarly indebted to Donald Elston of the Geological Survey's Flagstaff laboratories for the hours he spent explaining the nuances of paleomagnetism to me.

Cheryl Simon was the ideal review editor for this book. She brought to my preliminary draft exactly the outlook and talents needed—those of the intelligent and informed but scientifically untrained master grammarian. Her demands for clarity, simplicity, and brevity were a constructive counterbalance to the scientist's constant urge to qualify and elaborate.

At last, a paean to my long-suffering guardian angel and gifted wife, Jan, whose patience through five years of my near-total preoccupation with ancient history would have made Job seem querulous. She kept me functioning through crises and despair, feeding me at my desk, pushing notes under the door, coping with junk mail and other computerized flak, and even domesticating the family word processor.

My heartfelt thanks go out to all.

Preston Cloud
Santa Barbara, California
10 October 1986

PART I

INTRODUCING
THE EARTH:
A BEGINNING, AND
SOME FUNDAMENTALS

CHAPTER ONE

THE NEBULA
AND THE PLANETS

Given energy, matter, ample time, and space, anything can happen—almost. Put limits in the form of natural laws on what does happen and you have the visible universe. In one tiny corner of that universe is an inhabited planet, the only one known so far— Earth. The very matter of which that planet and its inhabitants are made is a minuscule fraction of the minute fraction left after the mutual annihilation of matter and anti- matter during the first catastrophic instant of this universe. It is the immediate by-prod- uct of the death of some ancestral star or stars, leading to condensation of the solar nebula and planets around 4.6 billion (4.6 × 10^9) years ago. Our Earth was one result of this condensation of nebular mass into earthy, icy, and gaseous components. The fact that it also happens to have the characteristics needed for life as we know it to arise and prosper is reason enough to probe its history. This chapter considers the context in which that history is set.

The Cosmic Connection

Planet Earth, galactic oasis, how bounteous it seems to us and yet how minuscule within the cosmos! Visi- ble only by reflected light, this tiny dark companion of our Sun has a total mass of but three one-millionths (0.00000302) of that unexceptional main sequence star—one of billions of such stars in the lonely reaches of the visible universe. Even within the planets of the solar sys- tem, Earth accounts for less than two- tenths of a percent of the total mass. Of that, its outer crust, source of all life- sustaining resources except sunlight, is less than half a percent of the total. How marginal our existence! How extraordi- nary that we should be here!

As it is with space, so it is with time. Our longest memories are aptly known as flashes. The pyramids were only yes- terday. *Brontosaurus* was just last week. If one could look back into the abyss of time where it all began, what might one see? By what miracle did Earth-stuff

1.1 CRAB NEBULA IN THE CONSTELLATION TAURUS, REMAINS OF THE SUPERNOVATION OF A.D. 1054. *[© California Institute of Technology 1959.]*

3

emerge from the embrace of the likely parental supernovation that squeezed a shapeless cloud of dust and gas into our ancestral solar nebula some 5 billion (5×10^9) years ago (similar to Fig. 1.1)? This book, as if restored from an old journal with many missing, torn, and scribbled-over pages, begins with that problem.

Our fascination with planet Earth, our launch-pad to space, arises chiefly from the fact that it is the sole cosmic body whose surface is yet known to support permanent liquid water (except superheated Uranus) and the peculiar properties collectively called life. Being the kind of life we are, we want to understand how this planet came to be. We care because we are *Homo sapiens*—man the wise, the compassionate, by desire and intent if not in fact. As beauty is only skin deep, so life, and the water essential to it, is only a veneer at Earth's surface. We cannot even represent them in a vertical profile of the planet at the scale of these pages without great vertical exaggeration. Nor can we do much better with the atmosphere and crust. Yet Earth's indigo seas, dazzling veils of vapor and ice, and kaleidoscopic patches of plant life reveal themselves from afar. Any observant space crew would instantly identify this planet among other solar satellites by its distinctive and beautiful pattern of hues, so faithfully captured by images from orbit.

To what does Earth owe these distinctive characteristics? They result from the fact that it happens to be just the right size and composition at just the right distance from a single parent star of just the right magnitude. Because of that it was able to acquire, retain, and evolve its blanket of atmospheric vola-

tiles, generate a radiation-shielding magnetosphere, and enjoy the range of surface temperatures within which water can remain liquid. Under such privileged conditions, the life we know arose and has persisted. Good things, it is said, come in small packages, and good workmanship takes time. Earth qualifies on both counts. Our tiny planet was a product of presolar and solar evolution that began far back in the dim reaches of the past, before any planets existed, more than 10 billion years ago, probably closer to 18 or even 20. That whole long and probably essential prelude is a cosmic saga of heroic dimensions, full of interesting twists and improbable characters like black holes and neutron stars (black holes from a British usage, because they imprison light). Here that saga is compressed to deal only with a central sequence of events and processes chosen to convey a balanced sense of Earth history.

The universe is even more peculiar than one might imagine. The part of which we know is mainly radiation. For every nuclear *proton* and *neutron*—not to mention the *quarks* of which they are made or the *gluons* that hold the quarks together—we know a billion particles of light, *photons*.

Imagine an average liter or quart of extragalactic space in which all motion is momentarily suspended. If you could see and count its contents, you would find more than half a million photons but not a single particle of *matter*—no neutrons or protons, no atoms, no molecules. Atoms of hydrogen, containing but one proton and one electron each, comprise up to 90 percent of all the familiar matter in the universe, the rest being overwhelmingly helium. Even so, one would have to dip hundreds of such containers of extragalactic space before coming upon a single hydrogen atom. Such a level of density has been compared with that of three bees in the whole airspace of Europe. By contrast, the "empty" space between the planets of our solar system averages about the equivalent of 5,000 hydrogen atoms in the space of a liter. A liter of ordinary "thin air" at sea level contains 27×10^{21} (27 followed by 21 zeroes) molecules of atmospheric gases, including nitrogen, oxygen, argon, carbon dioxide, and traces of others—a dense substance compared with interstellar or intergalactic space. An even more unusual perception is the growing belief among astronomers and cosmologists that some 90 percent of the expected total matter in the universe consists of invisible, nearly weightless particles unrelated to protons and neutrons (nonbaryonic matter).

Considering how thinly "ordinary" matter is dispersed throughout the universe, it seems little short of miraculous that solid bodies like Earth, its sister planets, and the rest of the solar system should exist. Science, however, may not call on miracles. If a miracle is assumed, there is no incentive to search for understanding. One must ask instead what natural processes may be responsible for all this matter and its aggregation into the Sun and planets. Where, indeed, is all the antimatter—differing from matter only in its opposite electric charge? (For quantum mechanics, symmetry principles, and Albert Einstein all tell us that antimatter should be present in the same proportions as matter.)

The best explanation so far is that the matter we know is the trifling residue that remained after the mutual annihilation of once-huge and nearly equal amounts of elementary particles and antiparticles that existed during the first few seconds of the present universe. The story is told with admirable clarity by astrophysicist and Nobel Laureate Steven Weinberg in his 1977 book *The First Three Minutes*, describing the theory of the Hot Big Bang.

Now, in these terms, hot is very hot indeed—so hot that everything is in the state of an intensely conductive fluid plasma. No chemical elements yet exist, even in the gaseous state, but only the basic particles from which, in later, much cooler times, elements were to be assembled. In fact, the range of cosmic temperatures is so great that it is measured in degrees *Kelvin* (°K), a system that is graduated on the same scale as *Celsius* (or Centigrade, °C) but begins at -273°C, or absolute zero, (0°K), where all molecular motion ceases. When temperatures level off to the range of more familiar things, the Celsius scale, where zero is the freezing point of water at sea level, is employed.

The universe, it seems, began at the mind-boggling temperature of 100 billion °K. After the first hundredth of a second, as the blazing universe expanded, it began to cool, receding to a mere 1 billion degrees at the end of the first 3 minutes. At that time the content of the universe was overwhelmingly radiation. By then also protons and electrons were able to combine to form elements of mass 2, 3, 4, and 7 (ordinary hydrogen is mass 1). The production of heavier elements was blocked by the absence of stable nuclei of mass numbers 5 and 8. This early universe continued to expand and cool until, after the first million years, the cosmic temperature fell to about

3,000°K. At such a temperature the previous unique state of radiation / matter was decoupled, the universe became transparent to radiation. Single protons and electrons could at last be joined to form *ordinary* hydrogen (H) of mass 1, and protogalaxies (or perhaps galactic clusters) could form. Hydrogen could now collect and burn to helium (He) at nuclear temperatures in the cores of stars like the one we call Sun. Under appropriate conditions other elements could also be made from such starting materials. For example, three helium atoms of mass 4 can join to form a single carbon atom of mass 12. And other elements up to iron (Fe), of mass 56, can be made in red giant stars. Still others, up to the heaviest known, are made and flung into space when a massive iron-rich star implodes and blows off its outer shells. Such stellar events are called *supernovae*.

This Big Bang theory of the origin of the universe is referred to as a gauge theory because it changes scale as well as direction. A close antecedent was proposed ahead of its time by the international figure George Gamow, in the late 1940s. As such an energetic universe or set of universes would also be expanding (and therefore cooling), Gamow predicted that present general background temperature in our universe, be it unique or one of many, should be about 5°K. Alas, no way was known in the 1940s to measure such a temperature. Validation of his prediction had to await a fortuitous discovery. While surveying background radiation for other purposes, Bell Laboratories radio astronomers Arno Penzias and Robert Wilson detected a weak general background radio static. It implied a universal background radiation of about 2.9°K. That figure was close enough to

the predicted number to satisfy most scientists that Gamow and his collaborators had been on the right track and to bring to Penzias and Wilson a Nobel Prize, too late for Gamow to share.

The Big Bang is where the action was. It cannot be heard, but it can be sensed. The 3°K general background radiation is one of several observations that warrant a high level of confidence in its reality. Observations of pregalactic lithium-7 of abundance close to Big Bang predictions, the fact that the substance of the universe we know is all matter, and the constant ratio of ordinary helium-4 to hydrogen wherever measured also support the theory. If, moreover, we believe astronomical evidence that distant galaxies were once more densely packed and that extragalactic starlight beyond our Local Group of galaxies is shifted toward the red end of the spectrum, then the present universe continues to evolve, expand, and cool. That *red shift* is, in fact, a Doppler shift of light similar to the shifting of sound to longer wavelengths as the horn of a passing car fades to deeper tones. It tells us that the universe is expanding in all dimensions, its galaxies like raisins in a baking loaf, at rates that increase with distance from the observer but are now known not to be as symmetrical as was once believed. This concept has been a major feature of cosmological thinking since the 1920s.

Omitting recently measured asymmetries in rates of expansion, the Big Bang explains why the cosmos is mostly photons, how it happens that the visible universe is one of matter to the essential exclusion of antimatter, why the universe is expanding, and why there is a red shift—an elegant and powerful ruling theory of our time despite proposed modifications and alternatives. It also explains why there is so much open space in the universe that the night sky is dark. Different lines of evidence indicate that this event occurred between about 13 and 18 billion years ago. The exact number is a subject of currently active disagreement and research.

With all these observations in its favor, it seems very likely that the idea of a Big Bang is close to the mark. Yet, as is commonly the case in science, some nagging uncertainties remain. Of the ordinary matter in the universe more of it seems to be hydrogen than the once-estimated 73 percent. The matter of the universe is not as uniformly dispersed as was once believed. Instead it is clumped into galaxies and galactic clusters, implying an initial or early lumpy state. The background radiation is not consistently uniform from all directions but shows unexpected minor variations. And distant galaxies 10 to 11 billion light years from us do not show the expected evidences of youth in the form of high concentrations of young blue stars.

The expansion of the universe in the form of consistent red shifts is seen only outside the Local Group of sixteen galaxies that cluster with our own—for we and it are moving *towards* the Virgo Cluster within the Local Supercluster. Perhaps the Big Bang applies only to one huge sector of a much larger universe? It has even been suggested that, rather than a single gigantic bang, there might have been a cascade of lesser bangs, with "local" universes then going through their own cycles of expansion and collapse. Perhaps similar explosive events in a number of giant stars could account for the same results? Right or wrong (and omitting strings, superstrings, and

bubbles) the Big Bang presents a grandly provocative idea, and one whose predictive success so far warrants a high level of credibility.

It also offers, for the first time, a philosophically satisfying explanation for the preponderance of matter in our universe to the near exclusion of antimatter. Matter, says that theory, probably existed from the beginning in the form of the tiny surplus of particles over antiparticles bequeathed to us by the Big Bang. The relatively massive *hadrons* (nucleons and other massive particles) and ghostly *leptons* (electrons and other near-massless particles) that make up everyday matter have been repackaged into different forms, at different times, in red giant stars and supernovae of different ages. The repackaged particles are then ejected into interstellar space as the dust and gas that make up the galactic nebulae. It is in those nebulae, somehow, that stars are born, mostly in groups of two or more. Where stars are single, stable, and long lasting (perhaps 1 percent of the total), planets *may* come into being as satellites of those stars. That requires much denser concentrations of matter than are observed in galactic nebulae. Of course it also takes a great deal more than the simple presence of matter to make a planet like Earth, write its history in stone, and evolve a creature that can decode, accumulate knowledge, ponder, and articulate that history.

Even given an Earth-sized aggregation of matter, a universally applicable law of physics—the second law of thermodynamics—would preclude the origin and evolution of life without a continuing source of disposable energy. The star we call Sun is Earth's life-sustaining energy source. It, almost alone, provides the power that keeps the life machine running, that adorns this planet with climate-moderating and nutritive vegetation, and that populates its seas and lands with animals. It interacts with the physical evolution of the planet to generate the succession of events we call Earth history. Without the Sun, there would be no Earth history of the sort we know and no one to decode and transcribe it.

Earth's history is also linked with that of its satellite and sister planet *Luna*, the moon. This was especially so during the early evolution of these bodies, when gravity and proximity were the shaping factors. Because of the gravitational pull of the moon and the Sun we also have tides on Earth, with all their manifold effects on living systems and sedimentary processes. Early lunar history is relevant to Earth history in other ways. It provides a window on the primitive Earth from a time before any records are found there.

Luna, reminding us by reflected solar light in the night sky that Sun will be back in time for breakfast, has a significance for Earth history that belies its bland image. A charmed observer, overwhelmed by the beauty of a full moon, may not reflect on its geologic significance. But a bit of study reveals records of lunar history that are clear to the sharp eye or the steadily held binoculars. The dark, nearly circular areas—the lunar maria—cut across the trends of the more extensive light-colored background of the lunar highlands. Like gouges across a design cut into soft clay, the maria plainly formed more recently than the truncated highlands. One may also often see, at the lower edge of the full moon, the conspicuously bright ray crater called

Tycho (after the early astronomer Tycho Brahe, predecessor to Kepler). From this crater, conspicuous white streaks radiate more than halfway across the lunar surface, like a pyrotechnic design. It looks as if the moon had been given a great whack with a giant sledgehammer that sent debris flying in all directions. And this impression is close to what actually happened—the whack was delivered by a large meteorite or asteroid. The rays from Tycho and other ray craters splash right across maria, highlands, and the boundaries between them, clearly signifying late events in lunar history.

So, without ever going to the moon, a keen observer with no equipment beyond field glasses can reconstruct three major historical episodes that affected the Earth-facing surface of the moon. First came the reflective lunar highlands. Second were the dark-colored, subcircular maria or lunar craters, cutting across the trends of the highlands as basaltic lavas welled up to heal the wounds of giant impacts. And third are the later ray craters whose conspicuous, though areally limited, linear streaks of light-colored debris extend far across the older lunar surface. We can see these features so clearly because our moon has no water or atmosphere and therefore no weathering or erosion to blur them, nor any haze or cloud cover.

We use the same and many other sequencing criteria to reconstruct a detailed history of the Earth and to place it in a quantitative time frame. That is a daunting task for a vegetation-covered Earth that has undergone so many episodes of weathering, erosion, tectonism, and veneering by younger deposits, and where its first several hundred million years of history are not directly recorded.

First Thoughts on an Aged Earth

Until the mid-eighteenth century, the principal accounts of Earth history were mythical or religious—Hindu, Egyptian, Sumerian, Babylonian, Hebraic, Druidic, and Norse, for example. In Europe, where historical geology arose, early concepts of Earth's origin were linked to Old Testament accounts. The first recorded biblically derived estimate of the age of the Earth (and universe) was made by Saint Augustine in the fourth century A.D. He counted about 6,000 years from biblical genealogies—a number that we find 12 centuries later in Shakespeare's *As You Like It.* That number was refined in 1598 by no less a figure than Johannes Kepler in his book *Mysterium Cosmographicum (The Mystery of the Cosmos).* Kepler, then professor of mathematics at Graz, calculated the date of creation to have been 3877 B.C., Sunday, 27 April, at 11 A.M. local time.

James Ussher (or Usher), Anglican Archbishop of Armagh and a respected biblical scholar, found a slightly different number in 1654. In his work, *The Annals of the World Deduced from the Origin of Time,* Ussher painstakingly calculated from Middle Eastern and Mediterranean history and biblical accounts that the universe was separated from the void in the year 4004 B.C. at 9 A.M. on Sunday the 23d of October. That date was widely accepted, it is reported, as a result of being recorded in an authorized version of the King James Bible in 1701. Attempts at greater precision, however, seem to agree only on the beginning of the divine workday. Another scholar, John Lightfoot of Cambridge University, had already set the moment of creation at 9 A.M., Friday, the 17th of September. Ussher further calculated that Adam and Eve were expelled from paradise only 18 days later and that Noah landed the ark on Mount Ararat on Wednesday, the 5th of May, in 1491 B.C.

In fact, these scholars could hardly have done better with the information available to them. Numbers in the thousands of years for Earth's age have not stood up simply because they are based on authority rather than on objective measurements from verifiable physical constants.

As we shall see in Chapter 4, geologic and cosmic time is now reckoned in much larger numbers, with a much higher degree of accuracy and repeatability. Methods available for determining it objectively are based on the constant rates of natural decay of radioactive elements to inert end products. A variety of time-keeping methods involving radioactive components of a dozen different elements provide a range of geologic and cosmic clocks. Their different but invariable "decay" constants comprise a set of self-checking systems whose utility depends on slight nuclear variations in chemically invarient atoms of the same element known as *isotopes.* Some of these isotopes are radioactive and waste away (decay), hourglass-like, to known and measurable end products.

Employing such systems, an age of about 3 billion years had become widely accepted as that of the solar system by 1956, when Clair Patterson nearly doubled it, utilizing a new and ingenious approach. At that time a postdoctoral student at the University of Chicago,

Patterson found that the ratios of radiogenic lead isotopes in a variety of meteorites all indicated the same age. Using oceanic sediments as the source for a thoroughly homogenized sample of Earth's metallic leads, he found that their isotope ratios also fell on the same trend. The isotope ratios of lead from Earth and meteorites alike were found to lie on a single evolutionary line denoting a common age of origin about 4.56 billion years ago. In addition, the duration of processes involved in the initial formation of our parent star and its inner planets has been estimated by computer modeling as about 100 million years, a number close to that estimated from the decay products of now-extinct short-half-lived radioactive isotopes. The true age of the solar system, therefore, appears to be close to 4.6 billion years.

Compared with the brief span of their own lifetimes, humans are likely to view as permanent things that have been around as long as the solar system. Nonetheless the "everlasting hills" eventually are carried to the sea. And neither Earth nor Sun has always existed or will continue to exist forever. Earth, its sister planets, and Sun all condensed somehow from a cloud of dust and gas known as the solar nebula fewer than 5 billion years ago. All will disappear when, in the normal course of stellar evolution, our Sun expands to become a red giant star in another 5 billion years or so, incinerating everything within its expanding periphery.

Tales Meteorites Tell

Earth and solar system history are but the most immediate sector of a panorama of cosmic evolution that stretches all the way back to the Big Bang 13 to (most likely) 20 billion years ago. What happened earlier? How shall we learn about the origin and evolution of galaxies, leading up to the separation and condensation of our solar system? Indeed it was only in the 1920s that astronomers first determined that there were other galaxies beyond our own and still more recently that research has focused on them. The resulting wealth of instrumental and theoretical advances now enables astronomers simultaneously to look farther out into space and thus farther back into the past than ever before. Questions about the origin and time of origin of the estimated 100 billion or more galaxies within the visible universe, therefore, are in a state of active flux. Until they are closer to resolution we must leave them to astronomers and astrophysicists and get on with the Earth and solar system.

For we do have some interesting if also controversial evidence that bears on immediate presolar history. It beckons us, ghostlike, from records of the long-vanished radioactive isotopes aluminum-26 and iodine-129 in meteorites kept by their inert end products. As if it were a sealed bottle wafting a coded message to some distant solar beach, a large carbonaceous meteorite descended into northern Mexico near the village of Allende in 1969. Inside was what seemed, after study, to be the first unequivocal evidence about events that occurred just as our ancestral solar nebula came into being. The *Allende meteorite* consists of fused, millimeter-sized siliceous spheroids called chondrules. It is said to be chondritic, and, being comparatively rich in carbon (about 2 percent), it is called a *carbonaceous chondrite*—in fact an important primitive kind known to planetary scientists as the type C-3.

Because carbonaceous chondrites disintegrate quickly under ordinary weathering in humid parts of the Earth, most of those recovered are observed falls or from dry Antarctic ice. Allende was an observed fall having an estimated mass of about 2 tons before breakup, of which about half was quickly recovered and preserved. The mass recovered far outbulked that of all previously known carbonaceous chondrites taken together. Allende attracted immediate attention because it was so big and had such a high carbon content. Moreover, some of that carbon was in the form of amino acids and other large organic molecules of kinds found in living organisms. Its presence established that such organic molecules formed on or within the parent bodies from which the carbonaceous chondrites came. It was exciting to think that such carbonaceous matter might be evidence for life or former life in other parts of the solar system, as some of the earlier reports on other carbonaceous chondrites proposed. Skepticism, in turn, had arisen from the fact that some carbonaceous meteorites were found to have been contaminated after their arrival on Earth. This was a fresh fall.

Another carbonaceous chondrite of similar size, the Murchison meteorite, fell in Australia in the same year. It brought new evidence before the court of scientific judgment. Keith Kvenvolden of the U.S. Geological Survey and his associates found that deep inside the Murchison meteorite, beyond the range of any likely contamination, amino acids found included only kinds that showed no optical bias. They rotated light preferentially neither to the right nor to the left. That quality is characteristic of amino acids that are produced nonbiologically in experimental work. Despite the presence of indigenous carbonaceous matter, therefore, such a result makes a biological origin for such molecules unlikely. It is important that unfavorable as well as favorable evidence come before the court of science. Like everyone else, scientists do make mistakes. The important thing is to correct them.

Allende brought another message relevant to presolar history in the form of isotopically anomalous white, pea-sized inclusions within the meteorite (Fig. 1.2). That evidence seemed to confirm the grand idea that an early nearby supernovation had enveloped and squeezed together the dust and gas that made the original solar nebula.

It is a charming story, if now disputed, and a good example of the triumphs and tribulations of the game of science honestly played. It begins with a study of the isotopes of magnesium (Mg) and can be told only by reference to those *isotopes*. They are all Mg, but they differ from one another in the number of neutrons in their nuclei and therefore in their mass number—the sum of neutrons plus protons. These differences can be measured with the delicate and precise instrumentation of geochemistry. Thus for example Mg, with mass number 24, is ^{24}Mg, more conveniently called magnesium-24 or Mg-24. Magnesium has isotopes of mass numbers from 24 to 29, with known ratios to one another except where Mg-26 exceeds its normal ratio because of the decay of radioactive aluminum-26 (Al-26) to inert Mg-26, with loss of one proton. When the number of protons in an element changes, it becomes a different element (see Ch. 4).

In the case of the Allende meteorite, Gerald Wasserburg and his associates at the California Institute of Technology (Caltech) found more Mg-26 than should have been present in proportion to the other isotopes of Mg. That implied a source of Mg-26 beyond what was made at the time of the original element-making or nucleosynthetic event—during which elements and their isotopes that comprise our solar system were produced in specific proportions according to precise governing principles. And that meant that the excess Mg-26 could be explained only as the stable decay product of radioactive Al-26, a short-half-lived isotope formed during a late nucleosynthetic event. Radioactive Al-26 was at that time known only from nuclear explosions and the expanding outer shells of recent supernovae, exploding stars (e.g., Fig. 1.1). It decays rapidly to stable Mg-26 by expelling one nuclear electron, with a half-life of only 720,000 years. And that implies that half the total amount of Al-26 at any given time decays to Mg-26 during the next 720,000 years. From the expanding growth of almost everything, we are familiar with positive exponentials, but this is a negative exponential. It means that a radioactive isotope decays to the vanishing point in about 10 half-lives and is barely detectable after 5 half-lives. In the case of the Mg-26 in Allende, it seemed to mean that no more than about 3 to 7 million years could have elapsed between the supernovation that created the Al-26 and the excess Mg-26 of those interesting inclusions. Otherwise too little Al-26 would have been left to account for the excess Mg-26.

That discovery implied exciting new insights to presolar history and the birth of stars. New stars can form in nebulae, in clouds of interstellar gas and microscopic dust grains such as in the pinwheel-like arms of our Milky Way and elsewhere (Fig. 1.3A). Most interesting for us are the long-lived sun-sized stars, called T-Tauri stars after the star T-Tauri itself (Fig. 1.3B), found in the constellation Taurus. Our Sun is believed to have originated as a T-Tauri star whose fierce stellar winds blew away its outer shell of gas, clearing the space around its satellites as it joined the "main sequence." It takes the thermonuclear reactions of a carbon- and iron-burning supernovation to generate heavy elements similar to those found in small

1.2 SLICE OF THE ALLENDE METEORITE. Showing calcium- and aluminum-rich inclusion containing excess magnesium-26. [Courtesy of G. J. Wasserburg.]

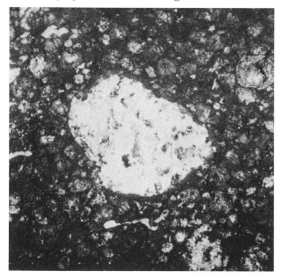

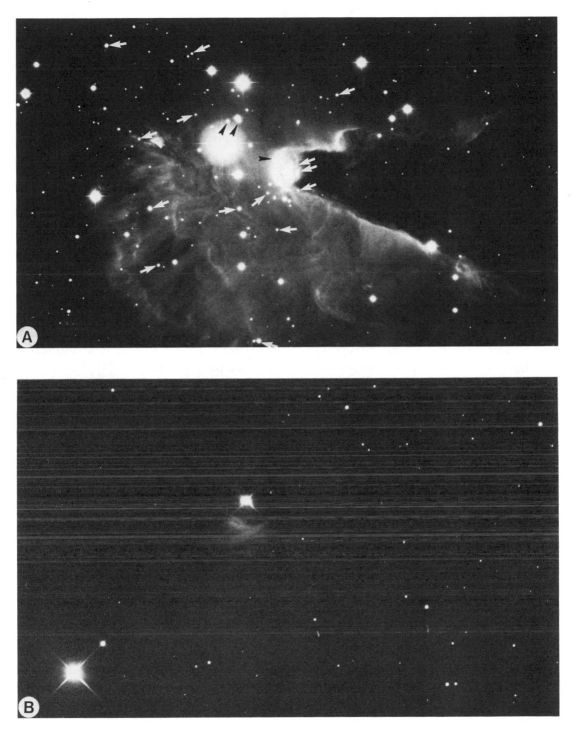

1.3 THE BIRTH OF SUN-LIKE STARS; T-TAURI AND THE CONE NEBULA. Arrows in *A* denote seven-teen T-Tauri stars of the Cone Nebula. *B* shows T-Tauri itself, astronomical up, or south, being to the right. Its light illuminates an adjacent small nebulosity. A recently discovered infrared source near the bright star at the lower left *might* be a planet. [A, *Palomar Observatory Photograph*. B, *Lick Observatory Photograph; by courtesy of G. H. Herbig.*]

quantity in Sun and the large outer planets and in proportionately larger quantity in the inner planets of the solar system. In addition, something may be needed to squeeze the dust and gas of such nebulae into compact-enough masses that the relatively weak forces of gravity take charge, leading to the condensation of new stars, with or without the satellites we call planets. Engulfment within a sector of the expanding outer shell of such a supernova (Fig. 1.1) could facilitate the pinching off, compression, and gravitational collapse of sectors of interstellar dust and gas into nebulae like that from which the Sun and planets of our solar system condensed. Thus it has long been hypothesized that one or more nearby supernovae preceded or accompanied such creative events.

The excess Mg-26 in those Allende inclusions, and the now-vanished Al-26 from whence it came, seemed to signify at least one such supernovation close in time and space to the ancestral solar nebula and very likely related to its final collapse. And Allende, a probable product of that supernovation, has been dated by lead isotopes as very nearly 4.6 billion years old.

Then, as so often in science, an unexpected new discovery interjected a note of uncertainty into that otherwise seemingly well documented conclusion. This discovery revealed a constant flow of gamma radiation, indicative of a source of Al-26 close to the same level as in Allende, but coming from the direction of the galactic center and apparently streaming from giant stars or their explosions.

Thus, although it is still possible, and in fact likely, that Allende records a supernovation nearly contemporaneous with the condensation of the solar nebula and even that such an event had something to do with solar condensation, this new evidence sounds a note of caution. Such is the way of science. As we so often discover after long and painstaking effort, more research is called for. Reconsideration of the meaning of the 1960 discovery by astrophysicist John Reynolds of excess xenon-129 in the Richardton (Minnesota) and other meteorites is now also called for. This excess Xe-129 could have come only from the decay of iodine-129, with a half-life of but 17 million years. And that also has been taken to imply an antecedent supernovation, which, on the same line of reasoning as for Al-26, would not have preceded the solar system by more than about 80 to 170 million years. All this implies that the age of Sun and its satellites cannot be much greater than about 4.6 billion years.

It remains increasingly likely, in any case, that supernovae were involved in the condensation of the solar nebula, both as a source of heavy elements and of the engulfment that triggered the final collapse of adjacent dust and gas to make the solar nebula. And answers as to just how that happened are implied by the isotope chemistry of meteorites, the results of planetary exploration, and discoveries that are now being made about those enigmatic, icy, deep-space probes called comets.

Without the Big Bang, the condensation of the galaxies, the supernovations that made the heavy elements and probably initiated the collapse of the solar nebula and its contraction, there would be no Earth, no us. We are made of star-stuff, processed through supernovae, concentrated from the contracting solar nebula, spun into biochemical aggregates with a difference, and graced, during our tenure here, by the ability to imagine, to conceptualize, to hypothesize, to create science, poetry, music, and works of art and technology.

A Plenitude of Planets

Our wet and teeming planet, scarcely 12,700 kilometers in diameter, is imbedded in a thinly discoidal solar system a million times or more as far across, through which it speeds at 30 kilometers a second. That is just the right speed to maintain its orbit against the pull of Sun's gravity and so to experience a range of temperatures appropriate for the origin and persistence of life. The entire solar system, in turn, moves at a rate of about 20 kilometers a second with respect to the neighboring stars around a spiral galaxy that is some 78,000 light years (7.5×10^{17} kilometers) across, 1,600 light-years through, and 117 million light years from its nearest neighbor. Such a litany of dimensions and distances can numb our sense of scale. The fiefdom of the Sun all by itself is quite big and important enough to challenge our capacity for comprehension and to focus our interest here.

Of the many solar satellites of all sizes only nine are conventionally designated as planets (Fig. 1.4). Conveniently enough, their common names are also their scientific names. It is also the practice in naming various subsequently discovered planetary bodies and lesser features to link related things and help our memories by adopting mythical, legendary, or classical names from various sources (except for the Apollo asteroids, Fig. 1.5). As we must also designate things unequivocally before we can discuss them unambiguously, the naming or numbering of new classes of objects and features is an important part of planetary exploration and mapping, as of other natural history.

To its inhabitants it is only natural that Earth should seem to be the center of the universe, especially when they daily see the Sun, the planets, and distinctive stars apparently going round it. Now that all recognize this to be an illusion, resulting from Earth's own rotation (as well as planetary motions), it is easy to chuckle at those who, in ancient times, could see no reason to question the commonsense evidence of their eyes. Nonetheless, the names for major stars, constellations, and planets have come down to us mostly from antiquity. Our home planet was *Gaia* or *Gaea* to the Greeks, hence *geology* for Earth science. *Earth*, a Middle English and Teutonic term, came later, but others accepted today are taken from the Roman deities. In the absence, before Newton, of a proper idea of gravity, moreover, it was necessary to explain how such objects, including Sun and our moon, retained their positions. And, toward that end, they were imagined as being imbedded in solid but transparent spheres or "heavens" that revolved about Earth. There were, depending on the scheme, a minimum of seven of these spheres—one for each of the five then-known planets beyond Earth, plus Sun and Earth's moon. That is how we happen to refer to heavens in the plural and to seventh heaven as the epitome of delight and perfection.

In that system, Sun occupied the fourth heaven out from Earth and Earth's moon the first. The heliocentric solar system changed that. Nearest to Sun was the planet that completed its orbit in the shortest time, Mercury, named for the swift-footed courier of the gods. Then the beautiful morning and evening star, called Venus, after the goddess of love and beauty. Next was Luna, our moon,

1.4 THE NINE CONVENTIONAL PLANETS, DRAWN TO SCALE AGAINST A HALF-SUN AT THE SAME SCALE, AND SHOWN IN THEIR RESPECTIVE ORBITS BENEATH. Order of increasing orbital distance from Sun is left to right. Symbols are astronomical shorthand for the planets, as indicated in top illustration. [Below, *adapted from* Earth, *3/E by Frank Press and Raymond Siever. Copyright © 1974, 1978, 1984 W. H. Freeman and Company.]*

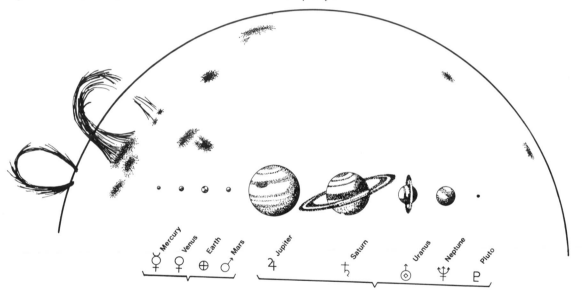

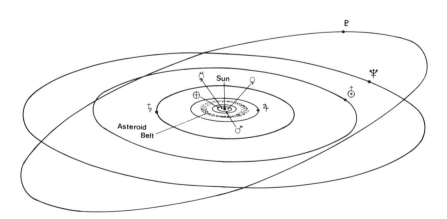

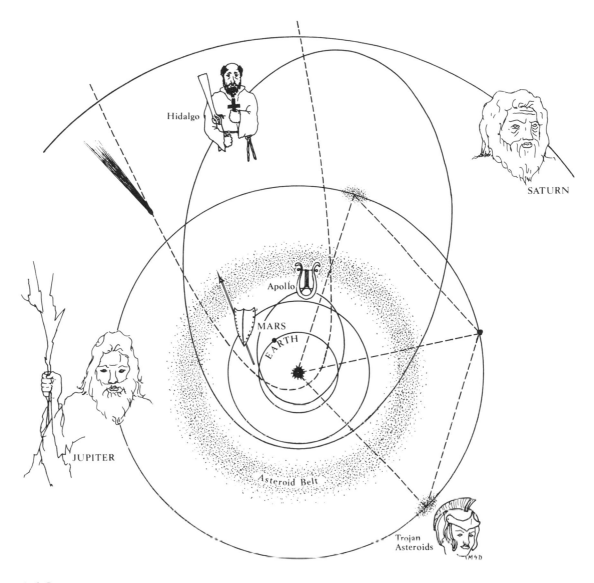

1.5 ORBITS OF SOME LESSER SOLAR SATELLITES COMPARED WITH THOSE OF SELECTED PLANETS. The Trojan asteroids precede and follow Jupiter at resonance points that make equilateral triangles with Jupiter and Sun. The Apollo asteroids comprise some fifty minor planets or planetesimals with strongly elliptical orbits.

and beyond Earth—the blood-tinged Mars, named for the god of war. Then came mighty Jupiter, in honor of the chief god, a bold planet that, unlike the brighter but demure Venus, rose high in the sky and shone throughout the night. Last of all was Saturn, the slowest moving of the planets. Named for Jupiter's elderly father, it needed almost 30 years to plod the whole way round the sky.

No other planets were known to the ancients. When new ones were discovered, the tradition of naming triumphed over attempts to honor political figures. Uranus, seventh out from the Sun, needed nearly 3 times as long as Saturn to circle the heavens and so was named for Jupiter's tottering grandfather. Neptune, eighth out, was named for the god of the sea, while Pluto, ninth and farthest from the light of the Sun, was more appropriately named for the god of the dark underworld. Finally, the largest of the bodies in the asteroid belt between Mars and Jupiter, sometimes considered to be a tenth planet, was named Ceres after the Roman goddess of agriculture. Were the planets to be renamed today, following the superlatively revealing grand tour of *Voyager 2*, the name Neptune would surely go to the one known as Uranus. For Uranus' molten rocky core is covered by 8,000 kilometers (5,000 miles) of superheated water, kept from boiling away only by the pressure of a heavy atmospheric lid.

The formally designated planets represent only the best-known and more conspicuous planetary bodies in the solar system. Their main characteristics are summarized in Table 1.1. In addition to these primary planets are some sixty known planetary satellites. Two of these,

Jupiter's icy Ganymede and Saturn's methane- and ethane-wreathed Titan, are larger than the planet Mercury. Some thirty-three asteroids more than 200 kilometers in diameter are known, plus innumerable lesser objects.

Indeed, there is so much stuff flying around out there that one might expect collisions to be common. And, in fact, the numerous tiny particles whose tracks in the night sky we call falling stars do collide often because they are so small, so numerous, and so readily attracted by larger objects. All of these objects, large and small alike, follow gravitationally prescribed individual orbits, except as deflected by proximity to larger objects. The so-called asteroids or "little stars"— which might more appropriately be called planetoids ("little planets")—mostly follow orbits within the broad asteroid belt between Mars and Jupiter. Some 20,000 asteroids more than a kilometer in diam-

TABLE 1.1 *Some Major Characteristics of the Planets*

	Radius × Earth's Radius	Mass × Earth-mass	Specific Gravity	Magnetic Field	Day × Earth Day	Distance from Sun (in AU)	Atmospheric Gases (in Bars)
Mercury	0.38	.06	5.4	Weak	59	0.39	He, CO_2 (10^{-12})
Venus	0.95	.8	5.2	–	243*	0.72	CO_2, H_2SO_4 (90)
Earth	1.0	1.0	5.5	+	1.0	1.0	N_2, O_2, Ar (1)
Mars	0.53	.1	3.9	+	1.03	1.5	CO_2 Ar (6×10^{-3})
Jupiter	11.2	318	1.3	+	±0.4	5.2	H_2 NH_3 (?)
Saturn	9.4	95.2	0.7	+	0.44	9.5	H_2 NH_3 (?)
Uranus	4.0	14.5	1.7	+	0.96*	19	H_2, CH_4, H_2O (0.4)
Neptune	3.8	17.2	1.7	?	0.92	30	H_2 (?)
Pluto	0.17	.006	.7	?	6.4	48.4	CH_4 (?)

Note: Asterisks indicate retrograde rotation.

eter occupy this belt, as well as perhaps another 100,000 big enough to see with the telescope, and a million or so smaller ones. And that is not all.

Two other groups of asteroids are known. One group includes the Trojan asteroids, individually named for heroes of the Trojan War. About a thousand of them cluster behind and ahead of Jupiter in the same orbit, traveling at the same speed as Jupiter and delineating equilateral triangles with it and Sun (Fig. 1.5).

Another group of about the same number as the Trojans comprises the Apollo asteroids (e.g., Apollo and Hidalgo in Fig. 1.5). These relatively large objects differ from others except comets in that they trace out strongly ellipsoidal and potentially collisional orbits that cross the paths of Earth and other planets. Some fifty of them have now been observed and named or numbered, and it is likely that most of the really large objects that have impacted Earth in the past 4 billion years were Apollo asteroids (or comets).

Properly speaking, of course, such objects are meteors or meteoroids only while they remain aloft. After landing on Earth they are called *meteorites* (the *-ite* suffix denoting a rock or mineral). The roughly 3,000 meteorites known so far are geochemically and mineralogically diverse. That diversity implies some sixty different larger parent bodies that presumably originated in different parts of interplanetary space and then broke up. When we include such hypothetical bodies, comparative planetology becomes a very large subject.

Finally, among natural objects in space, we find those celestial streakers, the luminous and ghostly comets—terror of early observers, fearful omens that coursed the skies with streaming hair, the cometary "tail." The tail is now known to result when their icy sheaths vaporize, blown outward from Sun by the pressure of a radiant solar wind of ionized particles traveling at speeds of hundreds of kilometers a second. Most comets are now thought to issue from a concentration of hundreds of millions of giant "dirty iceballs" that surround the solar system in a spherical cloud about 15 trillion kilometers in diameter—some 1,000 times the diameter given above for the whole solar system inside Pluto's orbit. This is the Öpik–Oort cloud of icy objects, described independently by E. J. Öpik in 1932 and by J. H. Oort in 1950 respectively.

When distances are so great, it is easier to think of them in astronomical units (1 AU, the mean distance from Earth to Sun, is about 150 million kilometers) or in light years (1 LY is not quite 9.5 trillion kilometers). The mean diameter of Pluto's orbit is 79 AU. The Öpik–Oort cloud is 100,000 AU or about 1.6 LY across. This vast cluster of potential comets has a total mass perhaps 100 or more times that of Earth. They become true comets when forced into independent solar orbits by the gravitational effects of passing stars or the massive outer planets.

In considering the origin and evolution of the solar system, therefore, comets and their icy cousins in the Öpik–Oort cloud are not to be overlooked. Rich in water (as ice), carbon, and other volatiles, these mysterious apparitions have potential implications far beyond that implied by their small mass, generally great distance from us, and rarity of appearance. Among these are their possible significance for the distribution of water and carbonaceous matter in the planets, their potential function in the sampling of space beyond the planetary orbits, their bearing on the distribution of angular momentum in the solar system, and the probability that a significant number of meteorite impacts were the products of cometary nuclei or their stony inclusions. Like Banquo's ghost, their influence may well exceed their present substance.

Indeed when we add up the whole mass of solar satellites we see that, after the hydrogen and helium of the giant planets, the next most abundant stuff is water (H_2O)—mainly frozen. *Voyager* satellites *1* and *2* have sent back signals in the form of data on density and reflectivity that ice is the most abundant rock in the solar system. Not only the 100 Earth-masses of the Öpik–Oort cloud, but also the two largest satellites of Jupiter and the seven largest of Saturn, are largely ice. Ice and icy slush may comprise the greater part of Neptune and its satellites, and may be important components of Jupiter and Saturn, not to mention the 7,000-kilometer-deep ocean of superheated water beneath the heavy atmosphere of Uranus.

The Nebular Nursery

How and when was the once thinly dispersed stuff of the planets, their satellites, and other bits concentrated, packaged into solid and densely gaseous bodies in specific parts of solar space, and gravitationally balanced so as to remain in orbit without more frequent collision?

The conditions under which a single star as small as our Sun could form and become surrounded with a bevy of satellites are evidently unusual. Beginning with the classic study of dark interstellar nebulae in our galaxy by E. E. Barnard in 1919, however, astronomers have come to recognize that a type of star found in the dark nebulae of the Taurus region of the galaxy makes a good prototype for our parent star. Such objects have masses close to that of Sun, are estimated to be about a million years old, and display temperatures and luminosities indicating that they are young stars. They are in a stage of gravitational collapse pre-

liminary to ignition and entry to the astronomical main sequence as hydrogen-burning main sequence stars. One of them, T-Tauri, has given its name to a whole class of protostars in the Taurus and other regions of the Milky Way (Fig. 1.3). The conditions of their origin and evolution are considered to reflect those experienced by our own star at an early stage in its evolution.

An interstellar nebula of sufficient density and size is inherently unstable, tending to collapse gravitationally into smaller and smaller pieces because, as its density rises with collapse, the pull of gravity increases. It is believed, therefore, that the formation of individual T-Tauri stars, including our own, involved the fragmentation of such nebulae into smaller protostars. Photographs of the Taurus region of the galaxy (Fig. 1.3) show that groups of T-Tauri stars averaging 0.5 to 2 solar masses are imbedded in dense nebulae of dust and gas some

30 trillion kilometers or more in diameter—one parsec (or 3.26 light-years) in astronomical terms. Still further fragmentation, however, would be needed to arrive at a nebula of the right size to condense to a single solar mass, and such fragmentation and condensation is where supernovation could have played a part beyond its critical role in nucleosynthesis.

Gravitational interactions in star clusters or large, turbulent, stellar nebulae of many solar masses explain why 80 percent of all stars in our galaxy belong to binary and multiple star systems. Double and multiple stars, however, are of little interest in the present context because their interacting gravity fields make it virtually impossible for bodies of planetary mass to find stable orbits around them. In addition, massive protostars evolve so rapidly that they pass quickly through the various stages leading to supernovation, with the return of parts of their mass to interstellar space. A star of 40 or more solar masses might go through its entire evolution in a million years or less—about half the length of the Pleistocene ice ages on Earth and not nearly long enough for the origin of a planetary system.

Fortunately for us, the one at the center of Earth's orbit is one of the long-lived, small, single stars, capable of providing stable planetary orbits for 10 billion years or so—presumably plenty of time for advanced forms of life to evolve from some initial concentration of potentially life-building organic macro-molecules. If, by analogy with the icy sheaths of some T-Tauri stars, we take the Öpik–Oort cloud of cometary nuclei as a measure of the initial solar nebula, a diameter of about half a parsec for the origin of a solar system (about 15 trillion kilometers) may be about right. As the mass of such a nebula would very likely have been less than 100 solar masses, shock waves from the supernovational implosion of large neighboring stars may well have been needed to bring about its condensation into the solar system we know.

Origin and Spacing of the Solar System

Here, then, we focus on the planets, revealed at last in all their mysterious beauty by NASA imagery and instrumentation. The planets comprise two sets. The inner terrestrial or rocky planets include, from Sun out, moonlike Mercury, torrid Venus with its CO_2 greenhouse and sulphuric acid clouds, Earth with its cool blue seas and multicolored lands, ice-capped Mars with long-dry rivers and giant extinct volcanoes, and the frigid and commonly carbonaceous asteroids. The outer Jovian or icy-gassy planets are hydrogen-rich Jupiter with sulfurous, volcanically active Io plus a dozen icy satellites, gassy Saturn with its equally icy satellites and prominent rings, and the three less well known outermost planets.

How then, and when, did so diverse a collection of planets come to be as they are? Over the centuries that humankind has pondered such questions, different hypotheses have been advanced, studied, and abandoned or modified. It now seems increasingly likely that the two sets of planets had different but interdependent origins. Probably the outer planets, or at least Jupiter and Saturn, formed first, as a result of the condensation of large gaseous protoplanets. Massive Jupiter then assisted the buildup of the terrestrial planets by deflecting planetesimal trajectories, including objects from the orbits of Mars and the asteroids. Because all of this activity is seen as taking place within the solar nebula, we speak of a nebular hypothesis, although not in the same sense as advanced by the French mathematician Pierre Laplace in 1796 (and by Immanuel Kant in 1755).

Laplace's original nebular hypothesis had the beautiful simplicity of all great ideas. It proposed that the planets condensed from successive rings of matter thrown off by a contracting and ever more rapidly rotating proto-Sun. Like a spinning figure skater, who turns ever faster as she lowers her arms, the early Sun was expected to conserve its angular momentum (a measure of rotational force) by rotating faster as it contracted, intermittently detaching rings of matter that condensed to form planets. Alas, this lovely idea runs afoul of a stubborn fact. Although it is a basic law of nature that angular momentum is conserved, calling, in Laplace's hypothesis, for a rapidly spinning residual star, Sun turns on its axis barely once a month. About 99 percent of the apparent angular momentum of the solar system is in the mass (318 Earth-masses), orbital velocity (13 kilometers per second), and orbital radius of Jupiter—and not in the rotation of the Sun. Even though recent research by solar physicists implies much faster rotation of the solar interior than of its surface, the problem of the missing

angular momentum remains a difficult one for the classical Laplacian hypothesis.

As in the ancient legends of our species, in which great achievements are often described as if the unique feat of some hero or heroine, usually dead, it would have simplified history to proclaim Laplace the winner and move on. In reality, it is more usual for the prevailing view at any time to grow by stages from accumulating insights and evidence to the point where some critical new discovery or realization suddenly restructures the paradigm and gives it a new name. In this way, for example, classic geological perceptions of continental drift gave way to plate tectonics in the 1960s.

So too has the Laplacian mechanism been replaced while holding to the broad idea of a nebular model. Earlier in this century geologist T. C. Chamberlin and astronomer F. R. Moulton tried to revive another old view calling for disruption of Sun by the gravitational attraction of a more massive passing star. They perceived such an event as responsible for chunks of matter called planetesimals that subsequently aggregated to form larger bodies, including the planets. The planetary satellites in this scheme were seen as large planetesimals that had drifted into orbital balance with the currently growing planets.

Although stellar disruption has also been rejected, the idea that planets grow by the gravitational accumulation of planetesimal-like aggregations of matter remains. That is combined with the concept of nebular fragmentation into gaseous protoplanets as the currently most favored mechanisms for building solar satellites of planetary dimensions. Current versions of planetesimal aggregation thus allow either for initiation at the centers of gaseous protoplanets, as in the Jovian planets, or for buildup within the solar nebula, starting with colliding dust particles (Fig. 1.6)—as in the terrestrial planets.

A word now about the interesting and perhaps significant spacing of the planets. At the center, of course, is the parent star Sun, then the planets in the order illustrated in Figure 1.7. Back in 1766 a German astronomer, Titius von Wittenberg (J. D. Tietz), noticed a striking regularity in the distance between the then-known planets. If one starts with Mercury at 0, gives Venus the number 3, and then successively doubles the number assigned to each planet from 3 outward, one gets the sequence 0, 3, 6, 12, 24, 48, 96. If 4 is added to each of these numbers and they are divided by 10, Earth becomes 1, signifying the distance from Sun to Earth as 1 *astronomical unit*

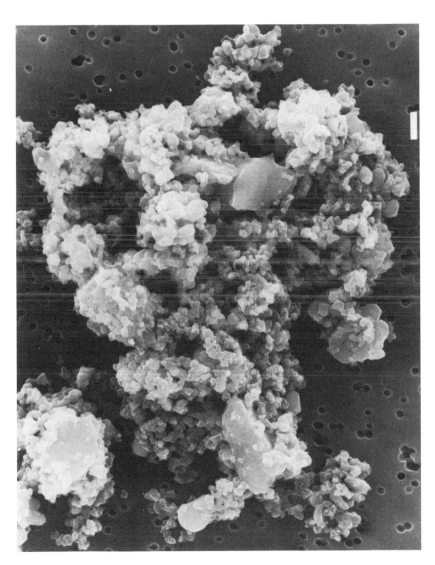

1.6 STARDUST: AN INTERPLANETARY PARTICLE. An electron micrograph of an aggregate of tiny grains less than 0.001 mm in diameter. The cluster is thought to be of cometary origin. *[Courtesy of J. P. Bradley; cover photograph,* Nature, *February 1983.]*

(AU). The distances from Sun to planets in AU are then Mercury 0.4, Venus 0.7, Earth 1, Mars 1.6, asteroid belt 2.8, Jupiter 5.2, and Saturn 10—very close to the measured spacing of the then-known planets. When, 15 years later (1781), F. W. Herschel discovered the new planet Uranus, the first since antiquity, at a distance of 19.2 AU from Sun, that number was so close to the 19.6 predicted by Titius' observations for the next planet out as to make the regularity seem fundamental.

1.7 ORDER OF THE PLANETS OUTWARD FROM SUN AND THEIR SPACING ON A LOGARITHMIC SCALE OF ASTRONOMICAL UNITS OR AUs. One AU, the average distance from Sun to Earth, is 150 million km. Symbolism as in Fig. 1.4, plus *A* for asteroids. Each planet's orbit is 75 percent larger than the next inner one. Kepler recognized the linear proportionality between the squares of the times of the orbits of the planets and the cubes of their mean distances from Sun. *[John A. Wood*, The Solar System, © *1979, p. 25. Adapted by permission of Prentice-Hall, Inc., Englewood Cliffs, New Jersey.]*

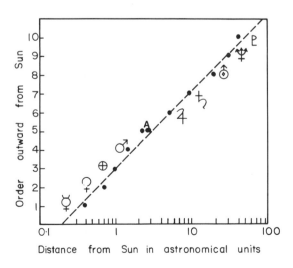

Poor Titius. Not only was his rule not substantiated by the subsequent first sightings of Neptune and Pluto, he rarely gets the credit for being first with the idea. If cited at all it is most commonly given as Bode's Rule (or, incorrectly, Bode's Law) for his countryman Johann Bode, who redescribed the spacing in 1772 without reference to Titius.

As is so often the case in science, it is important not only to have interesting ideas but to have them at a propitious time and in the right form. The Titius–Bode Rule is still with us but in a different form. If plotted in order against a logarithmic scale of distance on a slope such that the radius of each planet's orbit is 75 percent larger than that of the preceding planet going outward from the Sun, the planets fall on or close to a straight line (Fig. 1.7). A similar spacing is seen for the satellites of Jupiter and Saturn. It is most plausibly a product of the dynamic balance of gravitational, collisional, and centrifugal forces. For planetesimals, presumably, would tend to clump as planets in orbits that would have the greatest long-term stability in relation to the masses and gravitational effects of Sun and sister planets. The process is sometimes called dynamical relaxation. The orbital velocity and spacing of the planets and their satellites is just right to prevent their falling into Sun, while avoiding centrifugal effects that would fling them into space or one another.

Sorting Out the Planets

Another interesting relationship displayed by the various orbiting bodies of the solar system is their previously mentioned approximate arrangement in broad categories of decreasing density—rocky, icy, and gaseous. Rocky objects—the four inner or terrestrial planets plus the asteroids—are closest to the Sun. Icy and gaseous objects are farther out, although Jupiter's gaudy and fascinating pizza pie of a satellite, Io, is also rocky. That suggests that sorting of planetary components according to density, volatility, and distance from Sun might have something to do with planetary composition. The mainly icy and gaseous objects, however, do not show a clear segregation with distance. The giant planets Jupiter and Saturn, essentially of solar composition, are the gasiest. Their satellites and the comets are the iciest.

Such observations limit permissible conjecture about the origin of the planets. Any plausible model must cope with the following requirements:

1. It must satisfy the observed distribution of angular momentum.

2. It must explain the observed distribution of rocky, icy, and gaseous objects within the solar system, as well as variations in their proportions.

3. It must say how it happens that all of the planets lie in essentially the same orbital plane, why their rotational axes are inclined to that plane, and how it happens that some planets and satellites have retrograde rotations (rotate clockwise instead of counterclockwise like the others).

4. It should account for the small mass of Mars and the absence of a terrestrial planet of the expected size in the asteroid belt. It must be consistent with what is known about the evolution of main sequence stars.

5. It must explain, or at least be consistent with, geochemical anomalies such as the excesses in some meteorite inclusions of radiogenic xenon-129 and magnesium-26, as well as Earth's depletion in potassium (by about 4 relative to nebular abundances) and other variations that imply thermal irregularities, limited mixing, or undetected processes within or before the solar nebula.

Angular momentum continues to be troublesome. There seems to be a kind of consensus that it is not as serious a problem as once thought, but little agreement as to why.

The distribution of rocky, icy, and gaseous objects within the solar system, and variations in their proportions, can perhaps best be explained by some combination of the gaseous protoplanet hypothesis of the Smithsonian's A. G. W. Cameron and planetesimal accretion within the cooling solar nebula.

Although the observed rough gradient of outward-increasing volatility of condensing substances cannot be a simple consequence of distance from our Sun, it implies approximate equilibration of planetary components within local regions of the parent nebula. As solid particles at or near equilibrium for that part of the nebula accumulated toward its midplane, instabilities in the nebular disc would lead to a clumping of matter into

25

ever-larger planetesimals. The impact energy of objects in the same orbital path would remain low, and the cushioning effect of nebular dust and gas would favor continued increase in size with increased mass and gravitational attraction up to some limit. Above a certain size, especially if smaller bodies were gravitationally deflected into erratic orbit by previously formed larger objects (e.g., Jupiter), impacts would become more destructive. Only the largest bodies could continue to grow as lesser ones were broken by repeated collision into ever-smaller fragments, became assimilated to objects of planetary dimensions, or were ejected into asteroidal orbit or from the solar system.

The concentric structure of our home planet, with bilayered metallic core and multilayered silicate mantle, is interpreted in different ways. Some see it as a product of the outwardly successive accretion of shells of increasingly less refractory materials. Others interpret it as reflecting the conversion of aggregational energy to heat, leading to planet-wide partial melting and gravitational layering. Had such heat all been retained and not largely radiated away during the accumulation of Earth, for instance, our fully aggregated planet could have reached a temperature of about 29,000°C. As we shall see, that is not likely; but it is likely that melting had something to do with the concentration of iron in the cores of the terrestrial planets.

Omitting the asteroid belt, we come to the outer or Jovian planets. As already noted, the fact that Jupiter and Saturn are of essentially solar composition implies that they could be the condensed remains of giant gaseous protoplanets, left behind as the solar nebula con-tracted. In some ways they are like ministars, radiating more energy than they receive from solar sources and with planet-like families of satellites. They could also have aggregated initially from planetesimals and collected their thick gaseous envelopes later. The mass and average density of both Jupiter and Saturn imply the presence of rocky cores of several Earth-masses beneath the shells of metallic and liquid hydrogen that make up most of their bulk. Once grown to between 10 and 50 Earth-masses, they would have been massive enough to capture and retain hydrogen and helium from the T-Tauri and solar winds. To that they could have added veneers of solid and liquid H_2O, ammonia (NH_3), and gaseous hydrogen (H_2) until reaching their present size.

In the case of Uranus and Neptune, although their densities imply that they have rocky cores beneath their watery and icy mantles, they lack the mass to retain hydrogen and helium. The satellites of all these outer Jovian planets condensed so far from the center of the cooling solar nebula that they are mostly rich in ice of a range of crystalline forms.

Why do the planets travel in nearly the same orbital plane as Sun? And why are their rotational axes tilted? The fact that they lie in the same orbital plane shows that they condensed from the same discoidal solar nebula. Tilts of their rotational axes support an origin by planetesimal aggregation. Collision with a truly large planetesimal could have caused a planet to tilt or even to reverse its direction of rotation, as with Venus and Uranus. Uranus may have suffered a tremendous blow to account both for its reverse rotation and the near-90° tilt of its axis. Near misses and capture may also account for some of the planetary satellites. Others may have accumulated from peripheral rings of matter similar to those of Saturn but far enough out to exclude gravitational disruption. In fact, the currently favored origin of Earth's moon, attributed to A. G. W. Cameron in 1976, holds it to be the result of the impact of a Mars-sized body on the proto-Earth. Such an impact, it is proposed, blasted portions of proto-Earth and the impacting planetesimal into an Earth-centered orbit from which they then condensed as the moon. If that happened after the separation of iron cores in both bodies, it could explain the geochemical similarities of the moon to Earth's mantle, its lack of volatiles, and its depletion of iron, as the core of the impacting body settled out to join that of Earth.

The small mass of Mars, the absence of a body larger than Ceres in the asteroid belt, and the wide scatter of those asteroids may all be related to the presence of massive Jupiter. The gravitational effects of its great mass probably swept up or ejected so much of the solid matter in this part of the solar disc that there wasn't enough left over to make a larger Mars or a planet-sized body in the asteroid belt. It may also have deflected other planetesimals and cometesimals into collisional orbits with the growing inner planets, much as its gravity was used in 1980 to fling *Voyager 2* to the outer solar system.

The foregoing is consistent with much of what we think we know about the evolution of main sequence stars like our own. They are the comparative base for the interpretation of early solar nebular development. Although there is little agreement about the exact sequence of events in the early solar system, model experiments and other lines of evidence converge on 100 million years after initial condensation of the nebula as the likely time required to complete aggregation of the inner planets. All agree, though, that the condensation of the solar nebula probably involved a T-Tauri phase, during which a fierce plasma gale cleared that nebula of any remaining dust and gas.

Did the nebula-clearing T-Tauri wind precede or follow the aggregation of the inner planets, or did it "blow" during planet formation? To the extent that there is a prevailing judgment, it is that the main aggregation of the planets preceded the T-Tauri phase of solar radiation. That would have stripped the inner planets of any previous surface gases, leaving the origin of atmospheres to result from some process associated with the tapering off of the aggregational process and later outgassing events.

The final accumulation of the planets to their present dimensions was probably rapid after some critical size was reached. Debris from planetesimal collisions was rapidly assimilated by larger proto-planets. The age of 4.6 billion years found for the solar system dates that buildup. But planetary aggregation from preexisting objects evidently tailed off during another half-billion years into ensuing planetary history. Evidence for that is seen in impact craters and basins on Mercury, Luna, Mars, and other solar system objects all the way out to the Saturnian satellites and beyond.

Dated lunar rocks and crater counts on lunar surfaces of varying age indicate that a high rate of planetesimal, meteoritic, and probably cometary infall continued to shape the moon until about 3.9 billion years ago, when such collisions tapered off to something no more than perhaps 20 times the present infrequent rate. Similar crater counts and patterns on other solid bodies in the solar system imply a similar late meteoritic bombardment throughout the solar system.

Late infall of volatile-rich objects is discussed in Chapter 2.

Harvard–Smithsonian astrophysicist Fred Whipple, for instance, has calculated that, at some time following the major formative events of the solar sys-tem, the cometary flux in Jupiter's orbit was at the significant rate of about half an Earth-mass per million years. Although it has been argued that Earth's surface water results primarily from the outgassing of occluded water in its interior, we must now consider how much of it could be of cometary origin. A similar question must also be asked about carbon and the origin of the initial stock of organic molecules from which life was later to arise on Earth.

In Summary

Product of a long presolar evolution during which matter itself evolved and became locally concentrated, our solar nebula condensed from part of a larger interstellar cloud of dust and gas somewhat more than 4.6 billion years ago. The condensation of that solar nebula during the following 100 million years led to the formation of the Sun, planets, and a huge number of other objects, most of which occupy a common orbital plane. The processes that created the planets are believed to have involved the progressive aggregation of metallic, rocky, and icy components into planetesimal bodies and eventually planets, as well as the condensation of gaseous proto-planets. Although the solar nebula was generally well mixed, it was not truly homogeneous. Among other processes, the episodic injection of foreign matter from supernova shock waves saw to that.

Temperatures within this solar nebula probably ranged from hotter than the vaporization temperature of iron and silicates near the condensing proto-Sun to cold at the margins. What we know of planetary chemistry suggests that matter condensed outward along this temperature gradient in roughly the reverse order of volatility. Mercury, some 80 percent iron, aggregated nearest the parent star, the other metallic and rocky terrestrial planets at regularly increasing distances outward. H_2O is abundant as water on Earth, as ice on Mars, was probably once abundant on Venus, and fills a vast superheated ocean beneath Uranus' dense atmosphere. Objects in the asteroid belt are rich in carbon, although preponderantly silicates and iron oxides.

Ice is the common rock on the cold satellites of the outer planets, averaging nearly 50 percent of their mass. Jupiter and Saturn, of essentially solar composition, were probably the first planets to condense, perhaps from gaseous proto-planets. The huge mass and strong gravitational effects of Jupiter account for the small size of Mars and the dispersal of matter in the main asteroid belt. Jupiter was probably also a prime agent in the gravitational focusing of objects involved in the late meteoritic bombardment of the solid planets, including their veneering with the volatile stuff of carbonaceous chondrites and comets.

The process, however, was surely not smooth or homogeneous. Although the chemistry and petrology of the planetesimals presumably reflected the stable chemistry of the regions and times of their origin, bodies of somewhat variant composition joined to make the final planets. Most probably formed close together. Others may have grown deep in interplanetary space and were only later deflected to merge with the planets. At some point, the last of the interplanetary dust and gas was swept clear of the nebula by fierce presolar (T-Tauri) winds, carrying any planetary atmospheres then extant with them. This seems to have been a relatively late event in planetary evolution.

How does one explain local heterogencity within broad compositional trends observed? The subject of planetary origins is only partly understood. It is the subject of active research. About their very great age and nearly concurrent origin there is no doubt, but no such confidence is warranted concerning their mode and sequence of aggregation. It is a safe bet that future studies will find better and more complete solutions than those here proposed.

As for solutions offered, apart from being consistent with evidence available, it is more important that they stimulate and focus research than whether they are right or wrong. The hope is that research stimulated or provoked by them will eventually result in closer approximations to the truth, the ultimate goal of good science. Meanwhile, extant hypotheses provide a matrix within which we can view the likely primordial state and earliest development of our home planet—Earth.

CHAPTER TWO

THE HADEAN EARTH

The history of the first 600 to 800 million years of Earth's existence as a planet, the Hadean interval, is reconstructed from the clues of lunar exploration, comparative planetology, cosmochemistry, and computer modeling. They call on heat from aggregation, early radioactivity, and core formation to explain pervasive melting and the origin of Earth's concentric structure. They suggest early formation of an iron core with a solid interior and a liquid outer shell, source of Earth's radiation-shielding magnetosphere. They suggest how atmosphere, hydrosphere, and a supply of prebiotic organic molecules may have appeared later and concurrently from internal and meteoritic sources, including comets. They imply that simple microbial life may have arisen by the end of Hadean or the beginning of succeeding Archean history. But the evidence is indirect, the footing uncertain. Like the surface of a quaking bog, our vision of the Hadean Earth holds for the moment but may give way with the next step. It contains no clues to the eventful future that lay before our remote microbial ancestors once they arose.

Getting Started

What was Earth like in the very beginning? How does one get from a conglomeratic ball of stardust to a planet with a concentric structure—inner and outer iron core, silicate mantle, crust, oceans, and a layered atmosphere? When and how did life come into being? For life, from the moment of its first appearance, has played a major role in Earth's surface history. It would be hard even to think about Earth history without it.

The habitability of Earth, however, arises from attributes that may not be as common in the universe as in our livelier imaginings. In addition to the right kind of chemistry, a source of energy, and shielding from lethal ionizing radiation, life demands a continuously fluid medium in which its vital transactions can take place. As such transactions require both a continuing supply of fresh nutrients and volatiles and the removal of wastes, even the tiniest metabolizing body must be constantly flushed. The most probable mixing and transporting fluid is water (H_2O), although, given enough of it, ammonia (or perhaps some liquid

2.1 How sunrise of a short late Hadean day might have looked. A post-eruptive view inside Mount St. Helens, Washington, 1980. *[Courtesy of C. A. Hopson.]*

31

hydrocarbon) might work in an even more unusual environment whose temperature never fell below its freezing or rose above its boiling point at prevailing ambient pressures. Unlike water, however, which expands on freezing and thus makes an ice that floats and shields the liquid H_2O beneath, ammonia contracts on freezing and ammonia ice sinks, so that eventually no accessible liquid would remain.

Water, then, is the fluid of choice, at least on Earth, and the presence of life here assures us that, as long as living things have existed, liquid water has been continuously available somewhere on the planet. That, in turn, requires that surface temperatures were never so low as to freeze it all or so high as to vaporize it completely. As we shall see, microbial forms of life have very likely been present since 3.5 to 3.8 billion years ago. The record of sedimentary rocks over that same interval, moreover, tells us unequivocally that a shielding atmosphere and hydrosphere have existed on Earth for at least the last 3.8 billion years. That follows from the fact that an atmosphere is required to weather rocks to sediments and a hydrosphere to impose the observed sorting and stratification on these deposits.

What about earlier? When, in fact, did the first winds blow, the first clouds form, the first water flow, the first life pulse? Because a direct record of the first 600 or more million years of our planet's existence in the form of rocks is missing, we can visualize that history only dimly, through the mists of indirect evidence and our imaginations (Fig. 2.1). Although minerals older than 3.8 billion years have been reported and meteorites of even greater age are known, this part of Earth's history must be reconstructed from evidence in other parts of the solar system or inferred from the geochemistry of meteorites and later-formed Earth rocks. Such evidence tells us that Earth's primordial surface was probably too hot, too exposed to lethal radiation, too dry, and too battered by large planetesimals or meteoroids for life to emerge or survive during much or all of that long interval of unrecorded or only indirectly recorded history.

This part of Earth history, generally before about 3.8 billion years ago, is appropriately called *Hadean*. It is the first of the four great historical intervals, or Eons, with which this book deals (see Chapter 5), a time during which conditions were, by current standards, hellish. It was also a time during which our planet acquired such distinctive attributes as its layered structure, its magnetic field and radiation-shielding magnetosphere, a solid surface, our ancestral atmosphere and hydrosphere, and a stock of polyatomic organic macromolecules from which the magic fabric of life was to be spun. Hadean history closed with the formation of the oldest recognized solid crust about 3.8 billion years ago. When it is said that the solar system is about 4.6 billion years old, what is meant is that its oldest well-dated meteoritic and lunar components were then already present and that no older ones are known.

At this point the importance of the different nuclear components of an element such as lead (Pb) becomes clear. Although all atoms of a given element contain the same number of protons and possess the same chemical attributes, many have different numbers of neutrons. As mentioned earlier, atoms that differ from others of the same element only in the number of neutrons are called isotopes, and lead, with 82 protons and the same *atomic number*, has five different isotopes, depending on the total number of protons plus neutrons, representing *atomic weight* (Pb-204, 205, 206, 207, 208). Two, and only two, of them, Pb-206 and Pb-207, are the radiogenic decay products of two different radioactive isotopes of uranium that decay at different rates. The systematically varying ratios of parent to daughter isotopes, and of the different isotopes of lead to one another, provide three independent systems of determining rock ages. Employing sophisticated geochemical methods (discussed in Ch. 4), we have already learned from these isotopes that the solid components that aggregated to form the various rocky and metallic objects in the solar system condensed from the solar nebula around 4.6 billion years ago.

At the end of that primordial aggre-gation, the construction of the planets was essentially complete, but their history was only beginning. On Earth as on others, that history included substantial additions of meteoritic, planetesimal, and cometary matter during the following 700 to 800 million years of Hadean time—including, it seems, a late peak of very large objects about 4 to 3.8 billion years ago.

With so little direct evidence to limit speculation, however, one may visualize more than one plausible pattern for Hadean history. It is not known exactly how the initially aggregated planet acquired its broadly concentric internal structure, magnetosphere, rocky crust, and outer volatile envelope of gases and water during the next several hundred million years. Discussion, therefore, is limited to some general directions this history may have taken, and to the identification of salient problems and relevant evidence that may better inform the quest for answers.

Outlook from the Lunar Window

Gaea, mother goddess of Earth, guards her secrets carefully, and with increasing jealousy as they recede in time. The more distant the past, the more likely any records are to have been erased by erosion or obscured by more recent events. The sparse records of Hadean time are best viewed from our satellite Luna, a window on earliest Earth history. Although the familiar Apollo program of lunar exploration only scratched the surface of what might be learned, it made one important thing clear: A major part of the lunar landscape dates from Hadean times! The lunar highlands consist mainly of aluminum- and calcium-rich, silicate rocks of a type known on Earth as *anorthosite*—precipitated from an anorthositic melt, or *magma*. And two samples of it have yielded radiometric ages close to 4.6 billion years. These lunar highlands evidently crystallized at that time from a planet-wide ocean of molten rock at the surface of the primordial moon, a product mainly of the conversion of accretional energy to heat, perhaps including that from collision with the accreting Earth. The lunar highlands are intensively pocked by younger meteoritic craters, evidence that planetesimal impact continued to be an important lunar process long after solidification of the old highland rocks.

The infalling bodies responsible for these lunar craters included objects large enough to blast out the huge mare (már-ay) basins that are so prominent on the Earth-facing side of the moon. These basins in turn were flooded with very Earth-like basaltic lavas having ages of 3.8 to 3.2 billion years. A striking feature of these waterless lunar seas (maria) is that the lavas that fill them are only sparsely scarred by more recent impact craters. Between the late intensive cratering that excavated the mare basins around 4.2 to 3.9 billion years ago and the upwelling of their mare-filling lavas, the rate of conspicuous impact tapered off dramatically. The smooth mare surfaces of Earth's moon, Mercury, and Mars alike show average densities of 100 early craters larger than 10 kilometers in diameter per million square miles (2.62 million square kilometers). The rate of cratering thereafter drops sharply to barely 20 times the present, with really big ones then impacting only about once every 5 to 10 million years—a greatly diminished and episodic but still significant geologic process.

The inferred main geologic processes on the Hadean Earth, therefore, were impact cratering, impact-induced volcanism, and planet-wide melting and volcanism as a consequence of gravitational and radioactive heating. Between the main initial aggregation of our planet and the final tapering-off of meteoritic infall toward present rates about 3.9 to 3.8 billion years ago, Earth grew more slowly. As she did, she expelled great volumes of steam and other gases as products of extensive melting and vigorous volcanism, evolving toward a more habitable state.

Origin of Earth's Concentric Internal Structure

The infant Earth suffered severe growing pains. Her temperature increased with the conversion of gravitational energy to heat, thermal inputs from the radioactive decay of short-half-lived radioactive isotopes, and the descent of iron to the core. As a result, internal temperatures probably exceeded any calculated for hell itself. Pervasive melting was to be expected. That is considered a likely explanation for the layered structure of Earth's solid interior, somewhat like that of an onion but of outward-decreasing density and less regularity. The *inner core* is essentially solid iron, whereas the *outer core* is liquid iron. The multilayered *mantle* consists of silicate rocks and the surface *crust* of still-lighter silicates and sedimentary rocks of varying composition.

Early in the century it was widely believed that this layering was a product of cooling from a huge lump of molten or incandescently hot matter. Geologists widely compared Earth's folded mountain ranges with the wrinkled skin of a baked apple. Subsequently it came to be widely believed, and many still believe, that the planet was never hot enough to melt all the way through. Now the idea of a formerly molten or partially molten Earth is once again in favor.

Consider how the Earth may have become so hot, how the previously mentioned sources of heat may actually have worked.

First is accretional or gravitational heating—the conversion of the kinetic energy of infalling objects to heat. A single ordinary 10-ton planetesimal, for example, approaching from overhead at a likely speed of 25 kilometers a second, would carry with it more than 100 million megatons of kinetic energy—comparable to that in the nuclear arsenals of the planet. Geochemist George Wetherill of Carnegie Institution has calculated that, *if* Earth retained *all* of its accretional heat, its temperature at the end of aggregation would have risen to some 29,000°C. Although much of that heat would surely have been radiated off to space from the aggregating surface, some from very large planetesimals would presumably have been trapped within the planet. Only 10 percent of the computed heat of aggregation would have raised Earth's initial temperature to 3,000°C—far above the melting points of iron and the common silicate rocks.

A second source is radiogenic heating from relatively short-half-lived radioactive isotopes. Half of all the heat that could ever result from the decay of U-235 to Pb-207, for instance, was released during Hadean time. Even though this process is slow, it is especially significant over Hadean time because of the half-life effect (a negative exponential) and because it is generated within Earth, where conductive loss of heat is slow. This might have added another 800°C to earth's internal temperature during Hadean time.

A third heat source is compression of the earlier-accreted parts of the planet under the weight of its growing outer parts. Such compressional heating is also generated internally and thus released slowly.

The fourth, and last, important source is heat from the formation of Earth's iron-rich core itself (if that was not a primary event). Because iron begins to melt at Earth's surface at only 1,535°C, one may conclude that temperatures on the early planet exceeded the melting point of iron during much or all of Hadean history. The settling out of this initially dispersed iron, representing one-third of the semimolten planetary mass, would have catastrophically raised the temperature of the whole planet by a further 2,000°C.

Had Earth retained only its accretional heat, it would have vaporized—something that clearly didn't happen. The planet never approached temperatures such as would have resulted from efficient heat retention.

Given such a combination of heat sources, however, a likely explanation for Earth's internal structure is melting or partial melting followed by gravitational segregation and differential settling of now-layered components. Temperatures at Earth's surface of more than 1,600°C would suffice for that—hotter of course as pressure increased downward. Under such conditions heavy iron would have melted and trickled coreward through the lighter mush of partially melted silicates surrounding and beneath it (Fig. 2.2). Increasing pressure, of course, is why the inner core is solid while the outer core remains fluid, even today—assuring the persistence of our radiation-shielding magnetosphere.

Silicate minerals rich in iron and magnesium make up the outer two-thirds of the planet (Fig. 2.2). One may imagine them as separated into successive zones approximately along a gradient of temperature and pressure. Solidification of the outermost layer of these silicates would have resulted in the first solid crust of our planet. Alternatively, Earth might have aggregated in reverse order of the density of its silicate layers around a large, initially solid (or nearly solid) nickel–iron planetesimal similar to the large core of Mercury.

Earth's fluid outer core is, in any case, anomalous. It has the physical properties of a molten nickel–iron alloy with a density almost 10 percent less than that of metallic iron. Perhaps the light component (sulfur, oxygen, or hydrogen?) of this alloy played a part in lowering its

melting temperature enough for it to sift down through mantle silicates whose layering never reached such a degree as to destroy the mantle heterogeneities now being discovered by refined geochemical and geophysical techniques.

Other consequences of core formation at some stage of Hadean history are also of interest. The presence of this core explains why, although 35 percent of the total mass of the Earth is iron, that metal accounts for only 8 percent of the crust. The settling of such a mass to the core, if that is what happened, would also demand a large increase in Earth's rotational velocity in order to compensate for the shortening of the rotational radius of

2.2 HYPOTHETICAL RECONSTRUCTIONS OF THE HADEAN EARTH AS IT MIGHT HAVE APPEARED DURING AND AFTER CORE FORMATION. Cutout wedges at centers show inferred internal structure. *A*, the planet as molten iron settles toward Earth's center through a mush of upward-migrating lighter silicates. *B*, Earth slightly later in Hadean history. Core formation is complete, and the surface is cratered and scarred by volcanic rift systems. Water from volcanic outgassing and cometary sources accumulates in dark depressions between elevated areas.

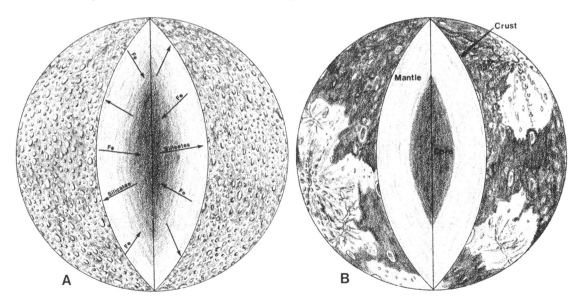

a third of its mass, as required by the law of the conservation of angular momentum. And that would rapidly reduce the previous length of day by half or more, with brief day–night and freeze–thaw cycles that may have been significant in chemical evolution leading to the origin of life. Finally, the dynamo effects of a rotating fluid iron core were responsible for Earth's magnetosphere.

How long after its initial aggregation did Earth first reach melting temperatures, or did that happen concurrently? How long then until it cooled enough to sustain a hydrosphere and atmosphere? Geochemical similarities between modern mantle-derived basalts and the oldest basalts we know show that Earth's core and mantle had already separated before about 3.8 billion years ago. Iron should be much more abundant than it is in these old basalts had most of it not already descended to the core. Terrestrial lead-isotope data suggest that core formation could not have been much later than about 4.4 billion years ago. And the currently favored theory of lunar origin calls for Earth's core to have already been present at that time. It was the first major event in Earth history, on the heels of or concurrent with planetary aggregation itself.

Estimates of the time required for a silicate mantle to cool enough to support a solid crust imply a duration of only about 100 to 400 million years. It now seems likely that our planet was largely molten by the end of its initial aggregation, and endowed with a core and magnetosphere by then or shortly thereafter. It first cooled to the point where it could sustain a crust and hydrosphere about 4 billion years ago. Rocks much older than the oldest now known are not expected. That age is 3.8 billion years, with hints as old as 4.3 billion. An ancestral atmosphere, of course, was also present by then. It only remained for these things to interact and evolve.

Sources of Air and Water?

Whether a planet is endowed with air and water, and what its atmospheric components may be, is a matter of gravity and temperature as well as potential sources. It depends on the mass of the planet and its atmosphere, its distance from the warming Sun, and the properties of the elements (Fig. 2.3) and their compounds.

Should one wish to go exploring in space, a spaceship must be set in motion away from Earth's surface at a speed that exceeds the escape velocity—the force required to break free from the grip of Earth's gravity. Although that velocity does not vary for a given planet, the power required to accelerate to it depends on the mass of the object being accelerated. Gases—which are naturally in motion and relatively light—are the only stuff that regularly escapes the present Earth, and only the lightest of gases do that. Their natural speed of motion increases with temperature, however, so that, if it is hot enough, all natural gases can escape. Gases can also be stripped from Earth's surface by the unshielded solar wind, which even today sweeps around Earth's deflecting magnetosphere at supersonic speeds.

Legend

ELEMENTAL SYMBOL — ATOMIC NUMBER

	3
Li	
LITHIUM	6.941

ATOMIC WEIGHT

D. I. MENDELEEV

ATOMIC WEIGHTS ACCORDING TO INTERNATIONAL TABLE OF 1971.
WHERE WHOLE NUMBERS ARE GIVEN IN PLACE OF FRACTIONAL
ATOMIC WEIGHTS THEY REPRESENT MASS NUMBER OF THE
MOST STABLE ISOTOPE. *SHADING INDICATES ELEMENTS
REGULARLY INCORPORATED INTO LIVING MATTER.*

Periodic Table (GROUPS / PERIODS)

Period	I	II	III	IV	V	VI	VII	VIII	VIII	VIII	VIIIa
1	H 1 / 1.0079										He 2 / 4.00260
2	Li 3 / 6.941	Be 4 / 9.01218	B 5 / 10.81	C 6 / 12.011	N 7 / 14.0067	O 8 / 15.9994	F 9 / 18.99840				Ne 10 / 20.179
3	Na 11 / 22.98977	Mg 12 / 24.305	Al 13 / 26.9815	Si 14 / 28.086	P 15 / 30.97376	S 16 / 32.06	Cl 17 / 35.453				Ar 18 / 39.948
4	K 19 / 39.09	Ca 20 / 40.08	Sc 21 / 44.9558	Ti 22 / 47.90	V 23 / 50.944	Cr 24 / 51.996	Mn 25 / 54.9380	Fe 26 / 55.847	Co 27 / 58.9332	Ni 28 / 58.71	
4	Cu 29 / 63.546	Zn 30 / 65.38	Ga 31 / 69.72	Ge 32 / 72.59	As 33 / 74.9216	Se 34 / 78.96	Br 35 / 78.904				Kr 36 / 83.80
5	Rb 37 / 85.4678	Sr 38 / 87.62	Y 39 / 88.9059	Zr 40 / 91.22	Nb 41 / 92.9064	Mo 42 / 95.9062	Tc 43 / 98.9062	Ru 44 / 101.07	Rh 45 / 102.9055	Pd 46 / 106.4	
5	Ag 47 / 107.868	Cd 48 / 112.40	In 49 / 114.82	Sn 50 / 118.69	Sb 51 / 121.75	Te 52 / 127.60	I 53 / 126.9045				Xe 54 / 131.30
6	Cs 55 / 132.9054	Ba 56 / 137.34	LANTHANIDE SERIES 57–71	Hf 72 / 178.49	Ta 73 / 180.9479	W 74 / 183.85	Re 75 / 186.2	Os 76 / 190.2	Ir 77 / 192.22	Pt 78 / 195.09	
6	Au 79 / 196.9665	Hg 80 / 200.59	Tl 81 / 204.37	Pb 82 / 207.2	Bi 83 / 208.9804	Po 84 / 209	At 85 / 210				Rn 86 / 222
7	Fr 87 / 223	Ra 88 / 226.0254	ACTINIDE SERIES 89–103	Rf 104 / 257	Ha 105 / 262	106 / 263					

Lanthanide Series (Rare Earths)

La 57 / 138.9055 LANTHANUM	Ce 58 / 140.12 CERIUM	Pr 59 / 140.9077 PRASEODYMIUM	Nd 60 / 144.24 NEODYMIUM	Pm 61 / 145 PROMETHIUM	Sm 62 / 150.43 SAMARIUM	Eu 63 / 151.96 EUROPIUM	Gd 64 / 157.25 GADOLINIUM	Tb 65 / 158.9254 TERBIUM	Dy 66 / 162.50 DYSPROSIUM	Ho 67 / 164.9304 HOLMIUM	Er 68 / 167.26 ERBIUM	Tm 69 / 168.94 THULIUM	Yb 70 / 173.04 YTTERBIUM	Lu 71 / 174.97 LUTETIUM

Actinide Series (Uranium Series)

Ac 89 / 227 ACTINIUM	Th 90 / 232.0381 THORIUM	Pa 91 / 231.0359 PROTACTINIUM	U 92 / 238.029 URANIUM	Np 93 / 237.0482 NEPTUNIUM	Pu 94 / 244 PLUTONIUM	Am 95 / 243 AMERICIUM	Cm 96 / 247 CURIUM	Bk 97 / 249 BERKELIUM	Cf 98 / 251 CALIFORNIUM	Es 99 / 254 EINSTEINIUM	Fm 100 / 257 FERMIUM	Md 101 / 258 MENDELEVIUM	No 102 / 255 NOBELIUM	Lw 103 / 257 LAWRENCIUM

2.3 PERIODIC TABLE OF THE ELEMENTS SHOWING THOSE REGULARLY INCORPORATED INTO LIVING MATTER.
[After Preston Cloud, 1978, Cosmos, Earth, and Man, Yale University Press.]

A little reflection tells us that Earth's present atmosphere necessarily evolved from one that was different. We know no primary source for the free molecular oxygen that comprises one-fifth of our present atmosphere. Compared with solar abundances, our atmosphere has only traces of hydrogen and helium but a disproportionate amount of nitrogen.

An important clue to the origin of our ancestral atmosphere is found in the abundances of the so-called *noble gases*—elements that, unlike oxygen, do not (or rarely) combine with others because they have the stable configuration of 8 (or 2 in the case of helium) in their outermost shell of electrons. As they do not ordinarily lose, gain, or share electrons with other elements, variations in their abundance imply different sources. Had Earth inherited its atmosphere directly from the solar nebula, the gaseous elements neon, argon, krypton, xenon, and radon should be present in approximately solar abundances, allowing for the addition of radiogenic isotopes. That is not the case. It has been repeatedly noted over the past half-century that all of the noble gases are grossly depleted in Earth's atmosphere compared with solar and cosmic abundances. They are depleted, in fact, by several to many orders of magnitude. This means either that Earth accumulated without an atmosphere of nebular proportions or that any initial atmosphere escaped its gravity field in some subsequent episode of heating that accelerated even the heavy noble gases to escape velocities.

Sources of heat during Hadean time were sufficient to account either for failure to accumulate an atmosphere that would be gaseous at present surface temperatures or for escape of any initial atmosphere during subsequent heating. Before Earth's fluid outer core was formed, there would have been no magnetosphere to prevent any expected atmospheric gases from being swept away by solar wind—like the gassy sun-swept "tails" of comets. Earth's ancestral surface volatiles probably came entirely from secondary sources—not only its atmosphere but also any H_2O gas or ice that condensed or melted to form its hydrosphere.

Two sources for these volatiles are possible. Either the atmosphere and hydrosphere were expelled by volcanic outgassing from occluded volatiles within the Earth, or they were brought here by partly or wholly icy objects that extensively pocked the lunar and other planetary surfaces between about 4.2 and 3.9 billion years ago. Outgassing has been the ruling hypothesis since W. W. Rubey's classic paper on the subject in 1951. The only serious debate has been over whether the volatiles arrived at the surface all at once as a huge initial outgassing, came as many biggish burps, or represent a continuing rumble of little burps—the big-burp, many burp, and continuous-burp hypotheses. Now, however, increasing knowledge of the solar system, particularly of the compositions of carbonaceous chondrites and cometary nuclei, suggests an alternative or supplementary mechanism—a peak of intensive late Hadean accumulation of volatile components.

This knowledge is of three sorts. (1) Growing evidence implies that objects in the asteroid belt are mostly carbonaceous chondritic meteoroids, in which inclusions of hydrated silicate minerals and magnetite (Fe_3O_4) are imbedded in a matrix of complex, low-temperature, organic compounds that comprise up to 5 percent of the mass by weight. (2) Icy objects from the Öpik–Oort cloud are episodically deflected into hairpin-shaped, sun-grazing orbit by passing massive objects as they and our solar system orbit the galaxy. (3) Recent computations and modeling studies imply that several Earth-masses of ice could have been gravitationally deflected into the region of the terrestrial planets from the neighborhood of the Jovian planets as the latter became massive enough to achieve such slingshot-like effects. Such sources would be more than adequate to account for all of Earth's surface and near-surface volatiles, including its carbon as well as its water.

It is also known that volcanoes, geysers, and hot springs constantly bring volatiles to Earth's surface from its interior, including those that make the sea salty. Although it now seems that most such volatiles are probably recycled, it is equally clear from the presence of primordial helium-3 in deep-ocean waters, some volcanic sources, and diamonds that some fraction of them have never known Earth's surface before. In fact, we find such *juvenile* inclusions in the crystals of even the most ancient rocks. And internal water is probably required to reduce the melting temperatures of rocks and otherwise promote the tectonic (mountain-building) activities ob-

served on our planet from the very beginning of the geologic record. It seems undeniable, therefore, that some of Earth's volatiles were initially products of outgassing from its interior.

Omitting further detail, it seems likely that both internal and external sources contributed to the initial terrestrial atmosphere and hydrosphere and that much of Earth's early carbon arrived with carbonaceous asteroids, meteoroids, and comets.

Consider now that a continuously liquid hydrosphere is essential to the origin and continuation of life. That calls for more than a planet massive enough to retain it gravitationally. It also requires that the orbital path of that planet be at an appropriate distance from its source of warmth—its parent star—depending on the luminosity of that star. If the distance from Earth to Sun were shortened by only 13 percent, water vapor would not condense and carbon dioxide (hereafter CO_2) would not be removed from the atmosphere by the formation of limestone and dolomite. Venus, for example, within that shorter radius, lacks water (H_2O) and has a CO_2-rich atmosphere, although it presumably started with a quantity of atmospheric H_2O equivalent to its CO_2. On the other hand, if Earth were much farther from Sun, its H_2O would be ice, as is that now remaining on Mars (Figs. 1.5, 1.7).

The Late Hadean Surface

Imagine the Hadean surface as it might have appeared perhaps 4.2 billion years ago (Fig. 2.2*B*). Separation of the iron core from the silicate mantle was complete. Earth had been cooling for a few

hundred million years. The cooling was a slow process. Heat could not have been lost fast enough by simple conduction. Convective heat loss was the only efficient way to cool, and convection in that hot early planet was undoubtedly vigorous. It would have manifested itself at Earth's surface by intensive volcanism.

Earth's surface at the end of Hadean history was extensively rifted, impact cratered, and probably shrouded in vapor and dust from late cometary and planetesimal impact. It was, perhaps, similar to regions of lunar maria, except that real water filled the maria and steaming volcanic vents and hot springs abounded (Fig. 2.1). The Hadean atmosphere, with its rich endowment of CO_2, its steam, and traces of other potential greenhouse gases, seems to have kept the surface warm enough for liquid H_2O, despite the fact that the faint early Sun was only about 70 percent as bright as now.

On this scarred and dismal surface were no animals, no plants, not even bacteria, algae, or lichens. Yet weathering surely took place under an atmosphere rich in CO_2 and volcanic gases. Rocks exposed to corrosive volcanic emanations were converted to clay that accumulated in pools already salty from various low-temperature condensates that arrived at Earth's surface with H_2O and other atmospheric gases. Complex (though nonliving) organic molecules were brought to this surface in carbonaceous chondrites and comets. An object no bigger than the earlier mentioned Allende chondrite could have concentrated as much as 100 kilograms of such organic molecules in a small sector of one of those claybanks.

A shielding magnetic field, a suitable range of temperatures, an absence of inimical *free* molecular oxygen (O_2), and a diverse supply of carbonaceous molecules seem a favorable set of conditions for chemical evolution leading to the origin of life. In fact, we know no primary sources of free O_2 that could have oxidized and destroyed any biologically interesting meteoritic macromolecules, or prevented reactions conducive to the formation of other such molecules or of living cells at that time.

Although some have considered the primordial atmosphere to be oxidizing because of its predicted content of oxygen-rich compounds such as CO_2 and H_2O, that is misleading. In the effective absence of free O_2 resulting from the continuing resupply of ample reduced volcanic gases, such an atmosphere is best referred to as anoxic, or neutral, or even anaerobic. For, although small quantities of transient secondary O_2 would surely have been introduced to it from the very beginning as products of the breakdown of H_2O and CO_2 by the energy of sunlight, no free molecular O_2 could have accumulated until all potential oxygen sinks were neutralized. Instead, immediate recombination of O_2 with its coproducts and other then-abundant oxygen-hungry reduced substances (e.g., reduced carbon, volcanic gases, and ferrous iron) would have seen to that.

For we whose metabolism depends on free oxygen and who perish without it, it may seem strange to think either of an Earth without it or of oxygen as a poisonous gas. Yet lethal it is to all forms of life above their particular tolerances, and anoxic the evidence of geochemistry certifies the primordial Earth to have been. Even the 20 percent of uncombined oxygen in the present atmosphere is destructive to creatures dependent on it in the absence of appropriate oxygen-mediating enzymes. The sites of our most essential vital processes are carefully shielded against exposure to any uncombined oxygen, peroxides, or superoxides.

Biological oxidation, in fact, is generally carried on by the removal of hydrogen rather than the addition of oxygen. And oxygen is transported through the body beyond lungs or gills in a fixed state as metallic pigments in specialized oxygen-transporting cells.

First Thoughts on the Early Biosphere

The gossamer web of life, spun on the loom of sunlight from the breath of an infant Earth, is nature's crowning achievement on this planet. The simplest organisms were already more complex than the most complicated physical systems we know. Although but a trivial fraction of Earth's total mass and volume, the realm of life interacts pervasively with the physical environment, creating its own important global realm of activity—the *biosphere*. Living things profoundly influence and are influenced by Earth's physical processes. They have played a major role in shaping and recording planetary history from time immemorial. The mutual feedbacks between life and its physical surroundings—biospheric and biogeologic processes—can never be far from our thoughts as we attempt to reconstruct that history.

The Hadean Earth would have been an unappealing habitat for most familiar forms of life. Nonetheless, it may have been the initial incubator. The oldest known sedimentary rocks of subsequent Archean history (Ch. 6), some 3.8 billion years old, contain carbon with relatively light isotope ratios consistent with a biological origin. They also include peculiar iron-rich silicate rocks, called *banded iron formation*, similar to younger ones with sedimentation that may in significant part have involved biological intervention. And sedimentary rocks of only slightly more recent age (3.5 billion years old) contain a type of layered structure that is widely taken as presumptive evidence for a biological presence—stromatolites, about which more is to come.

A number of experiments, using different plausible atmospheric components and likely primordial sources of energy, have yielded amino acids and other biologically interesting large molecules in experiments that had in common only the exclusion of free O_2. Hydrogen cyanide (HCN) and aldehydes (e.g., formaldehyde, CH_2O), known also from interstellar space, are common initial products of such experiments. They combine readily in suitable aqueous concentrations and the absence of O_2 to produce a variety of amino acids, the building blocks of proteins. And that seems to be especially likely if ammonia (NH_3) is present to catalyze the process.

Essential to the operation of such processes in nature are natural concentrations of the organic building blocks. As we have seen, one of the places where such molecular building blocks are known to be concentrated naturally is in carbonaceous meteorites. Such stony messengers have been found to contain some fifty different amino acids, of which eight are common protein-building types. Some have also been reported to contain all five of the basic structural components of the essential nucleic acids DNA and RNA (the nucleotide bases), found in all living things. Bury a large carbonaceous chondrite rich in prebiological organic molecules in a wet claybank and one has a concentration of organic building blocks in the presence of potential crystallographic templates from which something having the properties of life might be assembled.

Based on the foregoing and other evidence to be elaborated, it seems probable that life originated on Earth shortly after the planet acquired a continuous crust and water bodies of long duration. We see evidence for that in the oldest sedimentary rocks known, just at the beginning of post-Hadean history.

Finally, we should take note of the proposal that life did not originate on Earth but arrived here from some other place. Even were that true, it would not explain the origin of life; it would only transpose it to unknown conditions, place, and time. For life cannot always have existed when the very chemical elements of which it is made have not always existed. All are the products of nucleosynthesis in the wake of the Big Bang, and in main sequence stars, red giant stars, and supernovae. Life, wherever found, could only begin after the elements of which it is made existed. The weight of the evidence strongly suggests that, however often or wherever life may have originated, it was probably everywhere a carbon-based product of chemical evolution in an anoxic, hydrous environment of which significant parts did not range permanently below the freezing or above the local boiling point of water.

There is, in fact, no reasonable doubt that life now on our planet did begin here, probably as microbial ancestral forms not dissimilar to some of the old-est known and suspected fossils, and presumably within a UV-shielding hydrosphere. Its earliest chemical precursors, however, may well have been products of interplanetary if not interstellar space. The oxygen, nitrogen, and carbon that join with cosmic hydrogen to build the vital molecules are all products of stellar evolution. Even the basic molecules from which the life-stuff was initially assembled may have arrived on the late Hadean Earth in carbonaceous chondrites.

Although processes within our planet have had a profound influence on what transpires at its surface, any hypothetical history must be consistent with records kept by its crustal components. And these are all affected to some degree by the biosphere or its products. Crustal, atmospheric, and hydrospheric processes interacted with one another as they evolved to create the earliest historical records we know. Life added a fourth dimension—the biosphere. After that, Earth's surface processes could never be the same again. Although the earliest microbial forms were surely both anaerobic and dependent on external sources for nutrients (*heterotrophs*), life could not long have survived without giving rise to species capable of generating their own nutrients from sunlight or chemical energy (*autotrophs*). The gases they assimilated and gave off in the process, and their other biochemical properties and processes, interacted with and altered the chemistry of the hydrosphere and atmosphere, the kinds of sediments deposited, and the rocks produced.

With the origin of life, Earth history became a new ball game with new rules. Its complex unfolding was further complicated by convulsions of Earth's interior and the heaving and erosional rearrangement of its surface as a by-product of its continued venting of residual and radiogenic heat. The record of that history is a palimpsest—a dim and discontinuous record as of ancient directions to the location of buried treasure. Few clear signs stand out among many barely decipherable traces and scribbled-over antecedents.

The next chapter considers some of the principles by which we seek to decode this palimpsest—to impose order on Earth's history, to formulate fruitful hypotheses about the gaps, to reconstruct the story as it might have been.

In Summary

The first 600 to 800 million years of Earth's existence as a planet—the Hadean interval—is only indirectly recorded in events on its moon, in inferences from comparative planetology, in geochemical clues from meteorites and younger terrestrial rocks, and in computer-modeling studies.

Scant though this record is, it permits the provisional conclusion that our planet first began to aggregate from the solar nebula as a solid body about 4.6 billion years ago. Within a hundred million years or less it was essentially complete. Gravitational energy plus heat from compression and short-half-lived radioactive isotopes probably elevated temperatures to or above the melting point of iron right up to shallow depths. Iron trickled downward through a silicate mush to create Earth's bilayered iron core and the concentric shells of its silicate mantle. In response to requirements of the conservation of angular momentum, Earth's rotation accelerated. Dynamo motions in the molten outer sector of the core resulted in a magnetic field and a radiation-shielding magnetosphere.

During the next few hundred million years the planetesimal infall that created the initial Earth tapered off to a rate of only about 20 times the present. Earth cooled to form its first solid crust perhaps 4 billion years ago. Some evidence suggests a spike or temporary acceleration of meteoritic bombardment between about 4 and 3.8 billion years ago. That or earlier infall may have veneered Earth's surface with large quantities of carbon-rich debris from the asteroid belt and volatile-rich comets, perhaps deflected to Earth-crossing orbits by the gravitational effects of massive Jupiter. Such sources, plus outgassing of occluded volatiles from Earth's mantle rocks, probably account for the initial atmosphere and hydrosphere. They may also be the source of the carbonaceous macromolecules from which the first microbial creatures arose. They were anaerobic because there was no primary source of corrosive oxygen, heterotrophic because the synthesis of nutrients is a biochemically advanced process, and microbial perhaps because diffusion in and out of a small spheroid is the simplest energy-efficient metabolic system. Self-catalysis and radiant solar energy joined with more esoteric processes as probable driving forces for chemical selection leading to life.

The initial atmosphere from which the

present one evolved, and beneath which life on Earth arose, is judged to have been anoxic or neutral (with only transient traces of free oxygen), consisting of carbon dioxide, water vapor, nitrogen, and various reduced gases. Sedimentary rocks 3.8 billion years old tell us that such an atmosphere and a substantial hydrosphere was already present by the end of Hadean history. Microbial life, presumably, arose in the seas beneath this anoxic atmosphere near the end of Hadean time.

The story of the Hadean Earth is not one to be ignored or unduly abbreviated merely because we know so little about it. The beginning, being the most important part, must be dealt with as best we can. Limited and indirect though its records be, Hadean history was a highly significant sector of Earth's larger story. It includes the time during which the aggregation of our planet was completed, Earth's concentric structure and main chemical differentiation evolved, and the ancestral atmosphere and hydrosphere accumulated or emerged. At the end it probably saw the first separation of the planetary surface into relatively elevated lands and basin-filling seas, and, concurrently, the onset of chemical evolution leading to life. Such events set the stage for all future Earth and biospheric history.

3.1 HUTTON'S SICCAR POINT UNCONFORMITY, SOUTHEASTERN SCOTLAND. Here, in 1788, James Hutton and John Playfair gazed into "the abyss of time" revealed by the great angular unconformity between metamorphosed and upended Silurian strata and nearly horizontal overlying Devonian sandstones. *[Courtesy of Donald McIntyre.]*

3.2 VIEW NORTH ACROSS GRAND CANYON AND UP BRIGHT ANGEL CREEK FROM THE SOUTH RIM. Upended metamorphic rocks 1.7 billion years old in the inner gorge are capped unconformably in the foreground and on the right of Bright Angel Creek by the prominent flat-lying, cliff-making Tapeats Sandstone, Cambrian System. Above it follows a classic Paleozoic sequence. Upstream and left from a pre-Tapeats fault that follows Bright Angel Creek are billion-year-old sandstones and shales, then the same Paleozoic sequence, to the terminal sandstone and dolomite in the prominent, snow-dusted, distant cliffs. *[Courtesy of John C. Crowell; Betty Crowell—"Faraway Places."]*

CHAPTER THREE

ON READING
THE ROCKS

A subject becomes a science when it is brought within the bounds of natural law and a body of ordering principles that make its hypotheses testable—where evidence takes priority over authority and revelation. Such a perspective has now been reached from which the main outlines of a consistent and verifiable Earth history are visible. That is a consequence principally of insistence that natural phenomena be interpreted as the product of observable or lawfully consistent natural causes. Operating principles are needed to recognize events and trends, order them in a correct historical sequence, and correlate them with historically equivalent events elsewhere. The interpretation of planetary history draws on environmentally distinctive rock types, their preferred habitats, and the changes they undergo. By such means has the long history of our planet been extended, with varying levels of confidence, to include the 3.8 billion years of post-Hadean time. This chapter explores how the founders of historical geology groped their way from pursuit to science and a good outline of the last 15 percent of Earth's history—in the process recognizing discontinuities that have long defied explanation.

Sequence, the First Essential

History is both an art and a science. It aims to reconstruct past events, correlate them, and place them in an historical sequence. Bathsheba goes with David, Homer with the *Iliad*, and Cleopatra VII with Mark Anthony—in that order. Without sequence there is no history, only anecdote. Indeed a sense of time and sequence, of passing and recurrent seasons, of progression in human affairs, of past and future, is common to all mankind. It is central to perceptions of our relation to others, to our antecedents, and to our place in the universe. The conscious ordering of events was perhaps the most important conceptual advance in human evolution—leading our emerging bipedal ancestors to think beyond the moment, to learn from the past, to anticipate the future. It separates us from humbler forms of life. It inspires our vision.

Means for ascertaining duration and ordering sequence mark the advance of social order. Although time cannot be sensed or measured in any way separate from events that define its passage, all peoples have evolved a perception of its continuing flow, as well as some way of reckoning it. A range of scales and purposes have been served by systems based on the motions of planets, the drift of shadows, the flow of water or sand, ingenious mechanisms, the vibrations of crystals, the constant rates of natural radioactive decay, and finally the speed of light. Such observations, devices, and methods have made possible the planning of seasonal and daily activity, the equitable limitation of debate, accurate longitudinal navigation, and the calibration of history—including Earth history.

History and time, therefore, are closely linked, but they are not the same. Time flows, regardless. There is no history without events. Getting events in the right sequence is the key operation, especially for that vast reach of history that preceded the conscious keeping of records. What we know about the history of emerging humankind depends heavily on discovering and placing in temporal order such unintentional records as hearthsites, tent rings, tools, and others, as well as skeletal remains and burials. Reconstructing the history of our planet from earliest times to the present calls on fortuitous natural signals: the records of varying sea level, floods, volcanoes, avalanches, earthquakes, glaciers, and the changing manifestations of biologic, atmospheric, hydrospheric, and crustal evolution. A credible Earth history requires that the apparent jumble of rocks of different kinds, ages, and configurations—containing fossils and other signs of past events and states—be sorted out, placed in sequence, and interpreted as to conditions of origin.

The first step is to decide which way is up, the direction of "younging," as geologists say—oldest at the bottom, most

recent at the top. That is simple enough where the rocks are stratified and essentially flat-lying, like tiers in a layer cake, as in the Grand Canyon of the Colorado or the White Cliffs of Dover. But how to cope when the rocks are not layered? Or where they stand on end (Fig. 3.1) or lie upside down, as they often do in folded mountain ranges like the Alps, the Rockies, or the nuclei of continents—the so-called *shield areas*?

How does one identify the historical order when there is no obvious sequence? Suppose that the strata have been compressed into folds, like a wrinkled rug on a slippery surface, and then eroded so that the tops of the folds are cut away? And suppose that continued sliding and compression has caused these folds to rotate through the vertical until they again lie flat, to become *recumbent*, so that their properly under limbs are upside down? And what if they are then deeply eroded? In intensely crumpled regions of the Earth, rocks of the same sequence may even be repeated by stacking, one on another. If there is continuity of strata, even though deformed, the puzzle may be resolved by tracing or "walking out" individual layers in the field. But that is often prevented by discontinuities that result from erosion or physical displacement. Where erosion has removed all but the inverted under limb of an extensive recumbent fold, everything is upside down. Apparent lesser upfolds (*anticlines*) may actually be inverted downfolds (*synclines*) and the reverse. The apparent synclines and anticlines are then described as *synforms* and *antiforms*. To unravel the local history of a geologically complex area thus calls for an appropriate set of ordering principles, a trained eye, a sense of orientation, and willing legs. Until principles and procedures were discovered and enunciated for placing disturbed rocks in the correct sequence, relating them to one another, and interpreting their manner and circumstances of origin there could be no consistent Earth history and no science of geology.

In fact, until the nineteenth century, no organized science of geology existed in the sense we recognize as science today. Even the word did not exist until proposed by J. A. Deluc in 1778. The first organization to be called a "geological society" was that at London in 1807. The first *Introduction to Geology* was published by Robert Bakewell in 1813, and "Geology" first appeared in the Encyclopaedia Britannica in 1810.

Why did it take so long? Why particularly when the basic ordering principles had been articulated, by the man usually called Nicolaus Steno, in the sixteenth century? Able and energetic people were struggling to understand their observations of Earth's crust in terms of natural causes. Notable among them were eccentric polymath Robert Hooke and devout but truth-seeking Thomas Burnet in the late seventeenth century, insightful Benoit de Maillet (Telliamed) in the early eighteenth century—who

guessed Earth's age at 2 billion years—and the Compte de Buffon and James Hutton in the late eighteenth century. One problem was that their ideas of natural causes were constrained by dogma, conflicting, and often extravagantly conceived. The other was that the times were not right. Acceptance of new ideas is usually contingent on three preconditions: (1) the world must be ready for them; (2) they must be convincingly advocated by a persuasive person or group; and (3) they must be perceived as clearly superior to (or, at least, not in serious conflict with) other widely held beliefs.

The brevity of biblical time was a major obstacle in the European world. How could events observed have taken place during the 6,000 years then allowed unless processes of improbably catastrophic magnitude or miraculous order were involved? Contrary conclusions, if expressed, were likely to be advanced in allegorical or very guarded form. It took a long time for acceptance of Burnet's 1690 plea that no "truth concerning the Natural World can be an Enemy to Religion; for . . . God is not divided against himself." Indeed, that view, even today, is by no means universally accepted.

The Relevance of Ordering Principles

The first essential for geologists is to know or work out the historical sequence in a succession of rocks. In a flat-lying sequence of layered or stratified rocks, that usually means oldest at the bottom, youngest at the top, intermediate ages between. Geologists call that seemingly simple truth the *law of superposition*.

In circumstances such as are presented by the Grand Canyon, or other nearly horizontal or gently inclined sequences of sedimentary strata (Fig. 3.2), the relation is obvious or easily worked out. Indeed, observers as far back as the seventeenth century did well enough with flat and gently dipping strata like those of the Paris basin and the famed White Cliffs of Dover. But they ran into trouble with deformed rocks like those of the Alps and other complex or highly altered strata.

How could such extravagant deformation be attributed to causes now in operation when no one saw such things happening? It was clear that biblical time was insufficient, *unless* processes affecting the ancient world were of a vigor and intensity surpassing all experience. Thus, deep river gorges such as those in the Causses region of central France were long interpreted as preexisting catastrophic rifts instead of as the product of long-continued stream erosion. Mountain ranges, many supposed, were attributable to abrupt uplift to their present positions and configurations rather than to repeated or continuing stresses over millions or tens of millions of years. Spellbinding A. G. Werner, professor at the School of Mines at Freiburg, Saxony, for 42 years (1774–1817), the most influential teacher and "authority" of his time, taught that granites intruding the cores of mountain ranges were *older* than the strata they penetrated. In his view of Earth history, granites were the first precipitates from a global but shrinking ocean from which was deposited the entire succession of known rocks in brief and orderly succession.

Although several of the basic principles needed to resolve such relationships objectively had been laid out by Steno more than a century earlier and, properly applied, would have given different answers, Werner's charisma and persuasiveness overwhelmed his audiences. Because European and British students of the late eighteenth and early nineteenth century were likely either to be Werner's disciples or to be taught by them, Wernerian doctrine dominated prevailing views about the Earth in his time, at least in the German and English speaking worlds. Werner's great merit was that he stimulated broad interest in Earth history, made important contributions to mineralogy, and graduated some of the very students whose observations later contributed to the rejection of Wernerian doctrine.

Perhaps the reason it took so long for the prior ordering principles to penetrate was that they had appeared in an obscure volume in discursive Latin by a youthful Danish anatomist who was to become much better known as a Catholic churchman and bishop. This youth, Niels Stensen, later *Steno* (Fig. 3.3), on being ordered in 1669 to return from Florence to become royal anatomist at the Danish court, hastened to write out his geological observations before leaving. He published them at age 31 in a seventy-six-page book with a long Latin title that came to be known as *The Prodromus of Nicolaus Stenonius*. A prodromus is a preliminary discourse, and, like Darwin's *Origin of Species* nearly 200 years later, this one was billed as an abstract for a larger work that never appeared. Steno, presently a candidate for beatification, never returned to geology, but his slender book was as fundamental in its own nonquantitative way for geology as was contemporary Isaac Newton's *Principia* for physics. It laid the foundation for the later emergence of geology as a scientific discipline. The structure that has arisen is the product of efforts and insights by many later gifted contributors, among whom key luminaries are pictured in Figure 3.3.

Steno, more prophet than founder,

Nicolaus Steno
1638-1686

James Hutton
1726-1797

Georges Buffon
1707-1788

Georges Cuvier
1769 1832

Charles Lyell III
1797-1875

William Smith IV
1769-1839

William Whewell
1794-1866

A. D. D'Orbigny
1802-1857

Louis Agassiz
1807-1873

Charles Darwin
1809-1882

G. K. Gilbert
1843-1918

A. L. Wegener
1880-1930

Arthur Holmes
1890-1965

G. G. Simpson
1902-1984

H. H. Hess
1906-1969

3.3 GIANTS OF HISTORICAL GEOLOGY: FROM SUPERPOSITION TO PLATE TECTONICS.

recognized the fundamental truth that stratified rocks are usually the sedimentary "deposits of a fluid," and thus very likely formed in a sequence of upwardly decreasing age in an initially horizontal or near-horizontal orientation.

More broadly, this law of superposition, as applied nowadays, recognizes that strata in a given undisturbed sequence decrease in age from bottom to top, evidence seemingly to the contrary invariably being attributable to postdepositional intrusion or other disturbance of the original order of succession. That kind of basic "layer-cake" stratigraphy is well illustrated in parts of the Grand Canyon (Figs. 3.2, 3.4) as well as in many lesser displays of gently dipping to flat-lying stratified rocks.

It is not always so simple, however. A different situation exists beneath those colorful, flat-lying to gently inclined rocks of the canyon where Bright Angel Creek joins the Colorado River from the north. The somber-hued layers beneath the colorful flat rocks stand on end. These

3.4 ORDERING PRINCIPLES AND KINDS OF UNCONFORMITY ILLUSTRATED IN THE GRAND CANYON OF THE COLORADO RIVER, FROM OLDEST (*1*) TO MOST RECENT (*20*). The classic Grand Canyon sequence is an abridged primer on Earth history. It began with metamorphism and granitization about 1,700 million years ago (event *1*), followed by crosscutting igneous dikes (event *2*). Erosion during the next 430 million years is recorded by the *unconformity* at *3*. The sequence of deposits and discontinuities above that unconformity illustrates *superposition*, *original horizonality*, *original lateral extension*, *crosscutting relationships*, and the several kinds of unconformities, numbered in order of decreasing age. Event *9* is a *nonconformity* on the left and an *angular unconformity* on the right. Events *13* and *15* are simple erosional unconformities. About 140 million years of Earth history are represented only by erosion at event *13*, not so much at *15*, much more at *3* and *9*. The last event was canyon cutting by the river at *20*. Working out the nature and sequence of such events, ordinarily in less well exposed areas, is the foundation of Earth history.

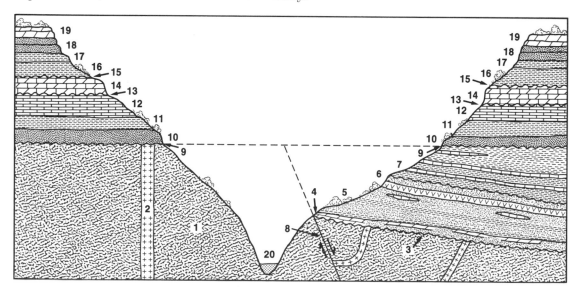

are the *basement rocks*, here about 1.7 billion years old. They were deformed, upended, metamorphosed almost beyond recognition of initial state, and eroded before younger but still very ancient and only slightly tilted blanketing strata were spread across their eroded edges. This is a classic discontinuity of the type that geologists call an *unconformity*. But that is not all. The colorful and nearly horizontal sandstones, shales, and limestones that fill the upper canyon walls and dominate the dramatic scene themselves rest unconformably on the gently dipping older strata that separate them from the upended crystalline rocks at the base of the canyon (Fig. 3.2). Here lie the records of almost half of Earth history, punctuated by metamorphism, deformation, and erosion, and interrupted by two great unconformities and a number of lesser ones (e.g., *events 13* and *15* of Fig. 3.4).

Sedimentary strata, of course, are deposited on underlying surfaces that were most commonly horizontal or nearly so at the time. So Steno recognized, and this has become a *principle of original horizontality*. Only keep in mind that either water-carried, wind-drifted, or airborne sediments may at places come to rest with initial slopes (*dips* to a geologist) up to 7° or more in the case of breaches or as much as 34° on volcanic cinder cones or other steep depositional slopes.

Steno also recognized that each depositional unit or *stratum* must originally have extended away from any point of observation, under present-day hills and across valleys, until it vanished. That perception is recognized as the *principle of original lateral extension*. It reminds us that any layer of sedimentary rock not only extends laterally away from its point of origin but also must eventually thin to disappearance (pinch out) or give way to something else in all directions. Where a rock layer is neither observed to pinch out nor to abut some obstacle, it must once have continued beyond the place observed, as now-distinctive layers in the buttes, pinnacles, and walls of the Grand Canyon can be matched by eye across the spaces between (Figs. 3.2, 3.4).

Where a barrier of natural or man-made origin causes an area like this to be flooded, as behind Glen Canyon Dam in the upstream Colorado Gorge, river-borne sediment may fill former valleys or canyons so that later erosion would reveal discordant relationships between canyon fill and former canyon wall. Compared to strata higher in the sequence, such relationships might appear to be exceptions to the law of superposition. On the contrary, they reinforce it. As in the case of the maria that truncate the lunar highlands, *the principle of crosscutting relationships*, formulated by James Hutton of Edinburgh in 1795, informs us that the truncated or crosscut deposits were there first. The crosscutting rock, sediment, or structure is always the younger, be it a lunar mare, an intrusive igneous granite, a basalt that truncates or inserts itself between older sedimentary strata, or a valley fill that impinges against walls of previously deposited strata. An unconformity is a special case of a crosscutting relationship, in which younger strata have been deposited across the eroded surfaces of older rocks. Figure 3.1 illustrates the famous *angular unconformity* at Siccar Point, Scotland, which Hutton (Fig. 3.3) called upon to demonstrate the necessarily great length of geologic time. This

and other kinds of unconformities are illustrated in Figure 3.4.

The elementary ordering principles discussed all involve relatively straightforward observation in undeformed strata. Yet they may become subtle and difficult to apply in regions of intensely deformed rocks such as are common along continental margins, in older parts of the ancient continental nuclei, and at other places where sectors of Earth's crust have been deformed during past episodes of mountain building. In such settings, strata may be strained by faulting and folding, stood on end, sliced tectonically into separate sheets (*nappes*) of rock and even completely overturned (Fig. 3.5). They may be transported tens or even hundreds of kilometers from their sites of origin, stacked in reverse or random order, and subsequently refaulted and refolded (Fig. 3.5*A*). Skilled students of rock deformation (structural geologists or tectonicians) routinely reconstruct how many episodes of deformation a rock has undergone. Sectors of Earth's crust may even be detached and moved horizontally for great distances (even thousands of kilometers) to end up in completely discordant surroundings (see Ch. 9).

Criteria are required for decoding crustal history. And so much of that history is stored in sedimentary rocks and their associations that they are in fact the focus of attention in this book.

Some well-known clues to the history of sedimentary rocks are illustrated in Figures 3.5 through 3.7. We will return to that subject as it becomes germane.

The Abyss of Time: The Principle of Natural Causes

Beyond the ordering principles of Steno and Hutton, two further liberating insights, both long "in the air" but still minority views, required broad acceptance before a flourishing antiquarian and commercial interest in rocks could mature into a science of geology. First was the recognition of the immensity of time. Second was the perception that all natural phenomena can indeed be explained as the product of natural causes. Together they can be summed up in the aphorism, "Given enough time, anything that can happen (that is consistent with natural law) will happen."

Hutton's observations at Siccar Point (Fig. 3.1) in 1788 convinced him, but few others in the European world of that

3.5 DEFORMED STRATA: (*A*) BORDER ZONE OF THE ALPS; (*B*) UPENDED STRATA IN THE ARBUCKLE MOUNTAINS, OKLAHOMA. Scene *A* looks southwest from an area known as the Glarus *window* in eastern Switzerland. Here the strata were stacked in reverse order of age by relatively north (*right*) moving low-angle *thrust faults* and then eroded, exposing overridden younger rocks beneath the horizontal "window" in which the viewer stands. The view is upward from within this window across the edges of detached sheets (*nappes*) of older rocks at various angles. Scene *B* is of upended older Paleozoic strata in the Arbuckle Mountains, near Stringtown, Oklahoma. The rocks were rotated through the vertical during a plate collision about 250 million years ago to a slightly *overturned* position. Up in a depositional sense is now to the right.

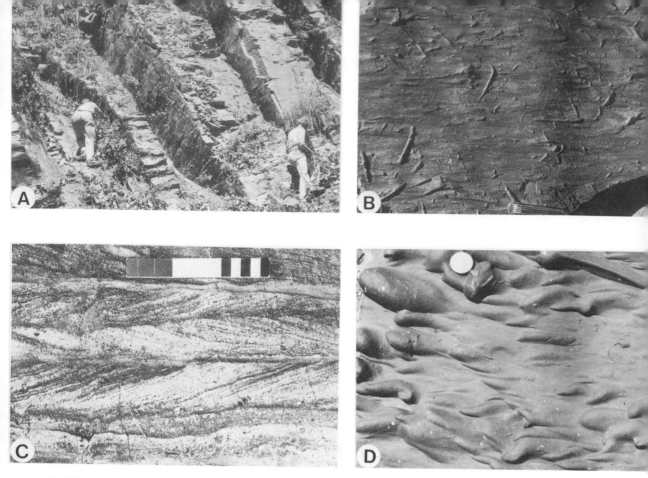

3.6 SOME CRITERIA FOR THE ORIENTATION AND DEPOSITIONAL HISTORY OF DEFORMED STRATIFIED ROCKS. View *A* illustrates upended 100-million-year-old Alpine flysch in southern Bavaria; abundant *sole markings* on the bottom surfaces show that the depositional top was to the right. Scene *B* is a closer view of a part of a bottom surface from the same locality, demonstrating current flow from left to right. Scene *C* illustrates a vertical profile of *chevron cross-bedding* in a 3.3-billion-year-old sandstone (scale 15 cm) from Swaziland, southeastern Africa, indicating by reversal of current directions a tidal influence. *D*, natural cast of tracks made by the descending eddies of a 300-million-year-old *turbidity current* in the Arbuckle Mountains, Oklahoma; flow direction shown by downstream broadening of these *flute marks*. [*C, courtesy of K. A. Eriksson,* © *Elsevier Science Publishers B.V., 1977,* Sedimentary Geology, *v. 18, p. 232.*]

time, that something was amiss with the idea of a 6,000-year-old Earth. How could the vertical strata below have been deposited, lithified, metamorphosed, stood on end, eroded to a level, and covered by younger stratified rocks, themselves now elevated, in so brief a time? On visiting the site with Hutton, his friend, clergyman and mathemati-cian John Playfair, was so impressed with the scene that he felt himself gazing into the "abyss of time." So do many visitors to the Grand Canyon nowadays; yet such a conclusion was bitterly opposed, even by renowned scholars, in the late eighteenth and early nineteenth centuries.

Thoughtful students of the Earth were

puzzled and inconsistent. They sought natural explanations but invented imaginary ones to escape conflict with biblical doctrine. Hutton himself earnestly sought reconciliation. The idea was afoot, nevertheless, that in order to be an acceptable scientific hypothesis any proposed explanation for phenomena observed must involve processes now in operation or demonstrably consistent with natural law—established or new. So one might deduce the likelihood of glaciation on a naturally ice-free Earth from the experimentally determined physical constants of H_2O. That is what I here call *the principle of natural causes*, described by Charles Lyell III (later Sir Charles) in 1830 in these words: "Our estimate of the value of all geological evidence . . . must depend entirely on

3.7 VERTICAL PROFILES THROUGH UPRIGHT (A) AND INVERTED (B) SLUMP BALLS. GRADED BEDDING (C). Slump balls, open at the top, reflect settling of homogenous sediments liquified by abrupt overloading or seismic shock. A, seismically shocked Pleistocene diatomite at Lone Mountain, Nevada, right side up. B, Triassic siltstone in Oregon's John Day basin, upside down. C, graded lamination in 620-million-year-old volcanogenic siltstone or tuff north of Durham, North Carolina.

the degree of confidence which we feel in regard to the permanency of the laws of nature. Their immutable constancy alone can enable us to reason from analogy . . . respecting the events of former ages."

Those words appeared in Lyell's justly famed work, titled *Principles of Geology: Being an Attempt to Explain the Former Changes of the Earth's Surface by Causes Now in Operation* (hereafter simply *Principles*). The subtitle, varying with the edition and not usually cited, accurately describes the emphasis of the work. The idea, in fact, goes back as a minority view to the time of Galileo and is foreshadowed in the writings of Hellanicus of Lesbos, a near contemporary of Herodotus, who attempted to eliminate appeal to the miraculous by offering naturalistic explanations.

It took courage, nevertheless, and finesse, in those times, to advance it vigorously and defend it successfully, as Frenchman Georges Buffon (Fig. 3.3) learned the hard way from contemporary cardinals with his "theory of slow causes." Lyell was lucky. His timing proved to be right. But he was more than lucky. He marshalled extensive supporting evidence and presented it persuasively. He supported the thesis of his subtitle with well-chosen examples from personal observation and wide reading. He wrote with a grace, clarity, and conviction that assured wide attention. *Principles* went through twelve influential editions during 45 years, with revisions and additions each time. Its initially controversial nature added to its interest. It was read not only by geologists but by the general intelligentsia of the day. In the contest with then-prevailing Wernerian views, which saw natural causes

in a more restricted light, Lyell emerged triumphant and has deservedly come down through the history of English-speaking geology as the central figure of his time in the field.

Above all, Lyell must be credited with having brought about a reexamination of outlooks considered basic in his time by insisting that all possibility of explaining phenomena observed by causes now in operation be exhausted before invoking unique explanations or inventing new laws. Moreover the latter, if invoked, must not violate already well supported natural laws.

Why did anyone question the invocation of natural causes to explain natural phenomena? Why did such a viewpoint require a champion? Inquiring minds of Lyell's vintage and earlier, who freely accepted naturalistic causes for the more recent aspects of Earth's development (e.g., France's Baron Georges Cuvier, Fig. 3.3), found it improbable that "causes now in operation" could explain the cyclopean deformities of the underlying crust, deep river gorges, and great mountain ranges. *The barrier to a consistent and universal acceptance of natural causes was insufficient time.*

One would have thought that Hutton's work would have dispelled that question, but, hydra-headed, it refused to go away. How was it possible for fossil seashells to be raised to the tops of high mountains, for deep gorges to be hollowed out by the streams within them, for once-horizontal strata to be stood on end? How could once-mountainous regions be worn down to flat or gently undulating surfaces by observable processes in the few thousand years of biblical time then widely (although by no means universally) accepted as the full quota allowed? How explain Siccar Point itself except by unimaginable catastrophe or divine intervention? That was the prevailing public view of the Judaeo-Christian world at the beginning of the nineteenth century, a view that came to be known as *catastrophism.*

Today no one other than a few politically active biblical fundamentalists, boxed in by the same self-imposed constraints of time, would argue with the conclusion that natural phenomena should be regarded as the products of natural causes. But, in 1830, when Lyell published the first volume of *Principles,* lines were being drawn for a big confrontation about time and process in which he was to play a pivotal role.

Before a particular cause can be accepted as natural, it must satisfy criteria mentioned. All valid scientific hypotheses make predictions about matters then unknown that can be confirmed or refuted by further inquiry, including past as well as present and future happenings. Processes either not presently known to operate in nature or not consistent with natural laws make no testable predictions unless they are put forth as new natural laws, requiring validation! Hypotheses that call on unknown processes to explain peculiar circumstances without saying how such processes might be discovered are called *ad hoc* hypotheses ("pertaining to this case only"). Unless one can say how their credibility might be weakened or enhanced by observation or experiment, such hypotheses are not scientific—even if proposed by persons who, in other circumstances, adhere to scientific principles.

The important and useful part of the principle which Lyell referred to as uni-

formity is its operational aspects, *not* rates: seek natural causes, eschew the miraculous, shun the supernatural. Despite his insistence on uniformity of "energy" (read "rates"?) as well as natural laws, Lyell clearly included among causes now in operation what most today would consider to be catastrophic processes. He mused on the extensive effects that could result from the abrupt release of Great Lakes waters by crustal rupture. He interpreted the puzzling erratic boulders of glaciated regions as ice-carried, and, by the third edition of *Principles* (1834), he was shifting continents from equatorial to polar regions and back to explain climatic changes of the past (71 years before Alfred Wegener of continental drift fame).

In fact, the view that natural processes have been, *on a long average,* uniform in no way conflicts with the current view that much or most geologic work is accomplished in short bursts of intensive activity: the release of accumulated stress, the coincidence of forces, the episodicity of events.

Lyell's well-warranted conviction that natural causes suffice to explain natural phenomena was undermined by his repeated reference to uniformity of processes as well as the laws that govern them. Perhaps it was necessary so to polarize the issues at that time in order to focus on them. In any event, that choice of words set the stage for the by no means entirely imaginary conflict between "uniformitarianism" and "catastrophism" that persists as a continuing dispute about rates.

The erudite philosopher, mathematician, and mineralogist William Whewell (Fig. 3.3), long master of Trinity College and briefly vice-chancellor at Cambridge, saw the problem in his 1832 review of the second volume of *Principles*. He asked: "Have the changes which lead us from one geological state to another been, *on a long average* [my emphasis], uniform in their intensity, or have they consisted of epochs of paroxysmal and catastrophic action, interposed between periods of comparative tranquility? These two opinions will probably for some time divide the geological world into two sects, which may perhaps be designated as the *Uniformitarians* and the *Catastrophists.*"

Whewell was the first to use those terms, not employed by Lyell. In another work Whewell correctly observed that, because the limits of gradation in geology must be left undefined, the distinction between "uniformitarianism" and "catastrophism" in a naturalistic sense becomes blurred. Whether by avalanche or soil creep or (characteristically) both, mountains are eventually reduced by natural processes of erosion and gravity to the general level.

The conflict was, in effect, creative, and the terms may once have been useful, but no longer is that so. What remains conceptually valid is better included under the *principle of natural causes*. With ample time at their disposal, whether spasmodic or gradual, the often-uneven effects of invariable natural laws will explain the phenomena observed.

Dispute continues about rates, degrees, and causes, but no longer is there any significant disagreement among informed geologists either about the antiquity of the Earth or about the explanatory validity of natural causes. We will see, however, that there have been changes both in rates and in prevailing trends, many rapid enough to be called events. Some, such as volcanic outbursts, were abrupt. Others, such as earthquakes and asteroidal impact, were, in effect, instantaneous. This is the stuff of Earth history. Without such irregularities it might be hard to punctuate the flow. With them we see an evolving planet, not a static or monotonously cyclic one. Indeed the debate about rates has taken a new turn. Enlivened by the ever-higher degree of resolution now achievable in radiometric age determination, it is being extended far back in geologic time in an effort to determine the rates of newly recognized geologic processes such as those of plate tectonism.

As in much current historiography, flaws are now being found in the classical view of Lyell as the white knight who single-handedly quenched the fires of the dragon of catastrophism. Regardless, he was the acknowledged and effective leader of the first significant flowering of geology as a science, and richly deserves his place in the pantheon of historical geology.

The Law of Biotal Succession

The most important discovery of classical geology was the recognition that particular layers or short sequences of stratified rocks are characterized by unique associations of fossils that allow fossiliferous equivalents to be recognized, wherever found. This had actually been suggested by the perceptive Robert

Hooke in 1667 in his *Discourse of Earthquakes*. His contemporary John Ray also hypothesized a great age for the Earth before yielding to ecclesiastical pressure to recant and agree that fossils were inorganic. It was also realized and courageously announced in 1779 by Gerard de Soulavie, Abbot of Nimes, that observed assemblages of fossils change systematically upward through the rock sequence. And other Frenchmen, the biologist Cuvier and collaborating engineer Alexandre Brongniart, announced a detailed sequence of fossiliferous strata in the Paris basin in 1808 and again in 1811, complete with map, vertical cross-sections, and illustrations of fossils. These observations exemplify what is now known as the *law of biotal succession.*

The codiscoverer of this law (with Cuvier), and for many the real hero of the story, was a remarkable Englishman of humble birth whose saga is at once inspiring and wrenching. Before and while the aristocratic Cuvier and his associate Brongniart were working out their elegantly detailed sequence of fossils in the relatively young strata of the Paris basin, plain William Smith, blacksmith's son and self-taught surveyor, was discovering a similar succession in older rocks in England. He found that a regular sequence of distinctive fossils in the Mesozoic rocks near Bath improved his supervisory activities in constructing the Somerset Coal Canal. He dictated his scientific findings to a friend, the Reverend Benjamin Richardson, in 1799 under the title *Order of the Strata, and Their Imbedded Organic Remains, in the Neighbourhood of Bath; Examined and Proved Prior to 1799.* Meanwhile he set out to prepare a geologic map of the whole of England and Wales, of which

fifteen sheets were published in 1815 before his publisher went bankrupt and Smith went to a debtors' jail in 1819. Smith's saga had a happy ending, however. He was awarded the first Wollaston medal of the Geological Society of London in 1831.

The mutually confirming achievements of Smith and of Cuvier and Brongniart (and, later, of D'Orbigny, Fig. 3.3) were monumental first steps in working out the progressively changing succession of fossils from ancient times to the present. This succession, more than anything else, establishes biological evolution as an observable fact calling for naturalistic explanations of processes involved. The idea of biological evolution—already enunciated in the late eighteenth century by France's Chevalier de Lamarck, but not widely accepted—was soon to undergo an interesting test.

In 1837 the English paleontologist William Lonsdale suggested from his studies of fossil corals in South Devon that they were intermediate in development and therefore in age between Paleozoic strata known elsewhere as belonging to the Silurian and Carboniferous systems. That is to say, they were more advanced than Silurian but more primitive than Carboniferous species. That conjecture, which lead to the foundation of the Devonian System by R. I. Murchison and Adam Sedgwick in 1838, has since been confirmed many times over. It was the first known practical test of evolution. And it came 22 years before Charles Darwin was to elaborate in detail the fundamental truth of systematic biologic change with time and habitat and to propose a convincing mechanism to account for it.

Subdivision and Correlation:
Central Tendencies and Unique Events

All things evolve. Nothing ever returns exactly to its ancestral form. The now-vast body of evidence supporting the principle of biotal succession permits historical geologists to array the fossili-ferous strata of the last 670 million years or more of Earth history in their proper sequence throughout the world. That is the central basis for *sequence geochron-ology*—the only kind of geochronology known until the discovery of radioactiv-ity and the realization that its regularity could be applied to the numerical mea-surement of time.

The goal, of course, is to divide history into manageable units, correlate them one way or another (Fig. 3.8), and cali-brate their ages and durations in years.

The simple order of central tendencies among varying rock types is widely uti-

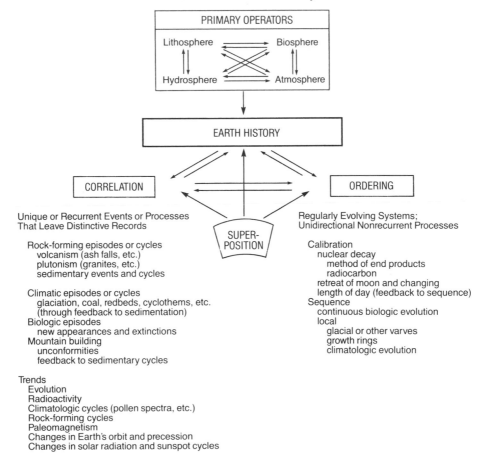

3.8 FLOW SHEET FOR EARTH HISTORY. *[From* Adventures in Earth History, *W. H. Freeman and Co., 1970, p. 8, © Preston Cloud.]*

lized in local correlation. If one observes the sequence limestone-sandstone-shale-dolomite, having distinctive thicknesses, on one side of a valley and then finds a similar order of similar strata on the opposite bank, it is usually safe to conclude that they represent the same gradational sequence. Scientific parsimony calls for the simplest naturalistic explanation that reconciles all the evidence. Although that is not invariably correct, it provides working hypotheses. To test such an hypothesis, the observed sequence then may be traced around hills and across other valleys for long distances, showing transitions from one rock type to another until it loses its identity. A new set of central tendencies comes to prevail. The crucial test of equivalence then depends on other criteria.

Suppose we find in the remote outcrops a thin layer of sticky greenish clay that has the properties of a weathered volcanic ash, similar to one that was earlier reported from the sequence we started to trace. In that case one could look for mineralogical or geochemical tracers that might support the conclusion that the two ash layers were, in fact, the product of a single eruption. Such a geologically near-instantaneous event provides a link among all rock sequences in which that ash is found.

Unique and widely recognizable events like volcanic ashfalls and paleomagnetic polar reversals are extensively used in geologic correlation. The Pearlette Ash of Pleistocene age can be recognized from Idaho to Nebraska. Huge volumes of ash known as the Bishop tuff erupted from an area north of Bishop, California, about 700,000 years ago. Hundreds of meters thick nearby, it tapered to only a single centimeter halfway across Kansas. Much lesser volumes of ash that erupted from Mount St. Helens in western Washington on 18 May 1980 left detectable traces as far away as 2,300 kilometers to the southeast at Norman, Oklahoma, and almost 2,000 kilometers east to Moorhead, Minnesota. The extent of distinctive ashfalls of this sort (Fig. 3.9) provides reliable evidence of near simultaneity wherever identifiable in the geologic record. Other less widely applicable methods of correlation are indicated in Figure 3.8.

The Natural History of Rocks

The prime witnesses in historical geology are the rocks themselves. Rocks do not just occur at random. They form well-defined patterns and trends, and, like plants and animals, different kinds of rocks are found in different places. By analogy one might say that they are adapted to those habitats. They even respond to changes in their local physical environment.

Igneous and metamorphic rocks that crystallize at depth within the crust, or under intense pressure, consist of interlocking crystals that are stable for their chemistry at prevailing temperatures and pressures of origin. As overlying deposits are removed by erosion, they respond to changing temperature, pressure, and moisture regimes. Reduction of load on such homogenous rocks sets up stresses that eventually result in concentric zones of weakness. Failure (strain) and

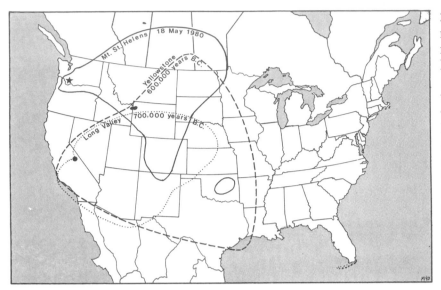

3.9 EXTENT OF
NOTICEABLE ASHFALL
FROM ERUPTION OF
MOUNT ST. HELENS,
1980, COMPARED
WITH LARGE PREHIS-
TORIC ERUPTIONS.
Despite its wide
extent, the volume of
ash from Mount St.
Helens was dwarfed
by that of the giant
Long Valley and Yel-
lowstone caldera
blowouts.

3.10 EXFOLIATION DOME IN THE HIGH SIERRA NEVADA, YOSEMITE NATIONAL PARK, CALIFORNIA.
Reduction of overburden pressure as a result of glacial erosion and melting, aided by ice
wedging, causes the massive domes to spall upward into a succession of thin slabs resembling stratification. *[Courtesy of David Pierce.]*

detachment along such surfaces may result in the spalling off or *exfoliation* of concentric outer layers as a result of further stresses set up by freeze–thaw and other cyclic temperature changes. That is the cause of the familiar onion-shell structure of the granitic domes of Yosemite National Park and elsewhere in the High Sierra Nevada and similar mountain ranges (Fig. 3.10). In moist, warm climates feldspar crystals in the granitic rocks may be altered to clay, softening to form "rotten rock," or *saprolite*, that crumbles to the touch and decays to a soil that is commonly red from the oxidation of iron. Metamorphic rocks, often rich in mica (the eisenglas window of great-grandma's kerosene stove), behave in similar ways. As mica is a clay mineral and feldspar weathers to clay, both generate gliding surfaces on which detached slabs of more quartz-rich rock slither downslope, like urchins

on sleds, influenced by that pervasive geological agent, gravity.

Limestone dissolves to make valleys in humid regions but stands up as ridges in deserts. Everything is eventually reduced to sediments or dissolved ions which, in turn, come to rest or precipitate somewhere. All eventually reenter the rock cycle (Fig. 3.11).

Certain kinds of rocks, therefore, tend to be characteristic of different parts of Earth's surface. Granitic rocks of various types and their metamorphic equivalents predominate in ancient shield areas, beneath stable continental platform deposits, and in the cores of major mountain ranges. Basalts and related volcanic rocks rich in silica, iron, and magnesium are characteristic of ocean floors (beneath their covers of pelagic sediments). They are also common in rift valleys where continents are pulling apart, as in East Africa. Sedimentary

3.11 MUSICAL ROCKS: THE EVER-CHANGING, EVER-RECURRENT ROCK CYCLE.

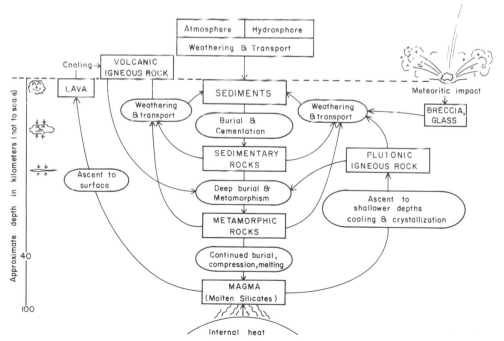

rocks of various types veneer the continental surfaces, grow thick on continental shelves and slopes and in subsiding basins of various sorts, and occasionally take the form of extensive coral reefs.

In this book sedimentary rocks receive special attention because they have so much to say about the surface history of our planet. Fossils entombed in them, their sources and modes of formation, distinctive sedimentary structures, and their geochemistry reveal when and where they were deposited, under what ecological and climatic conditions, and even whether and how they may have been transported after initial sedimentation to other locales. Sediments that collect in sand dunes and desert basins, on beaches and tide flats, in stream valleys and lakes or bogs, and on continental shelves and slopes all carry their own business cards. Those of estuaries and bays, of aerobic or anaerobic basins and inlets of the deep and shallow sea, and of the abyssal plains of oceans all exhibit characteristics by which they may be recognized. The main climatic zones advertise their former prevalence by distinctive fossils and weathering products, the textures and compositions of their sediments and soils, and sedimentary evidence relating to source of components and mode or modes of transport. Every rock has a story to tell, and their kinds are legion.

The variety of different rock types, in fact, is so enormous that no geologist knows them all by their first names. Individual geologists specialize on igneous, metamorphic, or sedimentary rocks and even special classes of rocks within those larger groups. Specialists on carbonate rocks may know little about shales and sandstones and less about granites and slates; but they can name the kinds of limestone or dolomite you bring in. And, with a bit of study, they can tell you whether it was deposited in fresh, brackish, hypersaline, or normal marine water, if it came from the deep or shallow sea, and otherwise when, how, and under what conditions it was deposited.

Happily, one can think about historical geology knowing little more about rocks than the distinction among limestone, sandstone, and shale, a bit about granitic and volcanic rocks, and a few common and important rock assemblages. Be on the lookout also, when observing rocks in nature, for lateral gradations between different kinds in the same stratigraphic space. Limestone may grade laterally to dolomite or shale, for instance, or to gypsum. Geologists call these different, stratigraphically equivalent aspects of the same rock body *facies* (Latin for "aspect").

Focus on Sedimentary Rocks

Now-solid sedimentary rocks were once soft or loose sediments. They were bound together (lithified) as a result of the precipitation of some kind of cement within spaces between grains, or by forces of compression or weight of overburden that locked the grains together. In such ways is sand changed to sandstone, silt to siltstone, mud to shale or mudstone, calcareous matter to limestone ($CaCO_3$) or dolomite [$CaMg(CO_3)_2$], and so on. In fact I will frequently refer to now-hard rock strata by the name of the sediment they once were (as geologists often do).

Limestone, except where detrital, is the product of chemical combination between oppositely charged dissolved ions (Ca and CO_3). Dolomite, also a mineral, is similar. They are the only common rocks in whose production dissolved carbonate ion (CO_3) plays an important role. As they are commonly associated in intergrading or interbedded sequences it is convenient to group them as *carbonate rocks*. Carbonate rocks are one of the geologist's favorite warm-water indicators because, other factors being equal, the likelihood of their precipitation increases with rising temperature. The real control, of course, is not temperature itself, but the concentrations and activities of component ions that vary with temperature and pressure, in particular the carbonate ion. These important relationships involve complexities of thermodynamics and kinetics that need not delay us here.

The most common sedimentary rocks are the familiar detrital rocks called sandstones, siltstones, and shales (including mudstones), products of the compaction or lithification of sands, silts, and muds. Although geologists usually imply a quartz composition when they use the terms *sandstone* and *siltstone*, they also employ *sand* and *silt* to indicate coarser and finer grain sizes of any composition, as in magnetic sands, gypsum silts, or even limesands (referring to sand-sized sediments of calcium carbonate).

Quartz (SiO_2), however, with a hardness of 7 on a scale of 1 (talc) to 10 (diamond), survives long and repeated transport where softer or more soluble minerals disappear. It is the most durable common rock-forming sedimentary

mineral. The debris of natural origin in rivers is mainly quartz and clay minerals. Limestone fragments quickly dissolve in the carbonic-acid-rich fresh waters of flowing streams. Shale breaks up to individual particles of clay, kept in suspension where currents flow and thus *bypassed* to the calmer waters of lakes, swamps, bayous, and the deeper sea. But sand goes on and on. It finally comes to rest in some slack-water setting, eventually becomes a rock, and at some still later time is again eroded and recycled. Who knows how many generations of recycling may have been required to achieve the sphericity of the sand grains in the beach beneath your feet or the rock beside your garden path?

Pick up a sample of sandstone and look at it with a magnifying glass, or, better, a binocular microscope. The grains are clues to how recently they initially separated from their parent igneous rock, perhaps a granite, when freeze–thaw cycles pried them free or weathering rotted adjacent feldspar into soft clay that washed away in the rain. Their degree of wear also gives clues as to how far they may have traveled and how many times they may have been recycled to make new sandstones. If the grains are angular, of unequal size (inequigranular), and mixed with a lot of easily fractured and weathered feldspar grains, they only recently became sedimentary particles. If they have rounded edges, irregular shapes, and feldspar is rare or missing, they may be second-generation sand grains. If they are well rounded and feldspar is absent, think of third- or fourth-generation sands. If nearly spherical and well sorted they are probably multigenerational or very long traveled.

Some Environmentally Distinctive Crustal Rocks

Some rocks are so distinctive, environmentally informative, and recurrent that they are familiar to every geologist who has traveled very far into the past. They are household words of historical geology—flysch, graywacke, molasse, ophiolite, tillite.

We know sands that travel rapidly to the deep seas, the product of catastrophic events that reflect naturalistic causes. Such sands may first pile up at shelf edges above deeper basins. From there they are thrown into suspension by seismic shock, storm, or flood—in effect liquified. This creates a turbulent current that picks up finer-grained sediment along the way as it is impelled gravitationally downslope in a near free-fall state, its load kept in suspension by its eddy structure. Such suspension currents, the *turbidity currents* of geologists, carry sands down the continental slopes at impressive speeds and with correspondingly great scouring power.

An unplanned and long-unreported experiment that may have recorded these speeds was carried out on 18 November 1929, two decades before the idea of such transport became widely accepted. The times and locations of successive breaks of undersea telephone cables south of Newfoundland, following the Grand Banks earthquake of that date, recorded themselves. When these records were obtained and studied by Bruce Heezen and Maurice Ewing of Columbia University in 1952, their findings implied speeds up to 80 kilometers per hour *if* the breaks were directly related to the heavily laden current itself. That conclusion was supported by later records of

similar cable breaks in the western Mediterranean, following a 1954 earthquake in Algeria. Lower current speeds are more likely *if* the cable breaks were products of liquefaction and collapse, resulting from seismic motion that triggered and was followed by slower-moving turbidity currents. In either case, the evidence implies that relatively fast-moving turbidity currents erode submarine canyons, fill near-shore basins, and spread sand far across the dark abyssal plains, where it surrounds and eventually buries any sea-floor elevations.

Physical evidence for turbidity currents is found in thick blankets of distinctive impure sandstone called *graywacke* that veneer modern abyssal plains and are interbedded with pelagic (open-sea) shales in deep-water deposits of all ages. In such graywackes the coarsest grains are at the bottom, grading upward to finer grained. The lower surfaces of such *graded* graywackes are often covered, like ancient papyri, with a variety of hieroglyphic *sole markings* (Fig. 3.6A, B, D) of both physical and biological origin that convey messages of source, depth, and direction of movement. Graywackes characteristically denote the deposits of former deep-oceanic depressions or trenches. Far travel, even on gentle slopes, implies substantial depths at sites of sedimentation. Associated phenomena are basaltic volcanism and growing mountain chains.

Such interbedded shale and graywacke sequences were long ago called *flysch* (Fig. 3.6A) from a dialectical term of Germanic origin referring to their propensity to slump downhill easily as they do in moist and mountainous Alpine terrains. Geologists have adopted the word worldwide. Although inconsistently defined, and considered enigmatic for generations, the implications of the flysch facies were clarified by the work, above all, of the eminent Netherlands geologist P. H. Kuenen in the 1950s (Fig. 3.6A). It, and its graded graywackes, are now among the best understood and most environmentally informative of sediments. At least we always know from the grading and sole markings which side is up, commonly from what direction the transporting sediment came, and where it was headed. Flysch is characteristically associated with the active growth of folded mountain ranges. It is commonly capped by thick agglomerations of coarse to fine, marine to continental, at some places coal-bearing, post-tectonic debris representing erosion of mountains at late stages of growth and afterward. Such are the sediments called *molasse*, from a French term meaning "soft."

We will take up other genetically significant kinds of rocks as they become germane to a particular problem. Two more, however, call for early introduction. They are the rocks called tillite and ophiolite—respectively indicators of climate and an important type of crustal deformation.

Tillites are, properly speaking, ancient glacial deposits—gravel-like glacial *tills* that have become lithified. In keeping with the scientific struggle for objectivity, the term should not be used unless a glacial origin can be demonstrated. That calls for analysis of variation in rock types, shapes, and packing, whether and how pebbles may have been scratched, and so on. Deposits sometimes mistaken

for tillites are those of avalanches, slope wash, and a kind of turbidity current deposit called pebbly mudstone. Where doubt lingers, the purely descriptive term *diamictite* may be used to refer to till-like mixtures of pebbly and sandy sediments. Convincing tillites and other ancient ice-related phenomena of continental extent are known locally and at intervals from about 2.5 billion years ago until now. Their episodicity and local distribution probably result from former sites of glaciation being rafted across the polar regions at successively different times throughout Earth history. Other likely influences were probably orbital changes, variations in solar activity, galactic motions, or increase in Earth's total reflectivity (albedo), causing reduction of solar heat retained. We shall have more than one occasion to return to that subject.

Ophiolite ("snake rock," denoting scaly surfaces), presently starring in the new world of itinerant continents and plate tectonics, is a word that once had two meanings. One, mostly American, was for a particular kind of rock, basically serpentine. The other, mainly European, was for a suite of rocks—an association of serpentinite (also "snake rock") and associated oceanic-type basalts. As now used, the term *ophiolite* refers to a succession of silica-rich oceanic basalts having structural (e.g., pillow lavas, sheeted dikes) and geochemical characteristics distinctive of mid-ocean spreading ridges, and capped by a thin layer of pelagic sediments. Such sequences denote places where fresh sea floor was being made and crustal plates were rafting away from one another like cargo on a flatcar.

Beginning of the Standard Geological Column

And so to the idea of a *standard geological column*, a central concept in geology. The intent of such a column is to designate as complete as possible an historical succession of rocks to serve as a framework for comparison worldwide—a kind of Rosetta Stone of historical geology. The idea of a regular order in the arrangement of rocks is an old one, long vastly oversimplified and only recently calibrated numerically. Early geologists and their forerunners had many names for the objects of their study and even simple, highly-inclusive systems of classification before the historical sequence now called the standard geological column. Competing schools in different parts of Europe, led by colorful champions employing picturesque terms and systems, jousted for men's minds then as now. And then, as now, the systems that emerged from this process of natural selection went through a long and continuing succession of debates and compromises as new or overlooked evidence came to light. The story is nicely told by W. B. N. Berry in his book, *Growth of a Prehistoric Time Scale* (1968).

As early as 1719, John Strachey worked out the local sequence of strata in the Somerset coal district of southwest England. Then, in 1756, J. G. Lehmann described thirty main zones of the copper-bearing sedimentary rocks of the Harz Mountains in Germany. Concurrently his fellow countryman G. C. Füchsel made such progress on a geologic map of Thuringia that, by 1762, he had projected his observations globally. Nor was Italy left out. Between 1760 and 1770, Giovanni Arduino, inspector of mines in Tuscany and later professor of mineralogy at Padua, set forth a grand division of Earth's rocks into Primitive, Secondary, and Tertiary, based on his observations in the Apennine Mountains.

Thus did the standard geological column begin to grow from an amalgam of local geologic columns. The ones that had the most influence on the eventual global synthesis were those of continental Europe and Britain—especially that of England and Wales. At first there were only descriptive names like (from older to younger) Old Red Sandstone, Mountain Limestone, Millstone Grit, Coal Measures, New Red Sandstone, the Oölite beds, the Chalk, and so on. Then, in 1794, while superintending the digging of the Somerset Coal Canal near Bath, in the same area studied earlier by Strachey, the aforementioned William Smith began the studies that were to bring him the sobriquet "father of stratigraphy." Although scholars 2,000 years earlier had recognized that fossils are the remains of once-living creatures, it was, as we have seen, Smith and Cuvier who first showed their regular order of appearance. In his 1815 *Geological Map of England and Wales*, Smith included a hypothetical vertical profile of rocks that began with ancient Mount Snowden in Wales and ranged stratigraphically upward and eastward through a succession of east-dipping strata into the Cenozoic clays of the Thames Valley at London (Fig. 3.12).

The concept of an orderly general stratigraphic sequence spread rapidly even in those days of exclusively written and personal communication. By 1842, barely four decades after Smith's seminal work of 1799, most of the main divisions of the conspicuously fossiliferous upper 15 percent of the now-known geological column had been named, defined, and placed in sequence. These divisions and others widely accepted up to 1879 are listed in Table 3.1 in downward order of increasing age.

In that early standard column the former Old Red had become the Devonian; rocks equivalent to the former Mountain Limestone, Millstone Grit, and Coal Measures had become the Carboniferous System; the New Red had become the Triassic; the Magnesian limestone had become the Permian; the Oölite had become the Jurassic System; and the Chalk, of course, had become the Cretaceous. One element of Arduino's 1760 nomenclature that (although fading) still survives, is the "Tertiary System," currently comprising all Cenozoic rocks up to but not including the last 1.6 million years. That bit at the end, the Pleistocene, together with the last 10,000 years, called Recent, comprise another anachronism, the Quaternary System of Desnoyers. That nomenclature needs revision and simplification. Increasingly nowadays a simpler twofold division of the Cenozoic Era into the older Paleogene and the more recent Neogene systems and periods is finding favor. Two other names that have gained wide acceptance as periods and systems in the United States are Mississippian for the Lower Carboniferous and Pennsylvanian for Upper Carboniferous. Where so used, the Mississippian and Pennsylvanian together may be called "the Carboniferous systems."

Little was known about rocks beneath the Cambrian in Europe in the mid-nineteenth century other than that there were some and that they were commonly highly deformed and metamorphosed. No one foresaw that they would be found to represent some 85 percent of Earth history. Nor did the early founders of geology foresee that the succession and history of these rocks might someday be deciphered. Instead they were called the "interminable graywackes," lumped under the informal term *pre-Cambrian*, and relegated to obscurity as something

3.12 WILLIAM SMITH'S 1815 GEOLOGIC STRUCTURE SECTION FROM NORTH WALES TO LONDON. Vertical profile of the tilted sedimentary cover, in which the brickwork pattern represents limestone. Other strata as labeled.

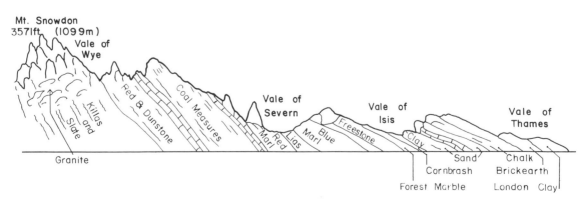

71

beyond the known and perhaps the knowable. As we shall see in Chapter 4, the discovery of radioactivity in the late nineteenth century and the radiometric measurement of geologic time changed all that. It showed not only that the geo-logic record of our nineteenth-century predecessors was several times longer than most of them supposed but also that it represented only a small fraction of geologic history.

TABLE 3.1 *The Standard Geologic Column a Century Ago (with Names of Founders and Dates of Founding)*

Main Historical Divisions (Eras)	Systems (Rocks) & Periods (History)		Series (Rocks) & Epochs (History)
CENOZOIC (J. Phillips, 1840)	NEOGENE (R. Hornes, 1853)	QUATERNARY (P. G. Desnoyers, 1829)	Recent Pleistocene (C. Lyell, 1839)
			Pliocene (C. Lyell, 1833)
		TERTIARY (G. Arduino, 1760)	Miocene (C. Lyell, 1833)
	PALEOGENE (K. F. Naumann, 1866)		Oligocene (E. Beyrich, 1854) Eocene (C. Lyell, 1833) Paleocene (W. P. Schimper, 1874)
MESOZOIC (J. Phillips, 1840)	CRETACEOUS (J. B. J. d'Omalius d'Halloy, 1822)		
	JURASSIC (A. von Humboldt, 1799)		
	TRIASSIC (F. von Alberti, 1834)		
PALEOZOIC (A. Sedgwick, 1838; Revised 1840 & 1841 by J. Phillips)	PERMIAN (R. I. Murchison, E. de Verneuil, & A. von Keyserling, 1841)		
	CARBONIFEROUS (W. D. Conybeare & W. Phillips, 1822)		
	DEVONIAN (R. I. Murchison & A. Sedgwick, 1838)		
	SILURIAN (R. I. Murchison, 1835)		
	ORDOVICIAN (C. Lapworth, 1879)		
	CAMBRIAN (A. Sedgwick, 1835)		
	pre-CAMBRIAN		

In Summary

To discover an uncharted Earth, to reconstruct a history for that seemingly unlimited time before there was anyone to make conscious records, demands that various kinds of proxy records be utilized and placed in sequence. The rocks must be read and the stories they tell arrayed in an order of before, after, and contemporaneous with.

Ordering principles are needed for that. They include the *law of superposition*, the *principle of crosscutting relationships*, and related principles, from which sequence is reconstructed for deformed rocks as first enunciated by Niels Stensen (Nicolaus Steno) in 1669. By the late eighteenth and early nineteenth centuries, the long-smoldering and cautiously expressed idea of a duration of geologic time far greater than the biblical 6,000 years was gaining ground. At the same time scholars were coming to appreciate the probability that natural causes suffice to explain all geological phenomena.

Although others played important roles (Fig. 3.3), Charles Lyell took the lead in enunciating these views as the basis for a truly scientific geology. Some confusion resulted, nevertheless, from Lyell's ambiguity about rates. What remains useful after 2 centuries of recurrent and, at times, heated discussion is the general perception that natural causes by themselves can account satisfactorily for all natural phenomena. The only assumption required is that natural laws (*not* rates) have remained invariant throughout Earth history—or, if they have varied, that they have varied lawfully.

Principles and criteria for establishing sequence and other forms of order in rocks have emerged during the course of intellectual history and continue to do so. Highly significant among these is the *law of biotal succession*, established in the late eighteenth and early nineteenth centuries. The now well-known succession of animal and plant fossils in rocks that range in age from around 670 million years ago to the present has two important consequences. It establishes the generally systematic change of life through time (*biologic evolution*) as a fact requiring explanation. And it provides an historical succession of fossils that can be used in sequencing and determining historical equivalence among fossiliferous rocks wherever found.

In this way has been established a broad global succession of rocks and, within such a framework, an Earth history. The general global succession can also be amplified and refined by criteria for dating and correlation other than the uniquely unidirectional systems of biotal succession and radioactive decay. Supplementary criteria include widely recognizable events such as mineralogically and geochemically distinctive volcanic ashfalls and paleomagnetism. By paleontological means alone, however, our predecessors of a century and a half ago, began to create the skeleton we call the standard geologic column and to add to it the flesh of history.

The action, the story of Earth's progression to its present state, must be elucidated by evidence wrung from the stubborn rocks. With a bit of persuasion, rocks do tell stories. They form under specific conditions and change when moved to others. This chapter provides some clues as to how those stories may be read.

CHAPTER FOUR

THE ALCHEMISTS'
REVENGE

To extend Earth's history much further into the past than that known to the founders of geology, to record the past before abundant, clearly visible, and stratigraphically distinctive fossils appeared, requires applicable criteria for time and sequence. They are found in the systematic transmutation of the elements. Although alchemists' attempts to transform lead to gold failed, the idea has been vindicated. Radioactive elements convert spontaneously to inert end products at rates that record the ages of rocks, of predecessor melts, and of subsequent thermal events. The passage of Earth history is thereby calibrated to an invariant scale of atomic time. And, although time alone is not history, there can be no history without it.

Paradoxically, time, like history, is recorded only by events. Related now to the speed of light, it is seen as an invariable, linear, but, in theory, reversible flow. History, a sequence of prevailing tendencies and punctuating events, is also linear, but variable and irreversible. It is calibrated but not defined by time. Including radiocarbon and the decay of uranium to lead, a dozen different radiometric timekeeping systems are now in use, supplemented by other numerical systems for more recent events. Earth's age is now unequivocally established as very close to 4.6 billion years, and its history from origin until now is calibrated within that range of radiochronologic years.

Historical Outlook on the Nature of Time

History is the succession of distinctive events and trends observed and recorded for given but variable lengths of time. But history without calibration is incomplete. Many aspects of Earth history could not be understood or even discussed unambiguously before it became possible to measure elapsed time preceding the present and the duration of events on an agreed-upon time scale. Perceptions of the great length of geologic time, its relation to history, and the methods of calibrating it are basic to all modern geology.

4.1 BRISTLECONE PINE, CREST OF WHITE MOUNTAINS, CALIFORNIA. *[Courtesy of C. A. Nelson.]*

What is this stuff we neither see, nor feel, nor sense in any way except duration? How are consistent measurements of it to be made, especially in prehistoric time?

Such questions are fundamental not only to history, but to practical human affairs. It is hard to appreciate their significance in ignorance of the progression of human thought. We can learn much about the progress of science and technology from an historical review of ideas of time. These ideas involve fascinating perceptions in their own right that ought to be appreciated before plunging into a narration of methods and numbers. There is more to geology than straight story. Think of the nature and uses of time.

Mankind—insatiably curious, ever the explorer—has spanned the Earth, mapped the geology of its moon, explored the planets to the edge of the solar system, and is searching by every means to probe the frigid depths of space. If we could but venture beyond our galaxy and look back at it, it would look something like the Great Galaxy in the constellation Andromeda, a blazing cosmic pinwheel of a hundred billion or so hydrogen-burning stars, one of billions of galaxies in the expanding cosmos. It is unlikely, however, that all of the galaxies strewn through space originated at the same time or that the stars in a galaxy are all of the same age. In fact, if we plot the stars in an old galaxy like ours according to temperature and luminosity, most of them are found to fall into a pattern, called the main sequence, that resembles an evolving system. As observed in Chapter 1, a distinctive class of small stars like our Sun are born as T-Tauri stars, ignite to enter the main sequence, grow more luminous, and

eventually expand, burn out their remaining nuclear fuel, and collapse into white dwarf stars—each according to its own individual and variable life-span. Bigger stars lead shorter, more violent lives.

To summarize and condense from Chapter 1, one sees a pattern of historical episodes and among them a sequence, a cosmic history of which Earth history is a part. Calculations from evidence and theory suggest: (1) an initial unimaginably hot and universal Big Bang, cooling to 1 billion degrees Kelvin (equals Celsius $+273°$) during the first 3 minutes; (2) an early universe of radiation only, lasting about 300,000 years, during which it cooled to about 30,000°K; (3) the condensation thereafter of elementary particles into ordinary light matter; (4) the clustering of that matter into galaxies, clusters, and superclusters, beginning perhaps 8 billion years later; (5) a continuing evolutionary cycling of stars and matter within the galaxies; (6) the roughly simultaneous condensation of the miniscule residues of star-born heavier matter into planets around favored single stars of near solar mass, including the condensation of our solar system some 4.6 billion years ago; and (7), most importantly for us, the ensuing sequence of events that have shaped and continue to shape our evolving Earth.

A sequence chronology, however—even a sequence chronology as well documented as that for the last 15 percent of Earth history—leaves much to be desired. As the late William Thomson (Lord Kelvin) of thermodynamics fame was fond of observing, the ability to measure something sharpens our perceptions of it. We want to number the years and bracket the trends and episodes of the historical record. We want to *count* the years behind the nearly 9,000-year record of the bristlecone pines (Fig. 4.1). We want to *measure* time.

Here we confront a problem as old as humankind. How does one measure something that is immaterial, that cannot be sensed in any concrete way? How is it possible even to describe something that has no detectable shape, weight, or other dimensions? By what standards do we perceive duration to occur, let alone measure it, when its very essence is transience?

Time, like space, is no more than a dimension of experience depending for recognition on material objects and events that give it substance. It is hard even to imagine empty space or uneventful time. The effort to define and standardize this elusive quality, therefore, has varied in interesting ways. All people in all ages have been aware of the passage of time in relation to events—before, after, contemporaneous with. Their commonsense perceptions of it and ways of estimating it, however, are strongly colored by climate, season, sociology, religion, and even age. The poor and the young, it has been said, have more time than the rich and the old. Time hangs on the hands or flies accordingly.

The coming to grips with this ephemeral stuff encapsulates the advance of science and technology. Across the last 10,000 years of human history crops have been planted and harvested, ships launched or beached, and military ventures undertaken or avoided according to the positions and movements of heavenly bodies and the resultant forecasts of weather or luck. Yet the first objective timekeeping devices of which record is preserved are the sundial or shadow clock

(known from about 1500 B.C.) still useful in principle to hikers and farmers, and the clepsydra or water clock. An Egyptian sample of the water clock survives from about 1700 B.C., and there may have been older ones in China. The Athenians put them to practical use in limiting debate. And the Romans commissioned handsome 24-hour water clocks like the great octagonal structure that stands in the Roman Agora at Athens. Sandglasses and candleclocks were of the same vintage. The oldest surviving gear-driven clocklike device, from a shipwreck in the Aegean Sea, was apparently made by some Rhodian craftsman about 87 B.C. as a gear-driven calendrical device, and by 1320 the mechanical clock was clearly in vogue.

The resulting world trade in mechanical clocks and the introduction of the Gregorian calendar in 1582 raised consciousness everywhere of the flow of time, reinforced the idea of the regularity of this flow, and perhaps overemphasized the idea of dividing that flow into segments.

Just as it had earlier been analogized with divine order, with the Renaissance the mechanical clock became significant in the emergence of modern science. In the early seventeenth century Johannes Kepler rejected former quasi-animistic concepts of the universe, comparing its operation with that of a clock. The earlier mentioned Robert Boyle, who also foresaw the use of fossils in geochronology, drew a similar analogy, as did other scientific figures of that time. Thus did the metaphysical significance of the mechanical clock evolve from that of a model of divine omnipotence to become the philosophical basis for the mechanistic view of nature that dominated emergent science from Descartes to Darwin.

Indeed, in abandoning design in favor of a statistical, mechanistic view of the progressive change in life through time, Darwin made the intuitive leap that predecessors like Descartes, Buffon, Darwin's grandfather Erasmus, and others had flirted with but not dared, in their religious climate, to advance except as poetic speculation or fiction. The eventual success of the unequivocally naturalistic views of Lyell and Darwin cleared the way to think of biological and geological history in non-predestinational terms.

The ultimate solution for geologic time was a triumph of insight over logic. The intellectual successors to the medieval alchemists, who thought to transmute lead to gold, were the present-day geochronologists, who found answers to the calibration of geologic history in the invariant natural transmutation of uranium to lead. The road to their liberating insights goes by way of the Julian and Gregorian calendars, precise navigation, the speed of light, and atomic time to intersect with radioactivity in the mass spectrometer—an atom-counting machine that grew in the minds of ingenious and disciplined dreamers.

Lost Days, Stretched Years, Relativistic Seconds

Time, then, is in some way measured in terms of events. Or, to reverse that perception, as some wag has said (approximately): "Time was God's way to avoid having everything happen at once." If, then, time is to be measured in terms of events, what events are to be chosen and how are they to be standardized so that the rich and the old can have equitable shares?

That is not a trivial question, as a little reflection on history will show. The day and year of common experience, based on Earth's rotation and orbital period, are hard to synchronize. Even allowing for leap years, as did the Julian calendar of 45 B.C., the cumulative discrepancy amounted to a loss of about 11 minutes a year. By A.D. 1582 the error had added up to 11 days and something had to be done to prevent the months, and particularly the religious holidays, from getting out of phase with the seasons. In

February of that year Pope Gregory XIII issued an apostolic letter accepting the results of a papal calendrical commission that solved the problem by rearranging the leap years and decreeing that the day following the 4th of October in that year would skip to the 15th.

Although the Gregorian calendar thus came into effect in countries under papal jurisdiction, Protestant societies were very upset at the idea of losing 11 days out of their lives. Worse still, the following year was to begin on 1 January instead of 25 March, the day of the revelation to the Virgin Mary that she was to be the mother of the Messiah and the proper beginning of the Christian year. Also in a year that began in January, the months of September, October, November, and December were no longer the 7th to 10th, as their names imply and were intended to be, but now the 9th to 12th. Everything would be topsy-turvy! Riots ensued and protestants refused to adopt the Gregorian calendar—the one in worldwide use today. It wasn't until 170 years later, in 1752, that the British Empire (including the American colonies) finally adopted the new calendar, again to the tune of much protest. That is how the birthday of George Washington comes to be celebrated on the 22d of February, when he was actually born 11 February 1732 of the Julian calendar. It took another 121 years for the Gregorian calendar to reach Japan, 45 beyond that to reach czarist Russia, and 11 more to China, where it was finally accepted in 1929, just 347 years after its adoption in Rome.

So much for technology transfer. Let us hope the metric system (adopted in France in 1791) reaches the United States faster than the Gregorian calendar came to China.

The Gregorian calendar is almost as good as was the Mayan calendar, but not quite. Even with a complicated formula of leap years and non–leap years it accumulates an error of nearly a day every 300 years. Great improvement was obtained, however, with the global adoption of ET in 1956—not the loveable alien hero of *E.T.: The Extra-Terrestrial*, but something called *ephemeris time*, invented by Yale's G. M. Clemence and based exclusively on the orbital motion of Earth and its moon without reference to Earth's rotation, whose rates are affected (and perturbed) by nongravitational internal forces. ET changed the definition of the second, formerly given as the 86,400th part of a day. The second was then defined as $1/31,556,925.9747$ of a year.

Yet the need for both precision and accuracy continued to grow with the increasing sophistication of navigational problems, such as those involved in directing the courses of spacecraft and missiles. Systems based on the common-sense application of Newtonian physics and Euclidean space were no longer good enough in a relativistic universe. Newtonian dynamics and Euclidean space are the stuff of classical solar and stellar (sidereal) time—ordinary clock time if you will. Relativistic (Einsteinian) dynamics, involving as its only accepted constant the speed of light, C (for celerity, in $e=mC^2$), gives atomic time, kept by vibrating crystals such as cesium or quartz. The cesium clock, invented in 1955 at England's National Physical Laboratory, is based on the fact that the stable isotope cesium-133 vibrates 9,192,631,770 plus or minus 20 cycles a second. It keeps time with an accuracy of plus or minus 1 second in 10 centuries.

Atomic time, the kind kept by radiometric geological clocks, has its own peculiarities. It gives a time scale for the present universe that has a beginning (the last Big Bang) and presumably will have an end. In a relativistic universe also, supposedly invariable Newtonian constants (e.g., the gravitational constant and any angular momentum) increase with time, while space, no longer Euclidean, becomes hyperbolic. As explained by E. A. Milne (brother of A. A. Milne of Winnie-the-Pooh fame), classical solar and sidereal time, the time of mechanical clocks, appears to be slowing down relative to atomic time. Planet Earth, with an atomic age of only 4.6 BY, would be immensely older than that in clock years, while the origin of the universe would recede toward the infinitely distant past.

Everything, including geologic time, is, in fact, now measured by the speed of light. Since October 1983 that has been internationally defined as 299,792.458 kilometers per second. The effect of this is that, whereas the length of the meter itself might be changed by better measurements of the speed of light, the velocity of electromagnetic waves, or photons, at all wavelengths is now invariable by decree, as well as (apparently) in fact. In a relativistic universe something must be set constant if anything else is to be measured. And on this planet C has now been chosen as the one absolute invariable.

The second, thus, is again redefined. It is now the time required for electromagnetic waves to travel a distance of 299,792,458 meters in a vacuum. Such measurements nowadays are routinely made by radar ranging to solar system objects or spaceships. The standard day and year now become appropriate multiples of the standard second, while the Gregorian calendar continues in use by virtue of periodic corrections. Most of us, nonetheless, will continue to measure time and space in traditional ways for traditional reasons during our brief individual tenures on this untidy little planet. As inhabitants of a particular conceptual world we will continue to regulate our lives by the *apparent* movements of the heavens.

Only remember that the radiometric time of geology is relativistic. Earth, under that theory, will have journeyed round Sun many more than 4.6 billion times over the past 4.6 billion radiometric years that geochronology reckons as its scientific age.

The Problem of Geologic Time

There was no lack of agreement, it seems, during the emerging years of geology in the late eighteenth and early nineteenth century, that a huge extent of time in terms of human experience would have been required to account for phenomena observed as products of natural processes. Disagreement centered on the matter of whether the probable explanation lay in a truly immense panorama of time or in an Earth whose early stages were the product of processes catastrophic beyond the tolerance of natural law and therefore miraculous.

The demonstration, during the early decades of the twentieth century, of the historic truth of immense time and its numerical calibration in radiometric years is, thus, the cornerstone of modern geology. It buttresses all subsequent advances in pure geology, including plate tectonics, paleomagnetic stratigraphy, the emergence of a defensible historical geology for the first 85 percent of the geologic column, and the impact history of the planets.

A procedural problem in geology, however, arises from the convenient but confusing tradition of treating history and time as if they were equivalent. Thus, although most geologists know better, geological statements and charts customarily imply that progression from one interval of history to another was instantaneous, whereas, in fact, most involved transitional intervals of variable length. Although there can be no history without time, therefore, time alone does not define history. History instead is a succession of related events that commonly begin and end at different times at different places. It is calibrated, even bracketed, but not defined by time. The centuries that bracket the Renaissance do not define that historical era. An instant in geologic time, even a volcanic ashfall or the dust from asteroidal impact, may take months to reach its farthest limits. The act of polar reversal requires some thousands of years. Perceptible modal changes in Earth history may demand millions, tens of millions, or even hundreds of millions of years.

Time is linear, but nondirectional (although radiometric decay is). Its equations are the same in either direction. History is nonlinear but unidirectional; it may "repeat itself" but it cannot reverse. The proper components of Earth and cosmic history in simplest terms are trends and sequences of events.

The basic principles by means of which historical geology is reconstructed were well understood by the turn of the current century, and the grand features of the most recent half-billion years were beginning to be understood. No one then knew how many years had elapsed, however, nor did they then know of any way in which the record could be extended to include events before the Paleozoic Era in any systematic way. Indeed, apart from inspired guesses by naturalists and philosophers such as the Comte de Buffon and Immanuel Kant in the mid-eighteenth century, there had been little discussion about geologic age until Earth processes, including evolution, made it a hot subject.

There then ensued great controversy about whether there was time for erosion and tectonism to do their work, and for natural selection to produce the living world. How old was our planet? Ingenious efforts were made to resolve those questions. One method estimated ages from inferred rates of sedimentation. Another method calculated how long it would take to account for all of the salt in the oceans, plus that in sedimentary salt deposits, assuming the oceans began fresh. Still others involved rates of erosion, rates of biologic evolution, and the cooling of Earth from a molten state.

The rate of erosion of the Niagara River from Lake Ontario to Niagara Falls was a classic indication that all was not well with estimates for the date of creation derived from biblical accounts. In 1789, in a paper read before Ben Franklin's American Philosophical Society in Philadelphia, a Niagara resident, Robert McCauslin, reported that such calculations fell far short of the time needed even to excavate the Niagara Gorge. Estimates since then have consistently yielded ages 2 or more times the roughly 6,000 years earlier allowed for the whole of Earth history. Carbon-14 dating has now confirmed a duration of 12,000 years for the cutting of the Niagara Gorge alone.

Clearly Earth is many times older than this, but how many? Darwin guessed about 400 million years (400 MY) as the minimum needed to account for the observed succession of animal fossils—surprisingly close to the results of modern radiometric geochronology. Turn-of-the-century estimates based on the salinity of the sea and the thickness of the sedimentary column generally came out somewhere between 100 and 400 MY. The methods weren't the best, but they were the best geologists could think of for some time.

Meanwhile Kelvin, cofounder of thermodynamics, calculated how long it would take Earth to cool to its present state from the incandescent origin then hypothesized. His first results, in 1862, gave possible ages ranging up to 400 MY, but by 1897 he had shortened that to between 20 and 40 MY. Not likely, said most geologists, smarting under his ridicule—that simply wasn't enough time. Something was wrong with the assumptions. In fact two things were wrong. One was that Kelvin had chosen the wrong constants for the cooling of a solid body by conduction. The second was radioactivity, discovered by Henri Becquerel in 1896. Geologist T. C. Chamberlin, (in 1899, in *Science* magazine) apparently independently, hypothesized atomic sources of heat to explain a slowed rate of solar cooling. And 3 years later Ernest Rutherford and Frederick Soddy published their findings on the generation of thermal energy by radioactive decay. That started the intellectual journey that has since led to nuclear physics, radiometric systems of geochronology and cosmochronology, and ages of billions of years for Earth and cosmos.

Such advances (eventually) made it possible for geologists to array the events of the earliest 85 percent of geologic time in historical sequence. They also provided a measure of control on cosmochronology as reckoned from shifts in the wavelengths of the stellar candles that light our expanding universe.

Introducing Geologic Clocks: Tree Rings and Varves

When taking the pulse of the universe or the Earth one wants a unit of time that is everywhere the same and that has not varied through the ages. That is the gift of radioactivity. Radioactivity involves the process of *nuclear decay* to stable end products—for example the now-familiar decay of uranium-235 to lead-207, with the expulsion of seven *alpha particles* (helium-4 nuclei). It records atomic time—akin to that kept by cesium-beam atomic clocks. And it proceeds with a statistical regularity that has, according to all available evidence, been constant during all of Earth history so far. By measuring the amount of parent isotope (e.g., U-235) remaining and that of *daughter isotope* or *end product* (e.g., Pb-207) accumulated, an age can be calculated. The rate of decay must be known, of course, and that has now been measured and internationally standardized for about a dozen different radiometric methods.

For the last few millennia, however, we know geochronologic methods that are simpler, more precise, and more directly related to human experience—tree rings and annually layered sediments. Tree-ring dating may have been known to Leonardo, and has been around at least since 1757 when Buffon and Henri Duhamel discovered that they could identify the same frost-damaged annual ring in a number of trees in their part of France. So much information is locked up in the bilayered annual rings of trees that timbers used to build ancient structures record not only the year they were cut but the time of year and the climate. In the dry southwestern United States, archaeological dating using this method is so reliable that other dating methods are rarely employed for ruins younger than 2,300 years unless suitable timbers are lacking. Specialists in tree-ring research (dendrochronologists) at the University of Arizona's Laboratory of Tree-Ring Research, using special coring tools, have matched distinctive tree-ring sequences, and compared their results with those of carbon-14 to prepare climatic maps for the last 400 years of western American history. Although the oldest living trees in that region do not exceed ages of about 4,700 years, dead but well-preserved trees extend the climate and tree-ring record back for more than 6,270 years. That is exceeded only by the west European sequence of 7,272 years and by California's bristlecone-pine record of 8,681 years (Fig. 4.1). These are counted, year-by-year records that reach back in time for some 3 millennia before the creation as calculated by Ussher.

Similar to tree rings are the annually layered or varved deposits of meltwater lakes and some other quiet aquatic basins. *Varves* are coded seasonal couplets of laminae or layers rarely more than a few centimeters thick. They form most typically where lakes freeze over in winter but are open during part of the year. Thin, dark intervals of organic matter and the finest clay particles settle slowly from suspension beneath the winter ice. Thicker, light-colored, silt-rich bands

represent the deposits of spring and summer freshets. Silty layers grade upward into the clay-rich portions of each couplet, but the clay interval is sharply separated from the coarser, lighter interval of the overlying couplet. From this graded bedding one reads not only the years but the direction of "younging"—as geologists say. The latter is not particularly interesting for the relatively undisturbed varved sequences of the last 10,000 to 14,000 years, but may be the clue to sequence in varved intervals that occasionally show up in ancient glacial deposits as far back as 2.5 billion years ago. Similar regular couplets known from calcareous and evaporitic sediments of quiet basins are also interpreted as annual layers and referred to as varves.

Tree rings and varves record Earth history year by sidereal year and season by season. They take it back continuously to the last major advance of Pleistocene ice. They also illustrate the advantages of quantification over simple sequence geochronology. Indeed such records can even be correlated by distinctive cyclic events from one continent to another. The Bronze Age varves of southern Russia, for instance, have been correlated with tree-ring records in Turkey and the bristlecone-pine sequence of the western United States. (Sherlock Holmes would have loved this.)

Radiometric Methods: Radiocarbon

Older varves yield year-by-year records for limited parts of the geologic column, but not ages. Actual ages in years before the present (BP) for these and other truly ancient rocks can be obtained only from methods involving

radioactive decay. About a dozen such *radiometric methods* are in common use—an activity called *radiochronology*.

Those methods are demanding and tedious but, in principle, easy to understand. They are of three main kinds.

First is the carbon-14 or *radiocarbon method,* based on radioactivity (actual counts of particle emission) itself and used primarily on plant matter or its products less than 40,000 years old. Second are the *methods of end products,* obtaining ages from the ratio of inert daughter isotopes to radioactive parent isotopes and applicable to rocks and minerals of a wide range of types and ages. Third is the use of *isotope ratios,* especially lead isotopes. Consider first the radiocarbon method.

Radiocarbon ages are found by measuring the rate of decay of the parent isotope (its radioactivity) rather than the accumulation of its end product. Carbon-14, or C-14, is produced by cosmic radiation in the upper atmosphere. It results when a proton is blasted from the nucleus of ordinary nitrogen-14, leaving

4.2 LIFE CYCLE OF CARBON-14. Cosmic-ray bombardment of nitrogen-14 converts it to C-14 at a statistically constant rate. It reverts to N-14 by emission of a nuclear electron.

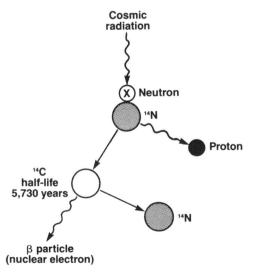

Cosmic radiation

X Neutron

^{14}N

Proton

^{14}C half-life 5,730 years

^{14}N

β particle (nuclear electron)

the impacting cosmic neutron behind (Fig. 4.2). It is like shooting marbles out of a ring when your agate stays behind, or other games where one object is displaced by the inertial energy of another. In the case of radiocarbon production, that act reduces the number of protons from the seven mandatory for nitrogen to the six of carbon, without changing the total number of nucleons (protons plus neutrons). So now N-14 has become C-14 instead of ordinary stable C-12, with only twelve nucleons.

C-14 is unstable, however. It wants to return to being stable N-14. It does this by emitting an electron (a beta particle) from one of its neutrons, converting it to a proton and restoring the magic number of seven nuclear protons that make it nitrogen again. Meanwhile other new C-14 atoms have been oxidized to CO_2 and dispersed by wind and water to become a miniscule but theoretically nearly constant fraction of the circulating carbon pool of the world, including the living biosphere. Inasmuch, however, as no more than a millionth of a gram of carbon contains some 5×10^{16} atoms, even a miniscule fraction of C-14 in CO_2 can make a lot of electronic blips on a radiation detector as it happily returns to being stable N-14. That C-14, meanwhile, entirely as C in CO_2, gets into plants and animals by way of photosynthesis, grazing, and predation, where a dynamic balance is maintained between incoming C-14 and decay so long as the host-creature lives. Upon death, the ratio of C-14 to C-12 immediately begins to decrease, as an isolated pool of radiocarbon in wood, bones, or shells converts back to ordinary nitrogen.

The critical measurement in obtaining any radiometric age is a negative exponential—the time required for half of any quantity of the parent isotope to decay to its final inert daughter isotope—its end product. That is the *half-life* (Fig. 4.3), which can be calculated from the *decay constant*, a factor in equations of age. The half-life of C-14 has been determined as 5,730 plus or minus 40 years—written 5,730 ±40. Unlike other radiometric systems, however, the ratio of parent-isotope C-14 to end-product N-14 cannot be measured because the latter and ordinary N-14 are in every respect identical. Radioactivity itself, therefore, is directly measured by special, shielded, counting devices. Other problems involve the introduction to the atmospheric pool of excess C-12 from the burning of fossil fuels and similar excesses of C-14 from nuclear tests. Happily, methods have been developed to correct for most such deviations from the norm.

Although the results are often quoted by users as whole numbers or round numbers, reports from well-run radiocarbon laboratories always quote an experimental limit of error—plus or minus some figure. For example, a 1959 analysis of carbonized bread from the A.D. 79 eruption of Vesuvius gave an age of 1,830 ±50 years BP. The meaning of that number is that at least two-thirds of additional analyses should yield ages between 1,780 and 1,880 BP for the bread, which in this case had an expected age of 1,880 BP—barely within those limits. How comforting for the rest of us to realize not only that those clever analysts don't always get truly accurate answers, but that burned toast has so respectable a pedigree.

Analytical error, however, is the least of the problems with radiocarbon dating. The accuracy of the method depends on the maintenance of a predictable rate of

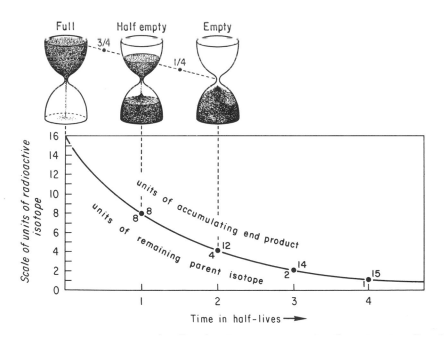

4.3 EXPONENTIAL TIMEKEEPING CHARACTERISTICS OF RADIOACTIVE DECAY COMPARED WITH ARITHMETICAL CHARACTERISTICS OF AN HOURGLASS. *[From Preston Cloud, 1978,* Cosmos, Earth, and Man, *Yale University Press.]*

production of new C-14, a function of the intensity of cosmic radiation which we now know has varied with solar activity and variations of Earth's magnetic field. That, plus industrial and military sources of discrepancy and geological error, are reduced but not eliminated by ingenious use of cross-linkage with tree-ring data and caution in selecting samples. Nonetheless, the *sequence* of events dated is not affected. Despite the fact that daylight saving time puts everyone out of phase with the Sun, most of us still arrive on time for appointments.

Radiocarbon dating is primarily employed in archaeology, paleoanthropology, and the dating of early historical and recent geological events. Because of the short half-life of C-14, however, its use is mainly limited to dating events younger than about 40,000 years. With dilution techniques and the use of particle accelerators in tandem with mass spectrometers for counting, that range can be doubled for appropriate samples. Some enthusiasts even predict extension of the method to as much as 200,000 years BP. That would be a boon for Pleistocene studies but still not interesting for events older than the last interglacial epoch.

Counting Atoms and Dating the Planet

For the dating of events older than about a million years, a number of radiometric systems are available, most especially methods of end products and isotopic ratios. Without radiometric techniques we would have only the crudest sequence chronology for truly ancient Earth history.

The basic requirement of radiometry calls for the transfer of matter from one pool to another at measurable rates—as in the dribble of sand glasses and water clocks. If one knows the rate of transfer and the amounts of beginning and end product, it is a simple matter to calculate the time elapsed. The calculation of radiometric ages is complicated only by the fact that the quantities are hard to measure with the accuracy and precision sought and that the rate of transfer (the decay rate) is a negative exponential (Fig. 4.3). That is really an advantage, however, because immensely long intervals can be recorded by very small quantities of end product at these incrementally very slow decay rates. Finding the ages of rocks by such methods is what geochronologists usually mean by radiometric dating.

Methods for estimating decay rates were slow to develop because the only usable system known for several decades was the transmutation of uranium to lead. No way was known to isolate and count the significant parent and daughter isotopes. Even after it was recognized that uranium decays to lead and does so at a measurable rate, the estimation of elapsed time was crude. A simple rule-of-thumb equation of age as time (t) in the 1930s was: $t = \text{Pb/U} \times 7,600 \times 10^6$, based on researches by B. B. Boltwood at Yale in 1907. Although Boltwood estimated a uranium half-life of 10 billion years (more than twice the present value for the half-life of U-238) and there were other problems, numbers obtained were consistent, large, and exciting. They were also surprisingly close to more accurate later values. By 1913 Arthur Holmes (Fig. 3.3), then at Imperial College, London, had published the first numerical time scale, finding a maximum crustal age approaching Boltwood's. And 4 years later Yale's Joseph Barrell was able to array events in a historically consistent sequence of numerical ages that culminated with Boltwood's estimate of more than 2 billion years for the age of the planet.

The mass spectrometer was the key to improved precision and accuracy. This instrument detects miniscule differences in mass or weight of isotopes. It literally counts and tallies the number of gasified atoms of different mass as they stream past its counting device (Figs. 4.4, 4.5). The principle of mass spectrometry had been known since its discovery by German Nobelist Wilhelm Wien in 1898, and was already suspected by Sir William Crookes in 1886. But the instrumentation only became applicable to the separation of isotopes as a result of refinements by Princeton's H. D. Smyth in 1932. It was soon found that the element lead included four isotopes of identical chemical characteristics but slightly different mass (Pb-204, 206, 207, and 208). All but one of them, moreover, resulted from the decay of uranium or thorium. Pb-204 alone was nonradiogenic, or *common*, lead. The first application of mass spectrometry to geology, therefore, was the elimination from further analysis of samples "contaminated" with Pb-204.

Modern radiochronology began in 1938 with the work of A. O. C. Nier, then a postdoctoral student at Harvard and subsequently the presiding instrumental genius at the University of Minnesota. Nier, then barely 27, was the first to measure lead and uranium isotopes (Figs. 4.4, 4.5). He placed electrostatic and magnetic separation units in series with an improved amplification system that, by measuring very small currents, provided the sensitivity to separate and measure the isotopes of lead and uranium. This enabled him to calculate the decay rate of U-235 to Pb-207, and thereby to determine the ages of minerals and rocks from very small amounts of material. That achievement pointed the way to all subsequent advances in geochronology, just when Hitler unloosed the dogs of war and uranium became a classified subject.

4.4 NIER AT THE CONSOLE OF HIS HOMEMADE MASS SPECTROMETER, USED FOR LEAD-ISOTOPE STUDIES AT THE UNIVERSITY OF MINNESOTA, 1939. *[Courtesy of A. O. C. Nier.]*

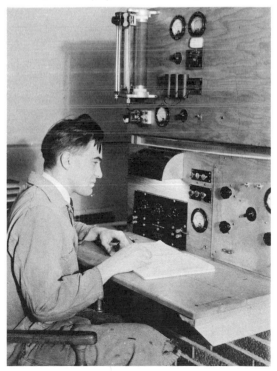

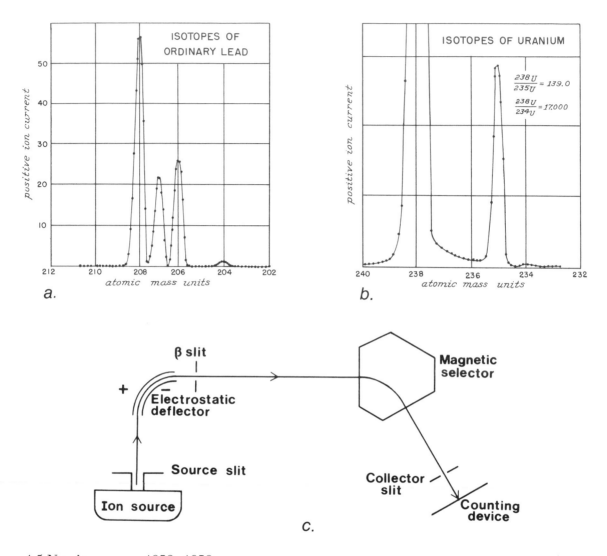

4.5 NIER'S ORIGINAL 1938–1939 RESOLUTIONS OF THE MASS SPECTRA OF LEAD AND URANIUM, AND THE ELEMENTS OF HIS ORIGINAL DOUBLE-FOCUSING MASS SPECTROMETER. [*Courtesy of A. O. C. Nier.*]

The crash program on uranium during World War II stimulated improvements in analytical methods and instrumentation that compressed what would otherwise have required decades of advance into a few years. When it was over, geochronologists knew how to measure other decay systems beside those of uranium. Now they could also measure the decay of potassium-40 (kalium or K-40) to argon-40 and of rubidium-87 to strontium-87: the K–Ar and Rb–Sr methods. New geochronological laboratories sprung up around the world like crocuses in the spring—staffed in the United States mainly by students of Nier and of H. C. Urey, who was then at the University of Chicago.

The next great advance involved the establishment of an age for the planet itself. Whereas Hutton had seen "no vestige of a beginning," the means were now at hand to seek just that. They arrived in 1946 with the simultaneous realization in England by Arthur Holmes and in Germany by F. G. Houtermans that instrumental advances had simplified the problem. One needs only to know the isotopic composition of three natural leads of widely different but accurately known ages to define a line that extrapolates backward to Earth's origin. That is because the differing half-lives of the parent uranium isotopes result in a trend in the proportions of Pb-207 and Pb-206 to nonradiogenic Pb-204 that increases toward the present but converges backward toward a lesser primordial ratio at Earth's time zero.

Different techniques were brought to bear on the problem by Holmes, by Houtermans, and by E. K. Gerling and T. Pavlova in the Soviet Union. All approached or exceeded Rutherford's 1929 age of about 3.4 billion years. As a result, a number upwards of 3 billion had become widely accepted as Earth's likely age, when, in the early 1950s, a group in Harold Urey's laboratory at the University of Chicago entered the friendly competition. Because all theories proposed for Earth's origin imply that all parts of the solar system formed over a very short interval, it seemed clear that the oldest meteorites should share an approximate age of origin with Earth. Thus Urey's students and younger colleagues began to look at the lead-isotope compositions of meteorites.

The effort culminated in 1956 in a classic paper by Clair Patterson. He graphed the ratios of nonradiogenic Pb-204 to radiogenic Pb-206 and Pb-207 for meteorites whose great age had been independently confirmed by the K–Ar and Rb–Sr determinations of others. All lead ratios were found to fall on or close to a line whose slope gave an age of 4.55 ±0.07 billion years. That works out to 4.56 billion years with decay constants now in use and other corrections. Analyses of thoroughly homogenized Earth lead from oceanic sediments also fall on or close to the same line, indicating a homogeneous initial source and identical age. All evidence then and now consistently points to an age of the planet-building components and solar system between 4.56 and 4.6 billion years, allowing for probable rates of accretion and other variables. That can be rounded off to 4.6. After a century of exponential growth (Fig. 4.6), estimates of Earth's age have reached a new and presumably stable plateau.

4.6 EXPANDING PERCEPTIONS OF EARTH'S AGE, 1664–1985.

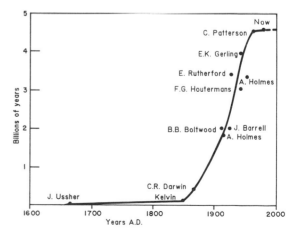

More Clocks in the Rocks

With all the different cross-checking ways we now know for dating rocks and different kinds of events within rocks, how can it be that so many in the modern Christian world still allow it no more than 6,000 to 10,000 years?

No one, on the other hand, should be expected to accept as true on their say-so alone, the great ages that are part of the everyday conversation of geologists.

Let me try to explain how some of the commoner cross-checking methods of geochronology work and the kinds of things we can learn about Earth history from them. The subject is not all that complex, but it involves symbolism and terms that will be unfamiliar to many readers and which require close attention to follow. If you find that boring or uncomfortable, don't hang up on it. Go on to the summary. But, on the way, have a look at the diagrams and try to appreciate the rigor and reliability of the effort that goes into obtaining the radiometric numbers you will see repeatedly throughout this book. They are the framework of modern geology.

What liberating insight enabled Patterson and his colleagues to break the code that until then had concealed the ultimate age of our planet and solar system? They realized that the isotopic composition of lead from metallic meteorites of vanishingly low uranium content should approximate the composition of *primordial lead*—lead having the isotopic composition existing at the time of origin of the oldest meteorites and Earth. Patterson measured the isotopic composition of lead in such samples of the Diablo meteorite of Arizona,

and the ratios he found have since been used successfully as *initial ratios* to correct for samples containing common lead. The very samples that used to be rejected because they contained common lead have now been found to be the most useful.

Three different and basically independent methods are now available for determining the ages of uranium–lead systems: (1) the decay of U-235 to Pb-207, the decay of U-238 to Pb-206, and the ratio of Pb-207 to Pb-206, customarily involving the ratios of both to Pb-204 (Table 4.1). Not only uranium isotopes, but all radioactive isotopes decay to different stable end products at different rates. Thus, agreement between results from different methods cannot be fortuitous. Instead it is elegantly self-checking.

First one needs to know decay constants, indicated in equations by the Greek letter lambda, written λ. Decay constants have now been determined repeatedly by actual measurement and were internationally agreed upon in 1976 at the Twenty-fifth International Geological Congress in Australia. Knowing λ, it is only necessary to know the *number of parent atoms* (P_t), the *number of daughter atoms* (D_t) in the end product, and the logarithm to the base e ($=2.171828$) to compute the age. Although an actual calculation is not made here, readers may as well see that the equation whose solution gives the age t when these quantities are correctly entered is not excessively complicated: $t = 1/\lambda \log e (D_t/P_t + 1)$.

Knowing λ, the half-life is easily computed. Table 4.1 lists these half-lives for different decay systems to the right of the relevant reactions. There one sees the systems now in use expressed in proper chemical notation. The varying lengths of the half-lives suggest the preferred methods of age determination for rocks of different expected ages. The method or methods chosen also vary with the mineralogy of the sample, its pre-vious thermal history, and the information sought. Indeed it is often possible to learn a great deal more about a rock than its time of primary origin by using different methods that date different aspects of its history.

On the same sample, for instance, one might learn the time of initial crystallization by applying the Pb–Pb method to one mineral (e.g., zircon) and find the

TABLE 4.1 *Standard Methods for the Measurement of Radiometric Time*

	Half-Life (in Years from International Decay Constants)
Methods of End Products	
1. $^{235}_{92}U$ ——————— $-7\alpha+1\beta$ ————→ $^{207}_{82}Pb$	0.704×10^9
2. $^{238}_{92}U$ ——————— $-8\alpha+6\beta$ ————→ $^{206}_{82}Pb$	4.68×10^9
3. $^{232}_{90}Th$ ——————— $6\alpha + 4\beta$ ————→ $^{208}_{82}Pb$	14.0×10^9
4. $^{206}_{82}Pb / {}^{207}_{82}Pb$	
5. $^{40}_{19}K$ ——— ┌——— $+1\beta$ ————→ $^{40}_{18}Ar$	11.93×10^9
└——— -1β ————→ $^{40}_{20}Ca$	1.397×10^9
6. $^{87}_{37}Rb$ ——————— -1β ————→ $^{87}_{38}Sr$	48.8×10^9
7. $^{147}_{62}Sm$ ——————— -1α ————→ $^{143}_{60}Nd$	106×10^9
8. $^{176}_{71}Lu$ ——————— -1β ————→ $^{176}_{72}Hf$	35×10^9
9. $^{187}_{75}Re$ ——————— -1β ————→ $^{187}_{76}Os$	46×10^9
Methods of Radioactivity	
10. $^{14}_{6}C$ ——————— -1β ————→ $^{14}_{7}N$	5730 ± 40
11. $^{10}_{4}Be$ ——————— -1β ————→ $^{10}_{5}B$	1.5×10^6
12. $^{36}_{17}Cl$ ——————— -1β ————→ $^{36}_{18}Ar$	0.31×10^6

Notes: Most commonly used of the above decay systems are 1–2, 4–7, and 10. Others have special applications discussed in the text. Kind and number of particles emitted or gained during decay are shown within long arrows indicating direction of reaction.

α = alpha particle (a helium nucleus) of mass 4.

β = beta particle (a nuclear electron) of negligible mass.

    For key to chemical symbols see Figure 2.3.

    Numerical superscripts preceding chemical symbols indicate mass number (total protons + neutrons), subscripts are atomic numbers (total protons).

    Not listed above are $^{26}Al \rightarrow {}^{26}Mg$ and $^{129}I \rightarrow {}^{129}Xe$ decay systems, special instances of the method of end products applicable only to presolar history (as discussed in Ch. 1).

age of the most recent thermal meta-morphism by applying the K–Ar method to another mineral (e.g., biotite). The latter might also be found from a so-called lead-loss line. An intermediate age of high-grade metamorphism may come from Rb–Sr analysis of the whole rock, and still other information from recycled minerals (e.g., zircons). Vari-ables once seen as obstacles have turned out to be rich in historical nuances!

Consider on Table 4.1 the isotopes of lead (Pb), uranium (U-235, U-238), and thorium (Th-232). Besides the lead–lead ratio whereby Patterson and others determined a planetary age of about 4.6 billion years, three systems that decay to Pb are shown in the top three lines of this table. You can see what happens from the chemical notation there. Alpha particles of mass 4 are energetically emitted—six by Th-232, seven by U-235, and eight by U-238—as those iso-topes decay respectively to Pb-208, Pb-207, and Pb-206 through a long series of unstable intermediates, including radium. You will also see that they have three very different half-lives—easily remembered because they correspond roughly to the age of the oldest conspic-uously fossiliferous rocks for U-235, the age of the planet for U-238, and the age of the oldest galaxies for Th-232. Because the half-lives of their parent isotopes are different, the ratios of their lead end products also change with time. Thus four different ages can be calculated from different lead methods, and, if all agree, one may feel confident that the number obtained is close to the true age of the rock.

A problem is that lead is a relatively soluble element. Lead loss, therefore, may skew results in the uranium and thorium decay systems. It does not affect Pb-207/Pb-206 ratios, however, which remain the same for any specified age regardless of lead loss (because all iso-topes of the same element have identical chemical behaviors). Where lead loss is indicated by discordant U–Pb results, therefore, the Pb–Pb procedure called the *concordia method* (or the concordia-discordia method) is used.

Concordia is the curved line of Figure 4.7 along which the simultaneous ratios of Pb-206/U-238 and Pb-207/U-235

4.7 URANIUM–LEAD CONCORDIA–DISCORDIA DIAGRAM FOR SEVEN ZIRCON SAMPLES FROM SODIC GRANITES OF THE MORTON GNEISS (*CIRCLES*), MINNESOTA. Had there been no loss of lead from these rocks, the ages of all samples would fall on the curved line, *concordia*. That they fall on the straight line, *discordia*, shows that only the times of its intersection with concordia are accurately dated. The lower intersection registers a time of lead loss 2.70 billion years ago; the upper implies the true age of the rock sam-pled to be about 3.66 billion years. *[Adapted from S. S. Goldich and J. B. Fisher, 1986, Chemical Geology, Isotope Geoscience Section, v. 58, p. 210, Fig. 7.]*

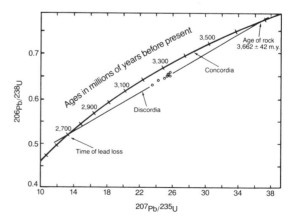

would plot had the samples experienced no loss of Pb. Samples that have lost lead will lie below concordia along a straight line (*discordia*) that intercepts concordia at two places. The upper intercept marks the age of the rock, in this case the 3.66-BY-old Morton Gneiss of southwestern Minnesota. The lower intercept usually gives the age of lead loss, a 2.70-BY-old thermal event in this instance.

Methods involving K–Ar and Rb–Sr are also in wide use. Ratios of radiogenic Ar-40 to radioactive K-40 are believed to give reliable cooling ages on igneous rocks where argon gas is likely to remain trapped in the crystal structure of potash-containing minerals. It does not give reliable primary ages on sedimentary and many metamorphic rocks for a variety of reasons. Ages found are sometimes (perhaps 5 percent of the time) too young because of argon loss or too old because of redeposited detrital minerals or absorption of excess argon at high temperatures by the mineral pyroxene. The K–Ar method has been useful, however, in dating thermal metamorphism. The argon clock is reset at different thermal thresholds, depending on the mineral used. If black mica (biotite) is the mineral studied, the argon clock records the time at which the rock temperature last stood at 275°C. If hornblende is the mineral, the time recorded gives the age at 500°C. With improvements by Berkeley's J. T. Evernden and D. E. Curtis, the method has given good results in dating terrestrial basalts associated both with sites of early man in Africa and with paleomagnetic reversal intervals. It does not work well on marine basalts because of their common entrapment of excess argon on solidification.

The K–Ar method has been widely used, especially in the Soviet Union, to obtain ages of sediments by dating the potassium-rich mineral glauconite, but all glauconite ages are in some doubt. Many are minimum ages because of argon loss. Others are too old because determined on redeposited glauconite from older rocks. Table 4.1 also indicates the decay of K-40 to Ca-40. This system, unfortunately, is not useful because Ca-40 is the common isotope of calcium. A new method that involves the ratios of Ar-40 to Ar-39 (made by irradiation of common potassium, K-39) has been found to elude some of the hazards mentioned above. It is now standard procedure in some age laboratories, employing laser-zapping techniques on parts of mineral grains.

As with K–Ar, results from rubidium to strontium (Rb–Sr) decay also require interpretation. That system utilizes the decay of Rb-87 to Sr-87 by emission of a nuclear electron (a beta particle), generating an increase in atomic number from the 37 of Rb to the 38 of Sr. The half-life of Rb-87 is the longest of any system in regular use—48.8 billion years. The method, therefore, is used almost exclusively on older rocks, where, unlike argon, strontium is tightly retained by enclosing minerals. In the case of Sr, as with Pb, there is an initial quantity or ratio to which radiogenic isotopes are added. If *initial strontium* is negligible, a simple determination of age can be made from Rb–Sr ratios. If initial strontium is present in greater than negligible amounts, it must be compensated for by standardizing Rb-87 and Sr-87 with reference to nonradiogenic Sr-86 and using the Rb–Sr isochron method. Figure 4.8 shows how this works.

In this illustration, the isochron method requires that the rocks and minerals dated have neither lost nor gained either rubidium or strontium as a result of weathering or other processes. If that is not the case they will not define an isochron. Moreover, if the samples do not have a common origin, or if the Sr *initial ratio* is not uniform, an isochron again will not result. The method is resistant to error when used with igneous rocks and meteorites. Its results are regarded with less confidence in the case of sedimentary or metamorphic rocks. The Rb—Sr clock can be reset by high-grade metamorphism, and detrital sedimentary rocks from a common source may yield isochrons that reflect the age of the source rather than the time of sedimentation. In addition, loss of Rb results in ages that are too old and loss of radiogenic Sr-87 gives ages that are too young. It is a sharp tool in experienced hands, but one to be employed judiciously and checked against the results of other methods where practicable.

4.8 A RUBIDIUM—STRONTIUM ISOCHRON DIAGRAM, SHOWING TIME-DEPENDENT EVOLUTION OF ISOTOPE RATIOS IN A CLOSED SYSTEM. Departing from a specific *initial ratio* of radiogenic Sr-87 to nonradiogenic Sr-86 at points *A*, *B*, and *C*, the decay of radioactive Rb-87 to Sr-87 results in an increase of Sr-87 and a decrease of Rb-87 relative to Sr-86. That causes the ratios at time zero for minerals *A*, *B*, and *C* to shift upward and to the left with time to points *A'*, *B'*, and *C'*, defining an *isochron* and with it the age of the parent rock at a given time greater than zero. The intersection of that isochron with the vertical axis gives the initial ratio, rich in its own historical implications. *[Adapted from Gunter Faure,* Principles of Isotope Geology, *Fig. 6.1, p. 80, copyright © 1977 John Wiley & Sons, Inc., with permission.]*

Note, finally, that a relatively new and marvelously simple procedure known as *fission-track dating* is giving *approximate* ages for the heretofore undateable interval between a few thousand and a million or so years (as well as older and younger ages). This method depends on the fact that, among those isotopes that emit alpha particles, the occasional atom undergoes a rare but statistically constant form of fission. The atomic physicist's *strong force* that holds the nucleus together is overcome so that the mutually repellent halves fly apart, stripping electrons from other atoms to produce destructive tracks a few thousandths of a millimeter long (Fig. 4.9). In uranium-238, for instance, approximately every 2 million standard emissions of alpha particles is accompanied by one such catastrophic fission event. When nonmetallic minerals, glasses, and pottery are etched with appropriate reagents these fission tracks became visible and can be counted under a microscope. The ratio of the density of fission tracks in a given space to the measured U-238 remaining, mul-

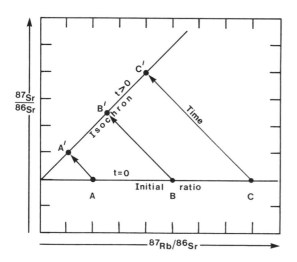

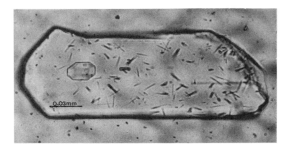

4.9 FISSION TRACKS IN A 0.14-MM-LONG CRYS-
TAL OF THE PHOSPHATIC MINERAL APATITE
FROM A SILURIAN VOLCANIC ASH NEAR LUD-
LOW, ENGLAND. The density of fission tracks
is a function of age, in this case about 420
million years. [Courtesy of C. W. Naeser;
U.S. Geological Survey.]

tiplied by the average rate of catastrophic fission for this isotope, gives the age. The method more than compensates for its lack of precision by two unique capabilities: (1) its wide range of applicability to rocks and artifacts of all ages, and (2) its versatility in recording relatively modest episodes of thermal metamorphism that obliterate the original fission tracks and reset local fission-track clocks to record a cooling age.

New Insights for Special Cases

Other radiometric systems listed in Table 4.1 are only now coming on stream, but they bring high promise of new insights to Earth and cosmic history in the hands of isotope geochemists. The rhenium–osmium system is a way of ascertaining probable ages of events in early Earth and cosmic evolution. Samarium–neodymium ratios, requiring the accurate determination of extremely small differences, are being increasingly used to date the ages of molten rock (magmas) in the process of becoming part of Earth's crust, antecedent to their actual crystallization. Sm–Nd ages are believed to be very resistant to metamorphism, but discrepancies are turning up that need to be worked out. Another method, the lutetium–hafnium method, is a new one for dating very old rocks and meteorites. And both beryllium-10 and chlorine-36, having relatively short half-lives, are being used on the one

hand to date polar ice and the deposits of saline lakes, and on the other to estimate how long meteorites have been in flight as separate bodies exposed to cosmic radiation.

Applications of Be-10 to the measurement of Earth processes now loom in the fact that some of it gets carried beneath the continents at places where they override the ocean floors. It survives in measurable amounts when host sediments are melted and returned to the overriding continental surfaces as constituents of volcanic rocks. It is also being utilized to estimate rates of erosion and their changes over historic times. Produced, like C-14, by cosmic radiation, ages given by the beryllium and chlorine methods are determined from radiation counts rather than by ratios of daughter to parent isotope. These systems, and the extension of the C-14 method to older rocks, depend on the use of particle accelerators for measuring very small levels of radiation.

Among the interesting things that are coming out of recent advances in radiometric geochronology is the possibility of extending it to check the ages of cosmic events. These ages are now based primarily on the shift to longer wavelengths of light and other electromagnetic waves as they recede into space, and there is disagreement about the factor (the Hubble constant) to be used in converting this *red shift* to time. Is the age of the universe 10 BY, or 20 BY, or some intermediate age such as the 13 BY implied by some estimates of the ages of galaxies? Limits to the age of our galaxy have now been based on Os-187 / Os-186 values, depending on whether or not the elements and their initial isotope ratios are considered to have originated simultaneously. The range of ages implied is between about 10.6 BY and 16.8 BY. The age for the universe where all methods overlap is between 15.5 and 13.5 BY.

There seems to be no end to the wonders of isotope geochemistry, including the use of stable isotopes as tracers of various dynamic processes in crustal evolution—a recurrent theme in this book.

In Summary

Time is a paradox. Like space, it is immaterial. It can be detected only by the passage of events. Its very essence is transience. There is scarcely a *now* in time, only a past and a future. Its equations work equally well in either direction. Yet we think of time as *flowing* toward the future, and it seems to flow at different rates for different people and situations. Its only manifestation is duration. Yet it is a property that we measure and that we measure with.

Perceptions of time vary with geography, social and political background, religion, the times, and age. All people throughout human history have endowed it with some meaning and kept some account of it. Beginning with the Julian calendar in 45 B.C. this account has been systematic. Calibration systems have changed, however, as the need for accuracy and precision grew. Now the basic unit of time, the second, is defined by its relation to the speed of light. The new relativistic or atomic second is the time required for light to travel 299,792,458 meters in a vacuum, and there are 31,556,925.9747 seconds in an atomic year. The seconds are recorded by the vibrations of crystals in atomic clocks—especially the cesium clock.

Geologists cannot hope to measure the aeons that have already elapsed with that kind of precision. Only for the last few thousands of years can we specify time to the nearest year by counting tree rings or annual layers in sediments. Before then, back to 40,000 years or so, the age in years is given by carbon-14, but only approximately, and within substantial limits of error. For older rocks the age is rounded off and notions of synchroneity are very broad. How could there have been precise global synchroneity? An ashfall, which may have taken several days to reach its limits, is considered geologically instantaneous. Even a paleomagnetic polar reversal, taking perhaps 5,000 years, is effectively instantaneous in a geological sense.

But geology has its own atomic clocks in the rocks, as well as in ice, water, and meteorites. They involve unstable atoms of ordinary elements that differ only in atomic weight from chemically identical iosotopes of the same element. These radioactive isotopes eject alpha particles (helium nuclei) and eject or acquire beta particles (electrons), thereby undergoing transmutation to stable isotopes of a different element.

In principle, the alchemists have been vindicated. But transmutation cannot be carried out at will. It goes on at invariable rates, except where U-235 becomes so concentrated that proton flux initiates a chain reaction. As the rates of decay are negatively exponential, total conversation is a slow process. Five to ten half-lives are needed to reduce the parent isotope to analytically negligible amounts. Each type of radioactive isotope also has its independent decay rate and half-life, so that similar ages found by different methods comprise a self-checking system. Different methods could not give the same age by chance.

Elegant procedures have been developed for reading these geologic clocks, correcting for potential errors, and utilizing seemingly anomalous results as sources of additional information. Loss of end products and parent isotopes, as well as the initial abundance of isotopes of the terminal element, need to be allowed for.

In the uranium–lead system the *concordia method* both corrects for lead loss and specifies the time at which it happened. The *isochron method* corrects for variations in initial strontium ratios in the rubidium–strontium system and is applicable to others. Using different methods, it is possible to get as many as four sets of different but informative ages on the same rock. Samarium–neodymium may date the melt that crystallized to become crust at that location. Uranium–lead or lead–lead usually date the time of crystallization of the parent magma. Rubidium–strontium might date subsequent high-grade metamorphism. And potassium–argon would confirm the age of the most recent thermal event responsible for lead loss shown by concordia. Other decay systems, properly analyzed, give a range of useful information on early man, the history of ice sheets, early events in solar and presolar history, and the origin of the elements and the galaxy. Fission-track approximations supplement all.

How was it ever possible to calibrate Earth history without these marvelous geologic clocks and geochemical tracers of the last 4 decades? Only crudely and fractionally, we now see.

CHAPTER FIVE

SYMPHONY

OF THE EARTH

Earth history, we shall see, is like an unfinished symphony. Neither has it come to an end, nor is there a score for its further evolution. The harmony simply unfolds in a grand syncopation of successive themes and movements, the evidence for which is teased from the mute rocks by the application of earlier-noted principles, techniques, and theory, livened by the occasional inspired insight. This chapter focuses on composition and organization: on geological notation, some instruments peculiar to the geologic orchestra, the distinction between time and history, and the means for putting events in order. It concludes with a comprehensive but condensed, revised standard geological column that reflects current progress and provides a framework for continuing discussion.

The Harmonic Elements

Adrift in a dimensionless sea of time, locked in its transient realm of solar space by the grip of gravity and the laws of physics, Earth's destination is unclear, its history an unfinished symphony. Yet our beautiful blue planet holds music for those who listen. With isotopes and biologic evolution to keep the beat and the rules of nature to maintain the harmony, one can begin to separate the main themes from the static in which all history is immersed.

Earth history is like the work of a gifted but absentminded composer-conductor, occasionally losing continuity or drifting off beat but all the while seeking harmony from the instruments of planetary evolution (Fig. 5.1). Those instruments are the winds that dry the deserts, evaporate the seas, and carry the waters to clement regions. They are the waters in all their states, chemical variety, and erosive moods. They are the forces of gravity, Earth's rotation and orbital characteristics, plate tectonism, and mountain building. They are the percussion of meteorites, the rumbling of volcanoes, and the manifold creatures of sea, land, and air.

The planet has its peaks of activity and intervals of calm in its passage from infancy to maturity. If Earth were now a

5.1 SYMPHONY OF THE EARTH. *[Drawn and elaborated by John Holden from author's rough design.]*

hundred years old, one might say that it grew rapidly during the first couple of years, achieved its full stature with a late burst of growth at the age of about 8, and thereafter matured slowly for the next 77 years before becoming a certifiable adult upon the arrival of animal life only 15 years ago. It will continue to evolve until the Sun dies.

Earth is a great heat engine. Its constant expenditure of nuclear energy maintains its internal temperature, but it would overheat if it did not also ventilate. Measurements of heat flow and the manifestations of global tectonics give evidence that Earth cools mainly by convection and has presumably done so from the beginning. Heat nowadays rises from within the *mantle* that wraps Earth's inner *core*, carried by and dissipated in the volcanism to which it gives rise and which, in turn, initiates the lateral motions of broad, thick slabs of crust (plates) away from volcanic oceanic spreading ridges (Fig. 5.2).

Heat also rises above *subduction zones* where these slabs return to the mantle. As the plates diverge or spread apart from one another, new volcanic sea floor fills the gap, solidifying in the wakes of the spreading plates. As sea floor cools and becomes denser, it glides ever so slowly downslope away from the spreading ridges, slowly rafting the continents with it to new positions. In this way, for example, the Americas have separated from Europe and Africa over the past 200 million years. It is as if Earth exhaled volcanic fluids of lava, steam, and gases along these divergent plate margins (e.g., the Mid-Atlantic ridge of Fig. 5.2), while, conversely, inhaling oceanic rocks, sediments, and water along the boundaries where plates converge or collide. There, relatively heavy, old, cold sea floor and underlying oceanic outer mantle are *subducted* (dragged down by gravity contrasts) beneath relatively light continents or thinner sea floor. At appropriate depths and temperatures this subducted sea floor

5.2 THE CONVECTING, COOLING EARTH OF PLATE TECTONICS.

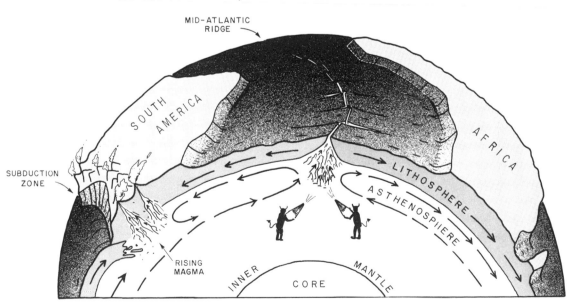

melts, or partially melts, and fuses with underlying hot, weak mantle (the *asthenosphere* of Fig. 5.2).

Some of these subducted materials become hot and light enough during their descent and mixing to rise again, as from the dead. Mixing still further with zones of partial melting within the continental lithosphere, they may become magmas that melt and return gravitationally upward. Such magmas, then, may reappear at or near the surface as intrusions of intermediate to sialic composition, hoisting with them chains of volcanoes and lofty sialic domes to grace the skies above places like the Andes and California's Sierra Nevada. In doing so they bear gifts of precious metals and porphyry copper. They add new volcanic and at places metal-rich belts to the belching Ring of Fire that follows the Pacific rim.

It is an important aspect of the production of new crust by sea-floor spreading that such spreading is approximately compensated for by the consumption of old crust as it descends along subduction zones. Grounds too extensive for elaboration here forbid an expanding Earth. Thus the cycle of sea-floor spreading, continental rafting, subduction, collision, mountain building, volcanism, and mineralization goes on. All are aspects of the geologically unifying theory of *plate tectonics.* Its actions constantly reshuffle Earth's crustal matter. They continue without notable change in its surface area or diameter. And although the oldest fully convincing evidence we have that this process was already in operation is not found until a bit more than halfway through Earth history, it is a typically recurrent symphonic theme thereafter. It expresses itself in the open-

ing and closing of oceans, the joining and breaking up of continents, and in a great variety of lesser rhythms that blend their own cadences into the broader harmonics. That theme recurs throughout this book. It will be elaborated in Chapter 9.

Suffice it here to observe that neither plate-tectonic nor other geological and geochemical rhythms or cycles are known to be metric in any exact sense. Earth evolves. But it sees an intermittent or episodic recurrence of similar progressions, and the term *cyclic* is almost universally used by English-speaking geologists in that sense. The rock "cycle" of Figure 3.11 is a good example. It is a repeated sequence of weathering and erosion, transport, sedimentation, conversion to sedimentary rock, metamorphism, more heat, fusion, and cooling. Then again crystallization into igneous rock, extrusion, more weathering and erosion, more sediment, and so on—over and over in its own peculiar beat.

Earth, reminds one, in poetic moments, of a living system. But it is not really alive. It does not replicate, mutate, or reproduce mutations. It does, however, react to maintain or regularly renew a harmonic balance among the ever-changing processes that affect our planet—a characteristic known as Le Chatelier's Rule, or, in the case of living systems, as homeostasis. The goal of the following chapters will be to discern the harmonic structure of that evolution, and its homeostatic mechanisms. They will attend to the comings and goings of the seas, the heaving up and wearing down of mountains, the assembly and dispersal of continents, the growth and subduction of ocean floor, and the succession of life and its interactions with the physical world. It will be seen that the broader cadences of Earth history are episodic, with soaring crescendos that rise above background and as often subside.

Getting Acquainted with Rocks

Rocks are the archives from which accounts of Earth history are transcribed. The record, however, is distressingly incomplete, like bits of brittle newspaper from an old trunk or fragmentary documents from the once-fertile crescent of Mesopotamia. To find, transcribe, and order its stories in stone, geologists go to rock country. The most complete and best preserved sequences of rocks are found in mountains, deserts, coastal zones, valley walls, excavations, and bore holes. There one studies how the rocks change upward and downward in sequences worked out from principles outlined in Chapter 3, and how individual rocks may change their properties when traced laterally along a ledge or *outcrop*. Transitions or boundaries between different kinds of rocks are observed, their orientation in space determined, and samples collected for analysis in the laboratory by a host of special techniques. Geologists with different skills and interests pool their talents to expand and refine the common understanding.

The oldest rocks we know are visitors from interplanetary space—meteorites. The weathering and erosion long ago of similar rocks on Earth and their fusion products resulted in sediments which, on consolidation, became the first *sedimentary rocks*. Alteration of either igneous or sedimentary rocks by heat and pressure as a result of deep burial or compressional tectonics is the source of *meta-morphic rocks*, found in the cores of mountain ranges and ancient continental shields.

Igneous rocks crystallize from melts to begin Earth's indigenous rock cycle. Hundreds of different kinds are known and named, but here let us reduce their numbers to a few broadly significant kinds. If such rocks crystallize slowly from melts deep within the Earth (magma) they form coarsely crystalline *plutonic* rocks like familiar granite or less well known gabbro. If they congeal quickly from melts at or near Earth's surface (lava) they form finely crystalline volcanic rocks like basalt or glassy volcanic equivalents of granite called rhyolite. The approximate mineral compositions of common igneous rocks, broadly defined, are indicated by the level at which they appear in Figure 5.3. An imaginary horizontal profile through the rectangle to the right of each name gives the approximate percentages of its principal components. Within that rectangle the dark pattern indicates dark, heavier minerals, whereas the clear areas signify pale-colored, lighter-weight minerals. Notice that for every coarse-grained plutonic rock there is a fine-grained or glassy volcanic counterpart.

Broadly considered and simplified, *granitic* rocks are continental and *basaltic* rocks oceanic. Alternatively, and even more broadly, one may speak of the continental types as *sialic*, combining the first syllables of their preponderant metal elements—silica and aluminum. Oceanic types are said to be *mafic* (or simatic), from the prominence in them of minerals rich in magnesium and iron. *Mafic* also denotes a lower percentage of silica than does *sialic*. Mafic rocks are not limited to oceans but also occur widely on continents. Ultramafic and intermediate rocks are self-explanatory variants.

Thus began Earth's indigenous and unending rock cycle (Fig. 3.11), often skipping steps. The rounded grains of quartz in the Coconino Sandstone, just below the rim of the Grand Canyon in Figure 3.2, are the product of a long pre-Coconino history. They started as angular grains in a plutonic granite, conforming to adjacent minerals as all congealed from the parent melt. After weathering loose and being transported and abraded, they were episodically reincorporated, with further rounding, in a succession of older sedimentary rocks before at last coming to rest where now seen.

Formations like the Coconino Sandstone provide the basic notation from which Earth's harmonic evolution is reconstructed. Their description and interpretation is an important element of Earth history, but it can become monotonous. Indeed, *stratigraphy*, broadly seen as the study, description, and sequencing of all kinds of rocks, is bedevilled by an excess of nomenclature and legalistics which, all too often, obscure the main themes. Here we shall minimize that.

The central factor in introducing order to Earth history is the ability to *correlate*, to establish continuity or equivalence between rocks of similar or different characteristics from one place to another.

Three sets of data are involved. First are the rocks themselves. Second are the historical sequence, conditions, prevailing trends, and events inferred from this record under the constraints of natural law. And third is the approximate numerical calibration in real time of rocks and events utilizing the geochronologic methods outlined in Table 4.1, or by equivalency to other rocks that have been dated by such methods—including bracketing between rocks or events of known age.

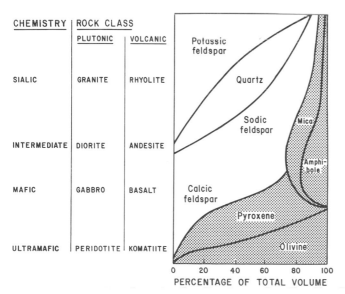

5.3 MAIN KINDS OF IGNEOUS ROCKS AND THEIR COMPOSITIONS. An imaginary horizontal profile to the right of names listed indicates typical mineral compositions. Stippling on far right denotes mafic minerals. [*Eicher/McAlester/Rottman*, The History of the Earth's Crust, © *1984, p. 42. Adapted by permission of Prentice-Hall, Inc., Englewood Cliffs, New Jersey.*]

103

Further Thoughts
on the Standard Geological Column

How did the idea of a long, decipherable succession of ancient strata ever arise at a time and in a part of the world where fossils were still widely considered to be sports of nature or works of the devil? How could this happen where the majority subscribed to the originally Chaldean view that the very same strata were all products of a brief universal flood?

As we saw in Chapter 3, geologists of the nineteenth century nevertheless did delineate major sequences of rock called *systems*, within which are lesser divisions called *series*, *groups*, and *formations*, in descending order of importance. Corresponding intervals of history were also recognized and assembled into *eras*. The stratigraphic systems and corresponding *periods* of the Paleozoic and Mesozoic eras were defined by their distinctive lithology in western Europe. Coal-bearing rocks became Carboniferous, chalks became Cretaceous, characteristic sequence gave the tripartite Triassic, and regional occurrence combined with paleontology and superposition account for the rest. Terms for systems and periods of the Cenozoic Era (the most recent) are still unsettled, but ever more commonly used are Paleogene and Neogene (Table 3.1). All of the *Phanerozoic* divisions (containing distinctively animal fossils) have long been recognized primarily by their characteristic biotas. Long suspected to be of very unequal length, they are now known to be so. The Quaternary (including the last 10,000 years, or Recent), sometimes considered a period or even a subera, includes only the latest 1.6 million years. The Cretaceous Period was 50 times as long.

Growth of the standard geological column was neither orderly nor systematic, and the rigorous separation of rock- and equivalent historical units, although grammatically correct, can make for stilted communication. Earth's history in any event is firmly linked to the rocks—the sole, if incomplete, source of hard data about it. And continuing research has supported the view that, allowing for some tinkering with boundaries, and albeit the work of human minds, most of the periods approximate recognizable natural intervals of very broad, if not literally global scope.

Even some time-limited rock units are remarkably distinctive, widespread, and persistent although none we know of is truly worldwide. Chalk, for instance, a pelagic marine sediment, is common among Upper Cretaceous rocks, from Texas through the U.S. Gulf Coast to Britain, the Middle East, and Western Australia. Coals, the carbonized remains of fresh or brackish water swamps, are found in different parts of the geologic column mainly younger than Devonian, but especially in the Upper Carboniferous and latest Cretaceous. The brownish-red, mainly nonmarine sandstones used to build the fashionable look-alike brownstone mansions of Victorian Europe and temperate North America are found among Devonian, Triassic, and late Proterozoic strata over large areas of the Northern Hemisphere. Flysch-facies graywackes—like the grim, gray *pietra dura* which sets the somber tone of Florentine public architecture—formed repeatedly from earliest times to the Mesozoic in similar deep-marine, volcano-tectonic settings.

Geologists of the early twentieth century, searching for synchronous global boundaries between natural systems, hypothesized explanations for the expected synchroneity. One widely favored hypothesis called on crustal deformation (*diastrophism*) as the "ulterior basis of time division." In this view, long episodes of deposition are separated by shorter intervals of diastrophism and erosion or nondeposition, resulting in extensive unconformities and changes in fossil biotas. Global mapping and geochronology, however, have shown that, although there are distinctive dia-

strophic episodes, diastrophism is going on somewhere most of the time as plates of Earth's crust (Fig. 5.2) collide with and grind past one another.

Nor did time stand still or sedimentation cease because of local nondeposition and unconformity. Yet, considering only the history of fossiliferous rocks, views that hypothesized brief transitions from one epoch to another are holding up well. Current work on the effects and potential effects of large asteroidal collisions—the heavenly nukes of prehistoric time—bear witness to the validity of catastrophic processes in nature, if never on the scale of biblical literalists. We need to know more about those effects and their frequency.

Extending History and Measuring Time

The founders of geology made a giant leap forward in creating a kind of sequence geochronology, but they were unable to calibrate it. In addition, their geologic column included only visibly fossiliferous rocks and those bracketed by them—rocks that were deposited during the *Phanerozoic*, the Eon of manifest animal fossils. The Phanerozoic includes only the last 15 percent of Earth history. Beneath its rocks lay the apparently unfossiliferous and "interminable graywackes" of the illustrious Sir Roderick Murchison, first director general of the Geological Survey of Great Britain and author or coauthor of three of the eleven original geological systems (Table 3.1). Anyone who has plodded through the thick, drab, monotonous sequences of impure, fragmental, volcanogenic sand-

stones usual to the graywacke suite can appreciate his designation.

Without fossils and no sign of a bottom, the best that could be done with these ancient graywackes in Murchison's time was to designate them pre-Cambrian or, more appropriately, sub-Cambrian, considering that the reference is usually to rocks rather than history or time. Few could have supposed in Lyell's or Darwin's day that this seemingly endless pile of rocks was but the local upper part of a succession whose cumulative thickness and total age was many times that of the then-known standard geological column.

The ancient metamorphic, granitic, and commonly mineral-rich rocks of the Scandinavian and Canadian shields, however, could not long remain

105

unprobed. The demands for mineral products set going with the industrial revolution grew ever more intense. The beginning of official geological investigations in the shield areas, particularly by the Geological Survey of Canada in 1842, initiated vigorous efforts to map, understand, and determine the sequence in space and time of these often highly deformed and metamorphosed rocks.

It was long supposed that degree of metamorphism might hold a clue to geologic sequence. The older the rock, the more opportunities it would have had to undergo deformation and metamorphism. Simple! But also misleading. Deformation and metamorphism are much more characteristic of places than times. They feature the margins of crustal plates, and their presence in continental interiors is evidence for former plate collisions, deep burial, or both. Intensively deformed rocks in the cores of young folded mountain ranges and little metamorphosed rocks of great age within stable continental platforms bear that out. Evidence for episodes of mountain building and accompanying rock deformation, however, are locally used even now, in the Canadian Shield and elsewhere, to separate younger rocks not affected by a given crustal deformation from older ones deformed by it.

A big problem with rocks older than Phanerozoic is that fossils, long thought to be absent from them, are even now rare, mainly microscopic, and of still limited, if growing, stratigraphic value. Radiometric ages are essential to the establishment of an historical framework within which these occasional microfossils can illuminate early biologic evolution and ecology.

It is now possible (Table 4.1) to determine radiometric ages within useful limits for appropriate rocks of all ages, often for very small samples or even single grains of the right mineral. There is a limitation even in radiometric chronology, however, where applied to very old rocks. A 1 percent error in the age of a 2-billion-year-old rock is 20 million. Three percent of 2 billion is the length of the Cenozoic Era. To have numbers even within such wide bounds of experimental error is invaluable, but the varying resolving power of different methods must also be kept in mind.

Thanks, nevertheless, to hard-won insights, improved resolution, and the labors of scores of geological timekeepers (*geochronologists*) the world over, we can now calibrate (within limits) the entire geologic column and all of Earth's recorded history. The uranium–lead method, employing durable, alteration-resistant zircons, has been refined to the point where it can resolve *relative* ages *within a sequence* of related rocks (not absolute ages) with a precision of 1 million years—about that of a good younger paleontological zonation.

The Calibration of Earth History

Interesting ambiguities have arisen from the very success of our predecessors in basing an *apparent* time scale on the sequence of historical phenomena during the Phanerozoic Eon (Table 3.1). They involve the meanings of time and history and the appropriate criteria for a standard geologic column, especially as concerns the historical subdivision of the first 85 percent of Earth's history. These ambiguities have not yet been resolved, even in the present golden age of conceptual and instrumental progress.

It is customary to refer to a geologic time scale and the geologic column as if they were the same, or nearly so. But they are not. A scale is graduated into equal divisions, something for calibration. The standard geologic column is a necessarily incomplete, globally composite sequence of rocks with historical implications. It defines (within limits) but does not calibrate. Both definition and calibration are needed, and they should not be equated.

What is meant by geologic time and how can it best serve Earth history? Although what has generally been called geologic time grew out of superposition and biotal succession, there is only one kind of clock that keeps something close to real time for materials much older than the agricultural revolution. That is the nuclear or radiometric clock—the measurement of nuclear decay in naturally radioactive substances as discussed in Chapter 4. The precision of such systems falls far short of the cesium atomic clock, yet their experimental repeatability is great. Like a good cargo net, they make up for their deficiencies of fine structure by their capacity to do the work called for, and their power of resolution continues to improve.

One would like, all the same, to find a method whose combination of precision and accuracy would facilitate testing hypotheses of synchroneity and diachroneity (time transgression) at an operationally more meaningful level of resolution. How common and how great is time transgression among our histori-

cal transitions? Very few rock layers are geologically "simultaneous" in the sense that a volcanic ash is. Aquatically transported sediments vary in age as the processes that favored their deposition encroached landward or retreated seaward across the depositional surface as sea level rose or lands subsided. Such strata (or, loosely, "sediments") are said to be *time transgressive* or *diachronous*. In the abbreviated language of stratigraphy, they *onlap* or *offlap*. Where the exposed edges of detrital strata can be traced for some distance toward or away from the ancient shoreline, one who walks such a route far enough can often see trends of coarsening toward the ancient shore and finer sediments off shore. That is sedimentary and time transgression. It characteristically follows subsidence or inundation of a land area long exposed to weathering and accumulation of sediments, the sediments then being redistributed by inundating seas. Transgressive sands denote an advancing sea or inundating land; regressive sands follow the sea offshore.

On a local scale, organisms also move about a lot with habitat change, but few so readily across oceans or latitudes. Darwin's defender T. H. Huxley was perhaps the first to call attention, in 1862, to the probability that the essentially simultaneous appearance worldwide of distinctive fossils and fossil communities would be a rarity, limited to species with unusual qualifications for dispersal. The same stratigraphic succession of organisms or communities at widely separated places means only that they arrived at (or evolved at) those places in the same order. Although appropriately taken as evidence for approximate synchroneity in the absence of evidence to the contrary, it *establishes* no more than historical equivalency or recurrence.

Hairsplitting? In some instances, yes. Indeed the extinction that separated the two last stages of Devonian history about 367 million years ago appears to have been essentially instantaneous in geological terms. The important part is to recognize that few major historical transitions are even approximately instantaneous and that "boundaries" should be defined by historical (preferably biotal) criteria, and *not* by some specific number of radiometric years. The supposed ages of events change with time, method, and materials. When they do, the numbers should not—and, in the Phanerozoic, do not—carry boundaries with them.

In historical geology, simultaneity and time transgression alike are hypotheses to be tested, and to be tested again for each supposed instance. When, and only when, one knows at what time an event or process began and how long it lasted can one have anything useful to say about rates.

It is clear that even the most breathtaking changes do not happen simultaneously wherever seen, and that they may never reach all parts of the world. The miniskirt, which burst on man's delighted vision in the 1960s, did so first in the more "liberated" metropolitan regions and only later dazzled the residents of suburbia. Alas, it took the best part of a decade for it to travel from the Côte d'Azur to Novosibirsk. It took the genus *Homo* well over a million years to attain a global distribution. How long may it have taken the first amphibians or reptiles? Perhaps more important, how long did it take those who did to become extinct?

Consider what we commonly call "boundaries" between variations in the less mercurial aspects of nature today. Looking seaward across the surfaces of certain equatorial Pacific reefs from a beach ridge at low tide one can see an apparently sharply defined color change where the landward zone of green algae gives way to a seaward zone of red algae. On close inspection, however, it proves to be not a sharp boundary but a transition, a belt some meters wide of mixed green and red algae between respective zones of dominance. Yet, where plants engage in chemical warfare (allelopathy), as they sometimes do especially in arid regions, equivalent transitions may be abrupt. We see that nature includes changes both transitional and abrupt, both evolutionary and revolutionary. Individual instances are to be analyzed and measured, not to be decided according to philosophical preference or authoritarian decree.

As it is with life and space, so it is with rocks and history. Changes in sequence may be transitional or abrupt. Discounting the briefer interruptions that account for so much missing history at the bedding surfaces between adjacent layers of rocks, we find sharp boundaries between sequences of rocks mainly where something dramatic has occurred. As we shall see, some of the important historical transitions found in older parts of the geological column are outrageously time transgressive. To assume synchroneity (or, for that matter, time transgression) is to forego opportunity to learn.

Although one may reasonably argue that most of the transitions observed in historical geology could be synchronous within the scale of resolution available, it would be mistaken to claim that an unproved assumption of synchroneity risks no serious conceptual consequences. The sailor and the surveyor can practice their trades whether the solar system be heliocentric or not. Astrophysicists have problems with a geocentric universe. "Common sense" gives us the flat Earth and the geocentric solar system. It is a poor guide to scientific reality. Time, the calibrator, keeps score but tells no story of its own. These pages will discriminate between time and history.

The Ambivalent Poles: Magnetostratigraphy as Scorekeeper

Consider now an instrument that promises to contribute simultaneously both to the making and the calibration of Earth history.

Like a fickle lover, Earth's magnetic field from time to time reverses orientation. That provides a powerful technique for correlating rocks and calibrating history. When that happens, the previously north-seeking needle points (and dips) south and the south-seeking needle points north. It has been calculated from the changing paleopolarity of Cenozoic basalts in Oregon that the time required to complete such a polarity shift is around 4,500 years, a transitional interval that is brief by geologic standards. No change in the orientation of the rotational poles or axis occurs—only their direction of polarity. In addition, plate tectonism rafts the continents about, so that, from any given point, the location of the poles *appears* to shift. To describe this shifting of continents with reference to the poles in a consistent, if imprecise, way, we refer to *apparent* polar wandering and to an *apparent polar wander path* or *APWP* (e.g., Fig. 5.4).

These APWPs track the apparent positions of the magnetic poles, which, in fact, do not depart far from the rotational poles. More to the point, they record the approximate track of continental migration as Earth's lands are rafted about by plate-tectonic motions, like cargo on a slow barge. The rotational poles stay in the same place and the magnetic poles wander only slightly compared to the drift of continents. Thus, in reality, it's plate wander, not polar wander. Like the geocentric Earth, polar wander only seems that way.

Polar reversals are something else. A good record of them has been logged in basalt and other magnetically susceptible rocks during the past 163 million years of sea-floor spreading. It provides a dependable indication of sequence and an indirect time scale, with a geologically modest range of error. In rocks older than 163 million years, the method is less widely applicable but still useful. Here, position in an APWP, coupled with paleopole orientation, holds promise, with continuing refinement, for relatively precise correlation—perhaps as much as a third of the way back toward the beginning of geologic time.

Information about polar orientation and location at any given moment in history is recorded in the rocks as if it were music on a magnetic tape or the price of groceries in your neighborhood market. Experimentation shows that this happens in lavas when tiny particles of metallic iron oxide cool below the *Curie temperature* at which they become magnetized as a product of the establishment of ferromagnetic order. This is 580°C for magnetite (lodestone or Fe_3O_4) and 675°C for magnetic hematite (another common iron mineral known as alpha Fe_2O_3). As basaltic lavas usually crystallize at about 1,000°C, their magnetic particles cannot move within the rock after magnetization has been fixed in them, immediately below the Curie temperature. The lava therefore acts as a fossil compass. To read this compass one must collect small, carefully oriented samples with a special drill designed for that purpose. Directions of the imposed magnetic field at the time of crystallization can then be measured in the laboratory, while their inclination from the horizontal records

how far the sampling site was from the contemporary pole when the rocks formed. Declination establishes the direction of the pole itself from the sampling site. A paleopole position can then be calculated using the methods of spherical geometry.

The particles that record the magnetic fields of sedimentary rocks (hematite in red sedimentary rocks, but magnetite in some others), were magnetized before they began their sedimentary history. Being able to swing like tiny, pole-seeking compass needles, they orient themselves in the existing magnetic field as they settle, or after settling but within still-fluid sediment. The oldest paleopole position yet recorded is that found for the 2.7-billion-year-old Abitibi green-

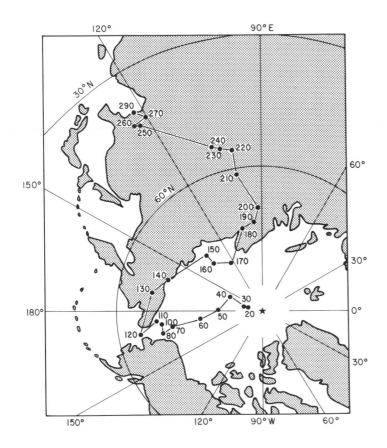

5.4 APPARENT POLAR WANDER PATH OF DRIFTING NORTH AMERICA FROM 290 TO 20 MILLION YEARS AGO. Error circles omitted. Eastward shift from 200 to 120 million years ago marks the beginning of the most recent opening of the Atlantic. Northwestward shift of paleopoles from 120 to 20 million years ago shows westward shift of North America during the opening of the Labrador and Greenland seas [*Modified from E. Irving, 1979,* Canadian Journal of Earth Science, *v. 16, no. 3, Fig. 1, p. 671.*]

stones of Ontario, indicating an apparent (probably north) paleopole at 20°S, 4°E. The most significant thing about this is that it implies that continents were then already in motion and that plate tectonics or something antecedent to it was already in progress.

Because polar reversal is irregularly episodic and because APWPs follow a separate track in time for each moving plate, rocks that yield identical paleopole positions and matching reversal patterns within an individual plate can be correlated historically with one another and placed in sequence. Given, then, a master set of dated paleopole positions and reversals, approximate ages of relatively high precision can also be assigned.

The distinction between precision and accuracy is important here. Approximate ages in years are available only from radiometric methods or, indirectly, from paleontological sequence. Rocks within an established pattern of apparent paleopole positions and durations, and having a specific location in an APWP, can be dated radiometrically, however. And once a sequence of ages is worked out, the location of other rocks within the same-dated APWP denotes their ages. Geologically speaking the precision is great, but the accuracy (from isotopic control) only fair. *Magnetostratigraphy* is like the minute hand on a clock. Only if you know the right hour can you meet your engagement at the proper minute. Its strength is in correlation, not age determination. What is correlated are paleomagnetic signatures, not the calibrating radiometric numbers.

Most important for precise correlation (and only indirectly for dating) is the magnetic orientation of a rock within a sequence of closely spaced polar reversal patterns, as for the last 25 million years BP—subject to great care that only the ages of primary (first) magnetism are utilized. Secondary magnetic overprints can be a source of confusion. When, for example, rocks, bricks, or mud in hearthsites or ovens are reheated above Curie temperatures, they acquire, on cooling, a secondary magnetic orientation (that of the then-contemporary Earth), like new sounds on an old tape. The same thing happens to old lava flows and other magnetized rocks that at some postdepositional time became reheated above the Curie temperatures of previously magnetized minerals (for example, if basalt that solidifies at 1,000°C is reheated above the Curie temperature of 580°C for magnetite).

It was, in fact, the imaginative use of paleomagnetism that finally demonstrated the reality of plate tectonics and, with it, continental drift. The breakthrough came with seaborne and aerial magnetic surveys that revealed a pattern of longitudinal north–south stripes of varying magnetic characteristics along the Mid-Atlantic sea floor. They were found to be arranged in parallel matching sets on opposite sides of the Mid-Atlantic ridge (Fig. 5.5). Similar matching patterns were found across other oceanic rises considered to be spreading ridges. At the same time, similar sequences of geologically young magnetic anomalies were being found and dated in volcanic rocks on land. When the patterns and ages on land were found to match with those at sea, they were seen to indicate rates of sea-floor spreading very close to those needed to open the North Atlantic in the estimated 200 million years or so since North America began to drift away from Europe and North Africa in early Mesozoic time (about 1 centimeter per year).

The utility of magnetostratigraphy is impressive for the past 163 million years (post–Middle Jurassic). With somewhat less confidence, it extends an additional 1,300 million years into the past. Apparent polar-wander positions and paleopole-reversal patterns are utilized to support the correlation of contrasting types of sedimentary and igneous rocks between 600 and 1,400 million years old along the whole of western North America. Astonishingly, they and other paleomagnetic findings imply an equatorial position for many regions of continental glaciation around 670 to 800 million years ago—a problem that will be explored further in Chapter 12.

Clearly "paleomagicians," as others appreciatively call paleomagnetists, are onto something big. Subtleties are to be appreciated and difficulties overcome,

however. Earth's magnetic field has at least two measurable components and one or more not measurable. The principal measurable component, called the *dipole field*, behaves as if it were a large dipole bar magnet at Earth's center (monopoles exist so far only in the imaginations of astrophysicists). Lines of force from this dipole field embrace the Earth and define its magnetic poles like iron filings scattered on paper above a bar magnet. The dipole field is, in fact, an idealized mathematical model that describes the behavior of the principal component of the magnetic field believed to result from the dynamo-like motions of Earth's outer liquid iron core. The magnetic field also displays large and variable irregularities collectively referred to as the nondipole field. These are the only components of terrestrial magnetism that can be measured from Earth's surface. Most theorists predict that the

5.5 Matching linear belts record sea-floor motion away from the mid-Atlantic spreading center during the past 4 million years and 4 polarity epochs. Shading denotes normal polarity, stippling reverse polarity. *[Adapted by permission of Smithsonian Institution Press from* Continental Drift *by Ursula B. Marvin. Figure 72, page 162. © 1973, Smithsonian Institution, Washington, D.C.]*

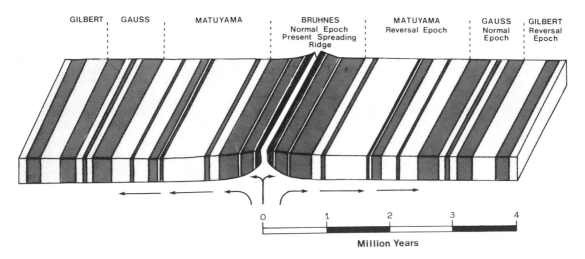

main component of the generating dynamo, probably gravity driven, is a doughnut-shaped structure within the circulating fluid shell that surrounds Earth's solid core. Because the self-enclosing lines of flux within this magnetic doughnut cannot be detected from Earth's surface, that sector of the magnetic field remains unmeasurable.

Changes in the orientation of the magnetic domains in a rock remain a problem. If reset by heating or later infiltrated by domains that record a postdepositional paleopole orientation, false conclusions may result. *Residual magnetism*, inherited from the initial magnetization (the *primary stable remanent magnetism*), is the field whose polarity and orientation one wishes to know. And, in finding that, paleomagicians have lived up to their nickname in discovering how to "clean" rocks of later magnetism while preserving the primary remanent magnetism.

Finally, the age even of the stable primary magnetization of a rock may not be the same as that of other early events in its history. In plutonic rocks, magnetization follows crystallization but precedes the setting of the potassium–argon clock at lower temperatures. Weathering and chemical alteration resulting from deep burial and thermal alteration can also reset polarity.

No one knows, as yet, why the poles reverse. It is known only that they do, not just on Earth and with a varying frequency (less often found in older rocks), but in our Sun and other stars as well. Sunspot pairs reverse polarity during their statistically regular 11-year cycle, and it is reported that the dipole field of Sun itself underwent a complete reversal of direction in the 5 years between 1953 and 1958. Such reversals clearly involve different materials and proceed perhaps 1,000 times faster in hot gas than they do in solid rock.

It was the reversal of poles itself—switching the orientation of magnetic poles without moving anything else—that earlier defied belief, a timely reminder that all things now accepted were once minority views.

Simple Framework for a Comprehensive Earth History

Where the founders of geology sought to discover and classify *the* natural order, it has now become apparent that there are so many ways of perceiving order that no single system can incorporate all. The goal, nonetheless, remains to identify the central variables and prevailing trends, to perceive how they may be interwoven into broad historical narrative, and to highlight the episodic crescendos, the punctuating events, of the unfinished planetary symphony.

Much progress has been and is being made toward the establishment of an objective framework of calibrating radiometric ages within which the various kinds and levels of natural order can be unambiguously discussed. Yet it is no small task to combine and interrelate all of the relevant methods, principles, facts, and variables of historical geology in a single framework, let alone to the satisfaction of all potential users. No knowledgeable and flexible-minded geologist believes there is a unique conceptual solution. Problems will continue to arise, for example, in tracking plate motions and integrating them with other aspects of Earth history. Disagreement exists about the particular time, duration, and causes of just about every important transition in geology.

All general schemes of historical classification, therefore, are progress reports. At the same time they provide frameworks for discussion. The one used in this book is the condensed scale of the major features of Earth history summarized on the front endpaper. It would be mistaken to suppose that this or any other framework is or will become the last word in a world of independent-minded humans with individual views of relevance. The levels and slopes of transitions and boundaries shown (especially on the left front endpaper) are all approximations and may be expected to change with reference to the column of ages on the left as better numbers are obtained or new data require revision of historical judgment. Alternative systems will be advanced. This one is intended to focus attention and thought on the main events of Earth history and their meanings, the major uncertainties, and to emphasize the apparently long evolutionary transitions between central tendencies among the older crustal rocks.

Scientific terminology and symbolism, to the extent deemed advantageous, should be clear, concise, and above all unambiguous. It should be comprehensible across nationalities and generations. The terminology of major features employed here is a working compromise between precedent and preference. It seeks to show the major features of Earth history at a glance and to scale.

For those already somewhat familiar with geological terms, a striking feature of the geologic column here outlined may be the absence of the widely sanctified term *Precambrian*. The word is an anachronism, impeding unambiguous discussion of the important transition from Proterozoic to Phanerozoic history. It has now served its purpose as a cover for ignorance and deserves an honorable retirement. Indeed, at least three historical intervals of the scope of formal historical eons in geology comprise the 85 percent of Earth history preceding animal life. Where a collective term is needed, *Cryptozoic* is available (since 1932). The Crypotozoic eons are also pre-Phanerozoic.

Among the major features summarized on that front endpaper, *eons* and *eras* are broad historical generalizations based on inference from the global succession of rocks. *Representative regional sequences* illustrate these global generalizations in regions of the planet where the main characteristics of the historical intervals are well developed, although usually to some degree incomplete. Most are supplemented by information from secondary events such as metamorphism and deformation, by inference from erosional gaps and initial and terminal deposits, by data from other regions, and by some controlled guesswork. Evidence may be incomplete or misconstrued, but the facts of nature do not change. As new observations and better measurements continue to be made, gaps will be filled or lessened, and interpretations will change. Future events will judge how closely the story here summarized approximates reality.

Indeed a contrasting proposal to eliminate further nomenclatural flux by fixing Cryptozoic historical geology within a framework of precise and invariable geochronologic numbers is, in fact, now in preparation by an *International Subcommission on Precambrian Stratigraphy*. When presented for consideration at the 1989 International Geological Congress, geologists will have to choose whether they want the perceived intervals of Earth history to be bounded by events recorded in rocks or by numbers.

Periods of history and *systems* of rock such as those of the original geological column (Table 3.1) are not specified in the summary of major features because none are yet agreed upon for the Cryptozoic eons. Such subdivisions for the Phanerozoic, however, are shown on the right front endpaper.

It is customary, as well, where major terms of history and rock sequence are employed, to capitalize prime modifiers where they are used formally. Where capitalized *Early* and *Lower* precede the name of a period or a system, therefore, they designate an explicit division of it. The same words, informally used, are *not* capitalized.

In referring to time numerically, terms for intervals of millions (10^6), hundreds of millions (10^8), and billions (10^9) of years would be convenient, but few euphonious words are available for ages greater than a millennium. I find *geocentury* tolerable for 100 million years, but nothing euphonious for 10^9, except billion, which in British usage means a million million (U.S. trillion, or 10^{12}). A nice crisp term, *aeon*, is available and increasingly used for 10^9 years, however, and I shall so employ it here, along with geocentury (for 100 million years), despite some trivial confusion of *aeon* with the little-used homophone *eon* (for very large

sectors of Earth history of unspecified duration). In practice *eon* is little used in geology. Geologists, in the fashion of journalists, also refer in conversation to the adjectival Phanerozoic, Paleozoic, Permian, etc., as if they were nouns, usually omitting the generic eon, era, and period—decriable to the classical grammarian, perhaps, but convenient and clear enough to be so utilized here.

The numbers with which Earth history is calibrated are similar to the B.C. numbers of historiographers, but are given as years BP—"before the present." Both conventionally *count backward* from some specified level. The "present" of geologists is defined as before 1950 for radiocarbon ages, that being the time when artificial radiocarbon of bomb origin began to enter the atmosphere on a large scale. Among really old rocks statistical limits of error are so broad that BP might as well mean "before the Pleistocene."

If time were reckoned forward from the beginning, one could say that Earth is now more than midway of its 5th aeon and nearing the end of its 46th geocentury. Both backward and forward counting time scales are graphed on the left of the front-endpaper charts to ease interpretation for the lay reader. In this book, in the tradition of the historical geologist, time is counted backward, but rocks and history are described from the beginning onward. That is similar to history preceding the Christian era.

Note also that the most recent, or Phanerozoic, Eon began only abut 0.67 aeons ago. Manifest records of animal life first appear then. The Cryptozoic eons account for the previous 85 percent of Earth history.

The eras are also of unequal length.

Eras are here added to include history before the Paleozoic, based on growing information about the origin and development of life during the Cryptozoic. At the very beginning, the Prebiotic Era was an interval not known to be represented by Earth rocks and one that probably preceded the origin of life. The Eophytic Era represents a scanty history of primitive microbial life. The conventional older Proterozoic (or, more descriptively, Proterophytic) Era was one of recognizably microbial but probably non-nucleate forms of life (procaryotic cells). Among them were some capable of oxygen-producing photosynthesis. And the conventional younger or later Proterozoic (or Paleophytic) Era saw the development and expansion of truly nucleate and aerobic but still mainly microbial life.

The knowledgeable reader will also notice that the Phanerozoic and Paleozoic intervals here begin around 670 to 650 million years ago instead of the 570 to 550 often given as the beginning of the Cambrian. That allows for inclusion of the fossiliferous Ediacarian System beneath the traditional Cambrian. For, in fact, the variety and extent of indisputably animal life (Metazoa) in Ediacarian rocks logically requires the extension of Phanerozoic and Paleozoic backward in time and down in the rock sequence to include Ediacarian or else to abandon their traditional biological implications. Recalling that, on two earlier occasions, Paleozoic was extended upward to accommodate improved understanding of Earth history (Table 3.1), so now it should be and here is extended downward to include the Ediacarian (see right front endpaper and Ch. 13).

In Summary

Having grown rapidly to near its present size, and having cooled sufficiently to support a solid crust and seas of liquid water, Earth still had to regulate the continuing temperature inputs within as a result of residual heat and radioactivity. Convective release of that heat through plate tectonism and predecessor mechanisms provided the energy for changing crustal alignments and generated interplay between life and its physical surroundings.

How can information be extracted from rocks about the nature and sequence of events and trends, their timing and duration, and the relation of the historical sequence in one region of the planet to that in others? What underlying rhythms may there be? It is interest in such matters and their connections that motivates live women and men to study inanimate rocks.

Terms and classifications are needed if planetary evolution is to be unambiguously discussed—names for rocks and for historical events and trends, if not for time. Although not precisely the same, the names of historical episodes and the record of the rocks that convey that history can be used interchangeably and the sequence of both worked out from superposition, biotal succession, and other clues to order. Time, although essential for calibration, neither defines nor is defined by history or rocks. The precision geologists seek in discussing explicit problems of duration, rates, and correlation, on the other hand, is better served in terms of radiometric years than by the broader terminology so useful in historical narrative.

In studying Earth's history, the recognition of equivalent but divergent types of rocks and events by *correlation* is critical. Correlation within the Phanerozoic eras is most widely achieved by a wide variety of paleontological criteria now powerfully supplemented by paleomagnetism—a body of principles, techniques, and data that broke to the fore in the mid-1960s and continues its advance. Paleomagnetic correlation is based on a pattern of magnetic anomalies that reflect apparent polar wander paths and polar reversals and is calibrated by radiometric time. It provides a good record of equivalence for appropriate rocks during the past 163 million years and is useful back to perhaps 1.4

aeons ago. Subdivision and correlation of older rocks and history depends on other criteria for crustal evolution, with a high level of dependency on time itself.

If time can be said to flow, history can be said to pulse, to surge, to syncopate. And, while it is true that the prevailing rhythms and punctuating surges of Earth's crustal and biologic evolution denote the flow of planetary history at a given place or region, time continues unaffected. In the end, the problem in Earth's history, as in other aspects of cosmic history, is to separate the signal from the noise. All progress in this direction depends on data from the rock record, but much data is simply irrelevant. It must be collected, sorted, and integrated with Earth's unfinished symphony.

The major movements of this symphony so far are summarized and denoted on the front endpaper.

PART II

EARTH
BEFORE ANIMALS:
THE NEXT
THREE BILLION YEARS

CHAPTER SIX

SEARCH FOR A PRIMORDIAL CRUST: THE OLDEST EARTH ROCKS

The familiar diversity of Earth's present surface is the product of an immensely long evolution. It is vastly different from any imaginable initial state, dependent as that was on the nature of the primordial crust and the processes that first began to shape it. The oldest Earth rocks so far convincingly recorded, ranging from about 3.8 to 3.5 aeons BP, have been so intensely and repeatedly metamorphosed, and are so limited in extent, as almost to defy reconstruction. Yet within small areas known in six to eight regions from Greenland to Antarctica are rocks of that age whose geochemistry and physical characteristics reflect their conditions of origin. Isolated mineral grains in southwestern Australia imply still-older rocks not yet discovered, perhaps as old as 4.3 aeons. The combination of evidence from these rocks with that from early lunar history and geochemical inference is a kind of case history in wringing maximal information from minimal data.

The Problem of the Primordial Crust

In the beginning, on the newly aggregated Earth, there was no water, nor soil, nor stable surface on which soil might form, nor any place where life might have survived had it been present. All of these things came later. A fierce solar wind of ionized particles, "blowing" at thousands of kilometers a second, stripped Earth of any gases that might have formed a permanent atmosphere or from which water might have condensed on cooling. Nor could processes leading to the origin of life have taken place in the absence of appropriate atmosphere and water. Here we come face to face with the penultimate mystery, with the

6.1 OLDEST EARTH ROCKS? ISUA REGION, SOUTHWEST GREENLAND. *Top,* view southwest from edge of ice cap over outcropping iron formation to engulfing Amîtsoq Gneiss. *Bottom,* massive ultramafic rocks to right, metasedimentary rocks at water's edge and across outlet of lake called Imarssuaq.

121

beginning of recorded Earth history, and to a search that has intensified in the years since 1973, when the oldest sedimentary rocks yet known anywhere in the universe were found by British and Danish geologists in southwest Greenland (Fig. 6.1).

Indeed, the nature of our planet during its first 600 to 800 million years as a satellite of the Sun is cloaked in mystery. Except for meteorites, samples from the lunar highlands, and a few apparently recycled 4- to 4.3-aeon-old crystals of the durable mineral zircon, no records so old are known. Yet there is a modest amount of indirect evidence about these times. Strong inferences about the nature and history of this Hadean Earth emerge from events on the moon and the impacted surfaces of those planets where the record has not been erased by weathering and erosion. Additional information comes from Earth's geochemical state at the end of this early formative interval, and from the requirements of gravity and the conservation of angular momentum.

To summarize, the ages of meteorites testify that rocky planetesimals were abundantly available for planetary aggregation some 4.6 aeons ago. Once a protoplanet of sufficient mass formed from such a cloud of planetesimals, gravity would assure its continued rapid growth. The aborted protoplanet Ceres, in the asteroid belt, illustrates that mass and freedom from gravitational disturbance by more massive bodies such as Jupiter are critical. The size and gravitational attraction of the proto-Earth and its greater distance from Jupiter were clearly enough for it to grow quickly toward its present size by the assimilation of neighboring and passing plane-

tesimals. Its internal temperature would have increased concurrently as a result of the conversion of gravitational energy to heat, core formation, and radioactivity. Accretional heat alone could have raised temperature by some 2,400°C, while core formation may have doubled that. In addition, half of all the radioactive heat ever to be generated within the planet by decay of U-235 was released during the first 7 geocenturies, strongly supplemented by similar heat from now-vanished shorter-lived radioisotopes.

As seismic and geochemical data reveal, our moon, Luna, became so hot from such effects that it melted. The results of the cooling and crystallization of this melt are seen in the rocks of the light-reflective lunar highlands, whose pockmarked surfaces and truncated edges so resemble, on a large scale, the Swiss cheese one gets with a continental breakfast. Planetesimal percussion reached a crescendo with the excavation of dry maria, triggering the upwelling of dark basaltic magmas to fill them, and then tapering off. The extension of such basaltic mass concentrations (*mascons*) at depth was profiled by early lunar orbiters—detectably curtsying to the gravitational power beneath as they passed overhead.

Earth, being several times the diameter of its lunar satellite, and having larger heat sources, could hardly have escaped a comparable thermal history. Surely it became hot all the way through—hot enough to melt at least its iron and some silicates. And liquid iron, being heavier than the silicate mass of the planet, trickled downward through the mush of upwardly segregating silicates, to become a core of nearly pure iron. The fluid outer core, however, contains small amounts of some lighter element, perhaps oxygen or sulfur. Thus, presumably, arose the somewhat heterogeneous, multilayered concentric structure of core, mantle, and lithosphere that characterizes the present Earth (Figs. 2.2, 6.2).

It follows from such a concentration of mass toward its center that Earth's moment of inertia (product of the sum of all masses multiplied by the squares of their respective distances from the rotational axis) would decrease. And, as in the case of the rotating skater when she drops her extended arms, it would spin faster. Faster Earth spin means shorter days, in that case perhaps as little as 8 hours. Frictional drag from the gravitational forces of Sun and moon since then, however, have gradually slowed Earth's rotation to the present 24-hour day and 365.25-day year, requiring another adjustment.

As in the case of balancing a budget, the law of the conservation of angular momentum (relating rate of rotation to moment of inertia) demands that any apparent loss (or gain) of angular momentum at one place must be compensated elsewhere in the Earth–moon system. In the case of a slowing rate of rotation in an Earth of constant mass and radius, the most likely adjustment is in the radius of lunar orbit about Earth. As Earth's day gradually lengthens under the retardational drag of solar and lunar gravity, its moon becomes ever more distant. Geophysical measurement and paleontological counts of days per year imply that the rate of this separation and the changing length of day have been essentially balanced for the past 4 geocenturies.

Although these are some of the processes that shaped our planet's early evolution, primary concern at this place is with the broad geochemical segregation that gave rise to its concentric structure. That this was achieved before extrusion of the oldest known basalts is certified by the fact that the average major-element geochemistry of basaltic rocks of presumed mantle source shows little change over the past 3.8 aeons (Table 6.1).

6.2 ELEMENTS OF CRUST, LITHOSPHERE, AND SUBLITHOSPHERE IN DIAGRAMMATIC VERTICAL PROFILE.

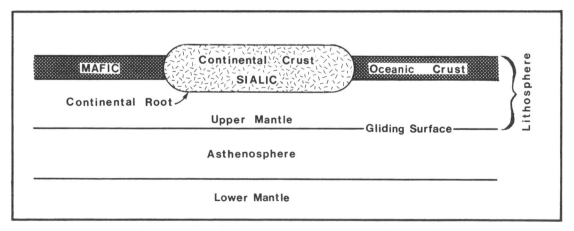

Despite mantle segregation, however, potential geochemical tracers remain. Certain mobile, rock-loving elements of large ionic radius (large-ion-lithophile or LIL elements)—such as potassium, rubidium, thorium, lead, and uranium—have become concentrated upward in the crust and thus depleted in the upper mantle with time, beginning about 3 aeons ago. These and the changing ratios of radioactive to inert sister isotopes provide important clues to the sources of the melts from which igneous rocks have crystallized and their relation to mantle history.

When Earth's barren initial surface first solidified, it was receiving about 30 percent less radiation than now from the faint early Sun but more heat by convection from the interior—perhaps as much as 4 times the present rate of outward heat flow.

In addition, because the rate of spin was probably thrice the present rate, centrifugal effects would have tended to concentrate sialic continental crust toward the equator. Our moon was closer, tides were higher, and the intertidal area proportionally more extensive. The shorter length of the day accelerated cycles of surface heating and cooling, perhaps stimulating early prebiotic chemical selection among organochemical stockpiles from cometary or other sources. In order for such selective processes to evolve toward life, however, Earth's surface would have needed to be insulated from the icy chill of the faint early Sun by some limited greenhouse effect. It has been calculated that a probable 100 times the present concentration of carbon dioxide could have provided a surface temperature of about 11°C (52°F). Higher levels than now of water vapor and the trace-gas ammonia (NH_3) would reduce the CO_2 requirement.

Questions flood the mind. How long did it take Earth to cool enough for the convecting mush of crystals beneath its unstable surface to solidify? What was the chemical and mineralogical nature of this initial solid crust? How and when did the primordial crust, of whatever composition, evolve into one where lighter sialic rocks could rise above the general level to become protocontinents, while heavier rocks sagged to floor contemporary oceans? How important was asteroidal and cometary impact? Was a generally sialic terrestrial scum similar to that of the initial moon ruptured by delayed planetesimal bombardment to

TABLE 6.1 *Average Geochemical Percentage of Basalts through Time, Omitting Volatiles*

	SiO_2	Al_2O_3	Fe_2O_3	FeO	MgO	CaO	Na_2	K_2O	TiO_2	MnO
3.8 aeons	50.6	13.5	2.9	10.7	6.3	10.7	2.2	0.44	1.1	0.26
2.7 aeons	49.8	14.7	2.6	9.0	6.1	9.0	2.4	0.34	1.1	0.20
1.8 aeons	48.8	14.1	2.4	9.8	7.3	10.1	2.5	0.22	1.1	0.22
1.1 aeons	48.0	14.7	1.9	10.3	8.1	9.8	3.3	0.20	1.1	0.19
Present oceans	49.3	16.9	2.1	6.8	7.5	11.6	2.8	0.18	1.4	0.18

create maria-like basins that became the protooceans? Did much or all of Earth's water and atmosphere come from a late influx of comets or was all or most of it outgassed? Did both sources contribute and, if so, over what time scale?

We cannot answer such questions with confidence at this time. They set goals for continuing research. But recognizing problems is the first step toward their solution, and some steps in that direction can be outlined.

Frontiers of Planetary History

Our neighbor Luna is a kind of fossil planet. Because of the absence there of weathering, tectonism, and extensive blanketing by younger rocks, we know much more about its earliest history than we do of Earth's. When that moon is bright and full, one can easily make out from crosscutting relationships, with simple equipment, the relative ordering of a half-dozen main events. To repeat, in order of age, these are: (1) light-reflecting highlands; (2) dark maria and peripheral debris blankets; (3) younger secondary impact craters; (4) ray craters; (5) still younger impact craters; and finally (6) more ray craters, culminating in the great Tycho. Tycho, named for Tycho Brahe, Kepler's teacher, is then the most conspicuous object on the moon. Its bright debris rays streak far and wide across the lunar surface—visible to the keen naked eye. From their reflectivity alone, geologists had guessed, long before the Apollo missions, that the light-reflecting old rocks were of feldspar-rich, continent-building types, whereas the dark, crosscutting rocks of the maria were of mafic (or basaltic) ocean-flooring types.

Now lunar rocks have been collected and analyzed, and seismometers have been installed to monitor moonquakes. The information so gained confirms and adds to what was already suspected. Luna's primordial highlands are indeed feldspar-rich rock, although, lacking quartz, they are not really granitic in a proper petrologic sense. They are of a type called *anorthositic gabbro*—a coarsely crystalline igneous rock consisting mainly of calcium- and sodium-rich feldspar of low-enough density to rise above the level of the maria basalts. The maria basalts, somewhat enriched in titanium compared with Earth basalts, lie low because they are "heavier" (denser) than anorthosite. Such rocks are familiar to everyone. Hawaii, like a number of other oceanic islands, is made of basalt, a common rock worldwide. Anorthosite is a beautifully iridescent, bluish building stone, sometimes called black "granite"—often seen as a decorative trim on public buildings. The Einstein statue, at the building of the National Academy of Sciences in Washington, D.C., gazes on a field of man-made "stars" set in a "heaven" of polished anorthosite from Larvik, Norway.

Lunar history is rich in implications for earliest Earth history. Highland rocks are consistently anorthositic, whereas mare (már-ay) rocks are basaltic. The ages of highland rocks range from 4.6 to 4.1 aeons, whereas mare rocks are consistently younger. Measurements of uranium and lead isotopes from highland samples by Caltech's Gerry Wasserburg fall on a straight line that intersects arcuate concordia (see Fig. 4.7) at 4.46 and 3.86 aeons BP. That means that some of the minerals of the highland rocks began to separate from a parent melt no later than 4.46 aeons ago. It also shows that some lunar clocks were thermally reset during an interval of intense asteroidal bombardment 6 geocenturies later. Studies of other radiometric systems in highland rocks indicate that crystallization of the primordial lunar crust had been completed by 4.35 aeons ago, perhaps dating the time of collision with Earth and core formation here.

Seismology completes the picture. It reveals a discontinuity in the physical properties of the lunar interior about 60 kilometers down. Above a depth of 60 kilometers these properties are consistent with anorthositic crust. Below that, they indicate denser mafic and ultramafic rocks.

Such evidence implies that the early moon became hot enough to melt most of its constituents, resulting in the upward segregation of lighter components to become a searing magma ocean some 60 kilometers deep. As this magma cooled and began to crystallize, the relatively light feldspar minerals that make up 80 percent of the lunar highlands floated upward to become rock islands in a hot magma mush. Eventually it solidified as anorthositic crust. That process, beginning during the first aeon following initial lunar accretion, seems to have taken only about a geocentury to complete.

What bearing has all this on Earth's primordial crust? Being much larger than its moon, and with proportionally larger heat sources, one would expect our planet to have reached substantially higher temperatures and taken longer to cool. It should also have experienced a similarly large thermal pulse from late asteroidal infall, perhaps lunar collision, and tidal stresses caused by the closer proximity then of Earth and Luna. Reasons have been found, however, to question whether Earth's initial crust could also have been anorthositic. Earth's size and gravitational attraction allow it to retain water, and the plagioclase feldspar that preponderates in anorthosite is said not to float well in a wet magma. Steep pressure gradients on Earth might be expected to convert the characteristic feldspars to the substantially heavier mineral pyroxene at depths below about 35 kilometers. Furthermore, had Earth melted completely, it should have expelled all its volatiles as did its lunar satellite, whereas helium-3, a primordial gas, is still escaping to Earth's surface in significant quantities.

How could Earth have failed to have a magma ocean when its (present) satellite did? Could it be that the reason Earth's surface seems to have taken so much longer to crust over than that of its moon is because its magma ocean was so much larger and deeper? Or, could it be, as the University of Saskatchewan's Euan Nisbet has proposed, that such an "ocean" was trapped somewhere within the upper mantle, from which ascended the Archean komatiites, so unusual in later geologic history? The evidence is so far entirely circumstantial. The jury is still out.

What are the oldest Earth rocks like? Where are they? What might they say about origins and sequence? How will we know when we find them? A goodly number of geologists and geochemists have long sought and still seek the oldest Earth rock. More than normal competitive urge drives them. The goal is to learn the nature of the planet's first solid surface, to find out what it was really like *in the beginning.*

Geochemistry and gravity assure us that this primordial crust was a silicate rock of some kind—with silica about or above 50 percent by weight, plus metallic oxides. Was the preponderant metallic element aluminum, the silica content well above 50 percent, and the crust sialic (Fig. 5.3)? Or was aluminum accompanied by significant amounts of magnesium, iron, and calcium, silica 50 percent or less of the total, and the initial crust mafic? Or were both mafic and sialic crust present from the beginning?

Beyond that lie important questions bearing on the origin of continents and ocean basins. Aluminum is lighter than iron. Sialic rocks, therefore, are lighter than mafic ones. That is the simple reason why there is so much granite on continents and so much basalt in oceans.

Throughout most of geologic time sialic and intermediate magmas have risen to occupy the more elevated crustal regions, tapering off, it seems, after the intrusive episodes of Archean and older Proterozoic history. Mafic rocks concentrate mass, resulting in basins which, when filled with salty water, are called oceans. Gravity sees to that. It is only coincidental, a function of distance from Sun and atmospheric physics, that on Earth our "ocean basins" are filled with water.

As to the oldest Earth rocks, it seemed well established by 1940 that they were about 3 aeons old. Then the 1950s brought recognition of a common age of around 4.6 aeons for the solar system. And, during the 1960s, Earth rocks from several localities within the ancient continental nuclei were found to have ages as great as 3.4 to 3.6 aeons—implying

that a floor for Earth ages may have been reached in a search so well focused that further exploration was unlikely to yield larger numbers.

The oldest then-known rocks, spanning a range in time from about 3.5 to as young as 2.5 aeons BP, are those of the Archean (see left front endpaper). They display a broad similarity of rock types and sequence the world over. Thick successions of preponderantly volcanic rocks of broadly oceanic types are followed by or interlayerd with immature sedimentary rocks consisting mainly of detrital volcanic materials. And all of these were squeezed into linear belts between rising, nearly contemporaneous *intrusive* granitic domes resembling those of the Yosemite region. The moderately metamorphosed basaltic rocks are called *greenstones.* With associated metamorphosed sedimentary rocks (metasediments) in the so-called *granite–greenstone belts,* they became the paradigm of Archean terrains.

It seemed clear at most places that the mafic or ultramafic volcanic rocks were the oldest. Ingenious models were invented to show how the initial crust started mafic and was later intruded by less dense sialic matter. That gave rise to protocontinents and eventually to the extensive, complex, deeply rooted masses of mainly sialic intrusive rocks and veneering sediments that make up the continents. Many were uneasy with that model, however, because, by parallelism with the moon, the light stuff should have floated to the top initially and stayed.

Gravity, it seemed, would be better served by a sialic initial crust, and there were geologists who thought there might be still older crust than the ancient greenstones. One was V. R. McGregor, who left idyllic New Zealand to take up residence in a bare and wind-swept coastal Inuit settlement near Godthåb (or Godthaab, or native Nûk) in southwest Greenland, not far from an uninhabited but geologically important place called Isua. The mostly gneissic and granitic rocks (gneiss is a kind of lineated, metamorphosed granite) of this region had been strongly deformed and disordered tectonically. Still, McGregor worked out their proper historical sequence by clues such as the presence in some of tabular, crosscutting, mafic intrusions called *dikes*. In 1973, a rubid-

ium–strontium age by Steven Moorbath of Oxford University showed that the oldest rocks in that sequence, now called the Amîtsoq Gneiss, were older than any previously known rocks, with a now well established (Pb–Pb) age of about 3.8 aeons on these rocks as well as associated metasediments. A new basement age was set. Similar discoveries elsewhere simultaneously improved the odds both that Earth's primordial crust might prove, after all, to be sialic rather than mafic, and that still-older rocks might turn up.

Beginnings of Continental Growth

Although the world today is roughly 30 percent land and 70 percent ocean, geologically speaking the continents represent more than 30 percent. That is

because an excess of water spills over the continent shelves and into low-lying regions like Hudson Bay and the Baltic Sea, blurring the boundaries between

6.3 DISTRIBUTION OF ARCHEAN ROCKS (*BLACK*) AS THEY MIGHT HAVE BEEN CLUSTERED (*STIPPLED*) 2.5 AEONS AGO IN A FRAMEWORK OF PHANEROZOIC CONTINENTAL OUTLINES. Numbers *1–8* indicate areas where rocks of the Isuan historical interval are exposed or convincingly inferred. Letters indicate Phanerozoic continental masses.

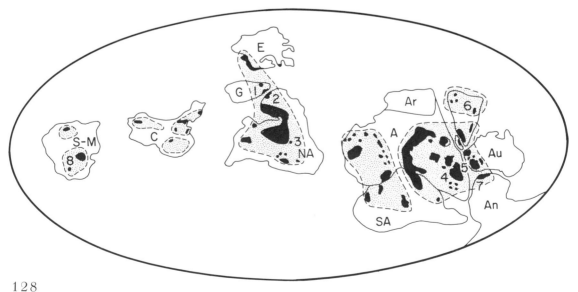

continent and ocean. The continental mass reached its present size as a result of the accretion and underplating of sialic matter, segregated by complicated processes from Earth's thick mantle. The sialic continents today float, as it were, within the mafic upper mantle, at levels some tens of kilometers above the still-deeper viscous asthenosphere.

Figure 6.3 shows the distribution of known Archean rocks as they may have been clustered at the end of Archean history in a framework of Phanerozoic continental outlines. The dark areas suggest where the oldest rocks clustered and may have joined to make the earliest continental shields (*stippled*), as well as how extensive these may have become by the time Proterozoic history was well underway. The suggested equatorial clustering would be a likely consequence of a faster rate of Earth spin.

Continental growth had to start somewhere, so the black areas of outcropping Archean rocks indicated in Figure 6.3 are good places to look for evidence of such beginnings. The numerals *1* to *6* on Figure 6.3 denote sites where rocks older than 3.5 aeons have been found, and *7* to *8* mark regions where there is good reason to expect them.

Rocks that span ages from 3.8 to 3.5 aeons old in southwest Greenland—the *Isuan* sequence—are among the oldest known. Being also the most extensive very early Archean sequence, they provide a useful reference model for initial Archean history. Although they include greenstones, they may not relate to a typical greenstone–granite belt. The most extensive outcrops consist of once deeply buried and highly metamorphosed, banded rocks of granitic composition called *gneiss*—in this case granulitic

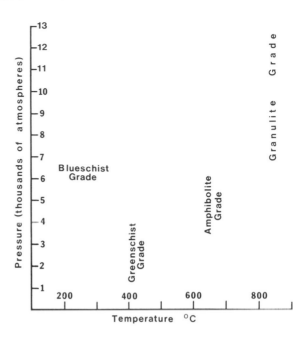

6.4 TEMPERATURES AND PRESSURES IMPLIED BY DIFFERENT GRADES OF METAMORPHISM. As rocks are buried, squeezed, and heated above about 200°C their original minerals change. *Metamorphic grade*, indicated by such mineral geothermometers and geobarometers, implies temperature, pressure, and depth of origin. The lowest grade of metamorphism is the *greenschist* or *chlorite grade*, characterized by a green, micalike, silicate mineral—chlorite. More heat and pressure yields shiny black prismatic minerals of the same chemistry called amphibole, defining the *amphibolite grade*. The characteristic minerals of the *granulite grade* crystallize at hot, dry depths. Very high pressure minerals such as glaucophane are found only in *blueschists*, distinctive of plate tectonism, formed under high pressures at modest temperatures.

gneisses, or *granulites*. Their mineral composition implies metamorphism at high temperatures and pressures, and thus burial to great depths, at some stage in their history (Fig. 6.4).

Undated metasedimentary and meta-volcanic rocks that surround these granulites at Isua may be still older. Lacking a bottom, with no clearly underlying surface on which they might have been deposited, they could have been engulfed by ancestral granite before metamorphism. The metasediments, moreover, also include quartzite-like pebbles of recrystallized chert and squashed fragments of sialic lavas, indicative of vigorous erosion, explosive volcanism, and flowing water at the time of sedimentation. In any case, the scene at Isua includes both mafic and sialic components. It even includes calcium-rich (*calcic*) anorthosites of lunar affinity, although in nothing like lunar abundances.

Could these Isuan metasedimentary rocks be reworked remnants of an early-intermediate or sialic scum, crystallized from the upper layers of a global magma ocean like that of the moon? Granulites are commonly regarded as the product of metamorphism at depths of 20 to 30 kilometers, but Mesozoic granulites from New Zealand imply that 15 is ample, and 8 has been considered enough at sufficiently high temperatures. Could the thermal gradient have been so high in the early crust that granulite minerals formed at Isua without burial to great depths? Or were dislocated slabs of such materials stacked under such a load by repeated low-angle thrust faulting during early plate collisions? If we really see here the remains of a protocontinent and if the protocontinents of the Isuan interval represented such small fractions of Earth's surface, what was the rest of it like? What was the volume, depth, and composition of the world ocean at that time?

From the foregoing, it seems likely that Earth's interior was hotter than subsequently and the rate of outflow of this heat perhaps several times the present. The lithosphere, therefore, well-qualified contrary judgment not withstanding, should have been, in theory, thinner than at any subsequent time—consisting essentially of crust with little solid outer mantle floating on asthenosphere. As a consequence of this, any continental roots would likely have obstructed incipient plate motions. Sediments and sedimentary rocks produced from weathering and erosion of such a crust would be immature in the sense of having had little opportunity to undergo rounding and segregation by physical properties. In addition, sialic crust, being lighter, would rise to higher elevations than mafic crust and would not be as susceptible to mantle recycling.

The Isuan Protocontinent of South Greenland and Coastal Labrador

Although high-grade gneisses occur in a number of different Archean and even younger settings, they are especially distinctive of Earth's oldest known metasedimentary rocks in the Isuan protocontinent (Figs. 6.5, 6.6) and other localities such as the north-central United States, southern Africa, and India (Fig. 6.3, *numerals 1–6*).

A good many of these gneisses are thought on geochemical or structural grounds to be recrystallized volcanic and sedimentary rocks and thus remnants of stratified rocks that accumulated on the surface of already existing crust—*supracrustal* rocks, they are called. In southwestern Greenland, however, McGregor found that the layering of the old Amîtsoq Gneiss was a product of deformation of an originally granitic rock, combined with the injection of layered intrusions. A primary derivation of the parent granitic rocks by *fractionation* (gravitational separation and concentration) from the partially melted mantle is similarly implied by their low strontium initial ratios, close to upper mantle values.

6.5 MINIMAL EXTENT OF ISUAN PROTOCONTINENT.

6.6 GEOLOGIC MAP OF ISUA SUPRACRUSTAL BELT, SOUTHWEST GREENLAND. Upended metavolcanic and metasedimentary rocks are here surrounded by the Amîtsoq Gneiss. Faults and dikes omitted. *[Adapted with modification from D. Bridgewater and W. S. Fyfe, 1974,* Geoscience Canada, *v. 1, no. 3, p. 9.]*

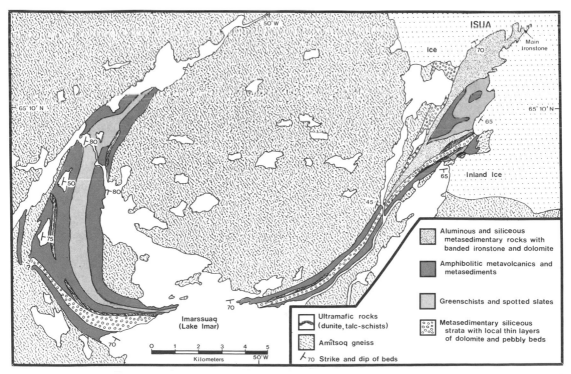

6.7 BANDED IRON FORMATION FROM PROSPECTIVE ORE BODY AT ISUA. Sample is 18 cm square.

The only surely supracrustal deposits of great age are the metavolcanic and metasedimentary rocks associated with the banded iron formation (Fig. 6.7), whose lead-isotope age of about 3.8 aeons BP has been interpreted as the time of sedimentation. This sequence, about a kilometer thick, borders and is partly covered by the Greenland ice cap, 150 kilometers northeast of the capital city of Godthåb, at Isua (Figs. 6.1, 6.6). Besides iron formation, it includes metamorphic equivalents of other sedimentary and volcanic rocks. The amphibolites are chemically similar to modern mid-ocean-ridge basalts.

In addition, traces of carbon are found as the metamorphic mineral graphite in glassy siliceous rocks called *chert*. The graphite is slightly enriched in the lighter carbon isotope (C-12), which would be *consistent with* microbial life by that time but is possibly a consequence of metamorphism.

Among Isuan rocks one also finds thin bands of dolomite containing reduced iron (Fig. 6.8), impure sands of volcanic origin of the type called *graywacke*, and cemented gravel or conglomerate. One such conglomerate includes layers of poorly rounded to angular, recrystallized chert pebbles, implying surface erosion and active transportation of already hard rock. Another consists of fragments and boulders of sialic volcanic rocks, some quite well rounded by stream or beach erosion. The volcanic fragments in such conglomerates are locally smeared into flat sheets by postdepositional tectonism (Fig. 6.9). These squashed pieces were originally porous, pumice-like fragments of explosive eruptive origin. Stratified sequences similar to but areally less extensive than that at Isua are also found as enclaves within the Amîtsoq gneisses over a large part of southwest Greenland.

Conclusions about early surface conditions that emerge from these observations are:

1. By no later than 3.8 aeons ago, a solid, emergent crust was already present and on it a variety of sediments and interlayered volcanic deposits.

2. An atmosphere rich in carbon dioxide and water vapor was also present at that time, moderating global temperature and accounting for the weathering manifested by the existence of sedimentary rocks.

3. A substantial hydrosphere existed, accounting, with other agents of weathering and erosion, for the production of sediments and their transportation and deposition in the sea. Ferruginous dolomite and banded magnetic iron formation, precipitated as chemical sediments, imply low oxygen levels.

4. Stratified rocks observed indicate that the contemporary surface consisted mainly of mafic volcanics, with sialic components.

5. Sedimentary structures and the presence of the typically marine carbonate rock dolomite in the sequence imply that the waters in which the Isuan sediments finally came to rest were shallow, near emergent land, and mainly or wholly marine.

6.8 VERTICAL FOLDS IN FERRUGINOUS DOLOMITIC AND SILICEOUS METASEDIMENTARY ROCKS AT ISUA.

6.9 METACONGLOMERATE AT ISUA: SIALIC VOLCANIC PEBBLES SMEARED INTO VERTICAL PLATE-SHAPED STRUCTURES.

6. Chemical similarities between the metavolcanic rocks and modern oceanic and mid-ocean-ridge basalt implies that the environment of deposition of the early Isuan supracrustals was that of an oceanic island, not yet truly continental but with sialic components in the form of associated highly siliceous lavas. The subsequent engulfment of these rocks by mantle-derived granites and their metamorphic conversion to granulite is our oldest record so far of a certifiably continental type of geochemistry.

The slightly carbonaceous chert from this area, in part recrystallized to a quartzite-like rock, may also place limits on contemporary surface temperatures. They could not have averaged less than 0°C, or water would not have been available for weathering, sediment transport, and deposition. An upper limit of 86°C to 146°C has been derived from oxygen-isotope ratios of the chert, and computer modeling suggests a surface temperature of 85° to 110°C. As the chert, however, is probably an altered volcanic ash, atmospheric temperatures were most likely not above boiling. The evidence for flowing surface waters given by the Isuan gravel deposits makes it unlikely that prevailing surface temperatures differed greatly from present ones, although it was perhaps uncomfortably warm by human standards.

Within these limitations, one may visualize that the original sediments and lavas of the supracrustal sequence now preserved at Isua accumulated in a shallow coastal zone surrounding or adjacent to a barren land of mafic and sialic lavas. Atmospheric moisture and runoff sufficed for weathering and transport seaward in multichanneled braided streams, unimpeded by vegetation. Cooling beneath the astronomer's faint early Sun was moderated by greenhouse effects, dependent on CO_2, H_2O, and probably trace gases. High surface heat flow favored abundant volcanism, igneous intrusion, outgassing of volatile substances, and a hydrosphere warmer than now. With a not-improbable geothermal gradient of 100°C per kilometer of crustal depth, the later engulfing granites (now granulites) might have formed at the base of a crust perhaps no more than 7 to 12 kilometers thick. In any case, as in that of the roughly 130-million-year-old granulites of southwestern New Zealand, they could have been upthrust and unroofed in a geocentury or so. A better analog for the developing Isuan protocontinent might be the oceanic island of Iceland, where an area of about 103,000 square kilometers is only 15 percent sialic.

An extension of the Isuan sequence is found southwest across ice-clogged Davis Strait from Greenland, in Saglek Fjord, near the northern tip of Labrador. There one sees an assemblage of granitic gneisses, the Uivak gneisses, similar to those of Isua. In 1975 these gneisses were found to have a Rb–Sr metamorphic age of 3.62 aeons. And they intrude older metavolcanic and metasedimentary sequences. The similarity in age and nature of the rocks at Saglek and Isua support a former connection. The combined outcrop areas of these ancient granulites and associated supracrustal rocks in Greenland and Labrador are the nucleus of the Isuan protocontinent. Although the remnants of this ancient land (Fig. 6.5) cover an area of only about 5,000 square kilometers, it apparently extended during later Archean time to occupy a much larger region (Fig. 6.3).

Global Extent of Isuan Rocks

As a result of the discoveries in Greenland, similar rocks in the Minnesota River valley were reevaluated and found to be roughly contemporaneous with those at Isua and Saglek. These high-grade gneisses, once thought from RB–Sr analyses to be only about 2.6 aeons old, were found by resourceful Sam Goldich and colleagues on the U.S. Geological Survey to have U–Pb concordia ages as great as 3.66 aeons (Fig. 4.7). Similar gneisses south of the Canadian granite–greenstone belts, in the northern peninsula of Michigan, were also found to have samarium–neodymium crystallization ages of about 3.6 aeons. These mid-continent gneisses resemble those of Greenland and Labrador in their mineral and geochemical composition. They may once have been a part of the Isuan protocontinent.

Other occurrences of a similar age and character, although few in number, are found worldwide. One of the better known comprises ancient metasediments of the Sand River Gneiss that occur along a southern tributary to the Limpopo River, barely 24 kilometers southeast of the only bridge connecting South Africa with Zimbabwe (Fig. 6.10). They are granulitic to amphibolitic and prevailingly sialic gneisses, almost concealed beneath a flat, brushy terrain of umbrella-like thorn trees and bottle-shaped baobabs, in whose distended hollow trunks, in times of tribal strife, legend hides the village virgins. These metasedimentary gneisses are highly deformed as a result of six distinguishable deformational events and five episodes of dike intrusion—smashed up,

but not beyond all recognition. In spite of this history, their stratigraphy has been unraveled by careful mapping, and the rocks described and dated. The oldest ages found were those of highly aluminous, quartzose gneisses that yield a Rb–Sr isochron age of 3.79 aeons—for all practical purposes the same as their historical equivalents in North America and southwest Greenland. As that may also be a metamorphic age, however, the time of origin of the parent rock could prove to be greater.

These sialic gneisses are apparently the oldest rocks of the region—truly basement rocks. They are intruded by early sialic dikes, which, in turn, are crosscut by two later generations of mafic dikes 3.5 and 3.1 aeons ago.

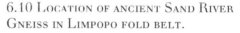

6.10 LOCATION OF ANCIENT SAND RIVER GNEISS IN LIMPOPO FOLD BELT.

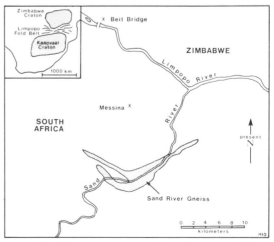

Beyond southern Africa the record of Isuan or older Archean rocks becomes more uncertain. Other rocks of similar age may include some 1,000 square kilometers of massive sodic gneisses near Champua, in northeastern India, and granulitic gneisses in the far-eastern Soviet Union. High-grade gneisses of apparently comparable age are also known from the Arctic tip of Norway, northeast Antarctica (Enderby Land), and the northwestern corner of the Yilgarn block in Western Australia. Southeast Asia (including China) and South America are the only large regions of the world where rocks of the Isuan sequence have not yet been recorded or strongly indicated.

Sialic rock was thus already globally known in earliest Archean time, 3.8 or more to 3.5 aeons ago. The building stuff of continents was beginning to accumulate.

The Elusive Primordial Crust: Where Is It, What Was It?

It does not follow, however, that still-older and perhaps different types of rock will not be found. What might lead one to expect such rocks? One thing is the presence of apparently older mafic and ultramafic rocks as inclusions in the above-mentioned high-grade gneisses. Another involves the fact that so many of the basement gneisses have been dated by rubidium–strontium isochrons. These isochrons could date the ages, not of initial origin but of metamorphism. It should soon become possible to discern beneath such metamorphic veils the predecessors of these old granitic gneisses. Given appropriate samples, the samarium–neodymium or lutetium–hafnium methods (Table 4.1) may be able to resolve such problems. A third line of evidence includes already reported older ages that have not as yet been unequivocally confirmed, some more than 4 aeons old.

These possibilities must be carefully studied. Omitting unrealistically great ages such as are now recognized to be products of argon absorption at high temperatures by the mineral pyroxene, we find the 1976 report by Soviet scientists of a lead-isochron age of 4 aeons for one of the previously mentioned granulites in Enderby Land. Alas, restudy by Australian geochronologists was unable to validate this.

More promising is the report by South African geochronologists of a lead-isotope age of 4.16 aeons for a bulk sample of the semiprecious mineral zircon and of ages of 4 to 4.1 aeons on sulfide minerals, all from the same granite boulder in a younger Archean conglomerate in Swaziland. Lesser ages found for selected zircons from the same boulder could imply more than one generation of zircon. Ages even greater than 4.1 aeons, therefore, may be found for other zircons in the sample and its parent rock—as yet not known to crop out in the region.

Finally, in 1984 William Compston and associated Australian scientists, utilizing an ion–microbe technique called by the acronym SHRIMP, found ages in excess of 4 aeons for a few individual zircon grains and fragments in an area in the south of Western Australia (Mount Narryer) where 3.6-aeon-old rocks were already known. In 1986 they reported finding (60 kilometers to the northeast) 17 such grains out of 140 analyzed, of which one gave the startling age of nearly 4.3 aeons. They are interpreted as having formed 4.3 aeons ago with later lead loss, or over an interval of 4.1 to 4.3 aeons BP. These zircons are much older than the quartzitic host rock itself, as well as other zircons in the same sample. The implication is that they came originally from some yet-unknown igneous rocks of probable sialic composition that weathered and eroded to yield some or all of the sediments that became the Mount Narryer Quartzite.

The discovery of real outcropping rocks that are verifiably 4 aeons old or older remains to be made. But it begins to look likely and the search is intensifying. What might such old gray rocks be like? How old might the eventual grandfather rock turn out to be? Or has an initial, mainly mafic crust long since disappeared beneath younger sialic continents in primordial subduction zones?

It is highly unlikely that any surface rocks predating the turnover of Earth materials during core formation could have survived. Because of its low melting temperature, most lead would have been removed from the mantle at the time of core formation, implying to some that the 4.6-aeon age of the Earth as measured by lead isotopes actually dates the time of core formation. The lead-isotope data, however, suggest to others that the formation of the core took place about 4.4 aeons ago, perhaps dating lunar effects. So we are looking for rocks that have radiometric ages between about 3.8 and perhaps 4.4 aeons old. One may note that the geochemistry of the oldest Isuan metasediments has been interpreted to imply derivation from undifferentiated igneous source rocks no more than about 200 to 300 million years earlier—around 4 to 4.1 aeons ago. Were such rocks to prove as old as 4.3 aeons, that would allow but 100 to 300 million years following initial turnover and core formation for Earth to cool enough to support a crust—a barely believable number.

The primordial crust may well have been basaltic. One can envision a truly Hadean scene, with fuming rifts, lava fountains, and volcanic buildups of mafic, ultramafic, and locally sialic lavas. Impact craters presumably abounded as well, filled with steaming pools of cometary and outgassed water that overflowed into other rifts and depressions, eventually blanketing a surface of low relief. Deposits like those of the Isuan sequence could have begun to accumulate adjacent to or on the drab volcanic islands or other emerged lands of such an initial earthscape. Meanwhile, the high heat flow required for venting residual and radioactive heat would express itself as thermally driven convection cells in Earth's outer mantle. These, in turn, might be expected to promote crustal growth by recirculating and assimilating portions of the protocrust, with upward segregation of lighter materials to initiate a permanent sialic crust. Could such a scenario be hidden in the metasediments, basaltic amphibolites, and high-grade granitic gneisses of the Isuan sequence (Fig. 6.11)?

Although geochemistry may favor mafic crust at the beginning, sialic protocontinents could not have been long behind.

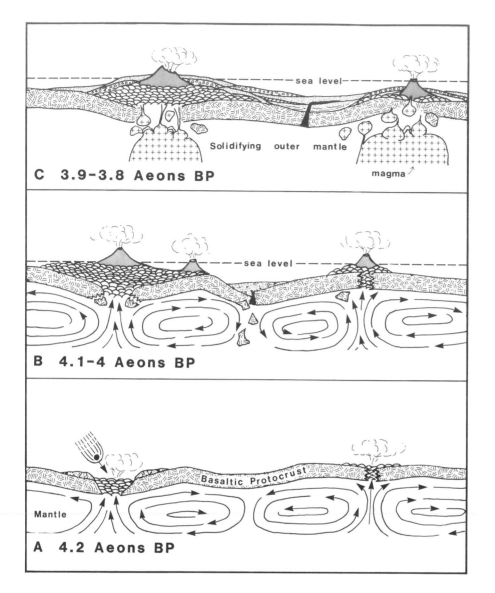

C   3.9-3.8 Aeons BP

Solidifying   outer   mantle

magma

B   4.1-4 Aeons BP

Basaltic Protocrust

Mantle

A   4.2 Aeons BP

6.11 Hypothetical vertical profiles through evolving late Hadean protocrust (A–B) and Isuan crust (C) in the region now southwest Greenland. Scene A, basaltic protocrust as it might have been about 4.2 aeons ago. Impacts on and opposing convection currents in the hot outer mantle disrupt the solidifying surface. Molten upper mantle wells up. Water and other volcanic gases mix with similar products of cometary origin. B, late Hadean protocrust. Broad swells and troughs result from sagging and undulations of convecting mantle beneath still thin, hot crust. Segments founder, and are assimilated in zones of partial melting. Volcanic islands grow. Continued outgassing and cometary infall adds volatiles to atmosphere and hydrosphere. C, early Isuan crust; beginnings of sialic accumulation. Continued recirculation within upper mantle results in intermediate and eventually sialic magmas that reach the surface locally as viscous lavas. Weathering, erosion, transportation, and sedimentation combine with further volcanism, intrusion, and sedimentation to yield components of the Isuan supracrustal sequence. Upwardly moving sialic magmas engulf older crustal rocks, and crystallize to form the parent granites of the Amîtsoq Gneiss.

Recirculation of the initial basaltic crust and its assimilation and further fractionation within viscous or fluid upper mantle (Fig. 6.11*B*) could give rise to ever more siliceous magmas. To convert average basalt with 50 percent silica (SiO_2) to a granitic rock with 65 percent SiO_2 requires the addition of 0.4 grams of SiO_2 to each gram of basalt, so the process of mixing and segregation must stew for awhile. One can imagine how, in time, similar to huge helium balloons rising through a thick and turbulent atmosphere, the increasingly buoyant sialic magmas slowly stoped their way to the primordial basaltic surface (Fig. 6.11*C*). Something like the processes and results imagined may have been involved in the granite–greenstone sequences of Chapter 7.

For the still-uncommon Isuan sorts of rocks, one must temper imagination with caution. Models that put sialic crust first are supported vigorously and on appealing grounds. The rock record reviewed here could be interpreted in their favor. Conversely, the interpretation of an initially basaltic crust finds support in the abundance of preexisting mafic inclusions in the oldest sialic rocks known. The initial question remains, perhaps for some inspired student or interdisciplinary minded physicist or chemist to solve: What was the primordial crust?

The record so far discovered clearly implies that Isuan history was a time of early sialic growth. The succession of events here envisaged calls on early ascent of sialic components through a mafic oceanic protocrust, giving rise to a scattering of increasingly sialic aggregates by 3.6 to 3.5 aeons ago. Evidence from the Isuan metasediments that parts of Earth's surface had by then emerged from the perhaps previously global sea means that the volume of whatever then served as ocean basins was large enough to accommodate much or all of the contemporary hydrosphere. Had ocean basins then already reached their present capacity, or did the volume of the hydrosphere continue to grow with increasing capacity of its oceanic reservoir? Geologists who work on such problems welcome all the help they can get from other sciences.

In Summary

The time of origin and the nature of Earth's first solid surface are uncertain. Much more was learned about the first aeon of lunar history during the 1970s than is yet known of Earth's, including useful parallels for aspects of Hadean and earliest Archean evolution. It is mainly from Earth itself, however, that we learn about its primordial crust.

It was long realized that the nuclei of all continental masses are characterized by distinctively Archean terrains, some as old as 3.4 to 3.5 aeons BP. Then, in 1973, still-older granitic gneisses were found in southwest Greenland and subsequently at localities in North America, southern Africa, India, Western Australia, and perhaps in Antarctica and the Aldan Shield of the far-eastern Soviet Union. These rocks range in age from 3.5 to 3.8 or more aeons ago. In Greenland they are associated with, but apparently younger than, water-deposited metasediments up to 3.8 aeons old, which they surround. Although the sediments were clearly formed at the surface of the Isuan (early Archean) Earth, the metamorphic mineralogy of the gneisses implies high temperatures and presumably elevated pressures at their sites of metamorphism, at depths of some kilometers within the Isuan planet.

These Isuan rocks raise a flurry of questions to which few widely agreed upon answers have yet emerged. The isotope geochemistry of the ancient gneisses shows them to have been derived from the metamorphism of primary sodium-rich (sodic) granites and not from parent sedimentary rocks. The characteristics of the Isuan metavolcanic and metasedimentary rocks assure us, however, that by 3.8 aeons ago there was already in this region a solid crust, including both mafic and sialic components. The presence of water-transported sedimentary rocks is conclusive evidence that a substantial hydrosphere and atmosphere were both present by then. Free oxygen remained low enough for ferrous iron oxides (FeO) not only greatly to exceed ferric iron oxides (Fe_2O_3) in the rocks but to be transported in solution. A depletion of C-13 with respect to C-12 in the reduced carbon of these rocks is consistent with (but does not require) the presence of life by then. The evidence given by sedimentary structures for the presence of liquid water implies prevailing surface temperatures within the range of present ones.

Isuan history was a time of earliest protocontinental growth, an ample but probably still-growing hydrosphere, and ocean basins large enough to accommodate that hydrosphere without completely submerging the occasional Archean lands. Despite the faint early Sun, temperatures clearly did not vary dramatically beyond the range of liquid surface water. The components of the heat-conserving greenhouse effect that assured this have been the subject of inconclusive modeling studies. Possibilities are CO_2, H_2O, and CO far above present levels, plus nitrogen and traces of lesser greenhouse gases.

Most Isuan rocks are high-grade granitic gneisses, characteristic of the interval up to about 3.5 aeons ago. Traces may be present on all existing continents. Because many earlier reported ages are Rb–Sr ages, these could date times of metamorphism (instead of origin) of rocks that crystallized much earlier. It is, thus, not surprising that older ages are being sought or that ages greater than any known exposed rocks (up to 4.3 aeons) are being reported from inclusions in South African and Western Australian rocks.

Where will the trail of the grandfather rock end? The smart betting is that, if found, it will prove to consist of relatively undifferentiated volcanic rocks including basaltic, ultramafic, and some sialic components that first crystallized little more than 4 aeons ago. It is likely that upward-floating balloon-like blobs of sialic magma, segregated by partial melting from mafic mantle, later stoped their way into this mafic crust, where they became sodic granites and the foundations of the original protocontinents.

7.1 LANDSAT IMAGE OF GRANITIC PLUTONS (*LIGHT*) AND DEFORMED GREENSTONE—METASEDIMENT BELTS (*STRIPED*) OF WESTERN AUSTRALIA. Part of a small shield of many separate plutons with intervening metavolcanic and metasedimentary rocks; the Pilbara block extends inland (eastward) from the Indian Ocean at Port Hedland. The circular area at top center is about 50 km across. The dark overlapping strata on right and left of it and in bottom quarter of image are Proterozoic. [*NASA Landsat image as processed by Division of Mineral Physics, CSIRO, Australia; image provided and publication authorized by Michael Horni-brook.*]

CHAPTER SEVEN

THE LATER ARCHEAN: MINIOCEANS, PROTOCONTINENTS, AND THE PHANTOM GLOBAL SEA

Earth's later Archean history spanned a time of seething mafic volcanism, emerging protocontinental crust, and high-level enrichment in precious metals. Buoyant masses of relatively light but viscous sialic magma (plutons) seem to have moved upward almost in tandem with sinking basins of heavier, mainly basaltic lavas and associated sediments in regular patterns of deformation. They appear now as domes of primarily sodium-rich granites and gray gneisses, surrounded by or engulfing deformed volcanic greenstones and metamorphosed sedimentary rocks.

Such somber terrains, spanning at their maximum a full aeon of geologic history, are typically Archean and everywhere broadly similar within the later Archean, along with extensive regions of preponderantly high-grade gneisses formed at deeper crustal levels. They are distinctive in their proportions of major elements, of rare-earth elements, of radiogenic versus nonradiogenic isotopes, and of metalliferous ores. Although similar rocks are found among younger sequences, only locally are their proportions so high. Archean geochemistry reflects that of the present lower mantle, in contrast to that of asthenospheric or higher levels. It is more oceanic than continental. The most recalcitrant problems have to do with the thickness and extent of sialic crust, the areal proportions of land and sea, and the prevalent modes of tectonism.

The Archean World

The later Archean (or Pilbaran) was like a hyperactive child—expending lots of energy with little clearly directed progress. It takes perspective to recognize that the foundation for all future development of our planet was there taking shape. Indeed the most notable characteristic of the preserved Archean rocks is their broad similarity from one place to another, a similarity that prevails over half an aeon or more of Earth history. Within that broad sameness, nonetheless, one discerns the progressive segregation of sialic crust, the concentration of precious metals, and, during the long transition from Archean to Proterozoic styles (left front endpaper), the roots of nearly everything else.

Alas, there were no dinosaurs, or even trilobites, to titillate the beginner's eye or pique the curiosity; no convincing evidence even of plate tectonics or other mind-gripping events beyond enormous sheets of flood basalts and the transfer to crustal levels of most of the world's accessible gold and radioactive elements. Archean things do not even offer the fascination of obvious rarity. For, although now but 5 percent of the visible continental crust, exposures of its rocks are not hard to find.

It is the rocks themselves that supply the challenge, with the profound but subtle questions they raise about Earth history and the insights there hidden about early environments, the origin of life, and the beginnings of biological diversity. Where was the sialic crust accumulating and how much was above water? How thick was that crust? Where, how deep, and how extensive were the oceans? By what processes did the planet dispose of its accumulated heat and the then still high additions from radioactivity? Such questions demand that one know a great deal about the rocks themselves before they can even be articulated in potentially fruitful ways, let alone answered.

What was it like in the Archean world, seen through the screen of folded and metamorphosed volcanic rocks and volcanic (*volcanogenic*) sedimentary strata within the broader terrain of sodium-rich (*sodic*) granites and gneisses (Fig. 7.1)? If one could travel backward in time (in a well-oxygenated space suit), the chances are one's landing site, if dry,

would overlook a volcanic foreground to an ocean view. Episodic seismic tremors would shake the ground beneath one's space boots as heaving magmas of different compositions stoped their way upward. No vegetation would have softened the view or assured an ample and steady supply of oxygen. No visible creature would have shared it. On a clear day one might have been able to see across the local pool or arm of ocean to its opposite volcanic shore, and perhaps to granitic uplands beyond. Not many days would have been that clear, however. During much of the time clouds of steam and dust from submarine eruption or from lavas entering the sea would have kept the atmosphere hazy.

To this haze, from time to time, a variety of sulfurous, chloridic, and nitrous volcanic gases would have been added, dwarfing the volumes of today's largest smelters and creating unparalleled and unalleviated conditions of smog and acid rain. An atmosphere of CO_2, H_2O, CO, N, and noxious trace gases would simultaneously have kept Earth's surface warm and its water fluid under the faint early Sun. That acid rain would have been an effective weathering agent. You would not have been tempted to remove your space helmet. There being nothing to hold the products of weathering and erosion in place to form stable soils, they would have washed directly downslope and into presumably yet-lifeless streams. Such debris would eventually have reached the narrow shelves of the marine embayments or engulfing oceans whose deformed floors one now sees as the predominantly volcanic greenstone belts so characteristic of Archean terranes.

The accumulation of sediments, consisting mainly of volcanic fragments, was occasionally dislodged from such shelves by one of the frequent lesser earthquakes of those times, thrown into suspension, and carried rapidly downslope. There it was dumped in calmer waters within or on top of the prevailing volcanic succession as layers of the impure volcanogenic sandstone and siltstone now known as graywacke. Graywacke layers, usually interbedded with pelagic (open sea) mudstone and shale, are commonly *graded*—coarser grains at the bottom, "fining" upward to gradational tops and then the sharp bottoms of following graded beds. The process and pattern are easily mimicked. Just dump quantities of mixed sand, silt, and clay into a big jar of water, shake it up, and, as it settles, observe the vertical sequence of textures; add another load and watch the sequence repeat. An oceanic setting for the greenstone sequences is demonstrated both by the common pillow-like stacking (Fig. 7.2*A*) so characteristic of basalts that have flowed into or erupted under water (*pillow basalts*) and by the associated, graded, volcanogenic graywacke layers (Fig. 7.2*B*).

Another feature of this interval of Earth history was probably a much shorter day than now. As noted earlier, the settling out of iron to form Earth's core would have set the planet spinning rapidly. It was the frictional drag of solar and lunar gravity over the following aeons that reduced Earth's rate of spin to the presently familiar 24-hour day, 365.25-day year. Shorter Archean heating–cooling cycles, then, would have affected weathering rates and probably chemical and biological evolution as well. Yet, as is customary, and salutary, whenever there is room for dissent in science, dissenters are to be found. The minority but

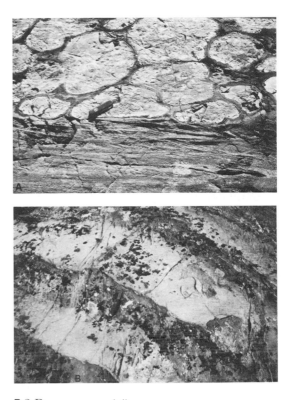

7.2 PILLOW LAVA (*A*) AND GRADED GRAYWACKE (*B*). Edge-on views of upended 2.7-aeon-old Archean rocks near Yellowknife, Great Slave Lake, Canada. The vesicular pillows of *A*, soft when extruded, conform to underlying surfaces. Their dark rims are devitrified glass, signifying abrupt quenching. They flatten against feldspar-rich cross-bedded sandstone beneath. The knife in *B* points up in the sedimentary sequence. Grading is denoted by the sharp bottoms and gradational tops of the strata.

improbable view (to be discussed) is that Earth's demonstrably slowing rate of rotation is at least partly explainable as a product of expansion since Archean time.

During the younger Archean also, a continuing process of recycling of cooling basaltic crust is envisaged. That process was likely one of engulfment and remelting within hot, viscous upper mantle, concurrent with gravitationally upward segregation of light sialic magmas as results of continuing partial melting within the mantle. Sialic mass was increasing even before there was enough of it to make conspicuously emergent continents.

Characteristic as the greenstone belts are, the equally distinctive sodic granites and gneisses of later Archean history may represent as much as 80 percent or more of the current Archean outcrop. In the end, granites, gneisses, greenstone belts, and the mainly volcanogenic sediments were all welded together locally to become one or a few continental masses. As implied by Figure 6.3, it seems likely (without clear evidence) that the growing protocontinental masses tended to aggregate in a few regions of the Earth, reflecting convectional patterns in the cooling outer mantle and the centrifugal effects of the rapidly rotating planet.

Few convincing signs of life existed on the Archean Earth. No animals trod its lands or negotiated its waters. Except for the mineralized or bacterial hues of local

hot springs and fumaroles there were only shades of gray, white, and black. Evanescent molecules of oxygen from photolytic dissociation of water and perhaps carbon dioxide at most places combined too quickly with reduced volcanic gases to oxidize and redden much of the ample, iron-rich, basaltic rock and ash.

No modern setting is quite like this. Southeast Iceland, with its basaltic lavas and short, broad, braided rivers flowing into the Norwegian Sea, is reminiscent; but it lacks a granitic core and stands in mid-ocean. Fiji, stripped of vegetation or other life above the simplest levels, might make a better comparison. The setting would not have seemed strange to any likely Archean creatures. Anaerobic bacteria would have found it cozy enough. Potentially damaging nascent O_2 from bacterial photosynthesis would have been neutralized by combination with sulfides, converting small fractions of them to sulfates.

Hostile as it would have seemed to us, the Archean world suited its microbial residents. It was also an essential prelude to those that followed.

Major Features of Later Archean History

The rocks tell us that the still-early Earth of later Archean times was a place of exuberant volcanism, continuing growth of sialic crust, evolving tectonic style, maturation of processes, and establishment of patterns for future crustal evolution. They record enrichments in gold, platinum, diamonds, and other minerals beyond the dreams of Croesus.

Important among the ways in which Archean history differed from that to

follow was its generation and loss of internal heat. Essentially all heat that could result from accretion, core formation, and decay of radioactive isotopes with significantly shorter half-lives than U-235 (Table 4.1) had already been generated when Archean history began. About half the heat that could ever come from U-235 had also been released. The rate of heat flow from Earth's interior has necessarily diminished since then, and that cooling history may have been the main event of Archean time—accounting not only for the long transition from Archean to Proterozoic characteristics, but even perhaps for the onset of plate tectonism.

Remnants of later Archean terrains today are somber regions of dark greenstones, gray gneisses, and sodic granites in which sedimentary rocks are consistently subordinate, generally late in the sequence, and locally derived. Although the total area over which Archean rocks are now exposed (Fig. 6.3) represents only a very small fraction of Earth's present continental surface, its subsurface mass may have been much greater. Certainly the volume of sialic rock continued to grow over the course of Archean history, and large additions to it occurred during the long transition from Archean to Proterozoic (front endpaper)—a time of massive granitic intrusion, by then increasingly potassium rich (*potassic*).

The gold-bearing Archean greenstones are also known for their high concentrations of other iron-loving (*siderophile*) elements—nickel, cobalt, and the platinum group of metals. The primary concentrations of those metals in Archean rocks links Archean geochemistry with that of meteorites and Earth's deeper mantle.

How good is the evidence for the widely held view that Archean crust was hot and thin? Not as good as it once looked, I fear. Major conflict between theory and evidence clouds the problems of early crustal thickness and rates of heat flow from interior to the surface. Theory, on the thin-crust side, is supported by the many early heat sources, some mineral deposits, geochemistry, and aspects of metamorphic mineralogy. Major ore bodies of this age include massive, high-temperature (350°C) copper–zinc deposits in layered volcanic rocks, gold-enriched intrusive veins, and high-temperature intrusive copper–nickel deposits.

Conspicuous, moreover, among the parent lavas of Archean volcanic rocks was an unusual ultramafic basalt, with a high melting point and a mantle mineralogy. Such rocks, *komatiites,* almost limited to the Archean, commonly show the pillowed structures indicative of submarine extrusion. Their textures are conspicuous for radiating clusters of a dark prismatic mineral (olivene or pyroxene) indicative of rapid early crystallization on cooling from outer-mantle temperatures of around 1,650°C. We see among Archean rocks broadly a prevalence of thermal metamorphism at a wide range of temperatures. Granulites are everywhere. They are laminated but mica-free metamorphic products generally considered indicative of dry, hot metamorphism (Fig. 6.4), but lacking high-pressure minerals.

On the other hand, this once widely accepted idea of a hot, thin crust for much of Archean time is vigorously opposed on the same and other evidence. Many find the hot, dry conditions of granulite metamorphism to be convincing evidence of burial to great depths, despite the absence of high-pressure minerals. Respected geologists report Archean crustal thicknesses in the range of 30 to 40 kilometers. If some kind of crust existed or accreted beneath stratified volcanic-sedimentary sequences locally reported to be as thick as 30 kilometers, a total thickness upward of 40 kilometers by the close of Archean history would seem likely. That is, unless apparent thickness were a product of later tectonic repetition. Extensive exposures of Archean granulites also require the subsequent erosional removal of the crust beneath which they formed. Even diamonds are called upon as evidence for great crustal thickness 2.2 to 2.3 aeons ago in Africa, although no one has explained why they could not be products of shock metamorphism from asteroidal impact.

What are the limits of early crustal thickness? Is there an irresolvable conflict between seemingly sound theory and reliable observation? Before one can even think constructively about such problems, plausible estimates of limiting factors are needed. What was the depth to the asthenosphere? How much, if any, solid upper mantle lay between asthenosphere and crust and what was its density? Could limited areas of crust have had roots that extended deeply into the asthenosphere without being reassimilated? Could such roots have reached depths as great as the 150 to 200 kilometers hypothesized by the Carnegie Institution's Francis Boyd and associates to explain the somewhat younger Kimberley diamonds (or might they be impact phenomena)?

Such problems are related to the mechanisms of plate tectonics. Those mechanisms could not have operated very well with scattered crustal roots extending deeply into the asthenosphere. What would hold a light sialic root at depths against the will of gravity? Physical constraints tell us that plate tectonism in the classic sense should not begin until oceanic lithosphere has cooled, thickened, and become dense enough to descend into the asthenosphere, dragging ocean floor behind it. Archean his-

tory seems to have been working up to that.

Turning ignorance into an asset, then, one might hypothesize that the long transition from the Archean to the Proterozoic mode may record planetary evolution from an internal rate of heat flow perhaps 3 times that now experienced to one approaching the present. During those transitional geocenturies (front endpaper), Earth's crust seems to have changed from a scattering of small, perhaps thick protocontinents to fewer but more extensive sialic masses of acceptably continental proportions (Fig. 6.3). Their surfaces thereafter became the main depositories for Proterozoic epicontinental sedimentation, marine and nonmarine alike. Meanwhile, what may have been a primarily convectional cooling system of numerous small cells with a prominent vertical component became one where lithosphere began to glide horizontally across the yielding upper surface of the asthenosphere, initiating a conventional style of plate tectonics.

If one accepts the findings of paleomagnetism, motions of that nature had begun by about 2.8 aeons ago. Evidence favoring such a conclusion is shown in Figure 7.3, where apparent paleopole positions are indicated for 2.8- to 2.7-aeon-old samples from the Canadian Shield and northern Idaho. They lie near today's equator, west from and in central Africa. A geocentury later the apparent paleopole was at today's 40°N, and 2 geocenturies from then the apparent paleopole was 60°N and more. Recall that this "apparent polar wandering" really tracks the wandering of continents, not of the poles themselves.

In truth, granite–greenstone complexes and high-grade gray gneisses are prevalent to the verge of monotony. The Archean crust does vary notably in prevailing rock type. Over large regions high-grade gneisses are so extensive that it is as if Isuan types of rocks simply extended upward. At a few places within the greenstone belts sedimentary strata outbulk volcanics. Such sequences are characterized by volcanogenic graywackes, dumped abruptly downslope from narrow coasts to become interbedded with slowly settling pelagic gray shales or mudstones (Fig. 7.2*B*). Along their landward margins are river-borne sandstones and conglomerates of more modern aspect, and locally small amounts of dolomite. The graywacke–shale sequences also are similar to the *flysch* sequences of younger Alpine regions, where they are associated with deepening seas and the growth of collisional, folded, and overthrust mountain belts.

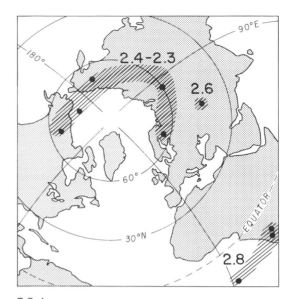

7.3 APPARENT PALEOMAGNETIC POLE POSITIONS FOR NORTH AMERICA 2.8 TO 2.3 AEONS AGO. Black dots indicate calculated paleopole positions. Error circles omitted. [*Simplified after E. Irving, 1979,* Canadian Journal of Earth Science, *v. 16, no. 3, Fig. 11, p. 683.*]

General Features of the Preserved Younger Archean Crust

The mafic basalts of the Archean greenstone belts are geochemically and mineralogically similar to those of modern oceans. Think of them as representing Archean proto-oceans or subsiding regions of oceanic geochemistry. They differ from sialic continent-building granites and gneisses, although they are now parts of the continents.

Crust-building processes, including plate tectonism, are strongly related to differences in melting temperatures and density of rocks. Although elemental iron has a substantially lower melting temperature than quartz, the fluxing effects

of bases such as calcium result in sialic melts at temperatures as low as 600 to 700°C. The density of average sialic rocks is about 2.8, compared to water with a density of 1, ordinary basalt at about 3, and outer mantle ultramafics at a density of around 3.3. Sialic melts, therefore, move upward to become parts of the more elevated continental crust, and (excepting small amounts entrained by subduction) they remain at Earth's surface after solidification. Oceans are floored with basalts, afloat on still denser outer mantle. Cooling from the molten state plus pressure as a result of burial converts sea-floor basalts to densities approaching or exceeding those of the outer mantle. They can then descend into it and even to the asthenosphere, where they may undergo partial melting and fractionation to yield a buoyant sialic fraction that rises and mixes with other sialic crustal components. At least those are aspects of how plate tectonics seems to have operated during the Phanerozoic.

As earlier mentioned, the ubiquitous granite–greenstone–metasedimentary belts of younger Archean history are broadly similar wherever found. They are, in fact, similar enough from one region to another (Fig. 7.4) that sequences of widely varying age have, in the past,

7.4 SCHEMATIZED MAPS OF GRANITE–GREENSTONE TERRAINS ON FOUR CONTINENTS AT ROUGHLY THE SAME SCALE. The idealized concentric pattern represents granitic and gneiss masses, stippled where potassium exceeds sodium. Shading denotes metamorphosed basalts (greenstones) and associated metasediments. Small open circles denote metasediments outside the main greenstone belts. Slanting lines mark post-Archean rocks. Paired, circled numbers *1–5* in *D* identify continuous east–west trending belts. Note that lower right quarter of *B* corresponds with Fig. 7.1. [A, *adapted from L. J. Salop, 1983*, Geological Evolution of the Earth During the Precambrian, *Springer-Verlag, p. 33, Fig. 2*; B, *adapted from A. H. Hickman, 1980, map of Pilbara Block, Geological Survey of Western Australia, Bull. 127, Plate 1*; C, *adapted from J. F. Wilson, 1973, Philosophical Transactions of the Royal Society of London, A273, Fig. 2, facing p. 408*; D, *adapted from F. J. Pettijohn, 1972, Geological Society of America, Memoir 135, Fig. 2.*]

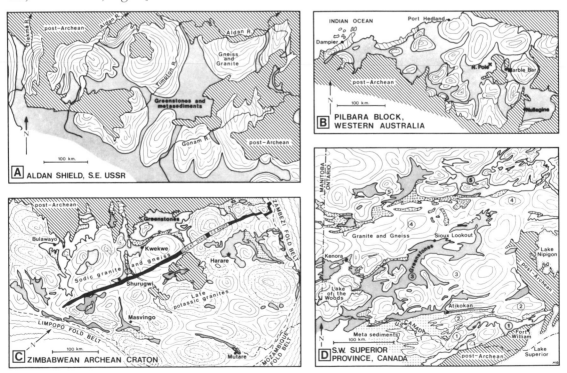

been considered to be time equivalents. In fact, younger Archean history, from about 3.5 aeons BP onward, is still very inadequately understood. The structure and stratigraphy of its rocks is complicated. Exposures are limited. And it is only since the early 1970s that reliable geochronologic information on time of origin as opposed to metamorphism has begun to be widely available. Surprises are to be expected still.

The areally largest greenstone belts are those of the Canadian Shield and its extensions into the United States (Fig. 7.4D), representing mainly rocks 2.75 to perhaps(?) 2.5 aeons old but with some as old as 3.2 aeons. A small, very old, classic, and not yet fully understood granite–greenstone belt is that of the Barberton Mountain Land in Swaziland and adjoining South Africa (Fig. 7.5A). But the most complete known sequence of greenstones and metasedimentary rocks is that of the Pilbara block of Western Australia (Figs. 7.1, 7.4B, 7.5B). The Pilbaran sequence, therefore, is here chosen as the most appropriate global reference standard for later Archean history, even though that sequence is as yet less widely known than others and may involve some structural duplication.

Distinctive though the granite–greenstone terrains are, younger Archean rocks globally are of two main modes. One is the often weakly metamorphosed granite–greenstone belts that dominate in the Archean of North America, southern Africa, Western Australia, central India, and the Aldan Shield of the far-eastern Soviet Union. The second comprises the intensely metamorphosed, and often somewhat older, sialic gneisses and granulites (with local metasediments) such as prevail in Greenland, Scandina-

via, parts of the Soviet Union, central and northern Africa, and Antarctica. Both modes are characterized geochemically by a prevalence of sodium over potassium and a relative depletion of rare earths and other large-ion-lithophile (LIL) elements as compared to post-Archean rocks (Fig. 7.6). Both record similar low initial ratios of radiogenic to nonradiogenic isotopes of the same elements (e.g., Fig. 4.8). The two modes probably reflect origin and metamorphism at higher (greenstones) and lower (granulites) crustal levels respectively; and they locally grade into one another through zones of intense deformation. They are found in all or nearly all old shield areas, worldwide.

A typical greenstone–metasediment sequence might consist of 40 to 50 percent pillowed, mafic, and ultramafic lava flows and intrusive rocks, about 25 percent intermediate flows and igneous debris, and a lesser volume of mainly sialic volcanic rocks dominated by airborne components. Capping such sequences may be as much as 10 to 20 percent of varied basinal sediments with a large volcanogenic component.

The geochemistry of the basalts implies derivation from a primitive upper mantle that has undergone little major-element geochemical differentiation since Isuan time. The abundance of pillow lavas, associated graded graywackes, and the extent of both combine with geochemistry to certify oceanic affinities. And where pillow lavas lack the vesicular rinds indicative of outgassing (Fig. 7.2A), they are judged to have been extruded at subsea pressures, thus depths, too great for volatilization of low-melting components.

Representative cumulative thicknesses of such sequences have been estimated

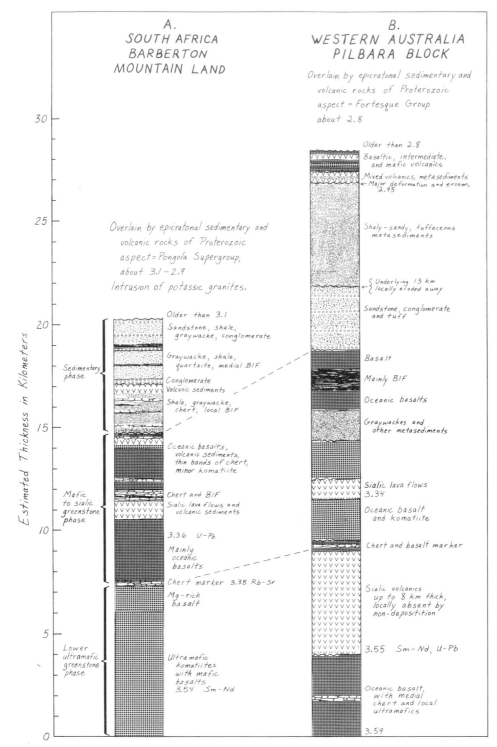

7.5 MOLE'S EYE VIEW OF THE VERTICAL SEQUENCE OF UPPER ARCHEAN ROCKS IN GREENSTONE—
METASEDIMENTARY BELTS OF SOUTHERN AFRICA COMPARED WITH WESTERN AUSTRALIA. Dashed
lines suggest inferred historical equivalence. Numerals at right of columns denote ages in
aeons BP, followed by dating method where known.

152

as 7 to 14 kilometers in Canada, perhaps 20 kilometers in South Africa, and 18 to 30 kilometers in the Yilgarn and Pilbara blocks of Western Australia, where they span a range of ages from 3.6 to 2.8 aeons ago. Their accumulation, however, by no means occupied the entire time represented. Intervals of nondeposition and erosion are represented by bedding surfaces and unconformities. And, in Canada, separate greenstone belts show repeated similar sets of rocks that become less mafic and more sialic upward, both within sets and upward within the sequence as a whole.

To sum up, supracrustal rocks of Pilbaran type, wherever found, are likely to comprise repeated similar sequences that start below with ultramafic to mafic and pillowed volcanics of oceanic geochemistry and grade upward to increasingly sialic rocks. They commonly terminate with graded graywackes of mafic volcanic affinities and other immature sedimentary rocks, then repeat the sequence, becoming a bit more sialic with each repetition.

Conspicuous subsidence during deposition in many such greenstone–graywacke terrains is implied by published vertical profiles that show huge thicknesses of tightly folded and faulted, near-vertical layers in linear downbuckled belts. Such volcanic-sedimentary downwarps are commonly associated with granitic intrusives that, at places, have shed dateable debris into nearly contemporaneous sediments as they ascended and became unroofed, concurrently with the subsidence of adjacent greenstone basins. Because of this, Archean tectonism was once generally thought to be primarily vertical, involving little horizontal compression or displacement other

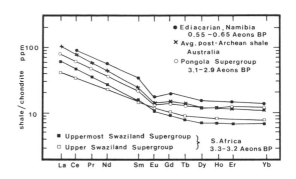

7.6 RELATIVE ABUNDANCES OF RARE-EARTH ELEMENTS IN AVERAGE ARCHEAN SHALES COMPARED WITH POST-ARCHEAN SHALES. The scale is in parts per million. The downward nick in the top three curves, indicating europium depletion, and the higher levels of the three upper compared to the two lower curves distinguish post-Archean (*above*) from Archean (*below*). [*Adapted with minor modification from S. M. McLennan, S. R. Taylor, and A. Kröner, 1983,* Precambrian Research, *v. 22 (1–2), composite of Figs. 2–4, 9, 11. © Elsevier Science Publishers B. V.]*

than local crustal shortening. Such an interpretation is still widespread, is accepted by many as the prevailing tectonic mode for Pilbaran rocks, and clearly involves some elements of reality. Right or wrong, it is now being challenged from several quarters. Evidence for large lateral motions, recumbent folding, and structural repetition of Archean sequences is reported from widely separated regions. Such motions, if extensive and verifiable, would suggest that reported thicknesses of 20 to 30 kilometers of layered Archean rocks could in part result from tectonic repetition—one more of the Archean Earth's inviting supply of yet-unresolved problems.

153

Origin of the Granite—Greenstone Terrains

There must be an explanation for the distinctive map pattern of granite–greenstone terrains. Do the granites that surround and seem to engulf the greenstones represent basement surfaces on which the latter were basin fills? Did they intrude the greenstones from beneath? Or were they mainly contemporaneous with the greenstones, the latter foundering as the granites stoped their way upward?

In fact the Archean granites and gneisses, for all their broad similarities, are a heterogeneous lot. The gneisses are products of metamorphism, primarily thermal. Some resulted from recrystallization of preexisting granites at depth or were layered sialic successions of sediments and acid volcanics before metamorphism. So many of the granites have been found to show mantle ratios of strontium-87 to strontium-86 that their direct crystallization from rising mantle melts in the lower crust is considered likely. Some such apparently primary granites have been found to show nearly concordant contacts with adjacent greenstone–graywacke sequences. At other places sedimentary cover was deposited directly on eroded granite surfaces (Zimbabwe). At still others the granites appear to intrude their greenstone–metasedimentary cover. And locally (northeastern Minnesota), pebbles and fragments of similar granites in associated sediments yield radiometric ages that are indistinguishable from adjacent granitic sources.

One does not need to be a geologist to realize that these variations imply a complementarity between rising relatively light granitic masses or *plutons* and sagging relatively heavy greenstone basins—remnants perhaps in many instances of once much more extensive flood basalts.

As the granitic plutons moved haltingly upward, overlying rocks were eroded. The granites were unroofed. And further erosion shed granitic debris onto the margins of greenstone basins to create partially time equivalent, terminal detrital sequences. All alternatives considered were probably to some extent realized. Some of the granitic and gneissic rocks are older than, some younger than, and some contemporaneous with the development of the greenstone belts. But none of that really explains the roughly quadrate pattern, which looks ever so much like that of intersecting fold belts (Fig. 7.4). Intersecting fold belts doubtless were involved at some point, but there is the question of when and how. Probably sedimentation in the greenstone–metasediment belts and erosion of the granitic domes continued in tandem as folding progressed.

A likely progression might be intersecting folds to initiate and continue the structural pattern, penetration of volcanic and sedimentary overburden by rising granitic plutons, and continuing sedimentation in the basins from both mafic volcanic and sialic erosional sources. Different combinations of such processes could account for the differences among the granite–greenstone–metasedimentary belts.

A Representative Upper Archean Sequence:
The Pilbaran of Western Australia

By all odds the most complete sequence of upper Archean rocks known is the 3.6- to 2.8-aeon-old volcanic-sedimentary sequence of the granite–greenstone belts in the Pilbara block of Western Australia (Fig. 7.1, 7.4*B*, and 7.5*B*). The sequence there attains an estimated cumulative thickness of 25 to 30 kilometers of layered volcanics and sedimentary rocks, stretching across an area of some 60,000 square kilometers.

Excepting flies, spinifex grass, and a breathable atmosphere, the Pilbara region has hardly changed since the last Archean sea withdrew and the folded metasediments were beveled by early Proterozoic erosion (Figs. 7.7, 7.8). Hot and dry most of the year, impassable during the rains, and remote, the region's main attraction is solitude, a kind of harsh beauty, and, of course, the unusually complete Archean sequence. Even geologists were scarce there before the early 1970s, when attention focused on Western Australia's mineral wealth and its impressive records of Archean and older Proterozoic history. From this has come appreciation of the fact that this huge pile of Pilbaran rocks, comprising some twenty formal stratigraphic formations in three groups, presents an excellent global reference standard for later Archean history (Fig. 7.5*B*).

Pilbaran greenstones and associated metasedimentary rocks once extended as a continuous thick cover over the entire Pilbara region before being segmented and shouldered aside by rising granitic domes (Fig. 7.1). That is demonstrated by the similarity between sequences of different, now separate, greenstone–metasediment associations within the region and the continuity across it of a distinctive chert–basalt marker zone—a

7.7 Upended upper Archean volcanogenic metasediments east of Nullagine, Pilbara region, Western Australia.

1,000-meter-thick repetition of low-grade banded iron formation and interbedded strata. The same is true of an equally distinctive 5,000-meter zone of basalts and intermediate sialic volcanics. Having been initiated perhaps by cross-folding, vertical ascent of mainly sodic but increasingly potassic domal sialic intrusions followed at some unknown intervals. Somewhere in this sequence came a great accumulation of flood basalts averaging 11 kilometers thick over the entire region, a total of about 660,000 cubic kilometers of basalt and ultramafic rocks. The final episode of deformation—with rising sialic plutons, tight compressive folding of the layered sediments in intervening basins (Fig. 7.1), and metamorphism—is dated as having occurred just under 3 aeons ago. That was followed by erosional leveling, more volcanism locally, and sedimentation up to the final erosional leveling that preceded the Proterozoic marine inunda-

7.8 FOLDED, ERODED, AND EXHUMED LATE ARCHEAN SURFACE (A) AND THE 2.8-AEON-OLD SANDS OF A STREAM THAT ONCE FLOWED ACROSS IT (B). Outcrop B is at the center foreground of A. Patches of similar cross-bedded sandstone allow the direction of flow of the old stream to be traced to the left (northeast) along the foreground slope of A, across the dry creek at the tall trees, then out of sight to the left of the low ridge across the road. The exhumed Archean surface of A bevels the uppermost Mosquito Creek sediments illustrated in Fig. 7.7.

tion of the barren, gently rolling terminal Archean surface some 2.8 aeons ago (Fig. 7.8).

Even though the total succession locally approaches a thickness of 30 kilometers, it is clear that there never was a depositional basin anything like that deep. Instead, the accumulation of volcanic and sedimentary layers almost kept pace with subsidence and may well have been repeated by gravitational slumping and compression. The prevalence of pillow basalts indicates submarine deposition, and other sedimentary structures imply episodically deep and shallow waters. The presence of both silica and barium sulfate replacements of gypsum (calcium sulphate) are evidence of local ponding, hypersalinity, and evaporation.

Despite the evidence that growth of sialic magma was taking place beneath older mafic crust and eventually invaded it, the primary sediments are volcanogenic—the product of erosional reworking of older volcanic rocks and the redistribution of contemporary volcanic sediments. The thickness of sialic crust by the end of this restless Pilbaran history is estimated by A. H. Hickman to have reached about 30 kilometers (not allowing for tectonic repetition).

A dramatic feature of Pilbaran history in its type region is that large areas there, plus a good part of the 600,000-square-kilometer and nearly flat Yilgarn block to the south, may comprise an exhumed erosion surface or surfaces of terminal Archean or earliest Proterozoic origin. The sparseness of vegetation in Figure 7.8 enhances the illusion that one is looking across a surface virtually unchanged from erosional leveling about 2.8 aeons ago.

Brief Notes on Two African Classics

On another continent, in another region where men have moiled and died for gold, the Barberton Mountain Land of the eastern Transvaal and Swaziland, has become a mini-classic and a mecca for students of granite–greenstone terrains. Incomplete by Australian standards and small compared with other well-known regions, its rocks are nevertheless well exposed, little metamorphosed, and accessible. Three generations of South African geologists and comparisons with historically equivalent sequences on other continents reinforce the idea of an Archean paradigm.

The complexly deformed deposits of the Barberton region comprise a dozen named geological formations in three groups, together comprising the Swaziland Supergroup. The salient features of this reputedly 20-kilometer-thick pile of volcanic and sedimentary rocks are summarized in Figures 7.5A, 7.9, and 7.10. Figure 7.5A is a mole's-eye impression of how this sequence might look if it could all be piled up in stratigraphic order from oldest at the bottom to youngest at the top, assuming any mole could burrow so deeply, and provided there is no tectonic repetition.

A classic reconstruction of the evolution of the Barberton greenstone belt is portrayed in Figure 7.10 (not to scale). That reviews its development as if in a set of vertical slices through the crust at successive intervals before the present.

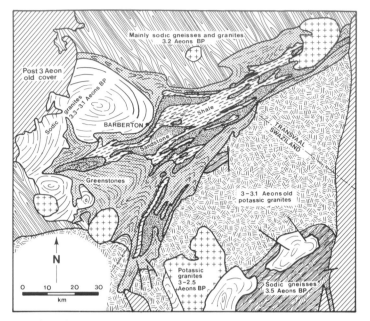

7.9 SIMPLIFIED GEOLOGIC MAP OF THE BARBERTON MOUNTAIN LAND. Within the wedge-shaped down-folded sedimentary basin light stipples outline the lower Onverwacht greenstone phase of the Swaziland Supergroup. Slanted dashed lines denote the overlying shaly Figtree Group. And the dark stipples mark the sandstone and conglomerate beds of the capping Moodies Group, all as generalized in Fig. 7.5A. Wavy lines indicate major structural trends. [Modified after diagrams in C. R. Anhaeusser, 1971, Geological Society of Australia, Special Publication No. 3, Figs. 1–2, pp. 204, 207.]

7.10 HYPOTHETICAL VERTICAL PROFILES THROUGH THE BARBERTON GREENSTONE BELT SUGGESTING A COMMON HYPOTHESIS FOR ITS PROGRESSIVE DEVELOPMENT FROM INITIAL SUBSIDENCE (A) TO THE PRESENT (D). Not to scale. [Modified after C. R. Anhaeusser, 1971, Geological Society of Australia, Special Publication No. 3, Fig. 3, p. 113.]

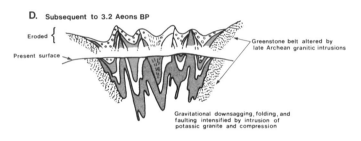

D. Subsequent to 3.2 Aeons BP

Eroded

Present surface

Greenstone belt altered by late Archean granitic intrusions

Gravitational downsagging, folding, and faulting intensified by intrusion of potassic granite and compression

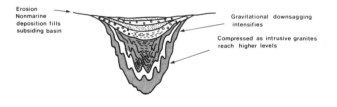

C. 3.3–3.2 Aeons BP   Sandstone–Conglomerate Phase

Erosion
Nonmarine deposition fills subsiding basin

Gravitational downsagging intensifies

Compressed as intrusive granites reach higher levels

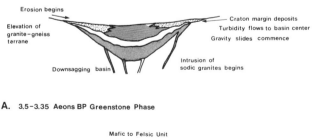

B. 3.35–3.3 Aeons BP Shaly Sedimentary Phase

Erosion begins

Elevation of granite–gneiss terrane

Craton margin deposits
Turbidity flows to basin center
Gravity slides commence

Downsagging basin

Intrusion of sodic granites begins

A. 3.5–3.35 Aeons BP Greenstone Phase

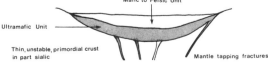

Mafic to Felsic Unit

Ultramafic Unit

Thin, unstable, primordial crust in part sialic

Mantle tapping fractures

Recent remapping of the region, how-ever, challenges that interpretation in favor of one involving much recumbent folding, flat thrust faulting, and struc-tural repetition of sequence during col-lisional plate tectonism. Whichever be closer to the truth, Pilbaran history in the Barberton region was terminated by extensive potassic granitization having a reported rubidium–strontium age of 3.07 aeons BP.

In every way that we can separate Archean from Proterozoic rocks by visi-ble characteristics, sedimentary deposits that overlie those granites are Protero-zoic. They are relatively mature sedi-ments with comparatively high ratios of potassium to sodium. They are enriched in radioactive silicate-loving elements of large ionic radius. And, as shown in Figure 7.6, their rare-earth abundances cluster with the Proterozoic pattern.

A second African classic lies north-ward from the Barberton Mountain Land, just across the great, green, greasy, hip-popotamus heaven called Limpopo River, in Zimbabwe (Fig. 7.4C). They wrap around the in-part contemporaneously ascending granitic intrusions immortal-ized as "gregarious plutons" by the deservedly famous pioneer geologist A.

M. Macgregor in 1947. The main histor-ical events of that classic sequence are summarized in Figure 7.11, a hypothet-ical vertical profile of the Archean crus-tal sequence in Zimbabwe. Incomplete but broadly similar to that at Barberton, this sequence covers a much longer his-torical span. The Zimbabwean green-stones and metasediments range in age from more than 3.44 aeons BP at the base to perhaps 2.6 aeons BP at the top, terminating with the intrusion of 2.6- to 2.7-aeon-old potassic granites.

All of these rocks were cut through 2.5 aeons ago, as if by a great cleaver from below, when the 500-kilometer-long Great Dyke intruded, dating everything penetrated as older. Would that all time markers were so specific and so sweep-ing! How continuous the original green-stone floor of this region may have been and what underlay it is unknown. The oldest granites cut the oldest green-stones, but, at one locality, 2.7-aeon-old sedimentary rocks of the Bulawayan Group are reported to rest directly on the eroded surface of 3.4-aeon-old gran-ites. At least locally, therefore, young metasediments of the greenstone belts were deposited on much older sialic crust.

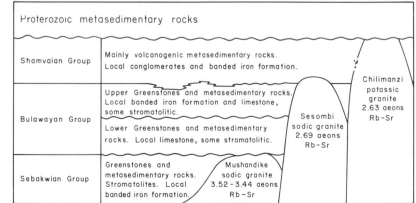

7.11 GENERALIZED VERTICAL STRATI-GRAPHIC SEQUENCE OF ZIMBABWE ARCHEAN CRATON. Not to scale.

Archean of the Canadian Shield

Most of the visible Archean in North America is in the Canadian Shield, including its extensions into the United States. Rocks of the granite–greenstone facies there appear later (2.9 to 3 aeons ago) and are reputed to last longer than those of other well-known Cryptozoic (pre-Phanerozoic) sequences.

In this complicated and often harsh terrain—inspiration alike for the hardy pioneers of Archean geology and the impressionistic artistry of Canada's beloved Group of Seven—a new generation of process-oriented geologists seeks to unravel its remotest history. They have found that distinct, bilaterally symmetrical, depositional greenstone basins can be recognized, their approximate

shorelines can be delimited, and contemporary islands and sites of volcanism can be identified by means of careful mapping of current directions, sediment-dispersal patterns, and otherwise discriminating sedimentology and sedimentary petrology. Despite the broad global similarities, one does seem to find differences in paleogeographic and tectonic style.

The pattern of the described beltlike distribution is shown in Figure 7.4D. Alternating greenstone belts there are marked by pairs of circled odd numbers (1, 3, 5) and granite–gneiss belts by even numbers (2, 4, 6). The even more generalized Figure 7.12 emphasizes the broadly linear style of these belts. They run right

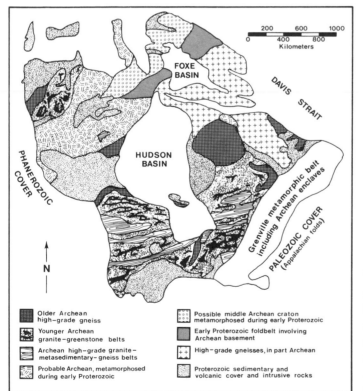

7.12 MAJOR GEOLOGIC PROVINCES OF THE ARCHEAN CANADIAN SHIELD. [Simplified after P. F. Hoffman, K. D. Card, and A. Davidson, 1982, Geological Society of America, Decade of North American Geology, Special Publication 1, Fig. 1, p. 4.]

across the Canadian Shield, disappearing under Phanerozoic cover to the southwest and abutting Grenville metamorphism to the northeast. Whereas most of the greenstone belts seem to be around 2.7 aeons old, ages as great as 2.9 to 3 aeons are known. An historical gap separates that sequence from 3.2-aeon-old gneisses elsewhere and from the Isuan gneisses of the Labrador coast. Because many of the new ages are by the lead-isotope method on stable zircons they are no longer likely to change significantly. Earlier reported Rb–Sr and K–Ar ages probably date metamorphic events.

After a century and a half of active inquiry and recent decades of unprecedented progress, one might think that understanding of the Canadian Shield would leave little to be desired. Reliable determinations of the time of origin as opposed to that of metamorphism, however, have until recently been sparse. Now that the whole region is being restudied by a group from the Geological Survey of Canada for a project called the Decade of North American Geology, that

will change. Pending the results of those studies, one visualizes the region as a cluster of minioceans and protocontinents, surrounded by a maxiocean somewhere beyond. Two classic and widely separated sequences exemplify the region. One is the Abitibi greenstone belt in the south, extending from Ontario eastward into Quebec and southwestward into Minnesota (*belt 1* of Fig. 7.4*D*). Second is the Yellowknife Archean province of Great Slave Lake, District of Mackenzie, far to the northwest.

A typical Abitibi sequence might consist of mafic to felsic, pillowed, volcanic rocks and associated volcanogenic sediments 10 to 14 kilometers thick, giving way upward to stream deposits of sands and gravels 3 to 3.5 kilometers thick along narrow protocontinental margins (Fig. 7.13). Directions of sediment transport imply island building in the same setting, along with recurrent subsidence, and the offshore sedimentation of shales and graded graywackes, representing relatively deep-water basins and corresponding to the earlier described flysch facies of tectonically active regions.

7.13 UPSTREAM IMBRICATION OF FLAT PEBBLES IN 2.7-AEON-OLD STREAM DEPOSIT AT KIRKLAND LAKE, ONTARIO. Current flowed to right.

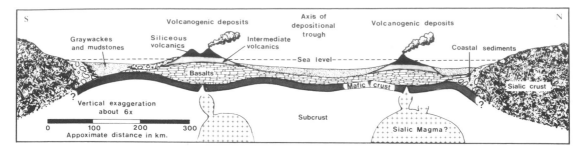

7.14 HYPOTHETICAL VERTICAL PROFILE OF ABITIBI GREENSTONE BELT, SOUTHERN ONTARIO AND SOUTHWESTERN QUEBEC. Suggested appearance before subsidence and compression; original width about 800 km. [*Extensively modified and elaborated after A. M. Goodwin and R. H. Ridler, 1970, Geological Survey of Canada, Paper 70-40, p. 20, Fig. 8. Reproduced with permission of the Minister of Supply and Services Canada.*]

A hypothetical vertical profile across this broad Abitibi depression before final granitic intrusion, uplift, and folding is shown in Figure 7.14. During its depositional history it may have been more equidimensional than beltlike. It might even have experienced a kind of protoplate-tectonic overriding of miniocean floor by opposing sialic protocontinents.

The margins of such basins in southern Canada are commonly characterized by local cherty iron formation, sialic volcanism, and subangular conglomerates; their interiors by basalt, sulfide iron formation, graywacke—mudstone sequences, and an abundance of sodic granite. An interesting question is: How could these greenstone sequences have been so similar to one another if they were deposited in isolated basins, now deformed into the patterns of Figure 7.4*D*? Their similarity implies similar sources and a high degree of interconnectedness. But, like other granite—greenstone complexes, their origin and deformational history remains disputed.

Pilbaran kinds of rocks at subarctic Yellowknife, goldmining center for Canada's 190,000-square-kilometer Slave Province, are full of conspicuous historical signals. Here, along the north shore of Yellowknife Lake, where woodlands give way to tundra and the howl of the wolf may still be heard, are great extensions of bare rock in a favorable setting for study—created by smelter herbicides, to give the devil his due. These superb exposures and the bonanza gold deposits of the region have made this one of the better-known Archean sequences in Canada. Much of it is accessible from the relatively modern city of Yellowknife, grown from the rustic mining camp of a half-century ago to an industrial enclave in the wilderness. These outcrops afford a classic instance of reading history and environment from the rocks at the very end of the Archean to Proterozoic transition.

Tightly folded 2.7-aeon-old rocks of the Yellowknife Supergroup start on the west of Yellowknife Bay with an estimated thickness of about 9 kilometers of mafic to intermediate volcanic rocks. Pillow basalts in this sequence, some vesicular (Fig. 7.2*A*), indicate underwater extrusion at moderate to shallow depth. They are thinly overlain in eroded depressions by locally derived sialic conglomerates and volcanics up to a hundred or more meters thick. The basalts are overlain by and apparently grade into a relatively deep-water assemblage of graded graywackes and mudstones that reaches a thickness of about 4.5 kilometers on the east of Yellowknife Bay (Fig. 7.2B). An eastward-thinning pile of intermediate to sialic volcanic ejecta and flows interrupts this sedimentary sequence, indicating temporary emergence. Renewed subsidence is then manifested by a continuation of graywacke–mudstone sedimentation, terminating the local Archean record. From such observations and the paleothermometry of mineral phases such as komatiites we read the pulse and temperature of the evolving planetary surface at a given time and place.

A favored interpretation of the Archean setting at Yellowknife is shown in Figure 7.15. Sialic basement rocks 3 to 2.7 aeons old on the left (west) were separated from a subsiding marine basin on the right by an arcuate chain of volcanoes. A sialic subcrust is implied by the local observation of greenstones in apparent depositional contact with underlying granites. Subsea eruptions of mafic to intermediate lavas ascended through fissures in the crust, loading the basin with volcanic rock and accelerating subsidence. Particulate matter from explosive eruption thickened the growing volcanic pile. Once it reached sea level, wave action and other weathering and dispersive processes generated local gravel deposits. As in other regions where the evidence is inconclusive, however, the evidence is inconclusive, however, plate tectonism now calls for reexamination.

7.15 SUGGESTED LATE-DEPOSITIONAL ENVIRONMENT OF YELLOWKNIFE SUPERGROUP, NORTH ARM OF GREAT SLAVE LAKE, DISTRICT OF MACKENZIE, CANADA. Scene pictured represents a time just before final intrusion and tectonism. Not to scale. [Extensively modified after J. C. McGlynn and J. B. Henderson, 1970, Geological Survey of Canada, Paper 70-40, p. 42, Fig. 1. Reproduced with permission of the Minister of Supply and Services Canada.]

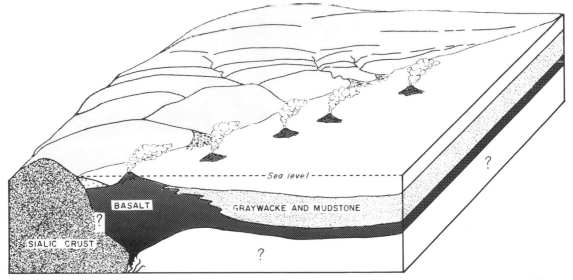

After the last suspension current left its calling card in the Yellowknife region in the form of a graded graywacke, we again lose the spoor of Archean events, this time for good. That history was terminated by the intrusion of 2.65- to 2.5-aeon-old potassic granites. Later granitic intrusion and compression, associated with crustal deformation called the Kenoran orogeny, conveniently squeezed the estimated 14 or more kilometers of volcanic rocks and sediments of the Yellowknife Supergroup into tight folds. Thanks to the erosion of their upturned ends it is possible to examine maximal thicknesses with minimal trudging. But one must traverse a brushy, boggy terrain, while being on guard against repetition of strata.

Later Archean Paleoenvironments and the Early Biosphere

Though it be the weakest of the four forces physicists seek to unify, the importance of gravity as a geologic agent cannot be overemphasized. It is a basic arbiter of all kinds of things from crustal evolution to surface processes. Nowhere was it more important than in the recycling involved in the growth of protocontinental mass and greenstone basins during later Archean history. The evolving granite–greenstone terrains may look so much alike wherever we see them because they underwent similar evolutions in unstable settings dominated by gravitational forces.

In the Pilbara block, the Zimbabwe craton, and the Canadian Shield we see evidence that a basal sequence of mafic to sialic lavas spread over regions of different sizes until reaching some critical thickness at which opposing gravitational effects took charge. Then came sialic intrusions, roughly concurrent basin subsidence, and chemically similar volcanic and sedimentary sequences until the basins began to collapse under the combination of their own weight, lateral compression, and stress fields between rising granitic plutons.

At that time there was no vegetation to hold fine-grained sediments in place on the land. Surface water trickled or rushed to the sea in broad, shallow, braided streams, as in many Arctic and desert regions today, or like the present North Platte River of eastern Colorado and Nebraska. Without the help of sediment-binding plants it could not meander sedately via well-defined floodplain channels. Coastal sedimentary accumulations joined effusive volcanism to keep narrow protocontinental shelves well supplied with mainly volcanogenic debris. Unstable piles of sediment from such processes needed only seismic shock or flood to place them in suspension and start them seaward as density currents, thereby adding to the accumulating pile of graded graywackes and pelagic mudstones.

An abundance of pillow lavas is evidence that much of the basalt was exuded beneath the sea. The presence or absence and type of vesicularity in the pillows informs us as to whether depth was moderate (vesicular) or great (nonvesicular).

Directions of current flow as indicated

by sole markings on the under surfaces of graywackes (Fig. 7.16), ripple marking, and cross-bedding join shrinkage cracks and other environmental criteria to tell of islands, shorelines, sources of debris, and current structure. The frequency of trough cross-bedding (Fig. 7.8*B*) among sands of Pilbaran age tells of braided streams. And the prevalence of feldspar-rich graywacke–mudstone sequences among the greenstones emphasizes the then-juvenile nature of depositional processes, dominated by volcanism and the local reworking and redeposition of volcanic debris.

Distinctive among other sediments were early but relatively small deposits of banded iron formation (BIF), a beautifully laminated rock that figures prominently in the story (told in Ch. 10) of the oxygen cycle on the primitive Earth. BIF had already made its appearance at the very beginning of Archean history (Fig. 6.7) and remained intermittently and locally conspicuous as that history progressed toward a great bulge of BIF abundance in older Proterozoic time (left front endpaper). Archean BIF forms an integral if generally minor part of many mafic to felsic volcanic sequences as well as associated flysch-facies sediments and cherts. Variations in the chemistry of iron formations with depth and distance from shore have been used to map the limits and relative depths of separate basins in the Canadian Shield.

The evidence of probably biogenic structures known as *stromatolites,* of microbial fossils, and of the stable isotopes of carbon, sulfur, and nitrogen are all consistent with an advanced state of biochemical evolution by about 2.8 aeons ago and the existence of simple life forms from 3.5 or more aeons ago onward. A

7.16 SOLE MARKINGS SHOW CURRENT FLOW ON BOTTOM SURFACE OF 3.2-AEON-OLD GRAYWACKE BED. Flow is from upper right to lower left on broken slab of Figtree shales in roadcut near Sheba Mine, Barberton Mountain Land, Swaziland and South Africa.

temperature of about 25°C for the shallow sea in which the younger Archean stromatolites of Figure 7.17 formed is implied by the oxygen isotope ratios of contemporaneous cherts—a warm sea, even by present standards.

The appearance of an Archean biosphere initiated a set of interactions among biosphere, atmosphere, hydrosphere, and lithosphere that marked an irreversible change in the surface processes of the planet. Whereas, on the

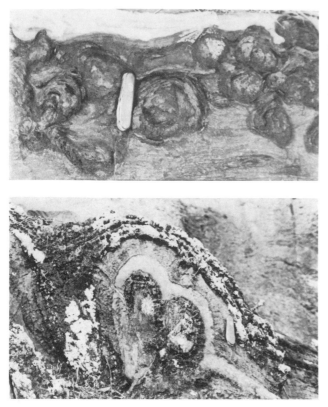

7.17 STROMATOLITES FROM 2.7-AEON-OLD METASEDIMENTS. Chesire Formation of Shabani district, south-central Zimbabwe. Here thirty-three well-defined stromatolitic zones occur within 30 m of sediment.

one hand, life on Earth is sustained and molded by its physical environment, the properties of Earth's air, water, and seemingly solid surface are themselves powerfully shaped by interactions with the contemporary biomass—the sum of which, with its environment, we call the biosphere. None of these would be the same without the others. Without life, Earth would be dramatically different from the planet we know. With it, it could never be the same again.

The Phantom Global Sea

The rather exaggerated area of out-cropping Archean rock indicated in Figure 6.3 represents only about 5 percent of Earth's present surface. To some extent these rocks, like giant trundle beds, continue laterally beneath younger covering deposits. Even so the estimated early Proterozoic land surface could probably not have been much more than about 18 percent of the present global surface (60 percent of the global land area). So what happened to the 82 percent, for which no records are known?

Two seemingly outrageous hypotheses have been proposed. One calls on plate tectonics and subduction since the Archean to recycle the ocean floor so completely that no trace of it is left outside the residual greenstone minioceans. The other hypothesis is even more outlandish. It proposes that an Earth of once much smaller volume has literally expanded like baking bread, extending its diameter to the present size. Preposterous as these ideas may seem, they meet the essential criteria of scientific hypotheses. They explain what we know, and they have verifiable consequences. They are testable. And here the law that governs the spacing and orbital characteristics of the planets—the law of the conservation of angular momentum—reenters the scene.

I must amplify an earlier allusion to this problem. If space for new crust were the result of Earth's expansion, a necessary consequence would be that the product of the components of Earth's mass, multiplied by the square of their respective distances from its spin axis, would be a much larger sum than they were at the end of Archean history. That is a way of saying that its *moment of inertia* has increased. In order to conserve angular momentum under such conditions, Earth's angular rate of spin would have to decrease toward the present by an equivalent amount. Slower spin means longer days and fewer days in the year—unless the year has also been getting longer, which is apparently not the case. Indeed, independent evidence from geophysics and paleontology indicates that Earth days *are* getting longer and that the present year has fewer days than that of early Phanerozoic time—an observation that may, on first thought, argue for expansion but which is consistent with either hypothesis.

In considering the consequences of Earth expansion, one must also think about the size and source of its water mass—the fluid H_2O, ice, and water vapor that comprise the hydrosphere. As we have seen, water was initially supplied either by sweating it out from Earth's interior (*outgassing*), or from melted comets, or both. All currently known evidence favors the view that both processes peaked during Hadean or early Archean history. That would have totally submerged a smaller Earth until it expanded enough to compensate. Yet, from at least 3.8 aeons ago onward to the present, emergent land has always existed somewhere on the planet. Either Earth was *not* significantly smaller during the Archean than now, or, contrary to other evidence, growth of the hydrosphere has kept pace with an expanding Earth. Moreover, a mechanism for expansion is hard to imagine in view of evidence that Earth's thermal budget has been decreasing rather than increasing.

If Earth has not expanded, and if new crust is being made at spreading ridges, then there must be places where old crust is shortened or vanishes at roughly the same rates as new crust is created. Both processes are observed and much evidence indicates that they have been active for at least a couple of aeons and at the needed scale.

If Earth has retained essentially the same diameter and surface area throughout post-Archean history, however, how does one account for its increasing length of day and apparent loss of angular momentum?

The answers are not complicated. Frictional drag by Sun and moon, reflected in the measurable tides of rock and water, slows Earth's rotation at geophysically measurable, if very small, rates. As for angular momentum, relevant scientific law requires only that it be conserved within the Earth–moon system, including as components Earth's rotation, lunar rotation, and the distance of both from their common orbital center. One way to conserve the angular momentum of the system is for the radius of the lunar orbit about Earth to increase by an amount just right to compensate for the observed slowing of Earth's spin rate. That would be consistent with evidence suggesting once significantly higher tides than now. For systematically higher early tides, with no evidence of variation in the length of the lunar month, demand a once-closer moon.

Subduction and, to a much lesser degree, the crumpling of folded mountain ranges are observable ways of making space for new crust. Although some 82 percent of Earth's Archean surface must be so accounted for, 2.5 aeons are available to do it in. Had subduction been active only during the last aeon at a global equatorial average of but 3 centimeters a year, that would have sufficed to eliminate the entire Archean ocean floor. Reduce that average to 1 centimeter a year, and the time available is just right. In any case, Earth expansion is not required and its assumption creates seemingly insurmountable problems with hydrospheric history.

Going back in one's imagination to the Archean, bits of sialic scum can be envisaged as rising to the surface here and there as buoyant products of partial mantle melting. The sweeping together and ultimate joining of such planetary scum by convectional processes could account for the growth of sialic nuclei around which continents were later assembled. Such a process is reminiscent of the growth of planets by planetesimal aggregation, but much slower. Except perhaps for greenstone mini-oceans or embayments, the Archean ocean floor is all gone now—a phantom global sea. It exists only in our mind's eye, constrained by what we can learn from the greenstone belts, the requirements of physical law, and the evidence of plate tectonics and Archean history.

In Summary

The still-youthful Earth, seen dimly through the mists of later Archean history, was the historical interval during which the basic surface activities leading toward the future world were being set in motion. The stuff of continents was being sweated out of a geochemical storehouse within the planet. Experiments in the regulation of surface temperatures were going on. Mineral wealth was being pumped within reach. Fundamental, if unspectacular, historical progress was being made. That history is recorded by prevailing rock types, geochemical characteristics, tectonic processes, and biospheric elements. To discern it, attention to detail is called for.

Two classes of rocks prevailed. One consisted of mainly sodic granites and granitic gneisses surrounded by mainly mafic metavolcanics and immature metasediments: the granite–greenstone belts. The second included almost exclusively high grade, gray gneisses representing higher thermal regimes and presumably lower crustal levels. Both are found in all large crystalline continental shields. The building stuff of continents was being assembled, but no large crystalline shields are yet discernible. Non-marine sediments show that land existed, but extensive epicratonal (epicontinental) deposits, commonplace later, are rare or absent.

Conflict between theory and observation clouds the problem of early crustal thickness. Theory, which expects high heat flow and a hot, thin Archean crust, is contradicted by observations that report upwards of 30 to 40 kilometers of crust and interpretation that predicts as much as 150 kilometers locally. It is clear that more detailed evidence as well as more explicit formulation of the problem is needed before it can be resolved.

Except for important differences in prevailing rock types and biological and atmospheric feedbacks, Archean depositional environments were not unlike selected modern ones. They occur, however, in significantly different proportions and with far less diversity.

The presence of probably biogenic stromatolites and carbonaceous structures implies a microbial presence from perhaps 3.5 aeons ago onward, while local depletion of the relatively heavy isotope carbon-13 in earlier Archean (Isuan) rocks is consistent with the presence of life as far back as 3.8 aeons BP. By 2.8 aeons ago, conclusively microbial structures are known. Some were, by then, surely autotrophic (making their own nutrients), photosynthetic, and perhaps oxygenic. But, of the latter, their O_2 was being rapidly neutralized, for the evidence favors extensive pools of reduced substances (O_2 sinks) and an anoxic atmosphere and hydrosphere at that time.

It is likely that large lateral movements or rotations, reminiscent of those involved in plate tectonism, began before the end of Archean history. Limited data for paleo–North America imply apparent migration of a magnetic paleopole, or, in real motion, parts of the present continent itself, through some 45 to 50 degrees of latitude during later Archean and earlier Proterozoic history.

For those keen to discern practical relevance, there are pots of precious metals at the ends of the Archean rainbow. Probably 80 percent of the world's gold first made it way to Earth's surface in Archean greenstone belts, even though much of it finally came to rest as paleo-placer deposits in younger sedimentary rocks. Ages of diamonds found as inclusions in the famous Mesozoic diamond pipes of southern Africa imply an Archean or earliest Proterozoic origin. Do they record deep mantle or asteroidal processes?

The approaching transition from Archean to Proterozoic history was signaled by extensive potassic plutonism, accompanied by compression of volcanic-sedimentary basins. That gave way to mainly potassic granites, to an epicratonal environment of better sorted and rounded, mainly sialic sediments, and to higher crustal concentrations of the rare-earth elements, distinctive of post-Archean history. The transition, although geologically rapid in any given region, varies in age at different places from about 3 aeons ago in the Kaapvaal craton of southern Africa to as little as 2.7 or perhaps 2.6 aeons ago in the Canadian Shield. Reported younger ages are probably metamorphic.

CHAPTER EIGHT

THE PROTEROZOIC
REVOLUTION:
HOW CONTINENTS GROW

The aggregation of Earth from stardust, the separation of core and mantle, and the origin of a primordial crust foreshadowed the distillation of sialic melts from mafic undermass to generate the building stuff of continents. The process continued throughout Archean history, culminating in a sustained interval of conspicuous continental growth. Beginning some 3 aeons ago in southern Africa and tapering off about 2.7–2.6 aeons ago in the Canadian Shield, a cascade of potassic plutonism and related tectonism joined previously separate sialic nuclei to create the oldest emerged regions of authentically continental dimensions and the first extensive surfaces of continental sedimentation. That long-lasting shift of prevailing tendencies, subject of this chapter, records the origin of a large fraction of Earth's total continental mass and the beginnings of extensive epicratonal sedimentation—perhaps even of plate tectonism.

The preserved record of Earth history was thereafter almost limited to the continental surfaces, including the continental shelves. The conversion was basically irreversible; continents might thereafter grow, lose bits down subduction zones, and change their configurations, but they could never return to the mantle. In contrast to the denser, therefore subductable, and much younger oceanic crust, continental matter and sediments remain mainly near the surface. Their recurrent passages through the rock cycle record most of what we know about historical processes and events. Proterozoic history began at different times in different places as that threshold was passed. The beginning of that transition is recorded in the Randian sequence of southern Africa.

8.1 Gold ore from Witwatersrand paleoplacer at Free State Geduld Mine, Welkom goldfield, South Africa. Sample is 25 cm high. Arrows point to detrital pyrite and Xs mark distinct quartz pebbles in the conglomeratic matrix. *[Courtesy of W. E. L. Minter; Free State Geduld Mines, Limited.]*

The Intervals of Cryptozoic History

As we take up the subject of the long and eventful transition from Archean to Proterozoic styles—the Proterozoic revolution—vexing questions about boundaries and transitions in the geologic record need to be addressed.

It is important here, as elsewhere, to distinguish between the indivisible continuum of time and the varying intensities of Earth's history as a succession of significant, but variably time-bracketed, events and trends, reconstructed from the rocks that record it. Boundaries do not exist in time and are perennial subjects of discussion and disagreement in history. Yet everyone realizes that there are recognizable intervals of history, transitions between them, and, at places, historical discontinuities (boundaries) in the record of the rocks. Important historical episodes and trends, thus, have been given names derived from regions where the rocks that represent them are well exposed (e.g., Devonian) or exhibit some distinctive characteristic (e.g., Cretaceous).

Although such events and trends take place in time and are properly discussed in a framework of radiometric numbers, they are not defined by time. To delineate them in terms of explicit time signatures, therefore, *as if* so defined has the effect of concealing our ignorance about important time-consuming transitions.

That's all very well for the 680 million years or so of Phanerozoic history, where fossils are abundant and distinctive historical markers, but troublesome when it comes to pre-Phanerozoic (Cryptozoic) history. For, although the record of fossils has now been extended almost to the beginning of the Archean, few of them are yet known to be widely distributed and stratigraphically distinctive.

An obvious, widely adopted, and eminently practicable solution to the problem among professional geologists is to discuss the history of the Cryptozoic eons entirely in a framework of radiometric numbers and methods. Geologists, however, also recognize very large assemblages of historically significant Cryptozoic rocks which it is convenient to discuss in collective terms. And, for the history thereby represented, the designations Archean (also Archaean, or Archeozoic) and Proterozoic have long been used—for the dominantly granite–greenstone belts and the dominantly epicratonal sequences respectively. Some named subdivisions of these eons are also variably in use.

So far so good. Archean and Proterozoic types of rocks can be visually separated by geologists in the field in terms of prevailing modes, with many uncertain intervals and, at places, with long transitions from one to another. They are hardly susceptible, however, to delimitation within well-defined "boundaries" of the sort geologists so yearn for. Indeed, an international subcommission, having considered that daunting task, now proposes to delimit Archean from Proterozoic history in purely geochronometric terms at the radiometric age level (method not specified) of 2.5 aeons BP—a number that recent radiometric dating may displace to 2.7. This ardently debated proposal, if adopted, could have the effect of stabilizing nomenclature, but at the cost of violating the most basic principle of historical geology—rocks are the final court of appeal. It would also, in effect, deny the existence of observable if dramatically time transgressive distinctions between prevailing Archean and Proterozoic lithologic and historical modes. Better, others conclude, that such important transitions be kept in full view than to be concealed behind neat but unreal time boundaries, perhaps delaying the search for improved historical resolution.

The solution here preferred (front endpaper) is to recognize this time-transgressive transition as a global historical event of its own and designate the Proterozoic as having begun at significantly different times in different parts of the world (about 3 aeons ago in Swaziland and South Africa, around 2.8 aeons in Western Australia, and not until about 2.7 to 2.6 in the Canadian Shield). Similar problems and differences, on a lesser scale, prevail thoughout pre-Phanerozoic stratigraphy. In another few decades, as stratigraphically distinctive microbial fossils continue to be found more widely in strata of the Cryptozoic eons, briefer transitions and a more refined global history may come within our grasp.

The Oldest Revolution

Although the transition from Archean to Proterozoic styles was gradational on a global time scale, it was revolutionary in terms of changes wrought. During successive waves of cratonization from southern Africa to Western Australia and onward to various parts of the present-day Northern Hemisphere, preserved rocks became more sialic—more continental—as Archean gave way to Proterozoic. A new generation of granitic rocks, rich in potassium and other "sial-loving" large-ion-lithophile (LIL) elements, flooded to the surface to open the Proterozoic scene. And as those rocks and their volcanic counterparts entered the weathering and erosional phases of the rock cycle (Fig. 3.11), sedimentary rocks with similar enrichments in LIL and rare-earth elements (Fig. 7.6) and a decreasing volcanogenic component accumulated on the cratonal surfaces. In contrast to the Archean, Proterozoic history was, above all, one of epicratonal processes and sialic geochemical signatures.

That transition from a conspicuously oceanic to a conspicuously epicontinental record of surface deposits was a turning point. For it was only as the distinction between these contrasting regions became pronounced that the mechanisms of plate tectonics as we know them today could begin to function.

The familiar fact that continents are high and mainly dry while oceans are low and full of salty water is only one of many ways in which they differ. The continental crust is thick (35 kilometers on the average today) and mainly sialic, with an average density of 2.8. The oceanic crust is thin (about 10 kilometers) and mafic, with a density (weight divided by weight of the same volume of pure water) of about 3 to 3.3, depending on temperature, water content, and mineralogy. Like blocks of aluminum and iron floating on mercury, continental and oceanic rocks rest on the mantle beneath in such a way that their differences in density and mass are compensated by the depths to which they sink within and the heights to which they rise above it. The continents stand high because their density is low. They have deeper roots than the ocean crust because they are thicker and of greater mass (Fig. 5.2). The forces of gravity and buoyancy are isostatically balanced. Moreover, the volume of water slightly exceeds the capacity of the present ocean basins. It spills over the edges of the continental shelves, making the continents seem smaller than they are. Continents also differ from oceans in the rate of travel of seismic waves through their crusts, the ubiquity on continents but apparent absence from oceans of folded mountain ranges, and the relative youth of ocean floors as compared to the antiquity of continents. These irreversible distinctions are products of the Proterozoic revolution.

The primary components of both continental and oceanic rocks are the results of melting within the mantle, followed by gravitational segregation. Oceanic rocks came directly from the mantle and are geochemically similar to it. Continental rocks, on the other hand, are mainly the products of repeated distillation. The contrast between Archean and Proterozoic is a contrast primarily between mantle geochemistry and continental geochemistry. That is recorded in the characteristics of the preserved rocks, their elemental and isotopic geochemistry, and the paleoenvironmental implications of their surface deposits.

Although intruded by and nearly everywhere underlain by mainly sodic granites and gneisses, the surface rocks

of later Archean age were, as we have seen, prevailingly mafic volcanics and immature sedimentary rocks derived from their reworking (e.g., volcanogenic graywackes and shales). Their isotopic and other geochemistry reflects mantle sources, and their depositional characteristics and geochemistry record prevailingly oceanic settings. Proterozoic surface deposits are either marine or nonmarine in depositional terms; but their marine geochemical affinities and depositional sites are preponderantly continental in the sense that they are derived primarily from sialic sources and deposited mainly on continental surfaces, whether or not submerged. They are likely to be quartz-rich sandstones and siltstones, micaceous shales or mudstones, and carbonate rocks (dolomites and limestones) such as are common on continental shelves and in interior epicontinental basins during all subsequent geologic history. Their geochemistry reflects dominantly continental sources and processes by their relatively high proportions of the LIL trace elements that tend to displace others in the internal structures of silicate minerals. They include the important heat-producing and timekeeping isotopes of uranium, thorium, potassium, and rubidium.

Although the change from Archean to Proterozoic was revolutionary in its results, it was hardly abrupt. Viewed on a human scale of time it was more comparable to the agricultural or industrial revolutions than to the American or French revolutions. It took geocenturies for the Proterozoic revolution to spread from the Kaapvaal craton of southern Africa, where it began, to the Canadian Shield, where it was completed. Gaps in the record and uncertain ages make it hazardous to generalize, but the transition appears to have occurred in most parts of the world between about 2.8 and 2.6 aeons ago, and its progress would have been imperceptible to anything then living.

What caused this transition? A plausible explanation is that it represents a threshold in Earth's cooling history. As long as convective loss of heat from Earth's interior was 2 or 3 times the present, mantle convection could have been vigorous, outer mantle hot and yielding, and mafic oceanic crust extensive and subject to small-scale convective recycling in the outer mantle. In contrast, light, sialic, continental crust seems to have been limited in extent and probably in vertical relief, while its initial ratios of radiogenic to nonradiogenic isotopes (Fig. 4.8) would have resembled those of the contemporary mantle. Epicontinental sedimentation and sedimentary rocks from continental sources would be restricted in volume and extent and not susceptible to mantle recycling. Such conditions do not favor plate tectonism as we know it today.

Persuasive evidence that plate tectonics was at last at work would be the presence of distinctive metamorphic rocks (blueschists) indicative of high pressures but low temperatures in subduction zones. Also sequences characteristic of migrated mid-ocean ridges—the so-called ophiolites. Neither blueschists nor convincing ophiolite sequences of classic type are yet known from any part of the Archean, although volcanic rocks similar to one component of the ophiolite suite are common. Could there still have been plate tectonism so long ago, as hinted at by Figure 7.3? Or plausible antecedents? Might there have been a major decline in rates of heat flow or other changes favoring it during the long transition from Archean to Proterozoic modes?

Had oceanic lithosphere by then become dense enough in its cooler parts and thick enough to undergo subduction? If so, a gliding surface should have existed beneath, along the surface of a warm, yielding asthenosphere below the deepest mountain roots. Such characteristics would most likely have resulted from reduction in the rate of heat flow in the mafic mantle as a function of the convective loss of residual heat—perhaps owing to the preferential movement of radioactive LIL elements upward from mantle to Earth's growing, sialic, continental crust.

The Golden Kaapvaal Craton: An Early Subcontinent

Among other geochemical distinctions that separate Proterozoic from Archean rocks is their respective abundances of primary (non-placer) gold. The primary source of most of the world's gold is Archean greenstones. Yet, by a quirk of weathering and erosion, the production of placer gold of Archean origin from the Proterozoic sediments of a single region (Fig. 8.1) accounts for more than half of all the gold so far mined worldwide.

That fabulous region is the tropical Kaapvaal craton of eastern South Africa and Swaziland, depositional site of the Randian basinal sequence—early Proterozoic if the rocks don't lie. It comprises some 600,000 square kilometers of mostly open and dry high veldt. That is slightly larger than Madagascar and not quite a third the size of Greenland. It is, nevertheless, a good deal larger than other then-emergent sialic masses we know of, and generally larger than features now called microcontinents. Only about 14 percent of that surface is Archean and that mainly granites and gneisses. Small greenstone belts and associated granites outline the craton and establish its age and continuity beneath younger cover.

The fever-ridden Barberton basin, described so briefly in the preceding chapter, is the largest and most complete of these marginal greenstone belts, rich enough in primary gold to have drawn its own quota of hopeful prospectors. Its folded and steeply dipping, layered rocks are bordered on the north and northwest by a variety of intrusive sodic granites and gneisses older than 3.1 aeons and on the south by homogeneous potassic granites 3 aeons old. Intrusive, characteristically pink granites of younger age, increasingly potassic and colorful with decreasing ages up to about 2.5 aeons BP, cut across the older gray sodic granites, gneisses, and belts of greenstone and lesser metasediment, literally welding them together into a single crustal block. Similar events over a much larger area produced the persistently stable region we now call the Kaapvaal craton or subcontinent.

If water were missing from Earth, as it is from Venus and Earth's moon, we would see our ocean basins as simply low, poorly reflective depressions, like the lunar maria and Venusian lowlands. But Earth has plenty of water, not only to fill its oceans but to lubricate tectonic motion. The floors of its oceans are heavy, mantle-like in composition, and prone to descend into the mantle under favorable conditions and become recycled. The continents, in contrast, as we have seen, are made primarily of rocks that were derived over long intervals from the relatively fusible, light components of a partially melted mantle. They have arisen piecemeal from that mantle because of their lighter density. Although they cannot again descend into it, they can and do change their positions and dimensions as a result of tectonic dispersal and microcontinental clustering and accretion as described in Chapter 9. The components of continents are complex, essentially permanent entities in contrast to the transient ocean floors.

The simplest of one of these complex continental components is the *pluton*, a usually sialic igneous intrusion, formed at depth by melting and moved buoyantly upward to crystallize within the crust as a distinct body. A group of associated plutons covering a sufficiently large area comprises a *batholith*, like that of California's Sierra Nevada. An association of plutonic, sedimentary, and metamorphic rocks that have been welded together by a later generation of intrusive rocks to form a stable, mainly sialic mass, is called a *craton* or a *shield*. *Craton* and *shield* are often used interchangeably, or a craton may be taken as a component of a shield. Here I use the term *craton* for the product of a single major (but not necessarily brief) cratonizing event (e.g., Fig 8.2). *Shield* is reserved for a generally larger complex of one or several cratons. Both are stable sectors of continents, which consist of one or more cratons plus post-cratonal

8.2 GEOLOGY OF THE
KAAPVAAL CRATON AND
ITS STRATIFIED COVER AT
END OF RANDIAN
HISTORY.

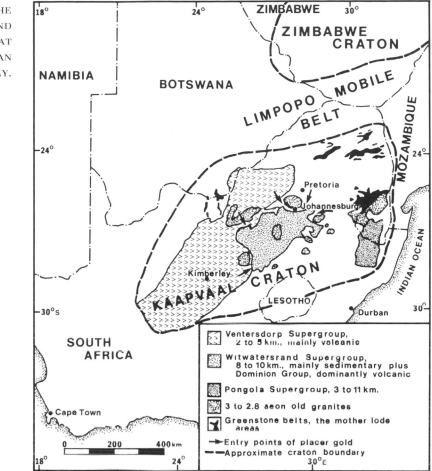

additions. If they are larger than about half a million square kilometers in area, they are considered to be subcontinents.

A thick sequence of little-metamorphosed sedimentary rocks and intermediate-to-sialic volcanic deposits rests unconformably on deformed Archean rocks in the middle of the Kaapvaal subcontinent. Those sediments, notable for their placer gold, are the river-borne and deltaic quartz-sandstones and conglomerates of the Witwatersrand Triad of major rock units (including the Witwatersrand and Ventersdorp supergroups), a household term in South Africa.

Although gold has been mined from about the third century A.D. at thousands of sites in southern Africa, it was not until after the often-tragic gold rush into the Barberton area in 1883 that the Witwatersrand deposits became known. Beginning with the European discovery of gold near Johannesburg in 1886 (Fig. 8.3) paleoplacers at sixteen different levels in these sedimentary rocks have yielded nearly 40,000 metric tons of gold, mostly from the upper part of the Witwatersrand Supergroup.

That is more than half of all gold produced worldwide until now, and the

Witwatersrand deposits are still producing. Current production is some 700 metric tons of new gold annually, with a value equivalent to about 10 percent of the U.S. gross national product in 1982.

The mother lode for all that gold has never been found, but geologists have a good idea where it is, or was. Gold (a *siderophile* element) has a strong affinity for iron. The ultimate source of most important primary gold deposits is in iron-rich rocks of mantle geochemistry that have presumably arisen as melts from low in or beneath the primordial lithosphere. Intervals of Earth history conducive to the formation of such gold-bearing rocks are times of strong mantle influence. The main interval is the Archean. Secondary ones are related to active subduction and resurgence of sodic granites, most recently beginning about 2 geocenturies ago with the present cycle of plate tectonism and subduction. But placer gold is found in coarse continental strata of all ages, and traces of dis-

seminated gold are known from virtually all mantle-derived basalts.

In southern Africa primary gold is found especially where mafic volcanics of the Archean greenstone belts drape over the crests of rising granitic plutons. Current directions of structures in the old placer-containing stream deposits on the Kaapvaal craton imply that most of the Witwatersrand gold came from the erosion of Archean greenstones and granites around the northern rim of the basin. Lesser sources were from the west, southwest, and east—from uplands now largely buried beneath younger covering strata. When this placer production is added to primary production from the greenstones themselves, one finds that the mother-lode terrain has, over the years, yielded something like 1.7 kilograms of gold per square kilometer of outcrop, more than all other known gold-producing rocks.

The payoff for geology, however, comes in the knowledge gained during the quest

8.3 Discovery site of Witwatersrand placer gold near Johannesburg, South Africa. Gold-bearing conglomerates found here in 1886 by prospectors George Harrison and George Walker led to the production of more than half of the world's mined gold.

for gold and other heavy metals. Because of it, we now know more about the sequence and depositional environments of these early Proterozoic strata than we yet do of many much younger and more widely represented sequences higher in the geologic column.

Indeed, we know rocks of Proterozoic aspect even older than the Witwatersrand Triad on the eastern margin of the Kaapvaal craton (Fig. 8.4)—those of the Pongola Supergroup. They tell us that events transitional from Archean to Proterozoic had already begun in southern Africa by about 3 aeons ago, perhaps involving plate collision from the north. That initiated the long succession of sedimentary and volcanic deposits of Proterozoic aspect known as the Randian sequence, while sodic intrusives and greenstone–metasedimentary belts continued to form at other places during the next few geocenturies—a long transition, but hardly a "boundary." Randian sediments ultimately overfilled the sub-

siding basins of the cratonal platform with a composite thickness of perhaps 28 kilometers of stratified epicratonal rocks of fresh water, marine shelf, and volcanic origin. The sedimentary components of this sequence were primarily quartz- and quartz-feldspar sandstone, siltstone, shale, and conglomerate with minor banded iron formation, a little carbonate rock, and what, *if* glacial, would be the oldest glacial deposits of record.

Such a sequence registers all the signals geologists associate with Proterozoic. It underscores the inadequacy of a distinction that would separate Archean from Proterozoic on purely geochronometric grounds.

The Randian (or Zuluan) sequence, then, including the Pongola Supergroup and the Witwatersrand Triad, followed on the heels of the massive granitic plutonism that joined the late Archean granite–greenstone terrains of southern Africa into the Kaapvaal subcontinent around 3 aeons ago.

8.4 OLDEST PROTEROZOIC. UPENDED LAYERS OF PONGOLA SUPERGROUP IN THE GREAT NSUZE DOWNFOLD, A SYNCLINE IN NORTHERN NATAL. *[Courtesy of Victor von Brunn.]*

8.5 SEDIMENTARY ROCKS OF THE PONGOLA SUPERGROUP IN NORTHERN NATAL. View *A*, quartzite at base of Pongola Supergroup rests nonconformably on older granites; thin lag gravel at contact. *B*, quartzites of lower Pongola along White Mfolozi River. *C*, low-grade banded iron formation from upper Pongola strata.

Rocks of the Pongola Supergroup are the oldest extensive stratified rocks of primarily sialic composition that were clearly deposited on continental basement (Figs. 8.4–8.6). They are found as scattered outcrops, suggesting an original depositional area of more than 30,000 square kilometers in hilly northern Natal province, adjoining parts of Swaziland and the Transvaal. These basal Proterozoic rocks include a lower mainly volcanic group (Nsuze) and an upper mainly sedimentary one (Mozaan).

They are here considered to be Proterozoic, among other reasons because the proportions of rare-earth elements in the shales are close to Proterozoic patterns elsewhere (Fig. 7.6). The characteristic maturity and preponderantly sialic composition of the Pongola sedimentary pile is also of Proterozoic aspect. The basal quartzites locally rest on and mingle with weathered and eroded residues of a potassic granite (Fig. 8.5A), and are overlain by interbedded stream deposits and volcanics having uranium–lead depositional ages of 3.1 aeons.

All of these rocks were intruded by 2.9-aeon-old intrusive rocks that date the deposition of the Pongola Supergroup as having occurred between 3.1 and 2.9 aeons ago. They attain thicknesses up to 11 kilometers—nearly half sedimentary, little metamorphosed, and extending over a large area. Their testimony joins with that of the Witwatersrand Triad to imply an early onset of the Proterozoic mode in this region.

Indeed, the Pongola Supergroup represents only the earliest phase of a would-

be marine epicratonal incursion from the southeast. The ocean was beginning to overflow the lands, or else the lands to sink. Its waters onlapped northwestward onto a tectonically unstable region that was to become, during the next few geocenturies, first a subsiding basin, then a shrinking one, and eventually an emergent platform of the Kaapvaal craton. The overlying early Proterozoic but post-Pongolan, Witwatersrand Triad is an extensive and diverse composite pile of lightly metamorphosed, mainly nonmarine but partly marine sedimentary rocks and mainly basaltic volcanics more than 14 kilometers thick.

Omitting detail and complicated questions about the age of rocks that cap and bound the Witwatersrand Triad, the gold-bearing siliceous sediments and occasional volcanics of the Witwatersrand Supergroup pique one's interest. They reached thicknesses upwards of 7,000 meters or more beneath the overlying

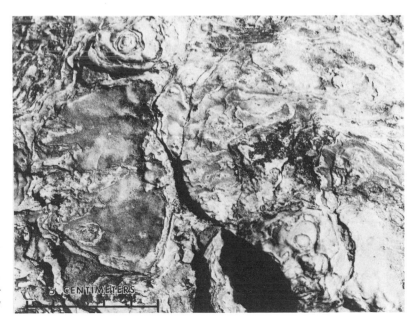

8.6 WEATHERED SURFACE OF 3-AEON-OLD STROMATOLITIC LIMESTONE FROM LOWER PONGOLA SUPERGROUP, SHOWING SMALL DOMAL STRUCTURES. *[Courtesy of T. R. Mason.]*

Ventersdorp volcanics that terminated the previously vast epicratonal basin. Some 70,000 square kilometers in area, this basin (Fig. 8.2) was flanked around the north end and partway down the western and eastern sides by upfaulted Archean rocks of the former mother-lode area. Its gold and associated heavy minerals, including uraninite, represent paleoplacer deposits, as is shown by their sedimentary characteristics and confirmed by the isotopic identity of lead from grains of associated detrital pyrite with that of gold deposits in the Barberton area. Detailed geologic mapping and discriminating sedimentology have shown that the goldfields are all in the form of alluvial fans or deltas where streams draining the ancient Archean highlands entered the subsiding basin.

That basin, open during Pongola history to the sea on the south, first closed to become a lake and then opened to let the sea in again before disappearing beneath the terminal volcanics. This shifting of levels also affected the alluvial fans and deltas of feeder streams. Combined with the final retreat of basin waters and basinward shifting of stream entry points, this explains the accumulation of paleoplacers at so many different levels. Although the gold surely came from the relatively uplifted ultramafic and mafic greenstones and associated granites of the bordering Archean uplands, the source of the uranium was evidently from younger intrusive Archean or earliest Proterozoic granites. Most of those placers are associated with ancient gravels and sands, now conglomerates and sandstones, that accumulated at unconformities in the sedimentary pile wherever vigorous erosion in the hinterland brought quantities of heavy metals to the stream ends before renewed basinal sedimentation. The setting is classically epicratonal, epicontinental, and Proterozoic. Little about it beyond the (reworked) gold would imply Archean elsewhere.

Beyond the Kaapvaal Subcontinent in Space and Time

Where might one expect to find a record like that of Randian history beyond the Kaapvaal craton? Perhaps in India, where gold has been mined for the last 2,000 years? Recent research by the Geophysical Research Institute at Hyderabad reports possible local parallels to the Kaapvaal craton, but auriferous deposits of such proportions are yet to be found.

The search for a second Witwatersrand gold province has been unsuccessful everywhere. Archean gold there was indeed, in Zimbabwe as elsewhere, but no great Proterozoic placers. Something about the history of the Witwatersrand basin and its mountainous hinterlands was different.

One might then consider the Archean–Proterozoic transition in Western Australia. There the Fortescue Group, consisting of 5 to 7 kilometers of basalt, intermediate volcanics, and interbedded sediments reminiscent of the terminal Witwatersrand deposits, is the Randian equivalent. There also a uranium–lead age of 2.768 aeons on zircon from a lower Fortescue rock favors an age near

2.8 for the onset of Proterozoic history. And different radiometric ages are found for similar transitions in other parts of the planet.

One sees, then, a distinctive shift from dominantly oceanic Archean to dominantly epicratonal Proterozoic modes that began at different times at different places. To recapitulate, a Proterozoic style began about 3 aeons ago in South Africa, about 2.8 aeons ago in Western Australia, and apparently 2.7 to 2.6 aeons ago in the rest of the world. Although those who, in practice, equate time with history have (until now) agreed on 2.5 aeons BP as "the boundary," there are only local boundaries. The global historical event is transitional. That transition was the time-transgressive Proterozoic revolution.

Randian Paleoenvironments and Paleogeography

The Randian sequence thus marks the time-transgressive beginning of epicratonal sedimentation on a large scale. It began with major cratonization that, over the next geocenturies, enlarged and welded together probably more than half the known continental basement rocks. Randian equivalents comprise our only source of information about what the oldest truly continental surfaces may have been like. Their deposits illustrate how much information can be wrung from rocks by careful attention to physical and geochemical details and analogy and existing environmental processes—particularly in post-Archean strata.

To visualize the Kaapvaal craton, picture a region devoid of vegetation or other conspicuous forms of life. Visualize it as about 6 times the size of Iceland, and with coastal lowlands resembling the barren sandur plains of Iceland's southeastern shores except in climate (Fig. 8.7). Without impediment to transport, sediments from Archean granite-greenstone uplands, raised perhaps by

8.7 PHYSICAL ANALOGY TO KAAPVAAL SURFACE OF RANDIAN TIME IS SUGGESTED BY BRAIDED DRAINAGE OF SAND PLAINS IN SOUTHEASTERN COASTAL ICELAND.

an early collisional event, would, as there, have moved seaward via broad, overloaded, multichanneled, braided streams. Such waterways would have merged along the coast to create vast areas of low relief where land and sea mingled across a broad intertidal zone.

The Randian sediments, mainly sandstones from decomposing Archean granites, consisted of quartz and feldspar, together with clays from the weathering of feldspar and the reworking of old shales. Although there were extensive volcanics in the region, volcanogenic components are limited. A mild climate is implied by the chemical disintegration of the granitic basement rocks that provided much of the sediment—a product mainly of weathering beneath a CO_2-rich atmosphere unaided by plant acids.

Southward dipping and pitching planar and trough cross-bedding in the basal sands of the Pongola Supergroup record the draining of southeastward-flowing, braided streams from the north and northwest. An initial marine inundation from the present southeast flooded across the weathered granitic surface of the barren Kaapvaal craton, locally reworking quartz and feldspar grains from the disaggregating granite into feldspathic sandstones, siltstones, and shales. At places the mixing and lack of sorting at the depositional interface results in a deposit so diffuse that it can be difficult to say where disaggregated granite stops and sediment begins. Soon thereafter, rifting introduced a mainly mafic volcanic regime that only intermittently yielded to quartz sands, occasional shales, and subordinate, chemically precipitated, cherty iron formation (Fig. 8.5C) until the total deposit piled up to a thickness of 6 to 7 kilometers. Some of the

basalts in this sequence are pillowed, indicating subaqueous deposition, perhaps marine, that apparently alternated with a fluviatile regime for much of later Pongolan history.

The upper Pongola, no more than 3 to 4 kilometers thick, is preponderantly cross-bedded quartz-sandstone, shale, and conglomerate of fluvial, intertidal, and shallow subtidal marine origin, with rare volcanics. "Herringbone" or "bow-tie" cross-bedding (Fig. 3.6C) in both upper and lower Pongola sandstones (Fig. 8.5B) indicates repeated current reversal and tidal influence, while flat-topped and doubled-crested ripples together with shrinkage-cracked mudstones, record their episodic exposure to the atmosphere. Intervals of locally stromatolitic limestone (Fig. 8.6) and dolomite up to 30 meters thick in the lower Pongola also record shallow-marine sedimentation, microbial sediment-binding, and perhaps local emergence.

Criteria indicative of tidal ranges almost double today's greatest imply greater gravitational effects and, therefore, closer proximity to the moon, a correspondingly shorter Earth day, and more days in the sidereal year than now. In addition to other effects, this could have been a contributing cause for the then generally greater volcanic activity.

As for the more recent but still early Proterozoic Witwatersrand Triad, that registers a more complicated history over a much larger area of the old craton northwest from the Pongola deposits (Fig. 8.2). Here, ringed by Archean highlands around the northern end, lay a subsiding inland basin reminiscent at its peak of the shallower parts of the Persian Gulf (less than 40 meters) or the Caspian Sea. The ancient sedimentary rocks of this basin contain records of deposition in rivers, deltas, and lakes, as well as intertidal and subtidal regimes.

Thanks to its gold, the eventful sedimentary history of the Witwatersrand basin has been worked out in great detail. It was an entirely epicontinental and mainly nonmarine basin that expanded and contracted in volume over time as it collected a great variety of sedimentary and volcanic deposits. The total area of its basin at a maximum was about twice that of Lake Superior. Even though the region was dry at times and the volume of water at best trivial compared with that great lake, the epicratonal Witwatersrand sedimentary basin at flood must have been an impressive sight.

The total composite thickness of Witwatersrand deposits includes over 12 kilometers of sialic detrital sediments and more than 5 kilometers of lavas, clearly indicative of rapid erosion of rising peripheral uplands, rifting, and regional subsidence and volcanism. Interruptions in sedimentation, recorded by unconformities, are associated with deposits of placer gold of impressive extent. The structures of the sandstones and conglomerates in the lower division of the Triad, imply mainly fluvial sedimentation at that time.

Among the interesting variety of associated sediments are matrix-supported pebble beds in which some pebbles are reported to display parallel scratches indicative of glacial transport. They may represent the oldest so-far-recorded evidence for glaciation, about 2.7 aeons BP.

The Witwatersrand gold and some forty other placer minerals were concentrated at places where tributary streams entered

8.8 WHERE THE GOLD IS. HYPOTHETICAL VERTICAL PROFILE THROUGH FAULTED WITWATERSRAND SUPERGROUP (*STIPPLED*) AND OVERLYING STRATA. Recumbent *V*s designate lavas and related volcanic rocks; sedimentary rocks as labeled. Vertical scale exaggerated.

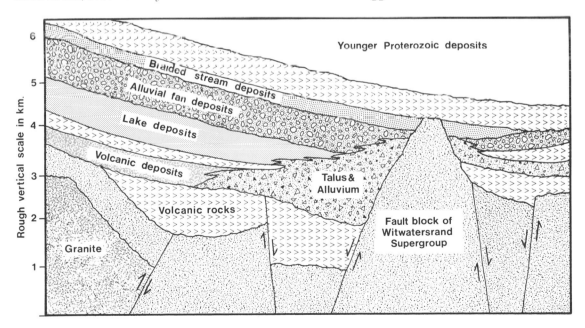

the basin, and one would expect heavy grains to lag or filter down in a strong current regime. Such minerals, therefore, are found on old erosional surfaces below conglomerates as well as in spaces within conglomerates (Fig. 8.1), at the toes of planar crossbeds, and in the scour troughs of trough-cross-bedded sandstones. Gold is about equally abundant in conglomerates and sandstones, although a few rich concentrations are associated with unusual thin carbon seams of disputed but not directly biological origin. Sediments of the gold-bearing fluvial fans and deltas entered the basin from every direction except directly from the south, recording the source rocks that almost surrounded it.

Witwatersrand history ended with shrinkage of its sedimentary basin to the vanishing point, followed by numerous offsets along or near vertical faults, development of a rugged topographic relief, and an extensive searing cover of basaltic lavas—*flood basalts* (Fig. 8.8). That and a lot of associated talus, gravel, and other debris made the search for terminal Witwatersrand gold a complex game of blindman's bluff.

The gold in this debris continued locally to be reworked into important second-generation placers in the fluvial environments of successive volcanic surfaces. As in more recent fault-block terrains, the surface that terminated the Witwatersrand Supergroup became the site of extensive and diversified continental deposits of everything from the coarse detritus of avalanches, rock talus, and debris flows to the finer-grained sediments of playas and longer-lived lakes.

Signs and Consequences of the Proterozoic Revolution: The Beginning of Plate Tectonics?

The long transition from Archean to Proterozoic crustal modes records the emergence of a new world of growing continental regimes and a new prevalence of mainly detrital sedimentary rocks of sialic composition. During some 3 to 4 geocenturies, beginning 2.9 to 3 aeons ago, huge quantities of new potassium-rich granites, the main cratonizing agent, ascended from the partially melted mantle to high crustal levels. There they invaded older mafic and sodic crustal components and welded them together, stabilizing large sectors of Earth's crust as cratons, the main components of shields and continents.

Growing subcontinents were weathered and eroded to produce first-cycle, feldspar-rich, quartzose sediments. On further weathering, transport, and sorting, such sediments became the kind of well-sorted, multicyclic quartz sands and silts we call *mature*. The feldspars concurrently weathered into clays, the components of shales and mudstones, while dissolved ions linked with those of opposite sign to precipitate as chemical sediments—as carbonate rocks and much banded iron formation. We see a new sector of the rock cycle in action.

The results of such activities are found worldwide at different times within the Archean–Proterozoic transition as thick, mainly sialic sediments and associated mafic to intermediate volcanic rocks that blanket surfaces and fill basinlike depressions on the partially submerged cratonal crust. Such are the conspicuous results. There were deeper and more pervasive geochemical results, with subtle feedbacks to Earth history.

One such result relates to the ionic radii of the elements and the control that exercises over how they associate with other elements and thus where they are found. Large ions do not fit well into the dense crystal structures of mafic minerals, which generally accommodate only smaller ions. The larger ions are more readily trapped in the more open lattices of sialic minerals. Thus the LIL and rare-earth (RE) elements substitute preferentially for others in the lattices of sialic silicates and especially potassic fieldspars. As we have seen, radioactive LIL elements like potassium, rubidium, uranium, and thorium are heat generating. They affect Earth's geothermal balance. They also decay to radiogenic isotopes of other elements whose ratios to nonradiogenic isotopes make sensitive and revealing tracers of long-vanished geological processes and events—such as rates of crustal motion and subduction.

As the potassium-rich sialic melts of early-transition rocks worked their way upward from the partially molten mantle beneath, they carried with them quantities of the LIL and RE elements, *except for europium*. Because of this, the mantle were depleted in those elements and the sialic crust enriched (except for europium). That is why the Archean and post-Archean profiles for RE-element abundances differ systematically from one another (as in Fig. 7.6) and why such differences are important. The geochemical distinction between *depleted* and *undepleted* (subasthenospheric) mantle

sources of post-Archean mafic intrusives can be critical for the determination of their origin, history, and even age.

When source rocks break down to become sediments they bequeath their distinctive trace-element-abundance profiles to those sediments. The post-Archean europium deficiencies result from the unusual retention of that element by the prevalent sodic feldspars of Archean rocks, resulting in its comparative depletion among the potassic feldspars and shales of Proterozoic crust where other trace elements are enriched (Fig. 7.6).

Such things relate to the Proterozoic revolution in several important ways. The flood of potassic granites that resulted in the initial cratonization of perhaps 60 percent of the sialic crust significantly depleted the outer mantle of heat-producing LIL elements. Why then?

Could it have been that earlier convective mixing was too vigorous for such geochemical distinction to be established and maintained? Could it have had something to do with the onset of plate tectonism? It would be consistent with such reasoning to propose that convective cooling and crustal transfer of radioactive elements by later Archean time had significantly reduced the thermal gradient from early Archean highs. Might that have cooled and thickened the underlying lithosphere enough so that subduction, plate tectonism, and turnover of the outer mantle could then begin, taking sea-floor basalts with it? Transfer of heat-generating isotopes from lithospheric upper mantle to continental crust just then could have further cooled that sector of the mantle while heating continental crust, increasing buoyancy contrasts, facilitating the growth of sialic

187

lithosphere, and perhaps triggering the onset of plate tectonism.

The previously mentioned ratios of radiogenic strontium-87 to nonradiogenic strontium-86 provide their own powerful insights to the relative importance of mantle and continental processes at different times in Earth history—but particularly to the transition here discussed. Because radioactive rubidium-87 decays to inert strontium-87 at an invariable rate, while non-radiogenic strontium-86 remains constant, the ratio of Sr-87 to Sr-86 increases predictably with time. That ratio began at 0.699 and has increased to a value of 0.709 in present seawater (0.704 in continental waters). That can lock in the ages of particular events in the past and trace sequences of events and processes through time.

Geochemist Ján Veizer of the University of Ottawa has used strontium-isotope ratios to trace important changes in ocean-water chemistry and sedimentary processes during the transition from Archean to Proterozoic. He finds ratios that imply a strontium input from continental sources 4 times that from the mantle. Mantle influences on strontium-isotope ratios result from interaction between seawater and oceanic crust, especially along the spreading ridges where hot new volcanic crust is being made from mantle sources. It depends on the intensity of heat flux and plate motions. Continental influence comes from surface drainage and rivers. It depends on the surface area of the continents and the intensity of weathering and runoff.

The two inputs vary with time and the intensity of plate tectonism. Mantle input was high during the Archean and, briefly, at later intervals—presumably coinciding with new episodes of plate tectonism and the tapping of mantle sources by volcanism. Continental imprint first became conspicuous with transition to Proterozoic continental influences. It has varied in intensity since then but has dominated the succeeding geologic record.

Strontium-isotope data clearly imply that the main shift from mantle-dominated to continent-dominated regimes took place during the Archean–Proterozoic transition. Following the Proterozoic revolution, the volume of continental crust experienced another large addition during a new cycle of intrusion and tectonism between about 1.9 and 1.7 aeons ago. By the end of that episode the judgment of knowledgeable Cryptozoic geologists and geochemists has been that Earth's sialic crust may have attained as much as 80 to 85 percent of its present volume, while the sedimentary mass was probably approaching its present size. Strontium-isotope data have been interpreted by Veizer to mean that sedimentary rocks since then have been largely recycled, while averaging close to their present volume. Sediments and sedimentary rocks, even now, are more mafic than their apparent upper-crustal igneous sources because the initial composition of the reworked Archean sedimentary mass has not yet been erased by 2.5 aeons or more of recycling.

Recent estimates of the current rate of growth of new continental crust by David Howell and Richard Murray of the U.S. Geological Survey, however, bring a new case to the rocky court of scientific judgment. Their estimates imply a net rate of addition (1.35 cubic kilometers yearly) more than enough to account for the

TABLE 8.1 *Prevailing Differences between Preserved Younger Archean and Proterozoic Crust of Different Kinds*

	Archean	Proterozoic and Younger
Granites and Gneisses	Na exceeds K. Feldspars usually sodic plagioclase. Color usually gray.	K exceeds Na. Feldspars usually potassic (ortho-clase or microcline). Color commonly reddish.
Volcanics	Commonly ultramafic. Low RE-element content. Low LIL content (K, Rb, U, Th). No europium depletion.	Less commonly ultramafic. Higher RE-element content. Higher LIL content. Conspicuous europium depletion.
Upper Mantle	RE-element content near 10 times that of chondritic meteorites.	Higher RE-element content.
Sedimentary Rocks	Immature (particles often angular, poorly sorted). Enriched in Na, Mg, and Ca. Relatively low in Si and K. Relatively low RE elements. Relatively low LIL content. No europium depletion.	Mature (particles commonly rounded, better sorted). Less Na, Mg, and Ca. Enriched in Si and K. Higher RE-element content. Higher LIL content. Conspicuous europium depletion.

entire continental mass (7.6 billion cubic kilometers) in the time available. A new challenge; a new problem. Will the growth of continents prove to parallel the episodic growth of sialic plutonism pic-tured on the front endpaper? Or will it come to seem a more sedate progress?

The salient characteristics of Archean and post-Archean crust are summarized and contrasted in Table 8.1.

In Summary

The Proterozoic revolution ushered in a new world of growing sialic cratons in which continental processes, a continen-tal geochemistry, and epicratonal sedi-mentation came to dominate the preserved historical record. The sodic granites and gneisses of the Archean world, together with its distinctive belts of greenstone and volcanogenic metase-diments, were transformed by massive granitic intrusions into larger cratons and subcontinents, predecessors to more familiar scenes.

With the potassium came other heat-generating, sial-loving elements, includ-ing rare-earth elements. That process simultaneously depleted mantle sources of those elements and enriched them in the growing sialic crust, perhaps shifting the declining global heat balance enough to contribute to the onset of plate tectonics. Related changes in the strontium-isotope ratios of sialic crust support the view that decrease in mantle heat flux, combined with the growing extent of continental surface and runoff, simultaneously reduced mantle influence on seawater chemistry while increasing that of the continents.

Although it took geocenturies for that transition to become global in extent, the Kaapvaal craton had already completed it 3 aeons ago. Western Australia followed about 2 geocenturies later (2.8 aeons ago) with its Fortescue Group, initiating the Proterozoic Eon and the Hamersleyan sequence there. But much of the present Northern Hemisphere seems not to have joined the Proterozoic world until around 2.7 to 2.6 aeons ago. The mantle-dominated Archean mode never again became prevalent on a long-lasting global scale.

The story of the Proterozoic revolution is encapsulated in that of the Randian sequence and its layered rocks, especially as developed on the Kaapvaal craton. Those rocks comprise a generally little deformed, composite pile of up to 28 kilometers of about 60 percent sedimentary and 40 percent volcanic deposits that accumulated first on the margins of the craton and then in a large, subsiding intracratonal basin. Following a long-lasting connection with the sea to the now southeast, the entrance to the basin closed, and, except for a brief and restricted later marine interlude, the sites of sedimentation were in an enormous lake and its tributary streams. Heavy-metal placers, containing more than half the world's known gold and large quantities of thorium-rich uranium, accumulated over large areas at many levels in debris fans and deltas near the entry of probably intermittent, braided rivers around the present northern, western, and eastern sides of the basin.

In general, Randian sedimentary rocks consist of well-sorted quartz-sandstones and siltstones whose rounded grains bear witness to the already prevalent recycling of ancestral sediments and the maturity of the product. Although younger sediments often differ in proportions, sources, and environment of deposition, the prevailing post-Archean pattern began here.

As much as 60 percent of the present continental mass may have come into existence by 2.6 aeons BP, when the Kenoran orogeny in Canada and similar episodes of potassic plutonism elsewhere finally concluded the transition to a Proterozoic mode. Later, intrusive rocks associated with other orogenies between 1.9 to 1.7 aeons ago added perhaps as much as another 20 to 25 percent of the continental mass. And it was nearing completion by the end of Proterozoic and Cryptozoic history.

The global sedimentary mass grew as a result of the weathering and erosion of large, areally extensive cratons and continents whose buoyancy assured freeboard above surrounding, commonly epicontinental seas in which sediments could accumulate. It attained a very large fraction of its present volume during earlier Proterozoic history, dwarfing the limited Archean record of mainly immature volcanogenic sediments. Beginning with the oldest post-Archean sediments we know, most are to some degree multicyclic. The same old stuff has been reworked, redeposited, relithified, and reworked again and again as it repeatedly moves through the rock cycle, becoming more and more mature in sorting, rounding, and segregation of friable and soluble components with each passage. Much of it has even, at some stage, melted, and then recrystallized before again becoming a component of the sample of sandstone, shale, or limestone one might break from some convenient or hard-won outcrop.

CHAPTER NINE

PLATE TECTONICS:
HOW CONTINENTS MOVE

Things are not what they seem. Terra is not all that firma. *The everlasting hills wear down. What was once tropic has changed places with the once polar. Sea floor is on a perpetual return trip to its mantle source, rafting continents and microcontinents with it, to be reassembled in other places, times, and patterns. The more we study the amazing world of plate tectonics, the curiouser it seems. Although ocean waters have apparently always covered the greater part of the post-Hadean Earth, the basaltic ocean floors are transient. Dense oceanic lithosphere, nowhere older than about 2 geocenturies, is regularly recycled through viscous upper mantle (the asthenosphere) at a current global turnover rate of about 110 million years. Lighter continental stuff remains at the surface like foam on the sea. Continents change outline, undergo erosion, locally increase or decrease in area and volume, but continental matter remains on top and never undergoes significant reduction in total volume. So goes the stubborn game of plate-tectonic chess, where pieces are moved without being taken from the spherical board and checkmate never comes. In this chapter we consider the manifestations, mechanisms, explanatory power, and beginnings of plate tectonics.*

The Diverse Manifestations
of Plate Tectonism

A creative surge of new insights transformed the prevailing geological perception of our planet between 1960 and 1965. After a half-century of contention, drifting continents were in and fixed continents were out, but with a difference. Earth is now seen as splitting open along a global system of interconnected rifts or spreading centers, like a wet softball bursting its seams (Figs.

9.1 MID-ATLANTIC RIFT SYSTEM COMES ASHORE IN SOUTHWEST ICELAND. *Above,* false color aerial view of main spreading center along northeast-trending breaks. Lake at center is Thingvallavatn, at Thingvellir ("Assembly Plains"), where the ruling Icelandic Althing formerly assembled. Reykjavik Harbor at left center. *Below,* scarp along the northwest coast of Thingvallavatn, in effect the crestal rift of the spreading center. [*Above, NASA, courtesy of Larry Carver.*]

193

5.2, 9.1). Continents are no longer supposed, as they had been by the original continental drifters, to wander through or over the floors of oceans. Instead they are ferried about, together with fringing chunks of adjacent seafloor, like cargo on a barge or a conveyor belt, but over courses constrained by spherical geometry. The barges are huge slabs of underlying mafic outer mantle. Along with their cargo of continents, microcontinents, and oceanic crust, they make up lithospheric plates that glide over the hot, viscous, shearing surface of the underlying asthenosphere at rates comparable to the flow of cold tar—about 2 to 2.5 meters (80 to 100 inches) per century.

The principal plates, thousands of kilometers across but only 70 to 100 kilometers or so thick, together with lesser plates and slivers, resemble a great ice pack. They are so tightly jammed together that individual plates cannot move freely. They jostle one another, generating earthquakes and volcanoes in the process. The major plate boundaries, in fact, are delineated on a world map of earthquake epicenters (Fig. 9.2). The Pacific rim, where plates collide with, slide by, and override one another, shakes the hardest and most often. Shallow-focus

9.2 EARTHQUAKE EPICENTER MAP OF THE PRESENT WORLD. Epicenters mark surface locations above the points (foci) where rupture is initiated and seismic waves begin. The linear arrays of shallow-focus earthquakes in the oceans denote the crests of spreading ridges. Dense clustering indicates convergent boundaries and transforms. Areas of deep-focus earthquakes (300–700 km deep) are circled. Most earthquakes outside such areas and some within are intermediate (100–300 km) to shallow-focus (less than 100 km) earthquakes. [*Data from National Oceanographic and Atmospheric Administration; earthquake epicenters from U.S. Geological Survey* Seismicity Map of the World *(data to 1974).*]

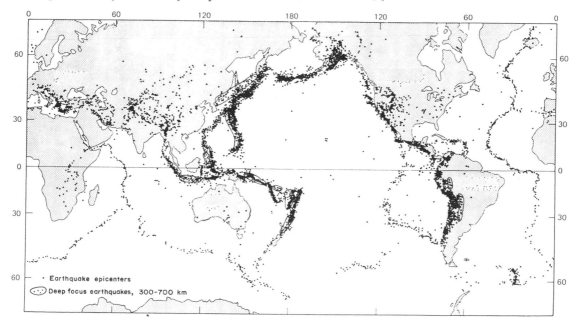

earthquakes of relatively low magnitude define a thin line along the geothermally active spreading centers or ocean ridges—divergent plate boundaries where new crust is made by upwelling from the asthenosphere, concurrently with lateral spreading. The epicenters of intermediate and deep-focus earthquakes define zones of convergence or collision between plates.

The continents themselves are quantitatively insignificant. The total mass of continental crust adds up to less than half a percent of Earth's mass. Earth's sialic continental stuff is conspicuous only because gravity dictates that it remain on top.

In contrast, the mafic oceanic lithosphere (ocean crust plus underlying rocky outer mantle) regularly descends beneath the continents or thin oceanic crust into the asthenosphere from whence it originally came, at a recycling rate of around 110 million years. There it may be remelted, mixed with more sialic fluids, and returned to the surface as the building stuff of volcanic-island arcs (as in the arcuate western Pacific island chains) or continental-margin arcs (like the volcanic Andes). The most ancient truly oceanic crust now known is only about 180 million years old, contrasting with the venerable framework of the continents, which everywhere is at least a dozen times as old.

Earth's surface is anything but static. (The most extensively recorded and damaging Earth movements in North American history since the arrival of European settlers began with the great earthquake at New Madrid, Missouri, in late 1811 and continued for the next 15 months during a series of aftershocks, two of which were as severe as the initial quake.) Like everything else in the universe, the face of Earth changes with time. It evolves. It constantly responds to the central driving forces of heat, planetary motion, the conservation of energy and angular momentum, and gravity. It ages. Although such responses are most readily observed in the properties and motions of air and water, they also drive the more stately motions of Earth's solid crust and mantle.

Such motions and the resultant large-scale changes in the location and configuration of continents and ocean basins reflect the general theory of crustal dynamics called *plate tectonics*. That theory had conceptual antecedents in the long-preceding interval of disputation about continental drift. It did not spring full blown to light as a single liberating perception. It did, however, emerge over a brief interval during the early 1960s from the heady ferment of new ideas and evidence that followed the intellectual frustrations and challenges of World War II. It has continued to grow in elegance and power with continuing research during its progression from hypothesis to well-supported explanatory theory. It explains alike the shaping of continents and oceans, the making and moving of mountains, the shaking and breaking of Earth's crust, the location and eruption of volcanoes, the biblical parting of the Jordan River (temporarily dammed upstream by quake-triggered landslides), and why the ocean floor is so young while the continents are old and wrinkled.

Perhaps the earliest reference to what we would now see as plate-tectonic motion is attributable to the prophet Zechariah in the King James translation of the Old Testament (Zechariah 14:4):

". . . the Mount of Olives shall cleave in the midst thereof toward the east and toward the west, and there shall be a very great valley; and half the mountain shall remove toward the north, and half of it toward the south." That describes what geologists today would call strike-slip motion along a transcurrent or transform fault.

It was motion along such a fault that so rudely awakened the residents of San Francisco at 5:13 on the morning of the 18th of April 1906. Frictional retardation of the upper 15 kilometers of the San Andreas fault, west of which the Pacific basin rotates northwestward (Fig. 9.3) at a rate of 5 or 6 meters a century, had long resisted motion. Stress then abruptly gave way to strain, offsetting two great plates of Earth's crust, fracturing gas lines and water mains, setting fire to buildings, toppling the feeble works of mankind (Fig. 9.4), and killing more than 800 people.

9.4 FAMED ZOOLOGIST–GLACIOLOGIST LOUIS AGASSIZ, OFTEN CONTEMPLATED IN THE ABSTRACT, SEEN "IN THE CONCRETE." Stanford University, 18 April 1906. [Photograph by G. K. Gilbert, courtesy of U.S. Geological Survey film library.]

9.3 THE PACIFIC PLATE AND ITS MOTIONS. [Adapted from P. J. Wyllie, The Way the Earth Works, p. 68, Fig. 5.5, copyright © 1976, John Wiley & Sons, Inc.]

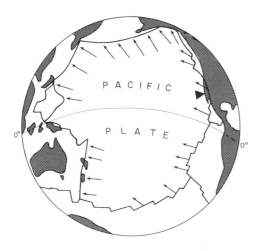

Thus was the sleeping city informed that it lay astride an active, 1,400-kilometer-long, northwest-trending, vertical fault of the same type as Zechariah described. In terms of energy released, the San Francisco earthquake was one of the greatest in history. It had an estimated *Richter magnitude* of 8.3 to 8.5— on a logarithmic scale in which the energy released increases 35 times with each full number up the scale.

As a result, crust west of the fault lurched as much as 6 meters closer to the Aleutian Trench, where it is expected to arrive about 50 million years hence. A small fraction of the energy released in such motions and their aftershocks is instrumentally recorded as fast-traveling compressional and shear waves that penetrate the Earth. Most of the energy liberated, however, is invested in moving rock. We feel it only as long and relatively slow-moving waves that shake the ground beneath our feet and account for most of the visible damage to structures. Slippage along the nearly straight San Andreas fault is sideways along the trace, or *strike*, of the fault.

Because motion along the strike of the San Andreas fault is lateral and slips to the right when viewed from either side (Fig. 9.5), it is called right-lateral strike-slip motion. The western side of the San Andreas fault moves northward relative to the eastern side at about 3 to 4 meters a century. An additional meter or two of motion on the part of the Pacific plate is compensated for by structural extension in the Basin and Range Province of the western United States. Only the part of San Francisco east of the San Andreas is geologically North American in the sense of being a part of the North American

plate; the area to the west is a part of the Pacific plate. Earthquakes result from motion along this fault because that motion is jerky. Stresses must accumulate before strain occurs. The fault is not quite straight, however. Stresses build up at places where irregularities in strike, or other sources of frictional delay, cause opposing sides of the cool, upper crust alternately to lock together and lunge ahead.

The shaken residents of San Francisco in 1906 were scarcely of a mind to appreciate either the scientific significance of the event they had experienced or the fact that the San Andreas fault is one of the three main kinds of plate boundary—a *transform fault*. Transform faults, abbreviated to *transforms*, offset and connect the ends of diverging plate boundaries, as in the short jogs along the right side of Figure 9.3. In 1906 no plausible explanation for such motions existed, and the sophisticated geophysical instrumentation needed to measure and date them would not be available for another four decades.

9.5 ROUTE OF SAN ANDREAS FAULT ACROSS CENTER OF PHOTOGRAPH SHOWS A KILOMETER OF RIGHT-LATERAL OFFSET ON A YOUNG STREAM. The air view is east toward Bakersfield across the Carrizo Plains of west-central California.

Once reasonably accurate world maps became available, however, imaginative people, observing the matching continental margins on opposite sides of the Atlantic, would likely have wondered whether the Americas might once have fitted against Europe and Africa. Indeed, global maps displaying such a hypothetical clustering and separation were published as early as 1858 by Antonio Snider, an American living in Paris. And Charles Lyell anticipated him in the third edition of *Principles* (1834), with a set of interpretative maps showing clusterings and rearrangements of the continents at equator and poles, with Australia at different times abutting North and then South America.

It, however, was not until 1912 that the intrepid German meteorologist Alfred Wegener (Fig. 3.3) clearly outlined— first in journal articles and later in book form (1915)—his view of the stately minuet of the continents that came to be known as *continental drift*. The 1929 English edition of Wegener's book, *The Origin of Continents and Oceans*, appeared shortly before his death on the Greenland ice cap at age 50. This remarkable work integrated evidence in support of his views from sources then known. Alas, the hypothesized mechanism violated physical laws, and many of the suggested consequences had plausible alternative explanations. Support came mainly from the Southern Hemisphere, where supporting field evidence was more compelling. It came particularly from two other outsized figures, Alexander Du Toit of South Africa, whose 1937 book, *Our Wandering Continents*, marshalled powerful support for continental separation, and later from colorful Sam Carey of Tasmania.

But science demands skepticism until inconsistencies are resolved and plausible alternatives found wanting. Skepticism about Wegener's ideas prevailed in the Northern Hemisphere, particularly in the United States and particularly regarding the proposed mechanics of drift. Wegener visualized the sialic continents as moving like gigantic icebreakers through an oceanic crust of mafic basalt. Harold Jeffreys, then dean of British geophysicists, epitomized opposing views when he correctly objected that such a mechanism was prohibited by the relative strengths of the materials involved. The seismologic identification of deep continental roots imbedded in solid outer mantle intensified the problem. As computations from physics once seemed to preclude a truly great age for the Earth, so now they seemed to prohibit continental drift. And, as before, mistaken assumptions invalidated the computations.

Matters remained at an impasse until 1956, when (too late for Wegener) paleomagnetic results from laboratories on two continents showed that Europe and North America, once together, had diverged to their present separate locations over the preceding 2 geocenturies. If continents can be, in effect, observed to diverge or drift, there must be a way to explain it. Minds were opened to the possibility of a different kind of motion, perhaps one driven by a system of convection currents such as had earlier been proposed by Britain's Arthur Holmes (Fig. 3.3) and America's David Griggs.

The time was ripe for new ideas, and they were not long in coming. The seminal works were those of Princeton's Harry Hess (Fig. 3.3) and of Robert Dietz, then of the U.S. Navy Electronics Laboratory at San Diego. In 1960, Hess circulated a preprint of a paper, the publication of which was delayed until 1962. Dietz, meanwhile, published in 1961 a remarkably similar paper in a faster-moving journal. A flurry of discussion arose (although not between Hess and Dietz) as to which of the two had priority—an event of the type that can be sometimes painful and sometimes ludicrous in a world given to simplification of history. In this case, the matter was settled, at least temporarily, by Dietz when he expressly conceded priority to Hess. In fact both men played important roles in the rehabilitation of continental drift, and Dietz's term for the motions that produce it, *sea-floor spreading*, has stuck.

The central perception, based first on paleontological and then on geophysical evidence, is that sea floor ages systematically in both directions away from spreading ridges, and that oceans, therefore, widen with time (before closing again). The liberating concept that grew out of this was that, instead of plowing through solid oceanic crust, the continents ride passively atop a thicker lithosphere (including solid outermost mantle), the motion of which is seen as a product of heat-induced buoyancy at the spreading ridges with cooling and increasing density away from them. Continents are rifted apart as oceans open. Hot, convecting, asthenospheric mantle expands, becomes lighter still as it incorporates water, changes mineralogically, and, nudged from below, expands buoyantly upward to rift the lands and elevate the spreading ridges. Such expansion simultaneously creates new oceanic crust and reduces the volume of ocean basins, thereby elevating sea level and inundating previously exposed continental shelves. The new crust cools, becomes denser, and therefore moves downslope away from the rises. Cooling and descent continue as oceans widen, until convergence of the spreading oceanic lithosphere with continents or with thinner oceanic lithosphere completes the cycle that began with sea-floor spreading and ends with the return of oceanic lithosphere to the asthenosphere. The elevated pressures and expulsion of water experienced by descending oceanic lithosphere result in continually increasing density until the subducted lithosphere completes its descent. There it becomes assimilated by the asthenosphere, or is mixed with continental lithosphere in a basal zone of melting and returned toward the surface, as in the Andes.

Far fetched as these ideas seemed, they were soon fortified by new observations. The insights and predictions of Hess were confirmed when sensitive ship-towed magnetometers revealed mirror-image sets of parallel bands of distinctively magnetized sea floor on opposite sides of the basaltic spreading ridges (Fig. 5.5). Each unit of those roughly matching paleomagnetic patterns separately acquired the signature of Earth's then-current magnetic polarity, as its parent lava cooled through the Curie temperature at which susceptible particles become magnetized. Again there followed concurrent insights by two sets of authors. (Clearly the subject was "in the air.") In 1963, two then-young English scientists, Fred J. Vine and Drummond H. Matthews, published the now almost universally accepted explanation. They suggested that the matching sets of magnetic stripes of alternating polarity track the successive polarities of the ridge crests, moving apart from one another as if on a pair of conveyor belts traveling in opposite directions. Like the checks that record your financial transactions, these stripes record earlier and different magnetic episodes that can be matched not only on opposite sides of the ridge axes but also with dated polarity reversals on land. They are a material record of sea-floor spreading and its average rates.

The north Atlantic Ocean is a key region. The Mid-Atlantic ridge, with its opposing but matching sets of magnetic stripes, runs right down the center, with a lot of east–west offsets (back endpaper, Fig 5.2). Canada's Tuzo Wilson suggested in 1965 that, if the Atlantic is really spreading apart in opposite directions at the ridge crests, the actual motions of those offsets (Wilson's *transform faults*) should be the reverse of those implied by the directions of offset. That was because Wilson saw the east–west offsets (the *transforms*) of spreading centers as initial irregularities in the original north–south break between oppositely moving plates. (Notice that the full arrows on the back endpaper, showing directions of spreading, point in opposite directions to the motions one would infer from the direc-

tions of offset along the transforms.) When the predicted motions were confirmed by the seismological studies of Columbia University's Lynn Sykes, most reservations about the reality of sea-floor spreading and the directions of its motions vanished. Yet in reality, the transforms are now better understood as adjustments to variable rates of linear spreading for the same radial spreading away from the poles of rotation.

Wilson's inspired misintepretation, and Sykes's "confirmation," of the predicted direction of motion along transforms had the effect of vindicating Wegener at last, while continuing inquiry served to validate his beliefs with a flow of new discoveries and insights. Continental drift became both a forerunner to and a consequence of plate tectonics. Continents do not really drift—they are ferried about on the backs of lithospheric plates that glide over the yielding asthenosphere, in patterns constrained by their tight packing and rotational axes. At the same time, evidence for the deep structure of continents took on a new twist. Continents do, in fact, have roots, although not as deep as was once suspected. The deep seismicity (from depths as great as

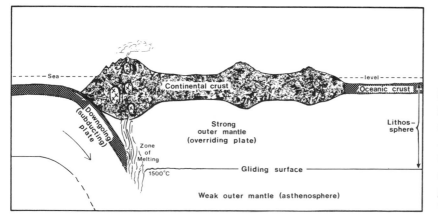

9.6 ELEMENTS OF PLATE MOTION AT CONTINENTAL BOUNDARIES. An *active* continental margin is shown on the left, a *passive* margin on the right.

700 kilometers) recorded along parts of some convergent continent—ocean boundaries comes not from continental roots but from downward extension (*subduction*) of the dense oceanic plate.

If sea floor spreads, continents must move, either because the plates that transport them are being pushed or pulled or both. Seismology establishes that the principal motion involves a gliding of the lithosphere over the readily sheared upper surface of the yielding asthenophere lying *beneath* the deepest continental roots (Fig. 9.6). While much is thereby clarified, much yet remains to be explained and integrated into the broader paradigm.

What Plate Tectonics Describes and Explains

In simplest terms, plate tectonics describes Earth's present lithosphere as comprising eight major and a variable but larger number of lesser plates that move horizontally with respect to one another and to Earth's interior, except where they converge and overlap. Their margins are delineated by the alignment of earthquake epicenters and volcanoes (Fig. 9.2, back endpaper), linear depressions at the crests of spreading ridges, and, more broadly, by folded mountain chains. The plates glide over the upper surface of the yielding asthenosphere, unimpeded by continental roots, at rates comparable to that at which an anvil might settle through solid asphalt.

Four main types of plate margins or boundaries are known. First are *divergent*, or *accreting*, boundaries (Fig. 9.7A), also called *spreading centers* (or spreading ridges or axes)—such as the Mid-Atlantic ridge. At such margins, plates move away from one another as hot buoyant mantle wells up from the asthenosphere and spreads laterally to create new oceanic crust. Second are *convergent*, or *consuming*, boundaries (Fig. 9.7B), where plates move toward and overlap one another—as along the western coast of South America, where ocean floor descends beneath the continent (Fig.

9.6). Third are *collisional boundaries* (Fig. 9.7C), where opposing buoyant continental masses strive to occupy the same space until their compressional shortening results in folded mountain ranges such as the Himalaya, the Alps, and the Appalachians.

Fourth is the group of *transform boundaries*. With the recognition that some transforms do undergo linear displacements of hundreds of kilometers in the direction of motion, offsetting plates to become plate boundaries themselves, the term has come to be used mainly or exclusively for that phenomenon. Thus transform boundaries are places where plates slide past one another on near-vertical fractures, not infrequently hanging up and generating great earthquakes. The San Andreas transform, for example, displaces the spreading axis of the East Pacific rise northwestward from the Gulf of California to its continuation west of northern California.

The boundaries of the American plates are thus divergent in the east (mid-Atlantic spreading center), convergent in the southwest (Peru—Chile trench), and mixed in the northwest, where the western edge of California is slipping northward along the 1,400-kilometer-long San Andreas transform while Oregon and

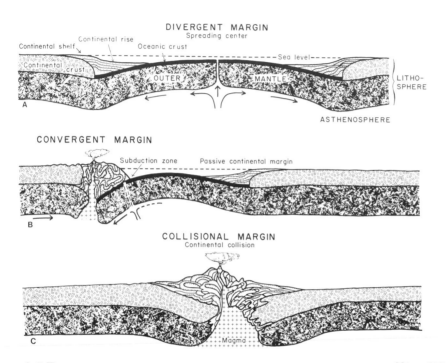

9.7 DIVERGENT, CONVERGENT, AND COLLISIONAL PLATE MARGINS. *[Simplified from R. S. Dietz, "Geosynclines, Mountains and Continent-Building,"* Scientific American, *v. 226(3), p. 37. Copyright © 1972 by Scientific American, Inc. All rights reserved.]*

Washington override seaboard micro-plates. Most plates include both continental and oceanic sectors. The Pacific, Nazca, and several smaller plates, however, include almost exclusively oceanic crust, while the Arabian and some other small plates are mainly continental.

Unlike the plates themselves, continents can be considered to have only two kinds of margins, active and passive (Fig. 9.6). Their margins can be said to be active where they coincide with convergent or transform boundaries, as along the Pacific coasts of the Americas. They are passive where they lie within a plate, as do the continental margins that border the Atlantic Ocean. Active continental margins are characterized by narrow continental borderlands and slopes, active seismicity, volcanism, the intrusion of sodic granites, distinctive metalliferous ore deposits featuring the precious metals and uranium, and (at transform boundaries) basinal oil deposits. Passive continental margins have wide continental shelves underlain by thick sedimentary wedges, salt domes (often oil reservoirs), little or no seismicity and volcanism, and mainly sedimentary ore deposits. Sulfide-rich metalliferous deposits of copper, chromium, platinum, nickel, and associated metals are found at oceanic spreading ridges and in active continental margins that override them.

The main Pacific spreading center, the East Pacific rise, differs from the Mid-Atlantic ridge in not being symmetrically positioned. It closely approaches the western margin of the South American plate and, in California, is actually overridden by the North American plate. Together with the Mid-Atlantic ridge and other spreading centers, it is part of a continuous, 75,000-kilometer-long, mostly submarine chain of ridges that completely encircles the planet. The Atlantic is widening, while the Pacific (a remnant of the former global ocean called Panthalassa) narrows. The East Pacific rise, however, seems to be spreading at several times the rate of the Mid-Atlantic ridge.

Spreading cannot continue unless new space is created elsewhere. How? As explained earlier, the space needed is most parsimoniously made at convergent plate margins as lithosphere descends along subduction zones, as well as by crustal shortening at sites of continental collision. In this way crust is consumed or shortened at rates comparable to the origin of new crust at the spreading centers.

The consumption of crust by subduction is well illustrated beneath the deep-sea trenches that now rim much of the Pacific Ocean and Southeast Asia (Fig. 9.8). At such places, cold, heavy oceanic lithosphere descends gravitationally beneath the opposing continental litho-

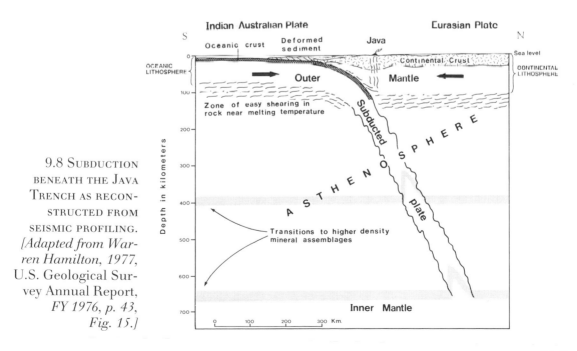

9.8 SUBDUCTION BENEATH THE JAVA TRENCH AS RECONSTRUCTED FROM SEISMIC PROFILING. [Adapted from Warren Hamilton, 1977, U.S. Geological Survey Annual Report, FY 1976, p. 43, Fig. 15.]

sphere or sometimes beneath thinner and lighter oceanic lithosphere. It melts, with the assistance of accompanying water, at appropriate temperatures and depths and is either assimilated or mixes with lighter sialic matter to produce magmas of intermediate composition and density. Such magmas arise to generate volcanic arcs, raise Andean types of mountains, and create deposits of copper and precious metals.

Plate boundaries are where the action is. Plates grow, disappear, jam against one another, and accrete transient microcontinents. Earthquakes and volcanoes concentrate there. There is no way active plate boundaries can escape notice.

The matching paleomagnetic stripes on both sides of the spreading ridges are clues to rates and direction of motions (back endpaper, Fig. 5.5). The correlation of the stripes with identifiable radiometrically dated reversals of the magnetic poles provides a quantitative foundation for plate-tectonic theory. By the late 1960s analysis of results in Columbia University's computerized data bank had shown that all of many spreading ridges fit the general pattern of the Mid-Atlantic ridge. In addition, the polar-reversal time scale had been carried back to 75 million years (and later to 165 MY).

All plates were found to be moving relative to all others, although there is evidence that the African and Antarctic plates may be fixed with reference to the deep mantle.

These motions, however, are not random. As noted earlier, they involve sectors of a spherical surface and are controlled by rules of spherical geometry. The direction of motion is away from spreading centers, parallel to transform boundaries, and around an axis of relative motion. Segments of a spreading ridge follow great circles whose relation to the spreading ridges follow small circles with respect to the pole of spreading, analogous to parallels of latitude. Relative linear motion thus increases with distance from the pole of spreading for the same angular motion.

Observed rates of linear spreading thus range from about 2 to 16 centimeters per year, in general slower near the spreading poles and faster away from them. An average rate of spreading of 3 centimeters per year would be more than enough to open the Atlantic in the time available since its last episode of drift began, while an aeon of subduction at the same rate would quite suffice to consume the entire global ocean floor. Disposing of the phantom ocean floor of Archean times by post-Archean subduction presents no problem.

Although long-term rates of motion between plates are given by distances between matching polarity stripes and their radiometric ages, precise rates, and whether motion is gradual or episodic, cannot be reckoned in this way. It seems likely from the thicknesses of the distinctive fissure-filling dikes at spreading ridges that later Phanerozoic motion has been spasmodic, in roughly meter-wide surges every 10 to 100 years. Rates and patterns of spreading are now being measured using laser ranging (by means of mirrors on fixed satellites and very-long-baseline interferometry between radio antennas equipped with high-resolution atomic clocks). Results available on eight such baselines connecting stations on different plates during the interval from 1980 to 1986 roughly confirm motions estimated from paleopolarity.

How Plate Tectonics Works:
Hot Spots and Hot Lines

Geophysicists long rejected continental drift because its proposed mechanism was physically impossible: "like trying to stuff whipped cream into cold honey," one remarked. Geologists, particularly paleontologists, were divided. Some found drift necessary to explain the observed distributions of organisms. Others considered that patterns could be explained given sufficient time and appropriate land bridges—for example, emergence of the Bering Strait. How is it that plate tectonics, resulting in continental drift, has become so widely accepted as a ruling theory in geology? What *does* move the plates? And how is one to explain tectonic features within plates as well as at their edges?

The critical mechanistic difference between the new plate tectonics and classical continental drift has to do with the gliding surface on which the plates move (Figs. 9.6, 9.8). Both hypotheses called on buoyancy to keep continents high, and most drifters liked convection as the driving force. But plate tectonism overcame the problem of anchored continental roots by having horizontal motion localized along a gliding zone of easy shearing *below* such roots, at the top of the asthenosphere—literally *within* the outer mantle. Instead of the continents traveling independently and randomly, driven but not directed by some mysterious force, they move only with their lithospheric plates, whose motions are constrained with respect to one another.

Some plate motion clearly occurs in response to the positive buoyancy of rising convection currents beneath the spreading ridges. Such motion is called *ridge-push.* The evidence for it is found in Earth's convective cooling and the resultant differences in elevation of the sea floor. Studies of heat flow at the planet's surface show that spreading ridges are hot, whereas trenches along convergent boundaries are cool. Even though the upper mantle was depleted of its heat-producing elements during terminal Archean and earliest Proterozoic time, their abundance in the lower mantle and core reheats Earth more rapidly than conduction alone can cool it.

Like air above a bonfire, therefore, the mantle convects, and the rising limbs of convection cells move asthenospheric matter upward beneath the spreading ridges like a toy sailplane above a bonfire. Hot rock expands and rises. Buoyancy increases as density decreases. Minerals of the outer mantle react with seawater as convecting asthenosphere approaches the surface, converting it from peridotite (with a density of 3.2) to serpentinite (with a density of 2.6 or less), so that it expands and elevates the ridge still more. Continuing reduction of pressure from load eventually results in partial melting and magma formation. The fraction that is to become sea-floor basalt is injected along the axis of the spreading ridge, adding to the forces that pry it another step apart. We see evidence of this last step in the linear depressions that everywhere follow the crests of more slowly spreading ridges.

Ridge-push is not the only source of spreading energy, however. A second force is provided at the opposite edge of the slab by *plate-pull*. Plate-pull results from the cooling, dewatering, and increasing density of the once-hot, new sea floor as it slowly creeps downslope from spreading ridge to subduction zone. Within the observable record the journey takes an average of about 110 million years. It was probably faster when Earth was younger and hotter. In a word, Earth's great heat engine, with gravity, is the driving force, linking the positive buoyancy of ridge-push with the negative buoyancy of plate-pull to maintain a balance between the growth and consumption of lithosphere.

A seismically delineated, vertical profile across western Java (Fig. 9.8) illustrates plate-pull. Here the oceanic lithosphere of the Indian–Australian plate on the south descends beneath the relatively thicker, lighter, continental lithosphere of the Eurasian plate mosaic to the north, pulling the floor of the Indian Ocean with it into the asthenosphere. In the upper sector of its descent, water carried with the subducted plate and squeezed out of hydrous minerals induces partial melting of oceanic crust and metal-enriched sediments as they descend to higher temperatures. That gives rise to magmas, which ascend into the continental lithosphere as the descending plate continues to sink, meanwhile reacting with it to generate a volcanic arc and sialic plutonism.

Some plates or parts of plates lose their seismic identity below about 250 to 300 kilometers, but not this. The Indian–Australian plate is observed to continue its descent toward the base of the asthenosphere until it eventually softens and disappears from seismic view at greater depths. In fact, subduction-zone earthquakes, until recently, have not been recorded from depths below 650 to 700 kilometers, implying that descending plates rarely find their way beneath the asthenosphere. As at the races, however, most records eventually fall. Seismic studies since the mid-1970s at MIT and the Carnegie Institution have now independently recorded penetration of the descending plates beneath the Kuril Islands in the northwest Pacific to depths of 1,000 to 1,200 kilometers—well into the presumably solid and geochemically distinctive lower mantle—a new observation, remaining to be explained.

The heat engine within the planet—assisted by water, density variations, and gravity—thus manages simultaneously to ventilate, to keep the continents and ocean floors moving, and to send a flow of new and commonly metalliferous magmas toward Earth's surface (not fast enough, alas, to meet industrial demands). It also keeps seismologists guessing, and provides an occasional surge of excitement, or terror, to the public in the form of earthquakes and volcanoes.

Subduction is the critical process. Were new space not created as fast as new sea floor is produced, spreading could not take place. Earth would have to find some other way to cool its radiogenic fever. Plate tectonics in the sense now known would not have started and could not continue.

How then does rock that once reached Earth's surface because of its positive buoyancy manage to descend again, as it is observed to do, creating plate-pull as it sinks? To summarize and amplify, estimates imply that more than half the heat produced within the planet is vented to its surface during the formation, cooling, and eventual reheating of oceanic crust. Molten magma that rises from the hot asthenosphere because of reduced density and load contracts and thins as it solidifies and cools again away from the spreading ridges. It glides downslope toward the ocean margins, becoming still denser as it continues to cool en route. On collision with thicker, lighter continental lithosphere or thinner oceanic lithosphere of an opposing plate, the cooling, shrinking oceanic crust continues to descend, lose water, and become denser still. The differences in density today—about 2.8 for continental crust, 2.9 for ordinary oceanic crust, and 3.2 or more for cold, descending, oceanic lithosphere—are enough, under the rule of gravity, to drive subduction into a weak, viscous asthenosphere of nearly the same or lesser density. Where collision is between opposing continents, significant subduction is forbidden by gravity. Chunks of sial may be engulfed and carried into subduction zones elsewhere and mafic rocks obducted, but that is not the pattern. When continents collide, they pile up into mountains with deep roots, as in the complex, multicollisional Himalaya.

Spreading ridges, therefore, are ruled by buoyancy. Subduction zones are places where "negative buoyancy" prevails. Transforms are vertical surfaces where plates slide by one another. And continental collisions show what happens when a nearly irresistible force meets a nearly immovable object.

Where spreading axes are known to be linear, the crests of convection cells beneath are probably also linear. How deep they go remains to be seen, but we have some clues in the form of geochemical distinctions between the melts that arise within plates and those that favor plate margins. Some forty instances of isolated volcanism or high heat flow, both within plates and along spreading ridges, have now been identified as *hot spots*, supposedly the surface expression of great plumes of high heat flow that arise from deep within the mantle. A few, such as at Yellowstone, are found within continents. Most, however, are oceanic. The Hawaiian chain is persuasively interpreted as the result of the northwesterly drift of the Pacific plate over such a mantle plume during the past 4 to 5 million years. About half of all hot spots, for example Iceland, are on or close to spreading ridges. Such a concentration supports the view that hot spots may be associated with *hot lines* beneath the spreading ridges, and some or all of the latter may even have been initiated by chains of hot spots.

Lava brought to the surface by hot spots differs geochemically from that of hot lines. The volcanic rocks of the spreading ridges resemble the present outer mantle in being relatively depleted in large-ion-lithophile, heat-producing elements such as potassium and uranium. They presumably come from asthenosphere depleted in such elements by reason of repeated episodes of sea-floor spreading. The volcanic products of the hot spots are not so depleted. Their abundances of the LIL elements are similar to those of Archean greenstones. The stuff in the hot spots, therefore, presumably comes from undepleted lower mantle, beneath the asthenosphere—from deep-mantle plumes. It suggests that the presently active convection cells beneath the spreading ridges are mainly limited to the upper 700 kilometers or less of outer mantle, ostensibly the asthenosphere.

In support of that interpretation, scientists of the Smithsonian Institution have found that the Mid-Atlantic ridge can be divided into segments between the equator and 72°N, based on the composition of lavas as well as surface elevation. Ten segments consist of typical low-potassium mid-ocean-ridge basalt in deep stretches of the ridge's axial valley. They are separated by short intervals of high-potassium intrusive activity on relatively elevated parts of the axial zone. That is strong evidence for a different, undepleted, and presumably deeper-mantle source for the hot spots and for a line of hot spots along the ridge axis.

Crustal morphology also suggests a connection between hot spots and spreading. Powerful hot spots may initiate spreading. As in lesser tension-crack systems such as columnar basalts and drying muds, they may generate triradiate cracks that define 120° angles. Such sets of plate junctures are aptly called *triple junctions*. Whether cause or effect, triple junctions imply coincidence with present or former hot spots and mantle plumes. It is not unusual in older structural patterns to find regions where it appears that one arm of such a triad failed to develop while the other two evolved to open a new ocean.

The New Paleogeography: Microcontinents, Exotic Terranes, and Collages

Just when Earth scientists were congratulating themselves on having agreed on a unifying theory, intraplate tectonics raised its quizzical head. Oceanic plateaus of ambiguous origin account for about 10 percent of the present ocean floor. Some of these, such as the Seychelles Bank northeast of Madagascar, show the basic structure of continents, with seismically identifiable crustal roots. In contrast with typical oceanic crust (only 7 to 10 kilometers thick), their roots descend to depths of 30 to 40 kilometers before reaching the crust–mantle boundary (*within* the lithosphere, Fig. 6.2). That is where seismic velocity abruptly increases from 7 kilometers or less per second to 8 kilometers or more beneath—the important transition known as the *Mohorovičić discontinuity* or *Moho*. Seismicity in the Seychelles records a 70- to 100-kilometer-thick continental type of lithosphere. What are such microcontinental structures doing out there surrounded by oceanic crust and lithosphere?

Microcontinents are common in the Atlantic, Indian, and western Pacific oceans but are rare in the eastern Pacific and Southern oceans. They are found most frequently in the marginal sectors of oceans but are rare in areas of new ocean floor near spreading ridges. That would be consistent with their having been rifted away from parent continents as new oceans first opened, subsequently to be separated from the spreading centers as the ocean basins widened. Where old continents override the ocean basins, such nonsubductable microcontinents would be swept against and annealed to the continents, or thrust over them (*obducted*). Such processes could explain alike the present rarity of microcontinents in the eastern Pacific and their abundance where the western Pacific is only marginally overridden, if at all, by the Eurasian plate.

That would also account for the prevalence of former microcontinental wedges and slivers now welded to the active Pacific coast of the Americas. Only since the late 1970s, in fact, have Earth scientists realized that the western 12 percent or so of North America is a collage of exotic crustal slices of contrasting, mainly pre–late Mesozoic rocks. Figure 9.9 shows the distribution pattern of these suspect *exotic terranes*. These far-traveled blocks and slivers of land docked against and became welded to western North America mainly during its late Mesozoic deformation. (Variation in spelling is used to differentiate such fault-bounded exotic *terranes* from the broader general use of the term *terrain* for other landscapes of any origin.)

The five patterned terranes in Figure 9.9 illustrate their variety. One of these, the Cache Creek terrane, is a classic example. Geologists a half-century ago were already puzzling over how to account for this anomalous mixture of late Paleozoic to Triassic sediments, ophiolites, and blueschists whose Permian microfossils were so like those of eastern Asia. The tiny Chulitna terrane of central Alaska, barely 50 kilometers long, is actually a composite of separate terranes, assembled somewhere far to the south before heading for caribou country. It includes a bewildering variety of rock types that range in age over much of pre–late Mesozoic history and are unlike those of other Cordilleran exotic terranes—*Cordillera* (cor-di-yér-a) referring to the great system of mountains from the Rockies to the Pacific coast in

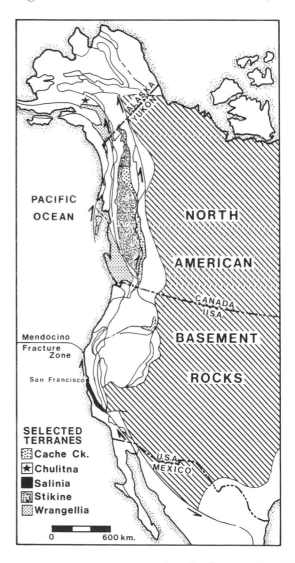

9.9 WESTERN NORTH AMERICAN EXOTIC TERRANES. Those discussed in the text are patterned; boundaries alone are shown for others. Arrows show movement of major strike-slip faults. *[Adapted by permission from Nature, Vol. 288, p. 330, Fig. 1. Copyright © 1980 Macmillan Journals Limited.]*

North America (as well as corresponding lands to the south).

The Stikine terrane is the largest. After a complicated Mesozoic history, during which it was folded, thrust eastward, stretched by strike-slip faulting, and, in effect, nailed to the North American plate by granitic intrusion, parts of it were somehow jammed both east and west over adjoining terranes. That was followed by the piecemeal arrival of Wrangellia, strung out over a north–south distance of more than 3,000 kilometers from the Wrangell Mountains of Alaska to eastern Oregon and rotated from its initial paleomagnetic orientation in the process. It consists of similar, little-deformed, late Paleozoic volcanic-arc deposits, overlain by a variety of Mesozoic rocks, all of which yield paleolatitudes that are equatorial with respect to early Mesozoic North America.

Finally, the Salinia terrane—wherever it may have come from—is of special interest because it is still moving northward along the west side of the San Andreas fault, carried by the rotation of the Pacific basin. Consisting entirely of metamorphosed rocks of undetermined age, it clings uneasily and temporarily to California, assisted by the stapling effect of late Mesozoic intrusive granites.

All of the five score or so of these suspect terranes now known in this region are far traveled. Paleomagnetic evidence implies that many came from the south or southwest. The subducting east paleo-Pacific plate scraped its buoyant microcontinents off against or onto the overriding North American plate like barnacles from a whale's back—annealing, stretching, and fragmenting them in the process.

The west American or Cordilleran collage is not the only such feature. The Himalayan–Tibetan region (to be heard from in Ch. 16) represents an equally or more complex product of plate amalgamation. China is a generally older assemblage of late Proterozoic and younger microcontinents and subcontinents brought together between about 1,800 and 50 million years ago. The stranger-than-fiction saga of the growth of the eastern fringe of North America also describes a collage. It records various stages in the opening and closing of the pre-Atlantic or *Iapetus* ocean during older Phanerozoic and probably latest Proterozoic history. Details are only now being worked out with the aid of seismic sounding, but bits of the Iapetus borderland range in time from early to late Paleozoic and in place from Florida to Newfoundland—even as far as northern Norway. As the Iapetus ocean narrowed and proto-Africa approached proto–North America, for example, two microcontinents (now the Piedmont and Carolina Slate Belt terranes) were trapped, shoved against, and sealed to proto–North America (Fig. 9.10). With continuing study of the region the whole Atlantic coast of North America is turning out to be a collage of similar ancient continental terranes.

Finally, all of the Paleozoic continents were swept together and joined to form a single universal landmass or megacontinent called Pangaea (from a Greek compound referring to a union of all lands). When Pangaea again rifted to initiate the present epoch of sea-floor spreading, the continents broke away nearly along their former boundaries. Several slivers of proto-Africa and Europe, as well as the Piedmont and Carolina Slate Belt, remained attached to present North America, along with what may be ghostly remnants of the subducting pre-Pangaean oceanic crust. It is estimated that, altogether, as much as 30 percent of present-day North America might represent Phanerozoic additions of exotic terranes—shuffled from their places of origin to new continental moorings like so many itinerant freighters.

Plate tectonism, including accretion of terranes, or microplate tectonics, has been an overriding factor in determining the shapes and positions of continents and oceans, the depths of oceans, and the sites and styles of mountain building during much of geologic history. It was also important, along with glaciation, in accounting for the apparent cycles of flooding and draining of the continents that are reflected so clearly in the sedimentary and paleontological records of the shallow epicontinental seas and basins of the past.

When a preexisting continental mass like Pangaea is rifted open, the expan-sion that elevates new spreading ridges simultaneously decreases the total volume available for storage of water within the sea. The more young, warm, shallow ocean floor there is, the less room there is for water. Sea level therefore rises with the newly elevated spreading ridges to flood the continental lowlands, shrinking the lowland continental habitats and expanding the shallow marine habitats available for colonization. It has been calculated that, because of higher spreading rates and larger ridges, the late Mesozoic sea was 100 to 150 meters above its present level—80 million years ago when present spreading ridges were young and hot, global climate benign, and when dinosaurs and coal swamps prospered.

Some time after such a relative elevation of sea level, the spreading cycle slows, the rate of heat flow declines, and sea level subsides again. Thus sea level varies, the flooding of continents waxes and wanes, and the degree of continentality and susceptibility to glaciation

9.10 DIAGRAMMATIC VERTICAL PROFILE ACROSS SOUTHEASTERN UNITED STATES TO THE APPALACHIAN FOLD BELT, SHOWING EXOTIC TERRANES. Fault-bounded Piedmont and Carolina Slate Belt terranes became annealed to eastern North America during three Paleozoic collisional events. [*Simplified and condensed from Jack Oliver, 1980,* American Scientist, *v. 68(6), p. 682, Fig. 5, and Zvi Ben-Avraham, 1981,* American Scientist, *v. 69(3), p. 297, Fig. 8.*]

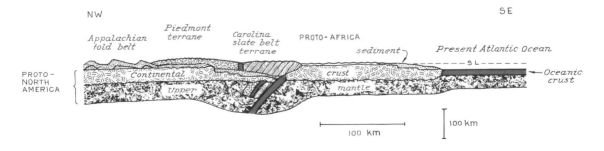

changes under the influence of plate tectonism, affecting both biologic evolution
and the geography and climates of former times. From its beginnings, plate
tectonics has been a force to be reckoned
with in Earth history.

Geologic Record of Plate Tectonism and Its Consequences for Earth's History

We return at last to the matter of how
and when the processes of plate tectonism may have begun and how they might
have affected earlier Earth history. Subduction, prime mover of plate-pull, is the
critical component. How far back in time
might subduction and a system of plate
tectonism similar to that of the present
have operated?

Consideration of that question involves
new language as well as new concepts.
Before plate tectonics it was widely
believed that folded mountain ranges
were caused by crustal overload and
unspecified compressive forces. They
arose, it was thought, from the former
sites of active sedimentation that so loaded
subsiding elongate basins or troughs—
called *geosynclines* (downfolds on a large
scale)—that they finally collapsed and
folded of their own weight and pressure
from some converging hinterland. Such
usually elongate, subsiding basins were
called *miogeosynclines* where they contained a preponderance of ordinary con-
tinental-shelf deposits like sandstones and
shales. They were known as *eugeosyn-
clines* where flysch, basalts, and graywackes were prominent.

As the sedimentary and deformational
consequences of plate tectonics came to
be recognized, however, it became clear
that the long-fruitful geosynclinal concept was flawed. "Miogeosynclinal" and
"eugeosynclinal" deposits were not those
of separate troughs but of continental
shelves and slopes on the one hand and
of oceanic basins and island arcs on the
other. Where oceanic crust of one plate
converges on the formerly passive margin of another, both are compressed into
parallel linear folds and low-angle thrust
faults as oceanic deposits peel away from
the descending plate. Although *syncline*
and *anticline* remain useful descriptive
terms for simple downfolds and upfolds,
the terms "geosyncline" and "geosynclinal" are out for large asymmetric
depositional sites and the tectonically
juxtaposed deposits of plate margins. The
purified new language of plate tectonics
declares that *syn* must go. Folded continental shelf deposits are henceforth to
be called *miogeoclinal* and oceanic ones
eugeoclinal. Subside they may, but it is
collision and not subsidence that causes
their deformation and it is sedimentary
environment that explains their differences. We now call other depositional
settings by descriptive plain-language
names—basins, troughs, rift valleys,
platforms, shelves, continental slopes.

The most widely accepted geologic
evidence for ancient plate tectonism is
the presence of detached masses of the
earlier mentioned distinctive rocks known
as blueschists and ophiolites. Other evidence comes from paired fold belts where
the sialic miogeoclinal sedimentary
wedges of formerly-passive continental

margins are folded against the eugeo-clinal belts of mafic oceanic sedimentation and volcanism.

Blueschists, those products of cold, high-pressure metamorphism, (blue from the mineral glaucophane) form deep *within* subduction zones. *Ophiolite* (once synonymous with the decorative stone called *verde antique*, or serpentinite) has come to refer to a sequence of rocks found only in connection with existing or former spreading ridges. The ophiolite sequence (once also called the "Steinmann trinity") is a key element in the reconstruction of ancient plate tectonics. From the bottom upward, a complete ophiolite sequence includes serpentinized ultramafic rock, a succession of sheeted vertical intrusions or dikes that grades upward into pillow lavas, and a capping of very-fine-grained, commonly siliceous, open-marine sediments or oozes (or their lithified products). During the last 4 geocenturies such oozes have consisted largely of the hard parts of pelagic microorganisms. The consistency and implications of this distinctive sequence are solidly rooted in tangible evidence. That evidence includes often far-traveled but well-exposed ophiolitic successions around the world and their petrographic similarity to demonstrably oceanic sequences and to one another. Observation and sampling of currently active spreading ridges by means of submersibles and from deep-sea drilling reveals the process at work.

The Mesozoic ophiolites of Cyprus and the Arabian Peninsula are classic examples of the sequence—even to the large calcareous tube worms and associated creatures that startled the scientific world with their discovery on widely separated parts of the East Pacific rise in 1979 and later and on the Mid-Atlantic ridge in 1986. Special circumstances, of course, are required for the preservation of ophiolites, which might be expected to return to the mantle in subduction zones. Although only a very small fraction of ancient oceanic crust has been preserved in this way, those bits have been diligently sought out, found to be remarkably widespread, and are intensively studied. In fact, a well-populated subscience of ophiolite geology now focuses on this problem.

Older still than the surprisingly extensive and well displayed Mesozoic and younger ophiolites of the Middle East and elsewhere, we now also know convincing and widely scattered Paleozoic ophiolites (back endpaper). Spot occurrences, as well as blueschists up to 4 geocenturies old, are found all along the Ural Mountains. This clearly implies that the Urals are a product of collision between proto-Europe and proto-Asia during Paleozoic time (probably late Paleozoic), an event that concurrently eliminated a paleo-Siberian ocean and created Eurasia. Although some senior Soviet geologists have expressed strong reservations about plate tectonism, mapping by their younger colleagues and a 1970 study by Warren Hamilton of the U.S. Geological Survey has accumulated compelling evidence in favor of a plate-tectonic origin for the Urals—now seen as a *suture* where an ocean closed and formerly separate continents became welded together.

In China also, some twenty-three belts of ophiolite have now been mapped, reportedly ranging in age from late Proterozoic almost to modern times, along with blueschists of Phanerozoic age in about half of them. Indeed China has experienced an unusually complicated microplate-tectonic history, the details of which are only now being worked out.

Ophiolites, moreover, provide evidence for ever more ancient plate tectonism. They are reported from rocks as old as 1.8 to 2 aeons BP west of northern Lake Baikal in the Soviet Union, and are fairly widely reported in rocks of the last aeon of Earth history. Their presence has convinced most geologists that plate tectonics was active at least that long ago. Support for plate tectonism that old also comes from paleomagnetics and from the presence of asymmetrically paired fold belts, where miogeoclinal continental rise and shelf sediments have collided with the eugeoclinal deposits of island arcs, ocean basins, or opposing continental margins.

Just when plate tectonism began to operate, however, remains unknown. Was it active before 2 aeons BP? The answer to that depends on less conclusive arguments for paired fold belts, on paleomagnetism, and on individual predilection. The early Proterozoic crust presents limited evidence of compressive deformation before about 1.9 aeons BP. It seems to have recorded a history—after the Proterozoic revolution—of broad stability, epicontinental sedimentation, stretching, and fracturing, accompanied by the intrusion of extensive systems of mafic dikes such as the Great Dyke of Zimbabwe (Fig. 7.4C) and the Widgiemooltha Dyke Suite of Western Australia.

My predilection is to accept, with only mild reservation, apparent paleopole positions that imply significant horizontal displacement, and therefore plate motions of some kind, back as far as 2.8 aeons BP. As plate tectonism without

subduction hardly suits current perceptions, however, we must consider whether and how subduction of such antiquity might be reconciled with estimated rates of heat flow back then.

Two suggested paths toward that goal call on the previously mentioned abundance of a dense ultramafic volcanic rock called *komatiite* in the Archean crust. Compared with a density of 2.9 for modern oceanic crust, and around 3.2 for solid outer mantle, the measured density of komatiitic volcanic rock is 3.1 to 3.2 grams per cubic centimeter—comparable to that of a lower segment of modern oceanic lithosphere 60 kilometers thick and 60 million years old. The Gordian knot of Archean subduction might be untied, therefore, say Euan Nisbet and Mary Fowler of the University of Saskatchewan, given plausible assumptions about the Archean oceanic lithosphere. If such a lithosphere 20 kilometers thick had consisted of 7 kilometers of komatiitic ocean crust on 13 kilometers of solid, ultramafic outer mantle, its total bulk density would have been about 3.23, comparable to the estimated density of the current asthenosphere. Because the eruption of komatiitic lavas implies an asthenospheric temperature of about 1,650°C, a molten density less than 3.2, and a low viscosity, subduction of such an oceanic lithosphere into such an asthenosphere would have been plausible.

Because much observational evidence implies a substantially thicker late Archean lithosphere, however, a second proposal, by Nicholas Arndt of the Max Planck Institute for Chemistry, is also appealing. Arndt suggests a different lithospheric composition, resulting in an Archean lithosphere as much as 40 kilometers thick and an easily subductable lithospheric density of 3.24.

In spite of remaining uncertainties, geologists are now busily attempting to rewrite Earth history in a plate-tectonic mode. That provides new outlooks on paleogeographic and paleoclimatologic evolution, new baselines for the relative motions of sea and land (both horizontal and vertical), and new perceptions of biospheric evolution. Previous ideas about all of these things need to be reexamined in the light of plate tectonism, cometary and asteroidal impact, and all their likely epiphenomena. The evidence becomes more fragmentary with distance into the past, however, and its connection with position on Earth's surface and in the historic time scale more tenuous. There will be plenty of interesting new facts to be discovered, new experiments to be performed, and new disputations to be resolved by future generations of Earth historians. The theory itself being incomplete, there will be new insights. And there will be surprises, even Nobel surprises, when that level of recognition comes to the Earth sciences.

In Summary

The ascendancy of a new general theory or paradigm transforms the science to which it applies. It opens new vistas and imposes new constraints. The emergence of plate tectonics from the heuristic if nebulous background of continental drift is a good example. It transformed the study of the solid Earth. Strongly held contrary views of the stability of continents and ocean basins gave way to fresh perceptions of the significance and underlying causes of previously mysterious aspects of Earth's crust and interior. Continents are stable (sort of), and they also drift (in a sense), but not freely. Instead, they ride on the backs of lithospheric plates as fixed and integral parts of those plates, their motions limited by tight packing and the rules of spherical geometry.

Happening, as it did, over but a few years in the early and middle 1960s, this conceptual transformation is appropriately referred to as a scientific revolution. It is as significant in its own way for geology as atomic structure is for chemistry and general relativity is for physics and astronomy. The mood of expectation when plate tectonics emerged was so keen, the predictions so explicit, and the tests so positive that it rapidly became central to all aspects of geology. The search it has stimulated for further assessment and applications has enriched and modified the body of geologic understanding in unprecedented ways. It has also led to the growth, elaboration, and extension of plate-tectonic theory to include aspects of intraplate tectonism and biospheric evolution that endow it with the qualities of a general theory of the last half to three-fourths of Earth's lithospheric history.

The 75,000-kilometer-long spreading ridge that encircles the planet like the seam on a softball is not everywhere a mid-ocean ridge, as first thought. It is extensively overridden by western North America and is far from central at other places. But it is truly close to being mid-oceanic in the Atlantic, Indian, and Southern oceans. And new ocean floor is being made at its crestal valley every-where.

The plates do move, grow, disappear, and glide past one another. Their growth in both directions from spreading centers widens oceans. In being consumed down subduction zones or shortened during continental collision they make equivalent space for new sea floor. Transform boundaries offset spreading ridges as horizontal motion adjusts to distance from spreading poles and rates of spreading in obedience to spherical geometry. Folded mountains are made in collision zones. Archean-type volcanic and intrusive mountains and volcanic-island arcs rise above the melting levels of subduction zones. Accretional mountains form by the assembly of drifting crustal fragments. And the "mid-ocean" ridge is a product of thermal expansion and hydration.

The epicenters of great earthquakes are found primarily above subduction zones or along certain transforms. Shallow-focus earthquakes of lesser magnitudes occur along all plate boundaries, and small ones are especially distinctive of the warm, less-rigid crust beneath spreading ridges.

Cooling of Earth's interior by convection is believed to be the initiating force of plate tectonism. Where the upwelling of a hot convection current brings asthenosphere close to the surface, reduction of pressure causes it to expand, take up moisture, emit basaltic lavas, and break through the crust. It may stop there, solidify, and abort further spreading. But where the fissure continues to widen and fill, ridge-push becomes active; sea-floor spreading and plate-pull begin; ophiolites are formed; and the magnetic record of Earth's reversing poles begins

to keep time in both directions away from the spreading axis.

Exactly how or where this is most likely to begin is unclear. But old rifts may reopen, and the features identified as hot spots may play a key role. The geochemistry and geometry of hot-spot basalt differs from that of ordinary spreading-ridge basalts in ways that would be consistent with a deep-mantle source for the spots and an asthenospheric source for the ridges. The spacing of hot spots along the mid-Atlantic spreading ridge, and the curiously triradiate 120° fractures where spreading ridges meet, reinforce this implication. It looks as if heat-driven, ascending, deep-mantle plumes might play a part in initiating or reinforcing self-propagating fractures that subsequently continue to open as new sea floor migrates in both directions away from spreading ridges, under the influence of gravity-driven ridge-push and plate-pull.

The principal limitation of plate tectonics as a general theory of the Earth has been its failure to account for intraplate tectonics, a deficiency now rapidly being remedied. Hot spots elucidate many isolated intraplate phenomena. The relatively shallow and repeated overriding of the former eastern Pacific plate by North America explains the extensive geothermal activity of the western United States. The concurrent sweeping up of microcontinents from the former eastern Pacific and the obduction of oceanic slices onto westward-moving North America between about 200 and 50 million years ago was a fundamental aspect of continental growth. It explains the collage of exotic terranes that dominates so much of the mountainous Cordilleran fringe of western North America—in effect, microplate tectonics. Comparable patterns have now been observed, not only in the Himalaya, but also (even older ones) in eastern North America. The collage of microplates that now comprises China had a geometrically different and much longer history.

The new plate tectonics, in all its variations, is the centerpiece of the new geology and a determining factor in crustal and therefore biospheric evolution. The present cycle of plate tectonism, now approaching a duration of 2 geocenturies, may be slowing down. But it has been repeated in different configurations at intervals during much of the geologic past and will take on new patterns in the future. By means of ophiolites, blueschists, the patterns of sediments and fold belts, and paleontology, its past caprices are being unveiled.

Plate-tectonic theory rests solidly on foundations established by prior advances in geochronology, paleomagnetism, and paleontology. It has been an historical process to be reckoned with for at least the past aeon or two. It, or some comparable antecedent process, may have been in operation as far back as 2.8 aeons ago. Plate motions now in progress affect in pervasive ways the geography, the climate, and the life of this planet (including our own).

CHAPTER TEN

THE OLDER
PROTEROZOIC EARTH
AND THE EMERGING BIOSPHERE

Conspicuous changes in Earth's prevailing surface processes accompanied the expansion of continental area as Archean norms faded into Proterozoic. Continental influences grew in importance thereafter, while mantle influences waned. Shallow marine, continental, and siliceous detrital deposits became more extensive and oceanic volcanogenic deposits and greenstones less so. Trace-element geochemistry reflects these changes.

The then already flourishing biochemical diversity of the microbial world transformed the planet in other ways as the Proterozoic continued to unfold. Earth could never be the same again. A teeming microbial biota of growing complexity interacted with physical settings in ways that have powerfully influenced the biosphere itself, hydrospheric and atmospheric chemistry, and even the solid crust. Appropriate biospheric starting compounds were present early on. They have been made experimentally under plausible, oxygen-free, simulated early Earth conditions and have been found in meteorites and interstellar space. The older Proterozoic saw the conspicuous flowering of simple non-nucleate cells from Archean beginnings. Higher forms, even of microbial life, had to await the evolution, during younger Proterozoic history, of energy-rich oxidative metabolism and an advanced hereditary apparatus.

Here we examine intermediate events and their feedbacks. Notable among preserved rocks was the record of epicontinental sedimentation, the deposition of the world's sedimentary banded iron formation, and the oldest evidence for glaciation of continental extent.

10.1 TYPICAL BANDED IRON FORMATION FROM THREE CONTINENTS. Bar scales 10 cm in *A*, 1 cm in others. *A*, Negaunee BIF near Ishpeming, Michigan; *B*, from Son River valley, Uttar Pradesh, India; *C*, from mine at Krivoi Rog on the Dnepr River, Soviet Union; *D*, from mine at Bjørnevatn, northeastern Norway. Age range of samples 2.4 to 2 aeons old.

Major Features of Older Proterozoic History

Like the Bayeux tapestry, the records of older Proterozoic history are incomplete, but they are a substantial improvement over those available for Archean history. They outline a broad historical framework of critical events in early Earth history between more than 2.8 and about 1.9 aeons BP, give or take a geocentury or two. Here I discuss only selected major features.

The transition from Archean to Proterozoic seems to have been a prolonged interval of major continental growth. The halting amalgamation of largely potassic cratons and their associated sedimentary and volcanic rocks between about 3 to perhaps 2.6 aeons ago (front endpaper) was an indecisive prelude to a long Proterozoic history. It was a record of extensive, stable platform and continental basin sedimentation and volcanism, cut intermittently by swarms of basaltic dikes and intersected by linear mobile belts between cratonal nuclei. This and apparent polar wandering at the time imply that plate tectonics had already begun. But older Proterozoic history is here more interesting as it pertains to climatic, biospheric, and atmospheric evolution.

It was then that continental influences first became widespread and conspicuous. It was a time of great enrichment of the outer continental crust in sedimentary iron deposits, of very early or even earliest continental glaciation, of extensive epicratonal sedimentation, of emerging biospheric complexity, and of transition from a generally anoxic or weakly oxic (oxygenic) atmosphere to one that offered prospect of eventually sustaining higher forms of life.

Because sedimentary rocks reflect feedback among hydrospheric, atmospheric, mantle, and continental influences, they record this history. As we have seen, older Proterozoic strata contrast with those of the Archean in many ways. Whereas Archean supracrustal rocks and geochemistry reflect a contemporary ocean and lithosphere strongly influenced by mantle processes, Proterozoic strata record a progressive increase of stream-borne and other sediments from continental sources. They reflect a notable expansion of shelf seas, continental platforms, river systems, and continental influences generally. It was no accident that expanding continents, growing maturation and recycling of sediments, continental glaciation, and extensive banded iron formations (*BIFs*) happened to be broadly coincident at that still-anoxic time. Detrital feldspars generally weathered to clays. Sandstones became more abundant and more quartzose, with better-rounded, multicycled grains repeatedly reworked from ancestral sandstones and other sandy deposits. Carbonate rocks, especially dolomites, became more common. Siliceous chemical sediments, called *chert* when lithified, became very extensive in the absence of biologic silica precipitation and in probably less alkalic ocean waters than now. Where iron was intermittently added, it became concentrated in the chert matrix of BIF.

BIF, largest and most colorful of the sources of metallic raw materials that underpin industrial society (Fig. 10.1), is a conspicuous feature of the older Proterozoic record. Deposits of BIF of this age on the order of a hundred trillion (10^{14}) tons are known from Brazil, eastern Canada, the Hamersley basin of Western Australia, the Transvaal, and the Ukraine—these five accounting for an estimated 92 percent of the world's minable iron. Next door to the Hamersley, incidentally, is the Nabberu basin, with 10^{13} tons of BIF—a tidy reserve, when added to Hamersley, for a nation that in mid-century thought it was running out of iron.

These great iron formations seem to line up along zones of incipient (and sometimes later) rifting, where iron from deep volcanic sources may have contributed to their growth. There is a hint that here one may be seeing a prolongation of the parallelism between the depositional histories of Australia and Africa, first encountered in the Archean (Fig. 7.5). When these continents are brought together, as they may have been from around 3.5 or more to perhaps 2 aeons ago, we see two of the world's greatest BIF deposits and their geologic settings juxtaposed (Fig. 10.2). The stratigraphy of these older Proterozoic iron-bearing sequences also shares broad similarities when carbonate rocks and chert are visualized as end members of a geochemical continuum (Fig. 10.3).

The oldest Proterozoic rocks in North America are of continental and partly glacial origin, but a somewhat younger set of iron formations and their associated rocks extends across eastern Canada and the Lake Superior region of the United States. Some 10^{13} tons of BIF are found in the lake and mosquito country of northeastern Minnesota, northern Michigan, and northernmost Wisconsin. The main features of this stratigraphic sequence are reconstructed in Figure 10.4.

10.2 MAJOR FEATURES OF OLDER SOUTHERN AFRICAN (*LEFT*) AND WESTERN AUSTRALIAN GEOLOGY COMPARED. [*Adapted from Andrew Button, 1976,* Minerals Science and Engineering, *v. 8, p. 262.*]

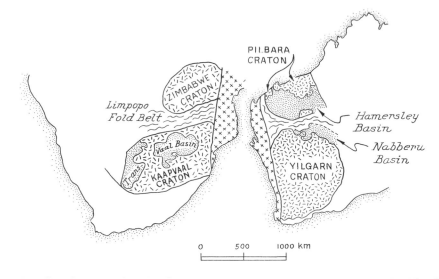

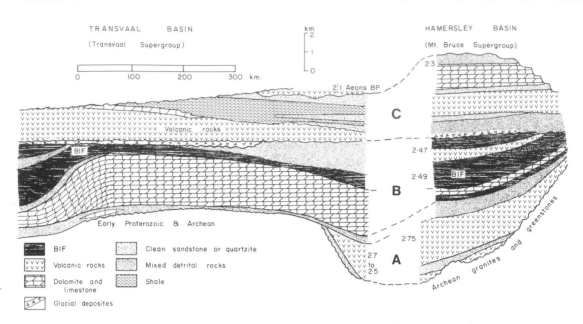

10.3 OLDER PROTEROZOIC IRON-BEARING SEQUENCES OF SOUTHERN AFRICA (*LEFT*) COMPARED WITH THOSE OF WESTERN AUSTRALIA (*RIGHT*). Sequence *A* is basal volcanic and sandy sequence; *B*, middle chemical sequence; *C*, upper volcanic and mixed sequence. Numbers are uranium–lead ages on the mineral zircon, given in aeons BP, except the estimated 2.3 aeons at the top of the Hamersley basin sequence. [*Prefolding profiles simplified after Andrew Button (right), 1976, Minerals Science and Engineering, v. 8, p. 262, and N. J. Beukes (left), 1983, Iron Formation: Facts and Problems, Elsevier, Fig. 4.4B. © Elsevier Science Publishers B.V.*]

10.4 DIAGRAMMATIC VERTICAL PROFILE THROUGH OLDER PROTEROZOIC SEQUENCES BETWEEN SOUTHERN ONTARIO AND WYOMING, WITH SPECIAL REFERENCE TO THE LAKE SUPERIOR REGION. Provisional prefolding reconstruction. Distances not to scale. Vertical scale approximate and highly exaggerated.

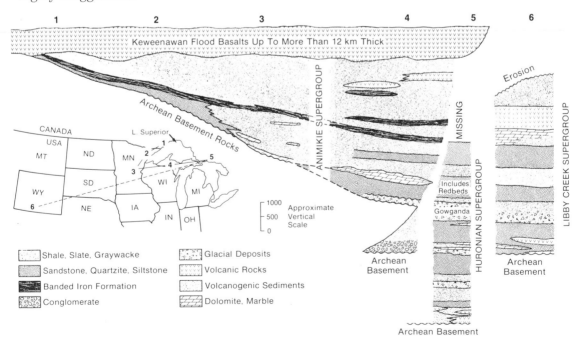

Oldest Extensive Ice Ages

Of special interest among the older Proterozoic rocks of North America is the presence of evidence for ice ages comparable in extent to those that so recently blanketed much of the Northern Hemisphere, but more than 2 aeons older and spanning a much longer time. Records of such a glaciation about 2.3 aeons old are found around pellucid Lake Superior, beyond there for another 1,200 kilometers northwest to the dwarf forests and muskegs of subarctic Canada, and far to the southwest in the Alpine meadows and coniferous groves of the Medicine Bow Mountains of Wyoming.

Similar deposits are known from rocks of comparable ages on other continents. They consist of dispersed, commonly angular, and occasionally parallel-scratched pebbles and large fragments of a diversity of rock types, suspended in a dirty, finer-grained matrix similar to modern glacial deposits called *tills*—hence *tillites* when lithified. Probable tillites and other glacial deposits of this age are widespread in southern Ontario.

The nonmarine to marginal-marine rocks of the Huronian historical interval between its glacial intervals represent a classic Proterozoic stratal assemblage of major geohistorical interest in North America. They extend 450 kilometers southwest across southern Ontario from Quebec to Lake Superior, thickening southward from a northerly source into a broadly subsiding depositional basin. Figure 10.4 suggests that extensions carry over into northern Michigan and thence westward, under different names. Although sedimentologists disagree about proportions of marine and nonmarine

deposits, both are present, along with intertidal equivalents. Torrentially cross-bedded channel sands, glacial deposits, and distinctively bedded quartzites betray nonmarine sources. Herringbone cross-bedding and casts of salt crystals in ripple troughs define intertidal settings. And dolomite combines with the salt casts to record marine influences. Reported wormlike objects have been shown instead to be sand injections into clay-lined curvate desiccation cracks in ripple troughs—a phenomenon observed in intermittently exposed sandstones of all ages.

In the Huronian sequence are three glacial and probably glacial intervals, each following above a sequence of mixed detrital sedimentary rocks of mainly continental origin and overlain by similar strata. The uppermost and most convincing glacial deposits are part of the Gowganda Formation (Figs. 10.4, 10.5). This roughly 2.3-aeon-old rock shows, in addition to tillites (Fig. 10.5B) and river-borne glacial materials, alternating clays and silts in graded bands a few millimeters to a centimeter or so thick, resembling modern annually layered glacial lake varves. The Gowganda varvites even contain isolated glacially scratched pebbles—*lonestones*, and in this case, *dropstones*—that advertise their ice-rafted origin by the distinctive local impact deformation, long before any other means that could account for such rafting is known to have existed (Fig. 10.5C).

When these Huronian glacial deposits are joined with others of like age in northern Michigan, an area west of Hudson Bay, and the Medicine Bow

Range of Wyoming (Fig. 10.4), a roughly 2.4- to 2.3-aeon-old ice sheet is implied, once covering a southwestward-apexing triangle about 2 million square kilometers in area. Similar glacial deposits found in 2.5- to 2.3-aeon-old strata in east-central Finland and the adjacent Soviet Union may belong to the same glacial epoch.

Because glacial deposits thought to be of equivalent age are also reported from the Hamersleyan sequence of Western Australia, from South Africa (Fig. 10.3), from central India, and from the Baikal region of Soviet Asia, older Proterozoic glaciation may have been very extensive.

If truly global or bihemispheric (as summarized by the University of Western Ontario's Grant Young), such glacial deposits might represent the shutdown of the greenhouse mechanism that, until then, may have kept the planet warm despite the faint early Sun of astronomers? Indeed, such a greenhouse would have been needed to keep the hydrosphere from freezing solid, and such a shutdown might have been a partial consequence of CO_2 assimilation by early microbial photosynthesis.

Warming climates following the Gowganda glaciation, perhaps due to equatorward drift of proto–North America, produced the oldest prominent redbeds in the upper Huronian of

10.5 GLACIAL DEPOSITS FROM 2.3-AEON-OLD ROCKS IN ONTARIO AND NORTHERN MICHIGAN. View *A*, dropstones in silty shales of Reany Creek Formation, Animikie Supergroup, west of Marquette, Michigan. *B*, Gowganda tillite with large granite blocks, Huronian Supergroup, west of Haileybury, Ontario. *C*, Gowganda varvite with granitic dropstone near Elliot Lake, Ontario. *D*, Fern Creek tillite, Animikie Supergroup, west of Loretto, Michigan.

Ontario. They also introduced the last of the great historical intervals of typical BIF deposition in North America, found in the mainly post-Huronian Animikie Supergroup and its equivalents (Fig. 10.4) between perhaps 2.2 and 1.9 aeons BP. It was the uppermost of the older Proterozoic sedimentary sequences in North America. Its history foreshadowed continental rifting and transition from low and transient atmospheric O_2 to concentrations probably high enough to generate a radiation-shielding ozone screen— about 1 to 3 percent of the present atmospheric level of O_2.

Also significant for older Proterozoic history and O_2 levels is the presence in the preglacial channel sands of the Huronian sequence in Ontario of large and extensive concentrates of easily oxidized, thorium-bearing, detrital uraninite. That (and equivalent deposits elsewhere) is independent evidence that atmospheric O_2 levels remained very low until at least as late as 2.45 aeons ago— the U–Pb zircon age of underlying sialic volcanics.

Older Proterozoic history was that of an evolving but still-archaic Earth.

Clues to the Nascent Biosphere

The Proterozoic seas harbored a teeming microbial biota, which was shielded from the harshest solar radiation by a sufficient 10 meters or more of overlying water, or by living within a thin protective layer of sediments that was precipitated and bound by their life activities. Successions of such microbial mats, upon lithification, are called *stromatolites*— laminated, usually calcareous, three-dimensional buildup structures of a variety of shapes and characteristically (but not invariably) of microbial origin. Members of such inconspicuous lilliputian communities left a discontinuous record of their presence from at least 2.8 aeons ago (Fig. 10.6) and probably 3.5 aeons onward. At the latter date, near the beginning of Archean history, one first encounters probable fossil evidence for earliest life and its interactions with an external world whose atmosphere would, to us, have made Los Angeles on smog alert look inviting.

In such a hostile setting, by some mysterious alchemy, there came to pass on this planet that indefinable stuff whose magic we share and whose source has been the subject of so much dispute. Turning to it, one can begin by excluding the highly improbable notion of the planet being seeded with preexisting life, either fortuitously or deliberately. It is unlikely that living things of the types we know, even the simplest, could have survived the vicissitudes of space travel unshielded. It seems equally improbable that a space-faring society would have bothered to seed the planet with such rudimentary forms as the oldest we know, or that advanced forms could have survived on the Archean Earth. Space-seeding, moreover, would not solve the problem, but only transfer it to another time and place.

Given the origin of life as the subject one must say what life is. What are the basic attibutes by which it is to be characterized? Probably all readers of this book will know (or have heard) that all

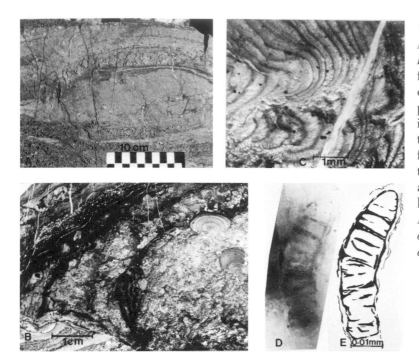

10.6 FOSSILS FROM 2.8-AEON-OLD ROCKS. Views *A–B*, stromatolites with fibrous structure. *C*, detail of fibrous structure. *D–E*, photograph and sketch of individual microbial thread from *C* above. All from Tumbiana Formation, Fortescue Group, along Newman–Port Hedland railroad, Western Australia. *[A–C, courtesy of Malcom Walter; D–E, courtesy of J. W. Schopf.]*

living things, from the simplest to the most complex, share a need for the same protein-building amino acids, the same special sugars, the same energy-transfering molecules, and the same genetic code. A simple aggregation of all the essential molecules of itself, however, is still not living. The components must be assembled and activated to interact in ways that depend critically on an interconnected system of internal fluids within which all of the basic molecules and others move about, directed by DNA and electric impulses, and through which otherwise poisonous wastes ultimately leave the system.

Life can arise and prosper only where appropriate conditions exist. Indeed it is hard to imagine life as a lone entity or even a single species. We know it only as parts of a larger biosphere so interwoven that Earth's outer shells themselves take on, in our imaginations, some of the attributes of a living system.

It is hardly surprising that we who are part of it share an enduring fascination with the intricate workings of this *biosphere*—the totality of living systems (*biomass*) and their interacting physical surroundings. How did the biosphere become so elegantly balanced? What keeps it that way despite a long succession of historical changes? How long will it last? What are the manifestations of life preceding the first clearly recognizable animal fossils that characterize the Phanerozoic systems? Above all, how did this complicated global ecosystem ever get started?

Challenged with such questions, and looking around us, most would agree that truly living things should be able to grow, to metabolize, and to reproduce, with or without the pleasures of cooperation. But these are minimal requirements. Something more is needed in order to discriminate unequivocally between life and nonlife at the simplest levels.

All living things are genetically programmed. In order to survive and continue their lineage—to evolve—they must also be capable of changing genetically from one generation to the next and of passing genetic changes on to their descendants. We expect more of authentically living things than that they merely be able to grow, metabolize, and reproduce. They must also be able to *mutate* (undergo spontaneous genetic change) and to reproduce mutations. Life, at a minimum, is something that grows, metabolizes, reproduces, mutates, and reproduces mutations.

The progression from nonliving to living, like so many other "boundaries" in nature, is, in fact, more logically viewed as a transition than an event. Beyond some point in a spectrum of happenings, all observers would concede the result to be living. Yet there remains disagreement about what might be the critical threshold. The British biologist N. W. Pirie models the process as something like an hourglass in reverse (Fig. 10.7). In the starting compartment (here the lower cone of the hourglass), radiation, chemical sources, lightning, or some or all of these, energize reductive processes that concentrate elements and compounds having appropriate self-catalytic, and other, properties. The neck of the hourglass represents the origin of life—called *biopoesis* by Pirie because "biogenesis" is preempted for a different meaning. The upper sector illustrates biologic evolution as a pattern of ever-widening diversification within the flow of time and energy. Starting from an initial and fundamental biochemical unity, living systems undergo an ever-changing diversity, first in the details of microbial biochemistry, and then, following the origin of the nucleate cell, in a diversification of body plan, shape, and behavior.

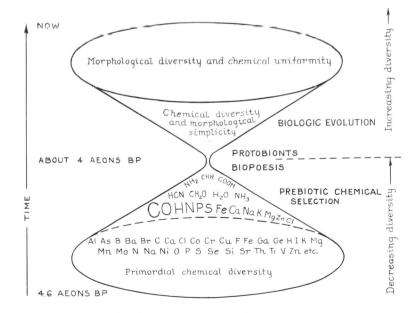

10.7 SALIENT ASPECTS OF PREBIOTIC AND BIOTIC EVOLUTION. Capital *R* in *CHR*, at center, refers to the variable defining side chain of the generalized amino acid structure. [N. W. Pirie; adapted by permission from Nature, Vol. 180, p. 887. Copyright © 1957 Macmillan Journals Limited.]

Concerning alternative chemistries, the only biochemistry available for study suggests prebiotic chemical selection of those elements best suited to the construction of the molecular building blocks of life. Inasmuch as these elements are also relatively common in the universe at large, one might expect life, wherever it arose, to have approximated the chemistry of Earth life in the major elements utilized.

Water is, in many ways, the key component: the vital fluid of the living planet. It is the universal solvent in which all biological reactions occur—the only natural fluid besides molten iron that expands on cooling, so that high-latitude lakes and seas do not generally become solid ice. It is the stuff whose removal or addition makes and breaks the bonds between many vital molecules. But, although living things are mainly water, life does not persist by juice alone.

Consider the abundances of the elements in dry protoplasm. Only a few are truly essential. The reader, for instance, thoroughly desiccated, would be about 48.4 percent carbon, 23.7 percent oxygen, 13 percent nitrogen, 7 percent hydrogen, 3.5 percent calcium, 1.6 percent phosphorus, 1.6 percent sulfur, and less than 1.5 percent of a dozen other things. The primary dry-weight constituents of organisms generally, all above 1 percent, are carbon, hydrogen, oxygen, and nitrogen (CHON is a handy mnemonic), plus phosphorus (making it CHONP) and sulfur (CHONPS). About thirteen other elements are regularly involved as secondary constituents in the construction of protoplasm, and another twenty-three are implicated variably.

What is perceived as more or less abundant, to be sure, depends on how one defines abundance—whether in dry or wet weight, volume, or numbers of atoms. What matters in chemistry is atomic abundance, activity, and ionic dimensions. Of the sixteen commonest elements in the universe, all except the noble gases (e.g., helium, neon, and argon), aluminum, and nickel are invariable constituents of living matter, despite the relative atomic rarity of many on Earth.

Carbon, keystone of the life-building arch, is present in inorganic matter in only relatively minute quantities, both on the planet and in the galaxy. Although nitrogen accounts for 78 percent of the present atmosphere, it is, in fact, three orders of magnitude more abundant in organisms than in Earth as a whole. Hydrogen and oxygen, with their striking reversals of relative abundance between Earth and the universe at large, are abundant and essential in organisms, but mainly as water. In dry protein they are about as abundant as carbon. Phosphorus and sulfur are also essential, for the construction of nucleic acids, some amino acids, most proteins, and many structural substances.

Of the major Earth-building elements besides oxygen, only magnesium and iron are essential to life processes and they, like phosphorus and sulfur, only in minute quantities. On the other hand, a clear similarity exists between the main electrolytic elements of seawater and those of the internal fluids of organisms. Just as the properties of the critical life-building elements imply a preceding interval of chemical selection, so do the elements involved in secondary metabolic processes imply beginnings in the vast, unresting sea. How delicate the balance! How precise the requirements.

The distinctive properties and positions in the classic periodic table of the protein-building elements and accessories are also illuminating. The CHON (or HONC) elements inhabit the first two periods of the periodic table (Fig. 2.3). HONC symbolizes the four smallest elements that regularly combine by accepting or giving up respectively one, two, three, and four electrons. That property is called *valence*, from the Latin *valere*, "to be strong." These elements make bonds with one another that are strong enough, but not so strong as to inhibit metabolic rearrangements and recycling. The HONC elements make negotiable molecules, tough sheaths, and long flexible chains, good for recycling and for building cell walls and (in animals) muscles. Oxygen attains stable but flexible proteinoid configurations with carbon, hydrogen, and nitrogen because no unfilled charges remain after they have combined. In contrast, when oxygen combines with silicon, a much more abundant terrestrial element than carbon, the result is a silica tetrahedron—the molecular component of common quartz, which grows into ever-larger masses or sheets as new unsatisfied charges at the corners of its identical tetrahedra become candidates for neutralization. The production of living systems involves all of these elements and others in very specific arrangements and pathways. CO_2, CH_4, NH_3, NO_2, and H_2O have very unlifelike properties until they enter more complex associations. Interaction, change, and eventually death are all properties of life.

Such things imply prebiotic chemical selection (as in the lower half of Figure 10.7), during which elements having suitable characteristics of charge and size were linked to form the essential large organic molecules of living things. Exactly how that first happened is something that can never be known with confidence, but no reasonable doubt remains that the steps to life were all the results of natural processes. From the classic 1951 results of Stanley Miller and Harold Urey onward, experimental work has supported such a view. It has been shown repeatedly that—given virtually any plausible early atmosphere and energy source, and in the absence of free oxygen—not only amino acids but most or all of the components of nucleic acids can be made from nonbiological starting materials. The more reducing the atmosphere, the broader and more ample the yield.

This has led many to hypothesize a highly reducing primordial atmosphere, and that may well have been the case during much of Hadean history. By the beginning of Archean history, however, a good deal of neutralization seems to have occurred, probably by means of transient oxygen released by sunlight splitting the water molecule—a process called *photolysis* ("breaking by light"). Minute bubbles in the fluid inclusions of mineral grains in 3.5-aeon-old rocks in Western Australia may be actual samples of early Archean atmospheric or dissolved gas. Crystals of the sulfate mineral in which this gas was trapped were formed in evaporating depressions that presumably interacted with the contemporary Archean atmosphere. Such gases are reported to be carbon dioxide, water, and hydrogen sulfide (CO_2, H_2O, H_2S), all expectable components of the Archean atmosphere. That is inconclusive, but it's all the direct data so far available.

Wide degrees of freedom are still permissible, therefore, in hypothesizing about early atmospheres. In fact, many interesting organic molecules are now known to be produced in small and presumably transient quantities in interstellar space. Likewise, amino acids and other organic molecules are known primary constituents of carbonaceous meteorites and are to be expected in chondritic asteroids and comets.

Experiments imply that the most important intermediates between plausible early atmospheric gases and amino acids, in the absence of free O_2, are hydrogen cyanide and formaldehyde—both general biocides. Chemistry, including biochemistry, although governed by strict rules, involves many subtleties. It is not obvious from the properties of elements or simple compounds how their marriage might turn out. Formaldehyde, for example, is "only" carbon and water joined chemically as CH_2O—but what a difference! Five formaldehydes, properly linked, make familiar glucose sugar. Hydrogen cyanide, formaldehyde, and water in the right combination make the protein-building amino acid glycine, while other seemingly simple combinations result in other amino acids. When water is removed in the right way, adjacent amino acids join to make peptides. The trick is to find the combination and generate or borrow the energy to make it work.

The processes that produce the interesting prebiotic molecules are thought to be similar to those involved in the industrial *Fischer–Tropsch process* for making hydrocarbons by passing a mixture of carbon monoxide and hydrogen over a heated iron-rich catalyst. Carbon monoxide, like hydrogen, was surely important in early chemical evolution and might have been both oxygen sink and molecular predecessor to much of the carbon dioxide in the primordial atmospheres. Important protein-building amino acids have now been identified in different laboratories from recently fallen carbonaceous chondrites. In addition, Australia's Murchison meteorite, a 1969 fall, is reported to contain potentially membrane-building lipid-like organic chemicals, as well as all five of the so-called *nucleotide bases.* They are the large molecules that, when strung on sugar–phosphate chains, make up the crucial stuff of life, the nucleic acids DNA and RNA (deoxyribonucleic and ribonucleic acid).

At least one of these nucleotide bases (uracil), along with other interesting prebiotic molecules, is now reported in three different meteorites by Dutch investigators. Thus it appears that the essential building blocks of proteins and genetic information systems can be produced in the absence of O_2 both in the laboratory and in nature. They were, therefore, presumably abundantly available on the prebiotic Earth—from meteoritic, cometary, or local sources.

From Molecules to Microbes

The central problem is how the appropriate building blocks might be assembled into some combination having the specified characteristics of living things. Properties of the only kind of life we know enter that assessment. Most striking, perhaps, is the fact that the twenty indispensable protein-building amino acids are all left-handed—they rotate polarized light counterclockwise. In contrast, biologic sugars are all right-handed. Amino acids and sugars of nonbiological origin contain equal numbers of right- and left-handed molecules. They are said to be *racemic.* One can't make spiral molecules of DNA out of racemic amino acids and sugars. So some not-well-understood process was involved in concentrating and arranging these molecules to yield the property of handedness (or *chirality,* from the Greek for "hand"), without which life does not exist.

The Australian cosmochemist Ron Brown of Monash University has reported that once as many as five amino acids of the same handedness are linked, they will preferentially add others of the same chirality from racemic mixtures. Assuming that, and given geologic time, might not initial linkages of five have occurred repeatedly, perhaps as a result of the cyclic thermal stimulation of short day-to-night temperature changes on a fast-rotating primordial Earth? A string of five left handed amino acids would be only one small step toward the origin of life, but a step in the right direction.

Another property that may seem odd to creatures as dependent on oxygen as we is the hypersensitivity of life to free molecular O_2 and its corrosive metabolic by-products such as ozone, the peroxides, and the superoxides. Even though the metabolism of advanced life forms requires O_2, any by-product peroxides or superoxides must be reduced to water by special enzymes (the catalases and superoxide dismutases) in order for the organism to survive and function. And all oxidative metabolism begins with an anaerobic step that is similar to the fermentative processes of bacteria, implying anaerobic beginnings.

The corrosiveness of uncombined, or *free,* O_2 relates back to the earlier observation that the experimental production and survival of life-building molecules works only in its absence. The message from all sources is the same. The beginnings of life were anaerobic, or at least *anoxic,* without free O_2 in other than trivial and transient quantities.

Considering that neither the requisite prebiotic chemistry nor the organization of its products into living systems could have taken place in the presence of O_2, objection to the idea that the accumulation of atmospheric O_2 was a later Proterozoic event runs surprisingly strong. Among geologists this may be a carryover from the often gradualistic "uniformitarianism" of earlier times. In fact, the real puzzle is not the scarcity of O_2 in the atmosphere and hydrosphere of the prebiotic Earth but its abundance on the present Earth. For we know no primary sources of O_2, and O_2 from secondary sources is highly reactive and characteristically short-lived. It quickly recombines, either with its coproducts or with other common and abundant reduced substances.

The mere production of amino acids, sugars, and nucleotide bases does not account for life, either. One amino acid does not a protein make—let alone a being. Amino acids must be joined to make proteins; and the sugars and nucleotide bases must be linked with phosphoric acid to yield nucleotides before nucleic acids can work their magic. Water plays a key role in these processes. As Nobel biochemist Melvin Calvin of the University of California has emphasized, the fundamental steps are *dehydration condensations*—the removal of water from appropriate adjacent molecules, causing them to bind together. Dehydration condensations are involved in linking amino acids to make peptides, polypeptides, and proteins. They link sugars to nucleotide bases and phosphoric acids to make nucleotides and join them as nucleic acids, genes, and chromosomes. And they are the removable links that bind and separate the phosphoric acid molecules of the adenosine mono-, di-, triphosphate (AMP, ADP, ATP) system, which oversees all biological energy transactions.

Life is sometimes said to run against the general downhill flow of usable energy in the universe. It is said to evade entropy, to be negentropic. In fact it runs on borrowed energy. It is an open system of flows for harnessing and temporarily storing external energy, mainly solar, and for using it to perform biological work. Like all other energy in the universe, however, borrowed biological energy degrades to unusable forms, and, when its credit fails, the energy bank forecloses. The organism dies. Entropy gets us in the end.

There are also many components without which biological energy storage and transfer could not function. Some of them are needed in almost vanishingly small but essential quantities. Phosphorus is a good example. It plays a critical role in the construction of the hereditary apparatus. It is essential in the metabolism of proteins, carbohydrates, lipids, vitamins, and inorganic salts, as well as in the synthesis of proteins themselves. The functioning of cell membranes, enzymes, photosynthesis, muscle, and nerves (including the brain) all require

phosphoric acid. It is the central energy broker of life and it runs that enterprise with enviable efficiency and utter ruthlessness.

ATP is the principle partner of this brokerage firm and it is present without known exception in the cells of all forms of life, wherever biological energy is utilized. The fact that appreciable quantities of ATP can be made experimentally under simulated prebiotic conditions implies that it has been available from the beginning. Together with the universality of the protein-building amino acids, the similar ubiquity of the amino acid–linking peptide bond, and the uniqueness of the nucleic acids as the basis for heredity, ATP bears witness to the unity of life and its ascent from a common successful prototype.

Informed perceptions of what that uniquely successful pathway to life may have been can be written, but the solid evidence is thin. Success in experimentally assembling a growing, self-sustaining, self-replicating, mutating system would only tell us of one pathway that could have been followed, not that it was *the* pathway. Nor is there any way to know how many unsuccessful origins of life there may have been before one triumphal biochemical lineage gave rise to, or emerged as, an organism that could manufacture its own nutrients from external sources of energy—solar or otherwise—thereby assuring the continuation of that genetic line.

Better, although by no means compelling, grounds are available for hypothesizing what the ancestral protobiont may have been like. The common analogy with tiny spheroidal, morphologically simple living forms bounded by a cell wall just may be too glib. Simple microorganisms that lack cell walls are known and there is no convincing reason to exclude the possibility that the protobiont might have been, for instance, a diaphanous film, ribbon, or disc of some sort into which essential nutrients could diffuse or be assimilated with ease. It is hard to imagine anything simpler than rickettsias (the tiniest of bacteria) or viruses (which are even tinier—1,000 to 10,000 of them could be stacked in the thickness of this page), but they are poor models for protobionts. Both function exclusively within host organisms (which they could not very well have preceded), and viruses are considered by some to be, in fact, nonliving organic crystals.

The oldest possible fossil remains known, however, have the sizes and shapes of bacteria, and the oldest indirect evidence of life suggests microbial agents. Moreover, studies in molecular evolution, based on amino acid sequences in the primitive universal iron- and sulfur-rich molecule called ferredoxin, imply anaerobic microbial beginnings and a long interlude of fermentative metabolism and biochemical evolution before even microbes learned to tolerate the presence of oxygen. That, of course, does not exclude anaerobic bacterial photosynthesis. Primitive porphyrins, the framework of the chlorophyll family, are entirely plausible early molecules.

The convergence of the evidence and the multidisciplinary "thought experiments," however, are most consistent with the prosaic conclusion that the primordial biomass was most likely microscopic, anaerobic, and dependent on external sources of nutrition—that is, heterotrophic as well as anaerobic. Its components may or may not have been bounded by cell walls, but, if living today, they would surely be identified as bacteria (or perhaps archaebacteria). A shallow, anoxic marine habitat, shielded from radiation by water or mud, or both, would have been a likely starting site. It could have begun with the landing of a carbonaceous planetesimal or comet in an early Archean or late Hadean mudbank.

What was the crucial initial step, the first self-replicating system? Many have been called but none has yet been chosen. New evidence and current opinion favors the RNA (ribonucleic acid) molecule, recently found to have not only its familiar information storage and transfer capabilities, but also enzymic properties that permit it to replicate itself—potential precursors to the combined functions of proteins and DNA. The actual setting and mechanism of life's origin, of course, can never be known in detail. The discovery of sulfur bacteria in the hot, hydrogen sulfide–rich waters of seafloor spreading ridges requires that even so strange a setting as that be considered as a plausible analog for the initial ecosystem. Who knows what surprises may await us on another planet?

The First Agricultural Revolution

Microbial heterotrophs were the hunter-gatherers of their time. Like early humans, they foraged for their food. The step from heterotroph to *autotroph*—an organism that makes its own food from sunlight or other energy sources—was the giant step to the continuation of life and biologic evolution. It was, in effect, the first agricultural revolution, preceding the so-called Neolithic and green revolutions by 3 aeons or more.

Creatures whose biologic energy came exclusively from external organic matter would have been like maroonees on a desert island. They would have had to become very inventive as available resources ran out. To win an initial contest of multiple origins, an evolutionary line would have had to be the first to give rise to a species capable of manufacturing its own nutrients by harnessing an external energy source. A creature at that level of advancement would have become a potential supplier of assimilable nutrients to others of its biochemical lineage.

All that one can say with a degree of confidence about that winner, therefore, is that it had previously been a self-replicating anaerobic heterotroph of some sort. Compared with known living organisms it could hardly have been called anything but a bacterium. Such wee creatures may not look very clever, but they are extraordinarily inventive biochemically. They have explored every conceivable habitat, and some inconceivable ones—like sulfuric and carbolic acid, smouldering coal, and perhaps 300°C jets of hydrogen sulfide–rich fluids at the crests of deep-oceanic spreading ridges. They outwit mankind's most elaborate efforts to exclude them from its company. And one may be sure that their ancestors were a match for the monotonously anoxic habitats of Archean history. Bacteria even evolved the ability to repair molecular dislocations in their DNA caused by high-energy ultraviolet radiation—persuasive evidence for their descent from lineages that existed before there was enough O_2 to create a shielding ozone screen. The meek may not inherit the Earth, but they staked the first claim and hold all records for survival and abundance.

The primordial autotroph, to be sure, might have harnessed any of a number of different sources of energy, from sunlight to a variety of chemical and geothermal sources. Living photosynthetic sulfur bacteria are able to oxidize hydrogen and sulfide to free sulfur and sulfates without the evolution or involvement of free O_2 at any stage in the process. A photosynthetic sulfur bacterium that learned to split hydrogen sulfide (H_2S) in preference to water (H_2O) as its source of energy-yielding electrons would have been a likely candidate for the microbial hall of fame. Had H_2S been in short supply relative to H_2O, that eventually would have resulted in selective pressure to split the stronger hydrogen–oxygen bonds of H_2O as an alternative energy source. Chlorophyll-a would have been needed to split the water molecule, realizing thereby the oxygenic step in photosynthesis now called photosystem II, which yields a twelvefold increase in the amount of biologic energy available from the more primitive anaerobic photosystem I alone. Thus the biologic production of free oxygen and oxidative metabolism, which opened a path toward the evolution of higher forms of life, was almost surely taken before 2.8 aeons ago and perhaps earlier.

Save for chlorophyll-a, the creatures envisioned had the characteristics of metabolically advanced bacteria. As England's Patrick Echlin and Boston University's Lynn Margulis have long insisted, such organisms may well be the remote ancestors of the oxygen-generating organelles or *chloroplasts* found in the cells of all true algae and higher plants. They, therefore, share traits both with algae and with bacteria, and have been called both. As they are neither, *proalga* (plural *proalgae*) seems suitably simple and equivocal. It is favored here.

They were the "blue-green algae" of yore, or their ancestors. A more advanced proalgal form, however, *Prochloron*, recognized only a decade ago by Richard Lewin to contain both chlorophyl-*a* and chlorophyll-*b*, may be an even more appropriate model for a chloroplast ancestor.

Free Oxygen and Its Interactions:
A Mixed Blessing

The oxygen these microbial wizards and their descendants released from combination is chemically unique. It is highly reactive. It is the third most abundant element in the universe, but mainly in combination with something else. The stuff is all around us but we don't see it. Chemically joined with other elements it accounts for more than a quarter of Earth's total weight and almost half the mass of the crust. Free or dissolved molecular oxygen, however, the kind of interest here, represents only 0.01 percent of the total crust, hydrosphere, atmosphere, and biosphere taken together. Only seven of every billion atoms takes the form of molecular oxygen (O_2), plus a neglible quantity as atomic or singlet O and ozone (O_3). Earth's present bounty of free molecular O_2, more than 99 percent of which is in the atmosphere, is not more than 3 percent of that which has ever been generated over the course of Earth history. The other 97 percent is locked in chemical combination with something else.

The fact that the present atmosphere is oxygenic is anomalous in a world that was so long without more than traces of that gas, and they transient. Even now our precious fund of free O_2 poses a cruel dilemma for all respiring organisms. Whereas it supports their lives at high metabolic levels, they are able to survive its corrosive presence only by reason of elaborate enzymic and other defenses. Their limits of tolerance are narrow. Either too little or too much can be deadly.

If the atmosphere contained much more O_2, it would be inflammable. Remove O_2 and only anaerobic bacteria could survive. We are the beneficiaries, first of the proalgal and radiative processes that put O_2 in the water and air to begin with, then of the oxygen-mediating enzymes that make it tolerable. Our lives depend on the imbalance between sources and sinks (Fig. 10.8) that brought this capricious gas up to its present level perhaps 400 million years ago and has somehow maintained it within tolerable limits through subsequent Earth history. Despite the hazards that go with it, the now-conspicuous morphological diversity of the biomass is a product of the high levels of energy-rich oxidative metabolism found among advanced cellular types.

Oxygenic photosynthesis (synthesis by light) is by far the largest and most familiar source of O_2. Upon its introduction at precariously trivial and fluctuating levels perhaps 2.8 aeons ago, O_2

began to play a role in the evolution of life and Earth's surface processes. After that, if not earlier, photosynthetic O_2 and perhaps plate tectonism joined sunlight, gravity, and water as lead players on the evolutionary stage. Yet, for another 6 to 8 geocenturies it remained at vanishingly low levels as a result of reactions with a variety of reduced substances.

Nor was oxygen-evolving photosynthesis the only source of O_2. Physical splitting of H_2O by photolysis was probably the prevailing initial process. Photolysis of CO_2, as well as the release of O_2 from metallic oxides as a result of microbial processes and chemical weathering, are potential but poorly understood sources, although the late W. W. Rubey estimated that weathering

alone, over time, could have produced some 6 times the amount of O_2 *currently* in the atmosphere.

All processes that result in the release of free O_2 must borrow external energy to break chemical bonds. The resultant thermodynamic compulsion to recombine with coproducts must be countered by segregation of the latter if O_2 is to accumulate anywhere. Hydrogen must escape Earth's gravity field for O_2 from the photolysis of water to survive. Carbon from the photosynthetic assimilation of CO_2 and its combination with H_2O to make glucose and O_2 must be buried in chemically equivalent quantities in soils or sediments for photosynthetic O_2 to survive. It is estimated that outgassed and cometary hydrogen alone would have

10.8 THE MODERN OXYGEN CYCLE. *[After Preston Cloud, 1978, Cosmos, Earth, and Man,* Yale *University Press.]*

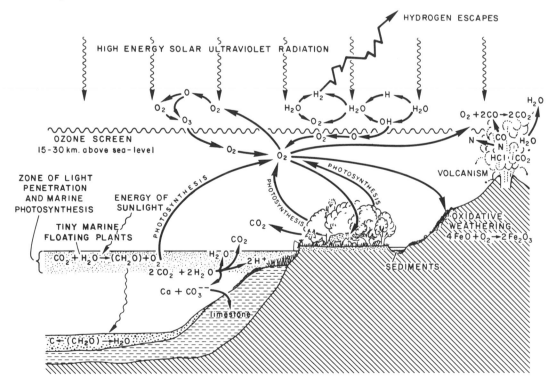

overwhelmed non-photosynthetic sources of O_2 in the primordial atmosphere.

Although evidence of extensive oxidation is often viewed as evidence of the presence of high O_2 levels, that is, in a sense, irrelevant. What oxidation signals is the former presence of an *oxygen sink* that could neutralize O_2. Whether free O_2 accumulates is another matter and one of great significance for Earth history and us. In fact, no significant quantities of O_2 from any source could have accumulated in either hydrosphere or atmosphere before contemporary oxygen sinks of associated reduced gases, surface materials, and dissolved substances had been neutralized. And then it could have accumulated no faster than coproducts were segregated.

Early oxygenic photosynthesis was, with little doubt, an exclusively marine process. Even after it began, significant quantities of photosynthetic O_2 could not likely have escaped to the atmosphere until after the marine hydrosphere became essentially saturated with it. And each later upwelling would have introduced new, O_2-consuming, anoxic waters from depths.

On reaching present levels, perhaps 4 geocenturies ago, oxygen was still consumed by new reduced volcanic gases, erosionally exhumed carbon, and reduced matter in the hydrosphere right up to the present. Levels fluctuate with rates of erosion and volcanism. The indefinite continuity of oxygen is not guaranteed. Indeed, Harvard's H. D. Holland has shown that its continued existence in the present atmosphere is a product of a 3-million-year lag between production and consumption. If photosynthesis were to cease today, photolysis would be our only source of O_2 three million years hence.

For the oxygen now in the atmosphere is just enough to neutralize the continuing influx of reduced volcanic gases and still-buried reduced carbon which will be exposed by continuing erosion during that interval, let alone carbon put there by industrial processes.

To grasp, if not to solve, the mystery of how O_2 managed to overcome the forces of reduction to establish and maintain Earth's now-breathable air, one must consider the sinks as well as the sources of O_2. Important sources of O_2 include both photosynthesis, with carbohydrates as a coproduct, and photolysis of H_2O, with hydrogen as a coproduct. Sinks are calculated as the sum of all reduced substances now or earlier at or above Earth's surface or in its waters. Estimates of the quantities of now- and once-free O_2 must be compared with similar estimates of the capacities of various O_2-scavenging systems over geologic times.

Rubey, in 1951, made the first attempt to calculate the amounts. He found an excess of carbon (calculated as CO_2) that implied to him either very large initial quantities of carbon monoxide and CO_2 or the evolution over time of quantities of O_2 from sources other than photosynthesis and equivalent to about half the calculated photosynthetic O_2. The first is likely. The second is possible but limited as far as photolytic O_2 is concerned both by the reductive power of outgassed hydrogen and by the fact that escape of hydrogen is temperature related and may not have been important under the faint early Sun.

Robert Garrels of the University of South Florida recalculated the quantities in 1983. He found a closer geochemical balance between organic carbon and O_2, implying that older Cryptozoic O_2 from simple physical photolysis could have been important only to the extent it was rapidly consumed in neutralizing juvenile H_2, CO, NH_3, H_2S, and other reduced substances.

Important early O_2 sinks other than the conversion of ferrous iron compounds to ferric, of sulfides to sulfates, of CO to CO_2, and of CO_2 to carbonate ion surely included large quantities of reduced gases from rampant early volcanism. The preponderance of volcanic rocks and volcanogenic sediments in the Archean sedimentary pile, and of exuberant cratonal volcanism in earlier Proterozoic time, bears witness to a corresponding abundance of contemporaneous gaseous O_2 sinks. The probable presence of CO, H_2S, and H from volcanic sources and in the primordial atmosphere could have neutralized appreciable transient O_2 from photolysis. And any methane, ammonia, or even reduced carbon from meteorites or comets would have added to the reductive capacity of the world our microbial photosynthetic predecessors set about neutralizing and then transforming to the oxic one we inhabit.

Canada's S. M. Roscoe had already drawn attention in 1969 to the fact that an interesting message about oxygen was being signaled by another historical sedimentary curiosity. He referred to the extensive survival of the earlier-noted large, detrital deposits of readily oxidizable, thorium-rich uraninite and pyrite not only in the basal Huronian strata of Ontario but in equivalent strata of Hamersleyan affinities at a dozen other localities on four continents. That age, once thought to have begun about 2.3 aeons ago, is now lowered by new uranium—

ages to about 2.45. It is coupled with the later appearance, perhaps 2.2 aeons BP, of the oldest thus-far-known significantly oxidized continental sandstones (*redbeds*) to suggest an age of about 2.2 aeons BP as likely for the first significant growth of atmospheric O_2.

The principal source of that O_2 was most likely its continuing photosynthetic production and upward diffusion following neutralization of oxygen sinks. Probable consequences of that were the final major sweep-out of stored marine iron— a trace element in modern seas—accompanied or followed by buildup of a protective ozone screen in the stratosphere and the beginning of significant microbial sulfate reduction.

As with O_2 in photosynthesis, biological sulfate reduction favors the light isotope of sulfur. It causes a decrease of heavy S-34 sulfide and comparable increase in sulfate. Exactly that relation is found in sulfates and sulfides of 2.4- to 2.2-aeon-old sedimentary rocks in the Huronian and Transvaal supergroups of the Lake Superior and Transvaal regions, but not, so far as evidence now known goes, in older sedimentary rocks. The histories of redbeds, sulfur isotopes, thorium-rich detrital uraninite, and natural nuclear reactors are all consistent with the view that free O_2 first became a significant atmospheric component perhaps about 2.2 aeons ago. Significant but not yet abundant. The prevalence of BIF before and redbeds after about 2 to 1.9 aeons BP, and the beginning then of stratiform copper ores, imply that O_2 may not have risen much above about 1 percent present atmospheric level before then. Increases above that level favored the emergence of eucaryotic life.

The Gunflint Microbiota

Brightest of the lamps that light the murky trail into the remote biogeologic past and that help to illuminate the oxygen story is the 2- to 2.2-aeon-old Gunflint microbiota. Its contemporaneity with the strata in which found is demonstrated by the concentration of its microbial components in the growth laminae of branching, fingerlike stromatolites but not in the adjacent black chert between branches. These microbial fossils are beautifully preserved, enormously abundant, and remarkably diverse—comparable to living microbiotas in every way except that ultramicroscopic detail is not preserved. Some thirty different kinds have been recognized, of which sixteen species in fourteen genera have been named and classified—as proalgae (six species), budding bacteria (four), and of uncertain affinities (six). A few are illustrated in Figure 10.9.

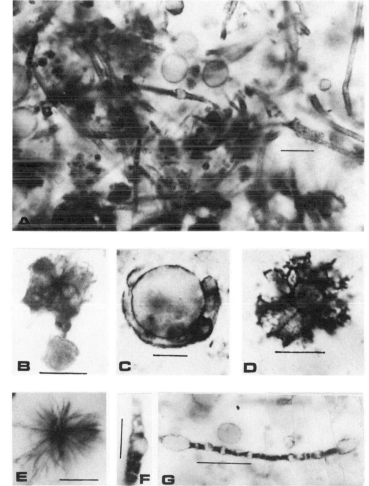

10.9 MICROBIAL FOSSILS FROM THE 2-AEON-OLD GUNFLINT IRON FORMATION. North shore of Lake Superior near Schreiber, Ontario. Bar scales all 0.01 mm (10 μm) long.

239

The discovery of this distinctive older Proterozoic microbiota in stromatolites of the Gunflint Iron Formation in southern Ontario cast new light on the old and obstinate problem of Cryptozoic life. It was also a classic example of discovery in the natural sciences, which often happens almost coincidentally when one is doing something else—like picking flowers to take home as a peace offering for a late return from the field. In 1953, at the end of such a long day, Stanley Tyler of the University of Wisconsin casually sampled an unusual-looking stromatolitic chert from the north shore of Lake Superior, a few kilometers west of the town of Schreiber, Ontario. On microscopic examination of thin sections of this rock, old Proterozoic hand Tyler quickly realized the significance of the swarms of minute septate threads and ovoid bodies inside it.

Making no pretense of paleontological expertise, he called on Harvard paleobotanist Elso Barghoorn to confirm and share his discovery. The ambiguous illustrations that accompanied their announcement in 1954 aroused little interest, but restudy by them and others led 11 years later to a pair of publications (by Tyler and Barghoorn on the one hand and by me on the other) that were to grow into what seems retrospectively a torrent of discoveries and publications about Cryptozoic life by a new generation of biogeologists and paleomicrobiologists. The key to further discovery was the superb preservation of delicate organic matter within this glassy rock. It had been saved from oxidative destruction by being, in effect, hermetically sealed within a probably primary silica gel.

Sporadic though such evidence still is,

it has removed all informed doubt that life was abundant on the primitive Earth, that it was then mainly microbial, or that it was continuous from 2.8, and probably from 3.5 or even 3.8, aeons ago onward.

Elements of the Gunflint microbiota itself have now been found in correlative or approximately correlative rocks farther northeast in Canada (the Belcher Islands, Hudson Bay), as well as in the iron ranges of northeastern Minnesota, northern Michigan, at least three different areas of Western Australia, and recently in northern China. The presently available or estimated ages of all lie within a range of about 2.2 to 1.8 aeons BP—a range that seems sure to narrow as more reliable ages become available. It supports the likelihood of a paleontologically distinct interval in early microbial evolution and the eventual application of distinctive microbial assemblages to the stratigraphic subdivision and correlation of Cryptozoic rocks. Its appearance is almost coincident with the important transition from essentially anoxic to oxic atmospheric conditions that marks the change from older to younger Proterozoic as recognized in this book (left front endpaper). At the same time, some of the species observed are so similar to living proalgae in details of form and probable habitat as to imply relationship, prolonged conservatism of shapes, and perhaps similar biochemistry and local ecology.

A few of the Gunflint filaments display an apparent variation in cellular morphology along their length that resembles the cellular differentiation of some living proalgae. One finds what resemble normal cells, terminal resting cells (akinites) capable of surviving desiccation to generate new filaments, and structures that resemble specialized oxygen-shielding cells (heterocysts) such as are dedicated in similar living forms to nitrogen fixation (Fig. 10.9G). Rare structures resembling short reproductive chains of cells (hormogonia) are also found (Fig. 10.9F). If the heterocyst-like structures are what they resemble (and there are those who question it) they add their bit to early atmosphere–hydrosphere biogeochemistry. The vital process of nitrogen fixation—limited to certain bacteria, archaebacteria, and proalgae—goes on quite happily in the absence of O_2 but shuts down when that gas exceeds a tolerable level. Nitrogen fixation can then continue only within oxygen-shielding heterocysts. The presence of cells so similar to heterocysts in the Gunflint filaments is consistent with the above-outlined view that sufficient O_2 had already accumulated in the 2-aeon-old hydrosphere to require shielding of nitrogen-fixing sites.

Even a confirmed skeptic, if not blinded by contrary predilections, would have to admit that the Gunflint microbiota and its congeners are conclusively biogenic, indigenous to the rocks in which found, and demonstrably *primary* (contemporaneous with the enclosing strata).

Banded Iron Formation and Its Bearing on the Oxygen Problem

The belle of the Cryptozoic rocks, beautiful much of the time and interesting all of the time, goes by the name of banded iron formation, BIF for short (Fig. 10.1). BIF is so interesting for students of early Earth history that it has become as much a household expression among them as DNA for biologists. It signals messages about the history and oxygen levels of the early atmosphere if we can but decode it. Its most distinctive feature is a striking alternation of relatively iron-rich and iron-poor, generally centimeter-scale bands and submillimeter-scale microbanding or lamination in a characteristically siliceous, often flint-like matrix called *chert* by geologists. A few deposits, as in the Hamersley basin of Western Australia (Fig. 10.10) and at Krivoi Rog on the Dnepr River in the Ukraine, even show a larger episodicity at a scale of meters. BIF is, in effect, symbolic of older Proterozoic history.

Because secondarily enriched BIFs are by far the world's most important commercial sources of iron, a great deal is known and hypothesized about them. They, and the products of their later oxidative enrichment, are distinctive among Archean and older Proterozoic rocks up to about 2 aeons ago. They are uncommon and atypical among younger strata (front endpaper, Fig. 10.11). The epitome of the typical, sequence-banded, siliceous BIF is that of the Hamersley basin, Western Australia (Fig. 10.10), conveniently accessible in this arid region except during the rains (Fig. 10.12).

Banded iron formation is simultaneously one of the most significant rocks for Earth history and the material foundation of industrial civilization. Its distribution, like oil, gold, and personal wealth, is very uneven (Table 10.1). About 92 percent of the roughly 600 trillion tons of BIF known worldwide is found in five

10.10 Macrobanding in Dales Gorge member of Brockman Iron Formation at (A) and north of (B) old blue-asbestos mine in Wittenoom Gorge, Western Australia. Numerals denote sixteen numbered microbands of volcanic ash seen as grassy slopes between seventeen ledgy microbands of BIF. The sequence shown in A is truncated by Cenozoic laterite at top right. Borehole correlation demonstrates continuity of banding at all scales across the entire 300-km diameter of the Hamersley basin.

10.11 Crudely banded, earthy, late Proterozoic iron formation of Yukon–Mackenzie boundary region, northwest Canada. More spotted than banded, but regularly bedded, the somewhat earthy Rapitan Iron Formation here pictured rests on cross-bedded detrital dolomite within a local iron-rich sequence.

10.12 GEOLOGISTS FROM
FIVE NATIONS AND FOUR
CONTINENTS COPE WITH
"THE WET" OF OUTBACK
WESTERN AUSTRALIA'S
HAMERSLEY BASIN,
AUGUST 1965.

TABLE 10.1 *Estimated Tonnages of BIF (>30% Fe) in World's Major Deposits*

Aeons BP	>2.5	2.5–2.2	2.2–2	1.9–1.8	0.7–0.55	0.37
Number of deposits	11	4	6	2	4	1
Continents	6 (plus Greenland)	4	4	2	4	1
Tons, main site	10^{13}	10^{14}	10^{14}	10^{9}	10^{13}	10^{12}
Total tonnage	3.3×10^{13}	4×10^{14}	1.2×10^{9}	2×10^{9}	1.2×10^{13}	10^{12}
Percent of total	6	71	21	0.0002	2	0.2

Source: Summarized from a selected list of 29 important or interesting BIFs in Harold L. James, 1983, in A. F.
Trendall and R. C. Morris, eds., *Iron Formation: Facts and Problems* (Amsterdam/New York: Elsevier), pp. 472–473.
Note: Tonnages of order-of-magnitude significance only.

huge older Proterozoic deposits that reach thicknesses between 100 and 1,000 meters and extend for hundreds or even thousands of kilometers laterally. Of the remainder, about 6 percent is Archean, found in many smaller and broadly lenticular deposits. And only about 2 percent is younger Proterozoic. Phanerozoic BIF is insignificant (front endpaper). The pattern of banding and areal persistence of individual laminae in much of the older BIF probably reflects early interactions among biosphere, atmosphere, and sedimentary lithosphere.

Although their total O_2 content is only about 10 percent of that consumed over geologic time in the oxidation of ferrous iron to ferric and ferro-ferric oxides in sediments, the problem of their geochemical transport, sedimentation, and concentration strongly affects our perceptions of atmospheric history.

Pristine BIF, before weathering, metamorphism, or enrichment, contains around 30 percent iron. Typical BIF from different parts of the world is so similar that unlabeled samples from different places could easily be confused (Fig. 10.1), and it is widely distributed in space and time (Table 10.1). The huge deposits are all older Proterozoic, and four of five are older than the oldest independent records of continental redbeds about 2.2 to perhaps 2.3 aeons ago.

Questions rise like a morning mist. Iron is readily oxidized. Dissolved iron is absent from the waters of existing seas and lakes except within sufficiently anaerobic local water masses and there only in the reduced or ferrous state. How did the BIF happen to occur so extensively and over such a range of history? How could it persist, in distinctive, chemically precipitated, submillimeter-thick, varve-like laminae such as have been traced for hundreds of kilometers across the shallow early Proterozoic Hamersley shelf basin of Western Australia? How did its predecessor solutions accumulate and where were they stored in such huge quantities in the soluble state pending introduction to the areas of precipitation? How was it carried in solution to its broad sites of deposition, and how was it then precipitated over the same region during brief intervals? If a pulse of soluble ferrous iron were abruptly introduced to an extensive oxic shelf environment, as by upwelling from an anoxic basin, how could the readily oxidized ferrous hydroxides have remained in solution over required distances? In particular, how could they have persisted in the presence of levels of free or dissolved O_2 in any way comparable to those now prevailing? Conversely, how could ferrous iron in those times have converted to the ferric state and precipitated except in the presence of ample free or dissolved O_2? Those puzzles concern the sources, geochemistry, and transportation of iron, silica (SiO_2), and oxygen.

The matter of sources has provoked continuing discussion. The conspicuously volcanic associations of the more restricted and perhaps deeper-basinal Archean sites imply primarily local volcanic sources of iron and silica, as may have been the case among some much younger BIF-like deposits. Such sources, however, seem inadequate to explain the generally much larger shelf and shelf-basin BIF deposits of older Proterozoic times, where associated volcanism was less conspicuous but perhaps also more subtle.

A view now in favor calls on the cumulative storage of ferrous iron from both anoxic weathering and volcanism in once-extensive anaerobic marine basins. Upwelling of iron-rich waters from such basins to sources of O_2 in shelf areas or shallower basins would have favored its oxidation and precipitation in the ferric state as hematite (Fe_2O_3), or ferro-ferric magnetite (Fe_3O_4), or both. In order for such precipitation to have occurred as thin iron-rich laminae over large regions, however, the iron would have needed to remain in solution above the surface on which it was accumulating until some basin-wide threshold level of O_2 was surpassed. And the thick sequences of rhythmically laminated and banded deposits that comprise typical BIF demand that such a process should have been repeated at brief intervals over very long times.

How could the levels of O_2 have been balanced in such a way that intervals during which ferrous iron was absent or suspended in solution alternated with intervals of conspicuous precipitation? So far, we have no better explanation for such rhythmically banded and laminated deposits than one that invokes the intermittent intervention of oxygenic microbial photosynthesis somewhat as fancied by A. M. Macgregor in his presidential address to the then Rhodesia Scientific Association in 1949 and more tentatively as early as 1927. Whereas photolytic O_2, at or above Earth's surface, probably did contribute to the long-term neutralization of oxygen sinks, it could hardly have accounted for the distinctive laminated and banded BIF structure. That episodicity is so regular and so extensive as to support strongly the inference that it was the product of some cyclic phenomenon. Seasonal

upwelling of iron-rich waters and related seasonal proalgal blooms may both have been involved, conceivably under glacial control.

How might that have worked? Recall that all forms of life are sensitive to O_2 beyond certain levels. A number of living proalgae find even very low levels stressful. In fact they like such low levels of O_2 that they prefer to live in the presence of hydrogen sulfide (H_2S), which simultaneously serves as a sink for excess O_2 resulting from their photosynthetic activities and a source of electrons for anaerobic photosynthesis. They have retained an ancestral ability to use H_2S instead of H_2O as a fallback source of electrons for vital energy under temporarily anoxic states that may last as long as 10 months out of a year. I have long supposed that some pre-Phanerozoic proalgal ancestor might have enjoyed a similar relationship with ferrous oxides. Now Israel's Yehuda Cohen has found that the charged ferrous ion itself can function as an alternative electron donor and that the presence of ambient ferrous iron actually stimulates procaryotic photosynthesis.

Given such relationships, the seasonal introduction of ferrous ion or hydroxides by upwelling under glacial or comparable cyclic control could have stimulated early proalgal photosynthesis. And that could have led either to the production of O_2 or to the direct conversion of ferrous to ferric oxides—either or both favoring the precipitation of insoluble ferric hydroxide and its conversion to hematite (Fe_2O_3). Reactions of that hematite with carbon could then have transformed it to magnetite ($6Fe_2O_3$ + organic C→$4Fe_3O_4 + CO_2$). Such a set of reactions could have accounted handily for the prevalent compounds of iron in BIF, the rarity of associated carbon, and the enrichment in light C-12 of associated carbonate rocks.

Of course no one knows what variations on the proalgal theme may actually have operated in those distant times. What can be said with a degree of confidence is only that something morphologically and perhaps functionally similar to proalgae had evolved by 2.8 aeons ago or earlier and was going strong by 2.2 aeons ago. It coped somehow with oxygen toxicity, and, when the major sinks were neutralized and O_2 accumulated to sufficient levels in the hydrosphere, the atmosphere also became oxic. Hypothesized consequences were first the BIF; second, the termination of extensive marine iron deposits for lack of dissolved iron in the sea; and third, the neutralization of any remaining atmospheric and surface O_2 sinks, the generation of a protective ozone layer, and the intensification of surface oxidation.

Oxygen could thereafter accumulate in the atmosphere to the extent that carbon was buried and new reduced substances neutralized. Consistent with such effects is the disappearance of all but rare traces of significant oxygen-sensitive detrital minerals such as thorium-rich uraninite and pyrite and the appearance of significant continental redbeds beginning about 2.2 aeons ago and becoming extensive from about 2 aeons ago onward.

The Enigmatic Sokoman Iron Formation

Finally, a word is in order about the huge (10^{14} tons), anomalous Sokoman Iron Formation of Quebec's Labrador Trough, 100 to 500 meters thick and reported to extend for hundreds of kilometers around the wet, black fly and mosquito infested northern peninsulas of the eastern Canadian Shield. Long considered to be substantially younger than other large BIFs, it underlies zircon-bearing rocks now dated by the uranium—lead method as 2.14 aeons old— about the same age as the iron formations of the Lake Superior region.

The Sokoman departs in several ways from more familiar BIFs. Very little of it is banded in the regular pattern of Figure 10.1. It does not fit neatly into categories made for other major deposits. However its deposits looked in the beginning, they have been so reworked by waves, currents, and gravity slumping that they now consist largely of chips, rolled pellets, and pebble layers that preserve little evidence of any former great regularity. They are conglomerates, angular magnetite sands (Fig. 10.13), oölitic or pelletal hematite sands, or slump breccias—the product of extensive reworking during the depositional process. The banded parts may be the only remains of the primary sediment. Its deposits, however they first started, last came to rest in an apparently shoaling, probably turbulent shelf and rift basin environment just before and during renewed subsidence (Fig. 10.14).

The enrichment in iron that made the Sokoman Iron Formation a direct shipping ore, however, has nothing to do with Cryptozoic history. It relates instead to a much later weathering interval that reached far across eastern North America during latest Mesozoic and earliest Cenozoic history—enriching parts of that surface in iron and locally in high-alumina clays. Plant-containing clays in iron mines and the conversion of former limestone layers in black shales beneath the Sokoman strata to ore-quality iron deposits testify to the effectiveness of this prolonged postdepositional event. In reconstructing Earth history, one must be wary of secondary processes.

10.13 MAGNETITE-BANDED DETRITAL AND CONGLOMERATIC SOKOMAN IRON FORMATION OF LABRADOR TROUGH, SHOWING PEBBLES AND RIP-UP FRAGMENTS.

10.14 HYPOTHETICAL RESTORATION SOUTH—WEST TO NORTH—EAST ACROSS THE PROTEROZOIC LABRADOR TROUGH BEFORE TERMINAL FOLDING, SHOWING GENERAL FEATURES OF KANIAPISKAU SUPERGROUP AND STRATIGRAPHIC SETTING OF THE SOKOMAN IRON FORMATION. Width about 100 km. Vertical exaggeration about 20 to 1. Location, Atlantic coast, about 55° present north latitude. Basal fluvial to deltaic redbeds and conglomerates were here fed into a coastal rift valley by longitudinal currents. They were capped by deeper-water shales, siltstones, and carbonate rocks across which nearly pure quartz sands then onlapped the crystalline Canadian Shield to the west (*left*). Iron deposits are interbedded with basalts to the east. The abutment of these pillow basalts against rift valley–like sediments and the capping deeper marine deposits above them implies east-to-west convergence of an oceanic sequence with the rifted continental margin. [*Modified after R. G. Wardle and D. G. Bailey, 1981, Geological Survey of Canada, Paper 81–10, p. 347, Fig. 19.10. Reproduced with permission of the Minister of Supply and Services Canada.*]

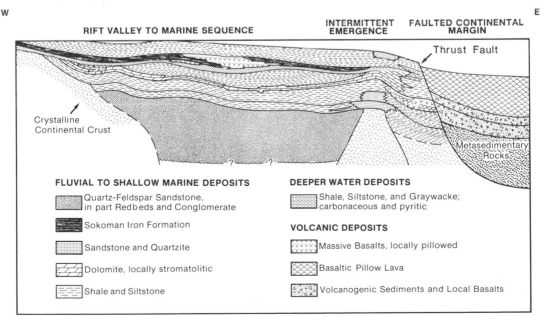

In Summary

Proterozoic history is a record of transition from the first demonstrably continental surfaces and the beginning of extensive cratonal sedimentation to the appearance of manifest animal life about 670 million years ago. Older Proterozoic history, until about 2 to perhaps 1.9 aeons ago, was an era of growing continental platforms and related epicratonal sedimentation. It was a time of conspicuous, rhythmically banded iron formation, of the first glaciation on a continental and perhaps a global scale, of emerging biospheric complexity, and of atmospheric transition from generally anoxic to detectably oxic states. Its preserved record began with the long Archean to Proterozoic changeover from a mantle-dominated geochemistry and sedimentary record to thereafter continent-dominated geochemistry and sedimentation.

Life during those precarious times records a diversification of biochemical pathways and early microbial body plans. Had Earth been a closed system, life probably could neither have originated here nor long continued to exist. But Earth is an open system. The life on it is energized by the Sun and will die with the Sun.

The big organic molecules from which living cells are made (and the little ones that make the big ones) need to be strongly and flexibly bonded to one another so that they can be broken and rejoined into new combinations without excessive energy costs. Carbon, hydrogen, oxygen, nitrogen, and bits of phosphorus and sulfur are the main life-building elements. Common and convenient temporary bonds are those made by removing water molecules, causing adjacent amino acids to bind together along carbon–nitrogen (or *peptide*) bonds to make proteins, just as carbon–oxygen and oxygen–phosphorus join toward other ends.

In this way early molecular structures were joined by dehydration. Such *dehydration condensations* were important in constructing the first blobs of protoplasm that could grow, replicate, mutate, and reproduce mutations—structures (with or without a cell wall) having the essential properties of life. Most likely they were simple, bacterium-like cells, intolerant of free O_2 and dependent on external sources of energy (i.e., were anaerobic heterotrophs).

The evolution of such early microbial consumers evidently gave rise at some point in some lineage to biochemically more complex, self-sufficient producers (autotrophs), else the stream of life we see could not have been realized. There may have been more than one unsuccessful origin of life before some triumphant line of descent became an autotroph whose organic residues could sustain biochemically similar *heterotrophic* cousins that depended on external food sources. The evidence for a single successful beginning rests on the basic biochemical identity of all organisms from the simplest bacterium-like creature to *Homo*—the same twenty protein-building amino acids, the same DNA and RNA, the same ATP.

The evolution of free oxygen exerted a major influence on the changing scene. Essentially all O_2 comes from secondary self-reversing reactions, driven by radiant energy. The unstable coproducts of these reactions must be segregated in order for O_2 to exist long enough either to combine with something else or to accumulate as free O_2. Carbon must be buried or hydrogen lost to space. O_2 could have accumulated only as fast as its photosynthetic and photolytic coproducts were segregated, and only after a formidable array of oxygen-scavenging sinks were neutralized. Hazardous as well as beneficial, O_2 exists as a free gas in today's atmosphere mainly as a consequence of a 3-million-year kinetic lag between initial production and later consumption.

The first molecules of free O_2 appeared briefly when water molecules were split by solar energy and hydrogen escaped the grip of Earth's gravity. But they did not last long. Reduced gases, then abundant, immediately neutralized those molecules and continued to do so as fast as they appeared for a long time. Photolytic O_2 was supplemented by photosynthetic O_2 when some anaerobic sulfur bacterium kicked the H_2S habit in favor of H_2O as an electron source, becoming a proalga with the acquisition of appropriate cytochromes and chlorophyll-*a*—later with both chlorophyll-*a* and chlorophyll-*b*. Hermetically sealed within 2-aeon-old carbon-rich stromatolitic cherts near the top of the older Proterozoic succession in southern Ontario is the classic proalgal microbiota of the Gunflint Iron Formation, so far our most impressive biogeological probe into the deep past.

Banded iron formations are distinctive among older Proterozoic deposits. These BIFs, comprising the world's greatest iron deposits, consist of thick and extensive successions of alternating iron-rich and iron-poor laminae and bands of a char-acteristically siliceous or cherty composition. About 92 percent of all BIF known is older Proterozoic. The sources of this iron were volcanism and weathering. It was probably stored in anoxic marine basins from which waters rich in ferrous ion welled up intermittently and probably seasonally to precipitate as ferric hydroxides and oxides on continental shelves or in shelf basins. It seems likely that oxygen-producing photosynthesis played a critical role. The main problem is to account for the extensive, regular, successively iron-rich and iron-poor banding and lamination of the siliceous bulk rock in a chemically plausible way. That could have been achieved by seasonal upwelling of ferrous oxides accompanied by seasonal proalgal (perhaps annual) blooming and temporary oxygenation, while silica maintained a steady rate of precipitation. That could also account for the probably biological light carbon now being reported from carbonate rocks associated with BIF.

Redbeds, indicative of atmospheric O_2, first appeared about 2.2 aeons ago. By 1.9 aeons ago there was enough free atmospheric O_2 around to support their extensive development, their preservation at appropriate sites, and probably a UV-shielding ozone screen.

Complex interactions during older Proterozoic history governed the evolution of the continents, continental volcanism, sedimentation, atmospheric and hydrospheric evolution, and the progressive elaboration of the biosphere itself. Feedback among all of these variables regulated the fate of each. An evolving steady state restrained potentially runaway processes such as an early atmospheric greenhouse and kept the whole of our planetary ecosystem in balance.

CHAPTER ELEVEN

YOUNGER
PROTEROZOIC HISTORY:
THE RESTLESS CONTINENTS

Younger Proterozoic history was an era of transition from a still-primitive to an incipiently modern Earth. Oxygenic photosynthesis and oxidative metabolism became firmly established at the cellular level. Microbial life differentiated, culminating in the central event of biologic evolution—the origin of the eucaryotic cell and hereditary apparatus, prelude to multicellular animal and plant life and the Phanerozoic Eon. Continental and oceanic realms reached nearly their present volumes and the hydrosphere its chemistry. Compared with the Archean prevalence of sodic intrusives and greenstone belts, the growth and stabilization of potassic older Proterozoic cratons, and the distinctive BIF depositories, younger Proterozoic history was one of growing continental influences and intracontinental tension.

Initial rearrangements of sialic lithosphere are thought to have resulted in large semistable shields and a small number of extensive continental masses. Heat accumulated beneath such sialic covers, peaking with the intrusion of nonorogenic plutons about 1.45 aeons ago and followed by later tensional rifting and intrusion of extensive mafic dike systems. Modern styles of plate motion became prevalent and new continental geographies evolved. The sialic component of volcanism increased with rifting and the ascent of subduction-zone magmas through continental lithosphere. Redbeds and other nonmarine modes of sedimentation became conspicuous. Calcareous sediments rich in stromatolites veneered shallowly flooded continental platforms. Save for its still-modest oxygen pressures and the resultant simplicity of the biosphere, the scene could have been Phanerozoic.

11.1 THE SUDBURY ASTEROID SCAR AND THE GRENVILLE FRONT IN SOUTHERN ONTARIO. October 1976, north up. The 1.85-aeon-old Sudbury *astrobleme* is the dim oval left of top center, said to affect a region about 140 km long. It was compressed from the southeast during the Grenville orogeny about 1.1 aeons ago. At the south is Georgian Bay, Lake Huron; at the northeast, Lake Nipissing. *[From NASA ERTS imagery.]*

The Maturing Earth

Earth history between 2 to 1.9 and 0.67 aeons ago was an age of experimentation, of growing, of maturing. In an odd way it calls to mind our own transitions from adolescence to maturity. In geologic terms it was a time of unrest, of exuberant volcanic activity, of nonorogenic plutonism, of crustal rifting, of increasing sedimentary variety, and of biological challenge. It was an interval of experimentation with hazardous oxygen, of the severance of old and the establishment of new connections, of reaction to the tensional stresses that grew with plate motions.

River systems and continental geochemistry continued to grow in importance, even as ocean water diffused through mid-ocean ridges, resulting in the buffering of seawater to near its present chemistry as a consequence of the changing kinetic balance between continental and mantle-dominated sources. In such ways did the maturing oxic Earth of younger Proterozoic times record its progress toward Phanerozoic physical characteristics, the passage from bacterial innocence to advanced microbial and eventually metazoan sexuality and complexity, the change from Proterozoic to Phanerozoic styles.

The younger Proterozoic record began about 2 to 1.9 aeons ago with the termination of typical, perhaps proalgally mediated, BIF as an important historical marker. At very nearly the same time, extensive subaerial redbed sedimentation became prominent. That record closed with the dawn of multicellular animal life (*Metazoa*), initiating the all-but-modern Phanerozoic Eon and its opening era and period—the Paleozoic and the Ediacarian.

The distinguishing features of younger Proterozoic history contrast with those of antecedent times (left front endpaper). Although younger Proterozoic history began with the last widespread intrusion of primary sialic plutons and their amalgamation into shields and larger heat-blocking continents, it later witnessed widespread intrusion of dike systems and opening rift basins—evidence of tensional stress. Extensive shallow continental platforms and shelves were veneered by marine sediments without Metazoa or signs of their activities. It was a time of unprecedented extension of redbeds above and a notable development of stromatolitic carbonate rocks at and below sea level. Its debut may even have been announced with a bang—the 1.85-aeon-old Sudbury asteroid fall (Fig. 11.1) whose scar (*astrobleme* or "star wound") is the site of one of the world's two major nickel deposits.

During this long younger Proterozoic interval, the ancient world approached its mature state in all respects except the presence of Metazoa and a level of O_2 comparable with later times. Before it closed, the growth of the sialic continental mass was so nearly complete that apparent later additions involved mainly remelting and recrystallization of old crust, plus recycling of younger sedimentary rocks together with small additions from the subcontinental asthenosphere. The North American continent was assembled or reassembled from a jigsaw of provinces and subprovinces (Figs. 7.12, 11.2), including slivers

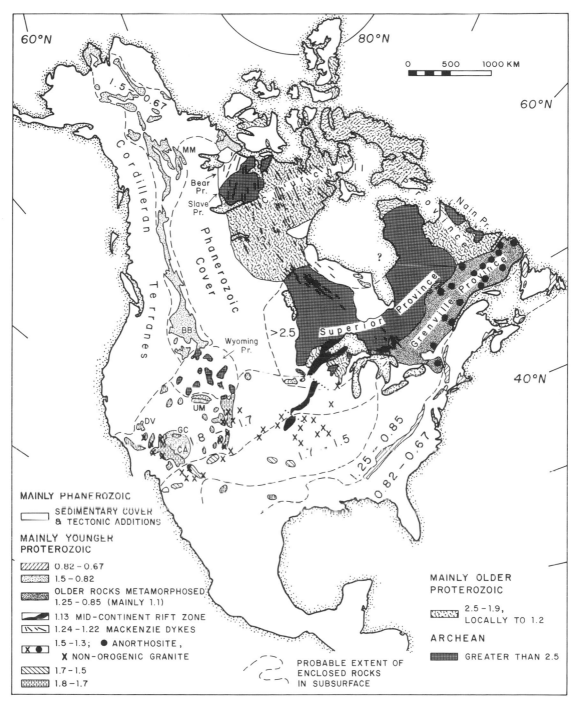

11.2 Distribution by age in aeons BP of Cryptozoic rocks in North America and their likely subsurface extensions. The 1.13-aeon-old mid-continent rift zone, (*black*) is a subsurface feature extending south from Lake Superior. Anorthosite (*dots*) and nonorogenic granites (*Xs*), cluster at 1.45 and range from 1.5 to 1 aeon BP. Ages include U–Pb zircon crystallization ages as well as Rb–Sr metamorphic ages. Letters indicate: *CA*, central Arizona; *GC*, Grand Canyon; *DV*, Death Valley; *UM*, Uinta Mountains; *BB*, Belt basin; *MM*, Mackenzie Mountains.

253

that were added to the now mainly eastern (then mainly southern) seaboard in a series of collisions with Africa and Europe (Fig. 9.10). A variety of interesting igneous rocks took part in this kaleidoscope of events. One of these was an unusual variety of high-silica igneous rocks wherein calcium oxides just balance potassium and sodium oxides (*calc-alkalic* rocks). They are believed to mark ancient collision zones where former mobile plates are sutured together to make the present shield areas.

Tensional Stress, Strain, and Volcanism

The north–south belt of black near the center of Figure 11.2 marks where North America almost split apart 1.14 to 1.12 aeons ago. A line of crustal weakness was here rifted open and filled with probably mantle-derived basalts and fluvial debris to a width of 60 to 160 kilometers along a mid-continent rift zone that extended 1,500 kilometers southwest from Lake Superior to Kansas. Phanerozoic marine sediments now conceal the rift to the south of Lake Superior, but we know where it was because its basaltic fill generates big gravity and magnetic signatures beneath the quiet cornfields—like the now-famous mascons found on the moon by orbiting Apollo spacecraft. It may be the world's greatest dike, but it is only part of the ample evidence for tensional stress in North America during younger Proterozoic history (Table 11.1).

The immense 1.13-aeon-old mid-continent rift zone was preceded barely

TABLE 11.1 *Records of Tensional Stress and Strain in North America between 2.15 and 0.67 Aeons Ago*

Age in Aeons BP	Present Trend	Extent (in km)	Nature	Location	Name or Explanation
0.67	ESE–WNW	2,000	Mafic dikes and sills	Across northern Canada	Franklin dikes
1.13–1.1	E–W	Superior basin	Flood basalts and sills	Lake Superior	Keweenawan volcanics
1.13	NE–SW	1,500	Basalts and mafic dikes	Lake Superior to Kansas	Mid-continent rift zone
1.23	SE–NW	3,000	Mainly mafic dikes	Lake Huron to Arctic Sea	Mackenzie dikes
2.15	E–W	360	Mafic sills and dikes	South Ontario	Nipissing diabase

a geocentury earlier by a less conspicuous but very widespread mafic-dike suite that splays northwest across the whole of Canada from Lake Huron to the Arctic Sea, proclaiming tensional stresses of continental extent—the Mackenzie dikes of Table 11.1. Counterparts of this event are seen in south Greenland and the Scandinavian Shield. And directly associated with it is the oddly funnel-shaped north-Canadian Muskox intrusion—a structure believed to have arisen from a depth of 120 kilometers and therefore was probably rooted in the asthenosphere. How that might be reconciled with plate tectonism remains open.

The history of younger Proterozoic tensional stresses began with the 1.9- to 1.8-aeon-old records of rifting followed by convergence in Canada's Wopmay *orogen* (a site of mountain building, or orogeny), just east of Great Bear Lake. Indeed, extensive tensional stress is signalled as early as 2.15 aeons BP when the mafic Nipissing diabase (Table 11.1) filled contemporaneous tensional rifts that transected Canada's older Proterozoic

Huronian Supergroup. And evidence from the eastern rim of the Labrador Trough implies that rifting, subsidence, and sedimentation on the continental margin of that region was terminated by convergence of plates along its present northeastern margin about 1.85 aeons ago, as the Sudbury object fell.

Perhaps there was a connection? If so, might it also extend to the apparent cut-off of the Huronian and Animikie supergroups (Fig. 10.4) along an imaginary line of that age that also continues northeast through Sudbury to behead the Labrador Trough near a major shift in metamorphic grade (Fig. 11.3)?

Continental *flood basalts* are first known from the younger Proterozoic. They covered areas of a million square kilometers or so with up to a half-million cubic kilometers of basalts in as many as hundreds of individual flows 10 to 30 meters thick. Known examples are found from younger Proterozoic to late Cenozoic times on five continents, and several of them are associated with rifting continental margins.

11.3 POSSIBLE LAURENTIAN PLATE MARGIN AT THE BEGINNING OF YOUNGER PROTEROZOIC HISTORY. This inferred plate margin may have radiated from a triple junction beneath Labrador at the end of older Proterozoic history, then converged and closed again 1.85 aeons ago. It may have been obscured by events during the Grenville orogeny which joined that province to proto–North America. See Figs. 11.1 and 11.2.

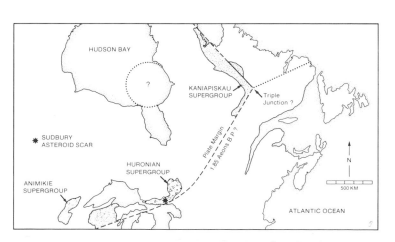

The oldest is at the north end of the North American mid-continent rift zone (Fig. 11.2), in the Lake Superior region (Fig. 11.4). Here some 400,000 cubic kilometers of mainly mafic, 1.1-aeon-old lavas 6 to 9 kilometers thick brought bonanza copper deposits to already iron-rich Lake Superior and, aeons later, temporary industrial prosperity to the region.

The eruptive scene is unimaginable! A miniocean of eerily fuming basaltic lava. A region the size of Maine engulfed in the acrid fumes of volcanic gases. The searing heat of a blast furnace across the whole of it. Literally hell on Earth.

The extent and cumulative volume of that lava compares with or exceeds that of the great Mesozoic and Miocene flood basalts and dwarfs the famous 8-cubic-kilometer fissure flow of 1783 at Laki, Iceland. Although the 200 or more individual flows of this Keweenawan Supergroup were nearly horizontal at the time of extrusion, their slopes increased with the total load, progressively deepening the subsiding terrain beneath until the volcanic overburden itself was buried under another 10 kilometers of 1.1-aeon-old upper Keweenawan redbeds and other detrital nonmarine strata.

Although continental flood basalts of such magnitude are unknown elsewhere among Proterozoic rocks, one finds petrographically similar tensional associations in both older and younger sequences. Basaltic dikes swarmed west to east through Scandinavia about 1.7 aeons ago and north to south through west Greenland at about 1.5 aeons ago. Similar rocks and associated redbeds developed widely in northern Canada and Europe, eastern Africa, and elsewhere as younger Proterozoic history waned (e.g., Table 11.1). Long-accumulating stresses gave way to active rifting and then more plate tectonism, beginning about 1.2 aeons ago.

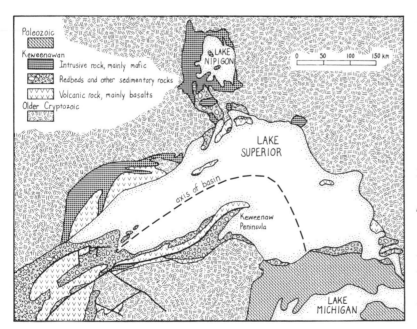

11.4 KEWEENAWAN ROCKS OF THE LAKE SUPERIOR REGION. *[Simplified after Henry Wallace, 1981, Geological Survey of Canada, Paper 81-10, p. 400, Fig. 22.1. Reproduced with permission of the Minister of Supply and Services Canada.]*

In addition to all this mafic intrusion and volcanism, intermediate to sialic lavas, often associated with redbed sedimentation, became increasingly important. The combination of such continental deposits with rift volcanism and stromatolitic shallow-water marine carbonate rocks without metazoan fossils is characteristically younger Proterozoic. Moreover, the wide distribution of redbeds, from the beginning of younger Proterozoic history onward, enhances the potential applications of paleomagnetics to paleoclimatic history and stratigraphic correlation during this time. Such rocks are excellent recorders of magnetic information. Moreover, new techniques for removing acquired magnetic overprints allow original signatures to be recognized and placed within a time-correlated apparent polar wander path.

Plutonism, Tectonism, and Crustal Evolution

Plutonism refers to the formation of magmas within Earth's crust or their intrusion from depths, followed by cooling and crystallization beneath the surface. It played a major role in Cryptozoic history. A column on the right of the left front endpaper depicts some relations of plutonism to other aspects of crustal evolution. Plutonic rocks often contain minerals—such as zircon, uraninite, and biotite—that are good for determining age by different radiometric methods. And the physical relationships of related plutons and other igneous rocks to otherwise undatable sedimentary materials often bracket their times of deposition. It is clear, for example, that any given sedimentary rock preceded any crosscutting intrusive rock, but that any sediment deposited on an eroded surface of the same intrusive rock came later—or, according to traditional systems of counting ancient time backward from some landmark date, it is younger (Figs. 7.11, 3.4). Thus, if a sedimentary rock is older in this manner of reckoning than one dated plutonic or volcanic rock but younger than another, its age is "bracketed" between those dates.

Plutonism is commonly associated with orogeny, and both are broadly episodic. That provides grounds for one system of punctuating Earth history by events that are usually referred to as orogenic, even though ages found most commonly refer to some aspect of plutonism. Problems persist. The intervals of plutonism (and orogeny) may be very long, rising to and descending from a peak of intensity that has no distinctive features except its age. In addition, each pluton has several ages: that of formation and introduction of the melt, the crystallization or cooling age or ages of minerals dated, and commonly one or more metamorphic ages. Each age is best determined by a different method. It is not clear, unless the method used and kind of sample dated are specified, which kind of "age" a quoted number refers to. In addition, numbers determined before the standardized decay constants of 1976 all require recalculation. It is only for smoother narration of well-established numbers that I quote undocumented ages here. All such numbers are open to challenge.

Measured ages of plutonic rocks from the Canadian Shield and Africa cluster into three main groups older than about 1.6 aeons, during which perhaps 90 percent or more of the total sialic crust was generated (left front endpaper, Table 11.2). Younger plutonic rocks result mainly from the remelting of preexisting ones, with nonorogenic exceptions.

Postdepositional numbers that presumably convey an important message about Earth history are the rubidium–strontium metamorphic ages that cluster almost globally around 1.9 to 1.7 aeons BP. They represent a very broad episode of thermal metamorphism at or very near the beginning of the younger Proterozoic. A major buildup of internal heat beneath the continents is implied, such as might result from extensive continental blanketing and which perhaps initiated the thermal and tensional stresses signalled by younger Proterozoic rifting, intrusion of dikes, volcanism, and nonorogenic plutonism—perhaps even the warmup to conventional styles of plate tectonism.

Because plutonism is usually associated with mountain building, nonorogenic plutonism is always cause for wonderment. Yet nonorogenic plutons

were abundant from about 1.65 to 1.1 aeons ago, and especially at around 1.45 aeons in two globe-spanning belts of limited width. One of these belts stretches northeast from California to the Adirondacks and Quebec (Fig. 11.2), then across the Baltic Shield and into the Ural Mountains and Siberia. The second is a similar but less voluminous belt that, in a pre-drift Mesozoic reconstruction of the Southern Hemisphere, runs from southernmost Brazil across Africa, to Madagascar, India, and Australia.

These plutons feature the bluish-iridescent rock called anorthosite (some 90 percent plagioclase feldspar). Associated with it, or in a parallel belt, is the also colorful and distinctive, potassic *Rapakivi* type of granite, with prominent, pink, sodic-rimmed feldspar inclusions, together with high-temperature metamorphic rocks. These younger Proterozoic anorthosites are well known as aluminum ores in Norway and as a decorative stone worldwide. All are high-temperature rocks, reflecting a deep source and probably thermal blanketing.

Of 140 or so anorthositic plutons known worldwide, well over half the total mass

TABLE 11.2 *Ages of Major Plutonism and Orogeny in Cryptozoic Rocks (Decay Constants Corrected to 1976 Standards)*

| Age Range | Peak Age or Ages | | Type of Plutonism | Associated Orogeny in Canada–United States |
	Canadian	African		
1.25–0.85	1.1	1.1	Mixed	Grenvillian
1.65–1.1	1.45	1.45	Anorthosite event	Nonorogenic
1.9–1.7	1.8–1.7	1.9	Orogenic	Hudsonian–Penokean
3.1–2.45	2.65	2.7	Mixed	Kenoran–Algoman
3.8–3.5	3.6	3.6	Mixed	Mortonian

is said to occur in the Grenville Province of southeastern Canada and adjacent regions: the circles on Figure 11.2 mark separate masses or clusters. Samarium–neodymium ages of anorthosites from eastern New York State cluster at 1.1 aeons BP. Other isotopic evidence implies that Grenville anorthosites of that crystallization age were derived from two hot-mantle-source regions dating to 1.65 or more aeons ago beneath eastern Canada. Isotopic similarity to present mid-ocean-ridge basalts suggests the asthenosphere as the primary magma source.

Some of these nonorogenic intrusives, at least, have a primary mantle source and are not simply products of remelting and mobilization of older rocks. Their greatest North American concentration at about 1.45 aeons BP may signal the approaching interval of plate tectonism that subsequently gave rise to the late Proterozoic ophiolite-like rocks whose distribution is shown on the back end-paper. This nonorogenic intrusive cycle was eventually phased out by the Grenville collisional event along the present-day eastern seaboard (Figs. 11.1, 11.2). Tensional dike systems were shut off. Rock temperatures rose to high metamorphic levels, and the Rb–Sr and K–Ar clocks were reset at 1 to 1.1 aeons BP. Except for the Phanerozoic addition of microcontinents to its margins and the veneering of the continental surface with shelf deposits, North America was essentially complete.

Interpreting the Sedimentary Environment

The sedimentary environment is an important route to Earth history at all levels, but understanding sedimentary processes and their products increases in importance with decreasing age, both as they come to play a larger role and as we seek a finer resolution.

Perhaps the most remarkable thing about the sedimentary record is that it is so well preserved on the continents. "Common sense" tells us that the continents are realms of erosion, from which all superficial accumulations should be carried eventually to the great oceanic storehouse, leveling the lands and filling the ocean basins. Yet the aeon or more of erosional history since essential completion of the continental mass has not accomplished this. Gravity and plate tectonism will not permit it. The bulldozing action of subducting plate margins scrapes off any accumulated sediments and plasters them against the continents or carries them downward to be assimilated at depths and regurgitated as components of volcanic arcs, bringing copper and precious metals with them. As erosion dissects young mountains, removing load from their tops, they rise in compensation (a process of seeking gravitational balance, known as *isostasy*). Thus, at some point before final erosional leveling, and perhaps for much of their history, the peaks of mountains are likely to be higher than their initial crests. Despite episodic marine flooding of their low parts, gravity, through isostasy, sees it to that continents maintain freeboard above the oceans. Continents may grow or decrease in area through plate-tectonic processes. The total mass of continental material may be increased from mantle sources.

259

But the total continental mass cannot decrease significantly.

On the other hand, ancient sediments and sedimentary rocks are preserved only when they are deposited below the level of erosion or on surfaces that subside faster than erosion can remove them. Thickness of a sedimentary pile is more a measure of subsidence than depth of depositional "basin." Indeed, the site of accumulation may be a basin only retrospectively. Other tectonic and erosional processes eventually bring what's left of its sedimentary record back to view.

As it happens, all pre-Jurassic stratified deposits now preserved (as well as sialic igneous ones) are found exclusively on continental surfaces, whether flooded like the seaward margins of continental shelves or exposed like their inboard areas and elevated surfaces. They include a surprisingly large number of intertidal, fluvial, lacustrine, glacial, and aeloian deposits—the most susceptible to erosion and thus the least likely to be preserved for long intervals. One begins to appreciate how remarkable the record of younger Proterozoic history really is. Considering that the older the rock the more likely it is to be either eroded away or concealed by more recent deposits, we find here an unforeseen variety of sedimentary, volcanic, and plutonic rocks. They, in turn, record an equally remarkable variety of environments, processes, and events.

Fascinating though the field of sedimentology is to its devotees, it is too esoteric and detailed a subject to explore in depth in this book. Three aspects of sedimentary geology, however, call for brief attention here.

First is the prevalence in many younger Proterozoic sequences of sedimentary structures and types indicative of continental influences, including extensive braided-river systems (e.g., redbeds, trough-cross-bedding, climbing ripples, festoon cross-bedding). Proterozoic oceans without animals can do little more with sediments than move them about, sort them into size categories, and precipitate dissolved components where the ambient chemistry is right. Excluding a volcanic minority, the detrital components and many of the dissolved ions are products of weathering and erosion on land and transport to the seas by river systems. The dominance of continental components, textures, and geochemistry over mantle influences stands out.

Second is the change in aspect of rocks, and especially sedimentary rocks, as outcrops or drill-core samples of the same stratigraphic level are traced from one locality to another. Recall that the patterns geologists use in their cross-sections (vertical profiles) to indicate kinds of rock change laterally as well as vertically (e.g., Figs. 10.3, 10.4, 10.14). They show that the kinds of deposits that become rocks change with place as well as sequence and time. Such changes are familiar—downstream in a river, between river channel and floodplain, and from sandy beach to muddy bottom in a lake. Also, in great variety, in the sea—from coastal marsh to rocky headland, to muddy lagoon, to continental shelf and slope, and eventually to abyssal depths. The different kinds of deposits laid down at different places at an equivalent interval in history represent the varying aspects, or *facies,* of the rock. It is possible to learn a great deal about the history of sedimentary rocks from the study of characteristic facies and their changes, including surface features, internal structures, chemistry, mineralogy, and fossils. Geologists try to identify sources of materials, modes and directions of transport; the flow regime if applicable; the depth, temperature, slope, and probable climate; and postdepositional changes. They study how such characteristics have varied with place and time.

The *third* sedimentary phenomenon calling for attention here is the abundance of stromatolites in Cryptozoic carbonate rocks. It lies so far outside the general experience and yet is so important for younger Proterozoic history as to call for substantially more discussion than the previous two.

DECODING THE CRYPTIC STROMATOLITES

Distinctive among the talismans of pre-Phanerozoic life and environments are laminated, ordinarily calcareous buildups of presumptive and sometimes demonstrable biogenic origin called *stromatolites*. They have a lot to say about ancient environments, especially in nearshore marine settings. They come in a range of sizes and eye-catching shapes—conical, knobby, potato-shaped, domal, simple and branching columnar, and others (Fig. 11.5). They consist characteristically of successions of microbially bound sedimentary laminae that reflect the growth habits and ecology of their builders. And they contain the most visible and most nearly continuous indirect account of the early biosphere. Now that we have a pretty good idea what most of these cryptic structures are, and have also found them to be helpful in ordering and interpreting Earth history, they have become everyone's favorite Cryptozoic "fossil." Although they occur in

calcareous rocks from far back in the Archean to the present, they are so abundant in younger Proterozoic carbonate rocks that it is appropriate to refer to that interval as "the age of stromatolites."

Stromatolites long went unnamed because they were widely misbelieved to be simple chemical concretions unworthy of attention. It takes an unusual mind to investigate what others consider obvious, and the beginning of stromatolite cryptography had to await such a mind. It waited until 1894, when then 83-year-old James Hall, state geologist of New York and the leading American geologist and paleontologist of his time, became interested in cabbage-like structures in Cambrian limestone at New York's fashionable Saratoga Springs. He gave them the formal genus and species name of *Cryptozoon proliferum*. "A proliferating cryptic animal" is what Hall's latinized words say; and cryptic it was, in ample measure, but not an animal.

In fact, even demonstrably biogenic stromatolites are neither animal nor plant, nor properly fossil in the conventional sense. They are mainly biogenic sedimentary structures of procaryotic microbial origin. Modern analogs come from a score of localities in marine, brackish, and freshwater environments worldwide (Fig. 11.6). They are the product of the precipitation of calcium carbonate as a result of the photosynthetic removal of CO_2. And such precipitates, along with trapped mineral grains, are bound together in distinctive shapes by sticky, sediment-binding mucus, secreted by proalgal communities of organisms.

Long before such modern analogs were recognized, Germany's Ernst Kalkowsky (1908) stressed their distinctively lami-

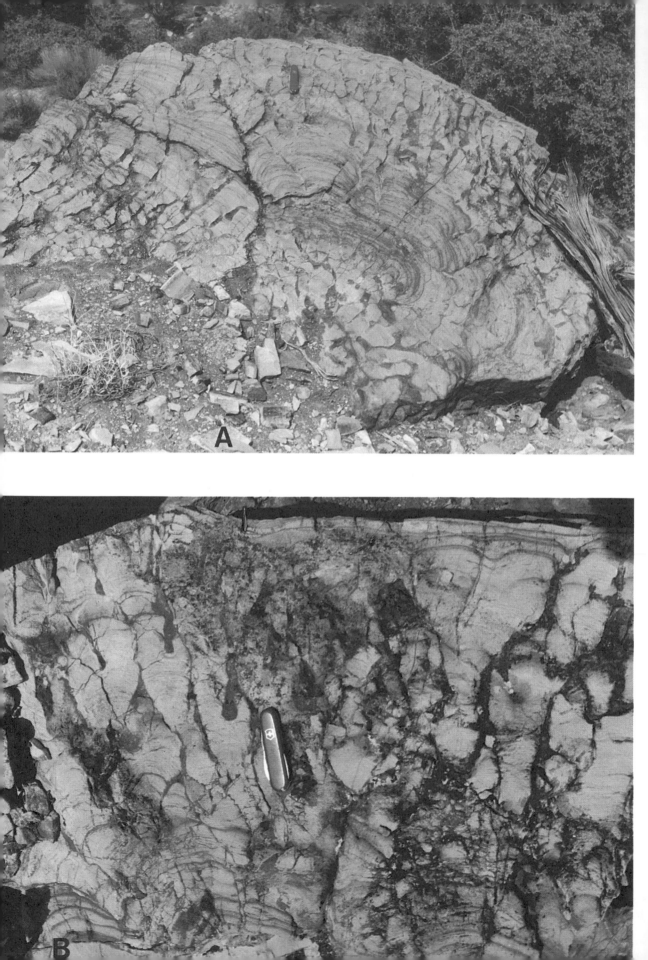

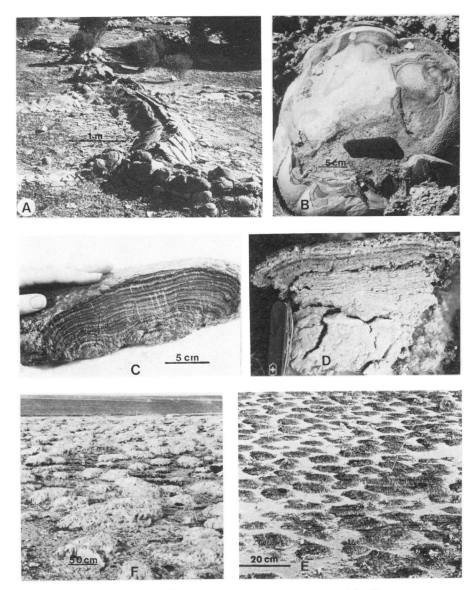

11.6 MODERN STROMATOLITES (*A–B*) AND STROMATOLITE ANALOGS (*C–E*) COMPARED WITH A FOS-SIL FORM (*F*). Scene *A*, modern stromatolites growing on edges of large contraction cracks where ground water resurgence has elevated margins; north side of saline Marion Lake, South Australia. *B*, larger stromatolites from same area, showing distinctive stromatolitic lamination. *C*, vertical cut through microbially bound laminae of a soft stromatolite analog at Laguna Mormona, Baja California. *D–E*, profile of an unlithified modern stromatolite analog and plan view of a colony of similar forms from a brackish marsh on Andros Island, Bahamas. *F*, Ordovician stromatolites from Arctic Canada.

11.5 A TYPICAL STROMATOLITE REEF, OR BIOHERM, OF PROTEROZOIC AGE. View *A*, vertical profile through a small calcareous buildup in the 870-million-year-old upper Chuar Group, Nan-koweap Butte, Grand Canyon, showing morphological responses to varying microenviron-ments. *B*, detail of the associated Middle Riphean form-genus, *Baicalia*.

11.7 ANCIENT STROMATOLITES (*A*) COMPARED WITH THROMBOLITIC MODERN ANALOGS (*B*) AT SAME SCALE; BOTH SHAPED BY TIDAL CURRENTS. Scene *A*, laminated, fine-grained, 1.89-aeon-old stromatolites at Great Slave Lake. *B*, stromatiform, coquinoid thrombolites at Shark Bay, Western Australia.

nated structure and recognized the probability that "plant-like structures of low organization" were their builders. He gave them their present collective name. Analogous modern structures are not properly stromatolites, however, unless lithified. In addition, many externally

similar Phanerozoic structures have distinctive clotted, coarsely fragmental, or coquinoid internal structures that differentiate them from typical laminated stromatolites and has led to their designation as *thrombolites* (Greek meaning "clotted rock").

That is the prevalent variety at the now-famous Shark Bay in Western Australia, otherwise strikingly analogous to 1.89-aeon-old stromatolites in the Wopmay region of Arctic Canada (Fig. 11.7). Although the genetic processes at Shark Bay were clearly similar, structurally more faithful modern analogs are the laminated stony stromatolites of Figure 11.6*A–B*, from the shore of a saline lake in South Australia.

These generally, but not invariably, biogenic sedimentary structures, the "termite mounds of the sea," epitomize the problem of levels of confidence in biogeology. Except where originally of smooth-textured black chert or early replaced by it, stromatolites rarely reveal any former microscopic builders. That is true even at high magnifications in transparently thin slices of rock (thin sections) or acid residues. Their detailed internal structure is often obliterated by metamorphism, and their microbial cadavers are destroyed by later oxidation. It is even the case that similar-appearing structures of nonbiogenic origin originate in caves and hot springs. On a rough scale of increasing confidence from permissive, through suggestive and presumptive, to compelling they are candidly rated no higher than presumptive evidence of life, *except* where found to contain microbial remains in a structural orientation. For those who separate descriptive from genetic terminology, *stromatolite* is a purely descriptive term.

Stromatolitic microstructures of convincingly nonbiogenic origin, however, are so uncommon that most knowledgeable students of these objects would accept their simple presence as strongly implying a microbial origin in the absence of evidence to the contrary.

The geologic record of these distinctively laminated buildups begins inconspicuously in the almost uniquely ancient sedimentary terrains of southern Africa and Western Australia, in carbonate rocks and cherts more than 3 aeons old. Truly abundant and diversified stromatolites are first known from shallow bank and reefal deposits, beginning on South Africa's Kaapvaal craton about 2.5 aeons ago (Fig. 11.8D–F). And they become nearly universal in younger Proterozoic carbonate rocks (Figs. 11.5, 11.8A–C). For revealing the utility of such structures in the correlation of Cryptozoic strata,

11.8 SOME FOSSIL STROMATOLITES. View A, *Conophyton*, exhibiting vertically stacked, linked cones. B, horizontal cross-sections of same. C, distinctive axial profile of same. D, inverted-canoe-shaped stromatolite whose growth morphology resembles a deformational anticline. E, horizontal profiles of three paleotidally truncated domes surrounded by edgewise rosettes of wave-stacked flat pebbles. F, internal structures of domes shown in E. A and C from 1.5-aeon-old Amelia Dolomite of north-central Australia. B from 1.4-aeon-old dolomite of Belt Group in Glacier Park, United States. D–F, from 2.5-aeon-old dolomite of Transvaal Group, South Africa.

265

however, we must thank the studies of Soviet geologists since 1950 in younger Proterozoic rocks of the Siberian platform and adjoining sedimentary basins.

Among stromatolites illustrated here, the oldest known to contain convincing microbial fossils in the form of tapering septate filaments are those from the 2.8-aeon-old Fortescue Group of Western Australia (Fig. 10.6), described by UCLA's J. W. Schopf, and M. R. Walter of Australia's Baas Becking Laboratory. These structures comprise the oldest known records of life about which there is no quibble. Whatever went before, the combined biogeologic evidence of stromatolites, paleomicrobiology, and biogeochemistry strongly implies that proalgae and the advanced aspect of photosynthesis called photosystem II were both functioning by no later than 2.8 aeons ago. Microbial communities, then and earlier, were building biogenic sedimentary structures in shallow, sulfide-rich, nearshore and intertidal marine environments. They were shielded from radiation within the successive layers of calcareous muds now seen as stromatolitic laminae, or by sufficient depths of water or both. Their life-style may be seen as an adaptation to early high levels of radiation.

Although living stromatolite analogs show a wide tolerance to variations in salinity, they are known only from waters shallow enough for light to penetrate often-murky suspensions of calcium carbonate. These are usually shallow to supratidal, hypersaline, lagoonal environments. Ancient stromatolites probably occupied a wider range of ecologic niches. In many Cryptozoic settings, such as the 1.9- to 1.8-aeon-old environments of the Wopmay orogeny, they built huge bank-edge reefs in presumably clear waters that reached depths of several hundred meters. These were the equivalent of modern tropical coral reefs in terms of volume of limestone produced and preferred habitat.

Few anymore would attempt to reconstruct surface processes on the Cryptozoic Earth without calling on the evidence of stromatolites. Many find them to be of value not only in attempting to relate ancient sites of sedimentation to their environment, but also for the correlation of younger Proterozoic strata.

Changing Life-Styles: The Modern Cell Appears

All advanced forms of life consist of structurally complex cells that contain a nucleus and other membrane-bound cellular components called organelles. They have paired rodlike chromosomes, instead of a single chromosomal loop as in bacteria, and are capable of sexual cell division. Such cells were designated *eucaryotic* by the French biologist E. Chatton in 1938 because they contain a *true* nucleus (*eu,* "true" + *caryon* or *karyon,* "kernel" or "nucleus"). Organisms made of them are called eucaryotes. The eucaryotic cell contrasts with the structurally simple non-nucleate or protonucleate cells of *procaryotes*—a group that comprises bacteria, proalgae, and other bacterialike organisms.

The appearance of the eucaryotic cell was a younger Proterozoic triumph—the main event of biological evolution after the origin of life itself. The decisive criteria are ultramicroscopic and have not been preserved among such ancient fossils, but the probable time of appearance can be bracketed. Inasmuch as the Metazoa are all unequivocally eucaryotic, their demonstrable presence 670 million years ago establishes that eucaryotes already existed at that time.

Evidence for an earlier debut is no better than presumptive, but it is all the geologic record offers, and it is persuasive to those who believe in the discriminating power of functional morphology. The most widespread and easily substantiated such evidence is simple cell diameter. Procaryotic cells, for instance, are generally less than 0.01 millimeters or 10 micrometers (millionths of a meter, μm or micrometers, or microns) in diameter, with a mean diameter of 5 μm. Eucaryotes record a mean diameter of 10 μm, and many are a good deal larger than that. A survey of records shows a prominent diametric increase (up to 60 μm) in a proportionally large number of spheroidal cell-like bodies, chains of such bodies, and branching filaments by 1.4 aeons ago.

The presence of eucaryotes by then is well supported. And probable levels of O_2 join that evidence to bracket the most probable time of appearance of the eucaryotic cell as between about 2 and 1.4 aeons BP.

The eucaryotes were surely all true algae then. And they contributed importantly to the growing supply of O_2, whose accumulation in hydrosphere and atmosphere eventually opened the way to metazoan evolution, to the emergence of vertebrates, and to us. Among younger Proterozoic microbial fossils, branching filaments and cells of large diameter as old as 1.2 aeons BP (Fig. 11.9*C,D*) are probably eucaryotic. And convincingly eucaryotic forms are also known from the well-described and well-illustrated 900-million-year-old Bitter Springs microflora from central Australia.

As for eucaryotic ancestry, three things seem well established: (1) they and their procaryotic ancestors shared the same biochemical forebears, (2) some of the membrane-bounded organelles of eucaryotic organisms are structurally and functionally similar to some independently-living procaryotes, and (3) both eucaryotic organelle and procaryotic forerunner continued to evolve independently of the eucaryotic host after the latter appeared. Possible ways of arriving at an ancestral eucaryotic cell, then, are by branching from the procaryotic mainstream (perhaps via an organism like *Prochloron*), by ascent from a common ancestor, or by linking independently evolved components into a kind of mosaic cellular structure.

More explicit clues to origin are suggested by biochemical and structural distinctions and similarities among eucaryotic organelles, the host cells they inhabit, and free-living procaryotes. Some eucaryotic organelles (e.g., plastids and mitochondria) are remarkably similar to independently living procaryotes in form, ultramicrostructure, and function. It is an inspired short step from that to the idea that the eucaryotic cell is a composite of host cell and its willing captive organelles—a mosaic, a classic case of the whole being greater than the sum of its parts. Although such a scenario—the hypothesis of *serial endosymbiosis*—has deep roots, it made little headway until given that name and vigorously advocated by Boston University's Lynn Margulis. It is now widely agreed that this hypothesis accounts well for many eucaryotic characteristics. The membrane-bound eucaryotic nuclei and their paired chromosome sets, however, have yet to find a convincing explanation.

How amusing to think of proud *Homo sapiens* as a sort of microbial composite.

We might prefer to consider ourselves as super-organisms of Proterozoic ancestry, whose continuing viability depended on the efficient functioning of captive but indispensable boarders and partners— rather like an oversized termite or a lichen. However defined, our vaunted self-sufficiency would be sorely tried without intestinal bacteria or ATP-producing mitochondria.

Among younger Proterozoic changes

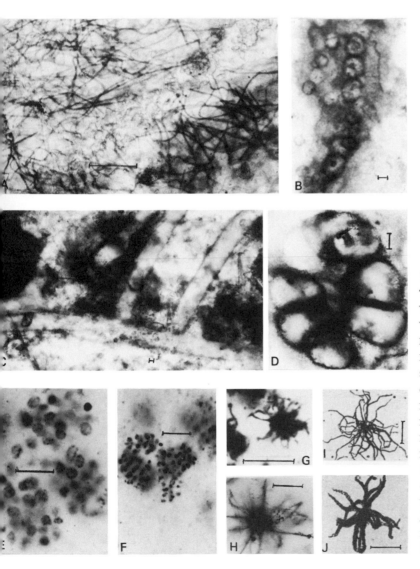

11.9 MICROBIAL FOSSILS IN TRANSLUCENT SLICES OF YOUNGER PROTEROZOIC ROCKS, WITH TWO VIEWS OF A MODERN FORM. ALL BAR SCALES 0.01 MM. *A–D* are from chert in supposedly 1.2-aeon-old dolomite of the Death Valley region, California. *E–G* are from roughly 1.6-aeon-old rocks of the Mount Isa region, Australia. Compare *G* with *H*, from the 2-aeon-old Gunflint chert, and *I–J* of the budding bacterium *Metallogenium* from a modern lake in Karelia, Soviet Union.

in life-style, there was one that might have been expected that either did not happen at all, did not happen until younger Proterozoic history was well advanced, or was so obscure as to be missing from the known geologic record. That was the appearance of a land vegetation—a development that, at advanced stages, was to become a major influence on the types and sorting of continental sediments. If there had been a continental flora during any part of Proterozoic time, however, it seems certain that it would have been a mere film of algae or proalgae, unable to retard or trap sediments to any significant degree. Indeed, the fluvial sediments of Proterozoic age continue to be mainly those of braided streams such as are common in today's subarctic and desert regions, where, of course, there is little vegetation. A remotely possible exception, consisting of silty-shaly overbank sediments at the base of the Apache Group in central Arizona, may symbolize something like a flimsy, sediment-trapping, algal or proalgal continental vegetation as early as about 1.35 aeons ago.

Impressive though the changes in late Proterozoic microbial life-styles were, there remained many steps to the Phanerozoic world. As one attempts to ascend them, it is important to separate the real and the probable from the spurious and the dubious—to make no more of the evidence than it allows.

Fossils, Pseudofossils, and Dubiofossils

Remains or traces of ancient life (*fossils*, from the Latin *fossilis*, "dug up") first became common enough during late Proterozoic history to be useful in interpreting past environments, ecology, and evolution. They have been sought eagerly by inquiring devotees since the recognition by the eccentric genius Robert Hooke (late seventeenth century) that they could probably serve as talismans of sequence and setting, and for their biological interest.

Many objects and markings made by purely physical processes, however, superficially resemble once-living objects and markings to the hopeful or desperate eye. Unfortunately, they have often in the past been described as fossils, so that the published record is, at places, embellished with descriptions and names of nonbiologic or questionably biologic structures. Genuinely biologic contaminants and suspected contaminants have also been described and named as fossils, as well as artifacts resulting from laboratory preparation. They make up a long list of *pseudofossils* and *dubiofossils*. The problem is uncommon among Phanerozoic rocks, where good fossils are plentiful. But it becomes troublesome among older rocks, where fossils are so eagerly sought and so highly prized that vigilance may relax.

It is severe in the case of the ardently sought but probably missing pre-Phanerozoic Metazoa, where many have been hopefully described but no convincing discoveries have yet been made in rocks convincingly dated as older than about 670 million years. To mention only a few examples, planar radiating crystal clusters and other discoidal or concentric markings of physical origin have commonly been misidentified as "jellyfish."

Cross-sections of filled shrinkage cracks, sedimentary dikes, and fluid evasion channels are often referred to as "worms." Modern termite burrows in weathered older rock and imprints of bee's nests and lichens dissolved into the surfaces of carbonate rocks have also been reported as Crypotozoic fossils.

The record of microbial fossils is better. It is especially good and growing for sedimentary rocks younger than about 1.4 aeons BP. But it has its share of pitfalls, and it has been diluted with doubtful or false reports. Modern fungal spores and pollen do get into sediments and even into laboratory preparations under the best of conditions. All too many microscopic particles of nonbiogenic origin have been mistaken for fossils. Caution is the only safeguard. It is called for wherever extensions of well-documented ranges in time or space are proposed.

Results have improved with experience and critical review. Progress toward a more refined Earth history is being made by means of a variety of cryptic, acid-resistant, organic-walled microscopic spheroids of distinctive structure and ornamentation and floating or *planktonic* habit (although some of them look wonderfully like the spores of living mosses). Such microorganisms, *Acritarchs* (Greek *akritos* + *arche,* "of doubtful origin"), are abundant in shales deposited between about 810 to 670 million years ago, in latest Proterozoic time. As with stromatolites, important, although by no means consistently reliable early work on these useful microscopic creatures was done in the Soviet Union. It has been widely extended and refined by others, particularly in Scandinavia, England, and the United States.

The Wopmayan Sequence:
Proto—Plate Tectonics Shapes the Canadian Shield

As we have seen, the restless younger Proterozoic Era in North America was one of tensional stress and active crustal mobility. Following much reshuffling of components, the general configuration of the continent we know took shape.

A striking and beautifully documented example of this restlessness is on display for the hardy in Canada's presently far northern Bear and Slave provinces (Figs. 11.2, 11.10). Studies by a series of field parties of the Geological Survey of Canada led by Paul Hoffman have made that region a classic of younger Proterozoic stratigraphy and early or incipient plate tectonism. They have revealed asymmetrically paired fold belts and west—east

directed thrust faulting in the Wopmay orogen (named for a river honoring the one-eyed World War I ace and bush pilot known as Wop May). This is a tectonically deformed belt that trends south from the Arctic coast of Canada to Great Slave Lake. Attempted subductions aborted—first westward, then eastward—and no oceanic crust has yet been found where the spreading center should have been. But the structural elements of converging plates are there.

Hoffman's interpretation of events is summarized in Figure 11.10. The Wopmay orogeny (processes leading to the growth of folded mountain ranges), whose stages are dated with high resolution and reliability by the uranium–lead method on zircon crystals, began about 1.91 aeons ago. It started in the region east of Great Bear Lake as a narrow Red Sea–like miniocean, grading eastward to a corresponding minicontinental slope that was locally emergent at the western margin of the Slave Province. During the next 15 million years, as much as 35 kilometers of volcanic and sedimentary rocks piled up within this rift basin and its eastward-extending, fault-bounded arms, including some 2 to 9 kilometers of passive margin, slope, shelf, and platform sediments (Fig. 11.10B).

These deposits record a well-dated history of tectonic evolution and related environmental change in three classic phases. First was the pre-orogenic miogeoclinal and basin phase; second, a mountain-building phase, during and following which mainly detrital sediments resulting from folding and erosion were dumped into and across adjacent basins; and third, a post-orogenic phase with alluvial fan deposits (*fanglomerates*) and immature, commonly red, quartz–feldspar sandstones (*arkoses*). Conspicuous among the miogeoclinal deposits of this interval are great reefs of

11.10 Plate motions in the 1.9- to 1.8-aeon-old Wopmay orogen, District of Mackenzie, Canada. [*Adapted from diagrams by P. F. Hoffman, 1980, Geological Association of Canada, Special Paper 20, Figs. 1, 5–6, 8.*]

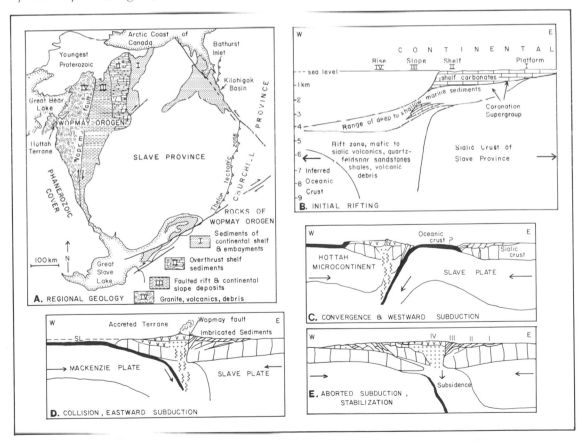

stromatolitic dolomite and associated deposits of the passive plate margin (Fig. 11.10*B*), grading landward into redbeds with impressions and casts of evaporative salt crystals. The initial paleomagnetic signature of the redbeds indicates an apparent polar position about 15°S, off the present coast of southern Peru, locating the Wopmay *orogen* (orogenic site or fold belt) itself at a subtropical latitude perhaps 25° from the equator.

Abortion of westward-directed subduction was followed by closure of the initial rift and intrusion of granitic plutons in the collision zone about 1.89 to 1.87 aeons BP (Fig. 11.10*C*). Cyclopean forces were at work. Previously deposited sedimentary rocks were moved eastward (in the present orientation) along multiple, now north—south striking thrust faults. They were folded and tectonically stacked. Renewed rifting, with reversal of attempted subduction, followed (Fig. 11.10*D*). With sialic lithosphere both east and west of the rift zone, and thin or no truly oceanic floor, neither plate edge was dense enough to subduct very far. In the end, renewed convergence and closure sealed the Wopmay orogen to what was then the Slave plate (Fig. 11.10*E*), while orogenic granites and volcanics completed the building of the present-day Bear Province.

In the final movement of this tectono-sedimentary ballet, the entire Wopmay orogen, over a 1,000-kilometer-wide swath, was cut by northeast-trending, high-angle faults. That has been interpreted as the product of collision far to the west of the mapped area about 1.81 aeons ago.

The entire history occurred within an interval of no more than 100 million years between 1.91 and 1.81 aeons ago (the bracketing ages are subject to the usual range of error, but the intervals between are precise). Ages found indicate that the Wopmayan sea floor could have been no more than 15 million years old when it first failed to subduct. Indeed, that was hardly time enough for even good mafic crust to thicken and cool to subduction density.

The Wopmayan sequence may exemplify the kinds of limited crustal movements and depositional processes that were going on around and within the region now called the Canadian Shield between about 1.9 and 1.7 aeons ago. Such movements, not necessarily involving great lateral displacement, could have been responsible for the final assembly and local deformation of the original plates and terranes that now make up this great composite shield.

How can one be sure that such motions were really manifesting plate tectonism when there are no blueschists, no convincing ophiolites, no evidence of completed subduction? One can't. One can say with confidence only that it seems, in other respects, to represent something similar to early plate motions.

The Wopmayan sequence comprises all of the sedimentary, volcanic, and intrusive rocks that are in any way involved in the history of the Wopmay orogeny and its successor sediments, including the now-detached partial sequences at the southeastern and northeastern corners and whatever sedimentation was taking place during late movements.

It is instructive to consider how much information about conditions and locations has been extracted from those nearly 2-aeon-old sediments. Deposits of the Great Slave Supergroup in the East Arm of Great Slave Lake (to the southeast) compare so closely with those of the principal Wopmayan shelf sequence or Coronation Supergroup as to make their affiliation clear. They comprise magnificent displays of stromatolitic reefs (Fig. 11.7*A*), as well as a complex set of detrital pre-reef, offshore, and coastal sediments—all accessible by boat and floatplane from the once (but no longer) primitive city of Yellowknife along a great length of scenic and still mostly unspoiled island coast. Similar orogeny-tracking sedimentation is seen in both the Great Slave and the more northerly Coronation supergroups.

Historically equivalent strata to the northeast, in the frigid haunts of the tundra-loving snow goose around Bathurst Inlet of the Arctic Sea, represent a different complex of environments. They are associated with the subsided, long-since-subtropical Kilohigok basin, a tectonic *foredeep* along a zone of miniplate convergence and westward thrusting at the northeast corner of the Slave Province (Fig. 11.10*A*). In this remote and demanding region, more than twice the size of Rhode Island, we find up to 7 kilometers of westward-thinning shallow-marine and nonmarine sediments of great variety. At the base of this sequence are shallow-marine platform deposits followed upward by deep-water deposits that grade westward to shelf sandstones and thin, shallow marine reef complexes of stromatolitic dolomite. Those strata grade upward to a variety of fluvial and deltaic deposits that represent the products of a complex of braided streams. These streams nearly 2 aeons ago drained westward from their rising sources near Canada's Arctic coast and east of Great Slave Lake.

They were deposited concurrently with rifting and subsidence in the Wopmay orogen, overlapped westward toward it, and were interrupted by discontinuities in sedimentation and the development of fossil soils. The Kilohigok sequence terminates with calcareous mudstones of evaporative origin, capped with a final flood of coarse fluvial sandstones and conglomerates.

Elsewhere in Canada the only other known sedimentary rocks of this once-extensive interval are scattered outcrops of 1.8-aeon-old redbeds and associated strata found eastward from Slave Province toward Hudson Bay.

Its equivalents in the United States are fragments and drill-core samples of extensive but mainly subsurface sediments and poorly understood metasedimentary rocks found at intervals from central Wisconsin to northern Arizona (Fig. 11.2).

Worldwide, this history is represented by extensive records of the earlier noted 1.9- to 1.7-aeon-old metamorphic event. Major additions of new sialic crust were also ascending at that time from depleted upper mantle, to judge from the evidence of neodymium-isotope ratios in sialic basement rocks of the Colorado Rockies. Sialic volcanism appears about then in parts of northern Australia. Redbeds of this age are also known from the Baltic Shield and the Transvaal region of South Africa.

The combination of redbed sedimentation and acid volcanism so distinctive of this age is exemplified by the deposits that fill the fault-bounded Ulkan trough, north of the Stanavoy Ridge, in the far-eastern Soviet Union. Here a thick sequence of red, mainly nonmarine, coarse, quartz–feldspar sediments with rare marine intercalations caps and intertongues with a 1.8-aeon-old pile of mainly sialic volcanics 6 kilometers thick.

To generalize, younger Proterozoic history begins with crustal unrest, orogeny, granitic plutonism, and a noticeable increase in sialic volcanism.

Continental redbed sedimentation was widespread for the first time. And stromatolites of great variety were abundant in carbonate shelf sediments where environments were shallow, subtropic, and characteristically marine.

Autobiography of the Wopmayan Successor Sediments

What happened during and after the terminal collision that fused the Wopmay orogenic belt to the Slave plate? What was the effect on surrounding environments? As sialic plutons arose in and west of the collision zone, they elevated a region that simultaneously underwent erosion. The debris it shed went tumbling and flowing gravitationally downward in scree or talus slopes, breaking up on the way. It broke up still more during stream transport to alluvial fans, was impacted and abraded into gravel, sand, and silt, and traveled northwest (by today's orientation) through what now is the home of the barrenground grizzly. It traveled in mountain torrents and then braided streams to a subsiding marine platform. There, in and near the sea, it accumulated as sheets of debris and chemically precipitated sediments above as well as below sea level, writing its own environmental impact report in the not-so-cryptic language of the rocks.

Here, in the barrens along and west of the lower Coppermine River, where forest meets tundra, and where Inuit and Indian once clashed over fishing rights, are 2,600 meters of nonmarine, intertidal, and marine sedimentary rocks. The story they tell is briefly and graphically recapitulated in Figure 11.11, illustrating how rocks record Earth history and how much they can communicate by mainly physical characteristics, even from very long ago.

That story, product of an imaginative group at Ottawa's Carleton University, is here condensed to thirteen lettered events and trends (*A–M*) that tell of intermittent exposure and erosion of granites and other rocks in the Wopmay hinterland. Episodic floods and fluctuating sea level are seen in a frame of time, space, and terminal volcanism. Evidences of life in these strata include abundant stromatolites, plus microbial fossils at five levels in the upper fourth of the sequence.

Accumulation of those deposits began about 1.66 aeons ago (U–Pb) with silicic volcanics and sediments from the erosion of the Wopmay orogen. They are penetrated by and are therefore older than the 1.24- to 1.22-aeon-old Mackenzie dikes (Fig. 11.2) and overlain without evident discontinuity by basaltic lavas of the Coppermine River Group of similar age. They record, therefore, a complex history of elevation and erosion in the south and east, subsidence and fluctuating sea level on the west, and intermittent rifting and volcanism between about 1.7 to 1.2 aeons BP.

11.11 AUTOBIOGRAPHY OF THE 1.4- TO 1.2-AEON-OLD WOPMAYAN SUCCESSOR SEDIMENTS NORTH OF GREAT BEAR LAKE, ARCTIC CANADA. *[Simplified and rearranged after Charles Kerans, G. M. Ross, J. A. Donaldson, and H. J. Geldsetzer, 1981, Geological Survey of Canada, Paper 81-10, Figs. 9.3 and 9.12, with additions. Reproduced with permission of the Minister of Supply and Services Canada.]*

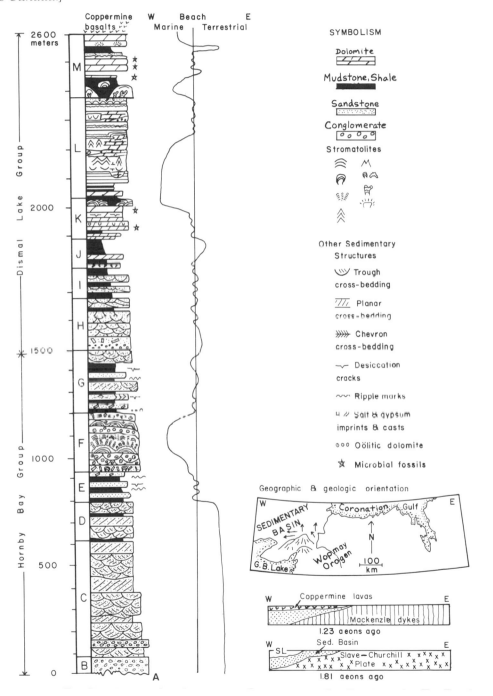

Take a good look at Figure 11.11 and see what you can make of it. Besides the rock symbolism and what you know of its environmental and climatic significance, note that the vertical profile in the center graphs the sedimentary setting with reference to sea level at the time of sedimentation. The sketches at the lower right denote present geographic, geologic, and directional orientation. Arrows point directions of sedimentary transport from upland sources.

From this symbolism, one can follow the travel of erosional debris north and west from the then-mountainous Bear Province somewhat before about 1.2 aeons ago and see where its journey ended.

Such debris was carried by braided streams across a broad, vertically oscillating shelf to empty into a shallow, westward-deepening, subtropical marine basin. The prevalence of trough- and steep planar-cross-bedding tells of fluvial and tidal flat environments. Herringbone cross-bedding (e.g., Fig. 3.6C) records tidal influences. Stromatolite morphology responds to depth, turbulence, exposure, and tides. Salt and gypsum casts, as well as shrinkage cracks and solution structures (karst), signify episodic exposure, evaporation, and groundwater movement. One sees how freely even very old rocks will relate their history to the attentive observer.

Riphean Sequence and History

The Riphean sequence spans most of younger Proterozoic history. It is named for outcrops in the rolling ridges, riverbanks, picturesque meadows, and birch–evergreen woods of the southern Ural Mountains, from an ancient term for the region (Fig. 11.12). Excluding, for present purposes, its conventionally uppermost strata, rocks of this sequence range in age from roughly 1.7 to about 0.9 aeons BP (left front endpaper), when late Proterozoic glaciation began to grip the world in icy embrace.

So delimited, the classic Riphean sequence is represented by largely siliceous detrital strata with occasional intervals of stromatolitic dolomite. Except for the absence of metazoan fossils and their traces, and the presence of an abundance and variety of stromatolites, its strata are very similar to Phanerozoic shelf-slope (miogeoclinal) sequences.

Riphean strata are the oldest to be regularly correlated from one part of the world to another on paleontological grounds. Scattered outcrops of Riphean and equivalent younger Proterozoic strata cover a vast expanse of Eurasia from Finland to the Okhotsk Sea and south (Fig. 11.13). And they are now related to one another primarily by means of stromatolites, confirmed by radiometric calibration and tested by the field studies of geologists of the Geological Institute of the Soviet Academy of Sciences. Their work became the basis for dividing rocks equivalent to the Riphean sequence beyond the Urals into three broad subdivisions—most widely called simply R_1, R_2, and R_3 (left front endpaper).

Although that system provides only a coarse, provisional *biostratigraphy* for younger Proterozoic rocks, it shows promise for refinement and extension into pre-Riphean rocks as paleomicrobial discoveries accumulate, broadening

11.12 Riphean and overlying strata of the southern Ural Mountains between Sterlitimak and Magnitogorsk, Soviet Union. [*Adapted from N. M. Chumakov and M. A. Semikhatov, 1981, Precambrian Research, v. 15(3–4), p. 247, Fig. 3. © Elsevier Science Publishers B.V.*]

11.13 Distribution of younger Proterozoic (mainly Riphean) and older Phanerozoic (Ediacarian) strata of eastern Europe and Siberia. [*Adapted from N. M. Chumakov and M. A. Semikhatov, 1981, Precambrian Research, v. 15(3–4), p. 230, Fig. 1. © Elsevier Science Publishers B.V.*]

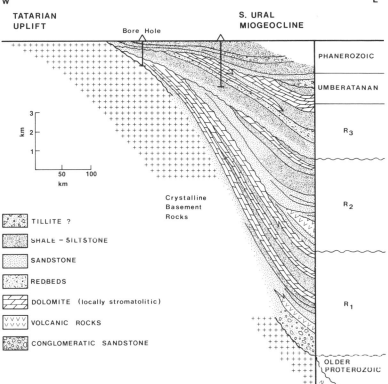

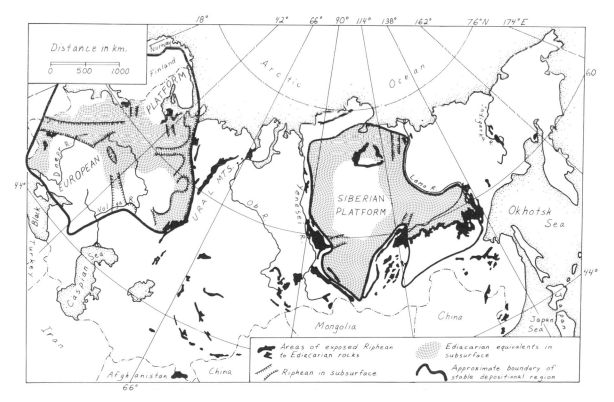

and extending the biostratigraphic base. It seems likely, for instance, that the wide extent of the Gunflint microbiota may approximate contemporaneity, as apparently does an assemblage of relatively large-diameter cells and branching filaments (Fig. 11.9C–D) now known from 1.2- to 0.9-aeon-old rocks in eastern California and South Australia.

Riphean history was the heyday of stromatolites and their mainly proalgal architects, objects of intensive study by Soviet stromatophiles since the 1950s and by subsequent converts around the world. The microbial creators of this earliest form of public housing flourished particularly in and adjacent to the shallow seas that flooded the European and Siberian platforms. Because their calcium carbonate—precipitating and sediment-binding activities depended on CO_2 assimilation, a function of their oxygen-releasing photosynthetic activities, their native seas could not have been too deep or too murky to transmit light of appropriate wavelengths. There is, however, evidence that stromatolites of Riphean age did form at depths as great as 100 meters (in eastern California) and locally built structures over 1,000 meters thick (in Canada's Mackenzie Mountains). It seems, therefore, that some Riphean seas were remarkably transparent and that, at places, they also underwent prolonged subsidence at average rates slow enough for the tiny stromatolite builders to stay within the sunlit zone.

In fact, the greatest extent and richest development of younger Proterozoic stromatolitic reefs and banks seems to have been in the Riphean dolomites that cross Siberia from the northwest to the southeast into Korea and north China. Across and beyond this beautiful but lonely land of larch and permafrost, the stromatolites bear witness to the fact that much of it, during most of middle and late Riphean history (R_2 and R_3), and locally during early Riphean (R_1), was bathed with clear tropic waters.

The carbonate rocks deposited from such waters commonly include casts or imprints of salt and gypsum crystals and are marked by shrinkage cracks. They also show other evidence of intermittent exposure to the atmosphere and evaporation of platform surface waters adjacent to erosionally beveled tropic lands. The cleanest and most richly stromatolitic of these then-tropic sequences is a roughly 2,000-meter-thick section of steeply dipping strata along the now usually icebound Kotuikan River, in the north Siberian platform. Here J. K. Korolyuk's hypothesis of the biostratigraphic significance of stromatolites survived its severest test when reexamination of an earlier discordant radiometric age showed that she had been right all along and the geochronology premature. Other miogeoclinal sequences that reach thicknesses of 6 to 7 kilometers and contain up to nearly 2 kilometers of stromatolitic limestones are found in eastern Yakutia, and widely throughout Siberia (Figs. 11.14, 11.15).

11.14 FOLDED LATE PROTEROZOIC (MIDDLE RIPHEAN) SHALES AND STROMATOLITIC DOLOMITES ALONG THE KHANDA RIVER, FAR-EASTERN SOVIET UNION. The Khanda or Belaya River is an eastern tributary of the north-flowing Aldan, which joins the Lena River on its way northward to the Sea of Laptev. The sequence is classic for the remote Yudoma–Maya region.

Despite the fact that the stromatolitic carbonate rocks are far outbulked by siliceous detrital rocks (Figs. 11.12, 11.15), their abundant presence and evaporative inclusions imply a Riphean equator right across today's Siberian platform (Fig. 11.16*A–B*).

The huge former expanse of these mainly platform and miogeoclinal deposits spans the whole of the Soviet Union and beyond. It exceeds the area covered by the most extensive Phanerozoic epicontinental seas and approaches them in terms of the diversity of environmental conditions represented and land areas bounded. Riphean and equivalent rocks outcrop or are penetrated by deep boreholes intermittently from the White Sea eastward to the Lena Delta and beyond. They are found south to the south end of the Ural Mountains, to the Tien Shan Range in Middle Asia, to the Soviet far east, and into Korea and China. Over much of this enormous region these ordinarily little deformed rocks are either sharply separated from older metamorphic and cystalline rocks or rest on an initial younger Proterozoic sequence of sialic volcanic rocks, redbeds, and intergrading marine to nonmarine platform sediments.

It seems clear, nonetheless, that this now-continuous region was not all one great platform sea. Figures 11.12, 11.13, and 11.15 show that the sedimentary sequences thinned shoreward and at times disappeared against emerged surfaces from which detrital sediments were being eroded. Were the marine basins thus defined within a greater continental region? Or were they but the upper slopes of miogeoclines that descended into intervening oceans? It is well established that a paleo-Siberian ocean lay between the European and Siberian platforms, later to subduct beneath the Urals. Pres-

11.15 Riphean and younger strata flanking and blanketing the Aldan Shield in southeastern Siberia. Vertical profile runs east to west 1,100 km across the southern Aldan Shield from Aldan River to the Patom Highlands near Lensk. [*Adapted from N. M. Chumakov and M. A. Semikhatov, 1981*, Precambrian Research, *v. 15(3–4), p. 248, Fig. 4. © Elsevier Science Publishers B.V.*]

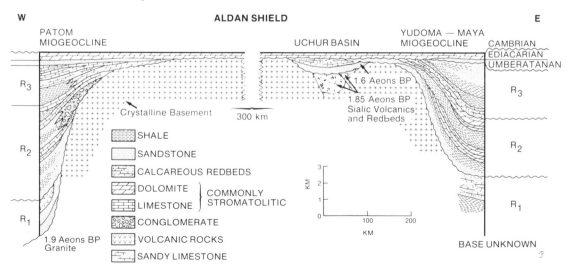

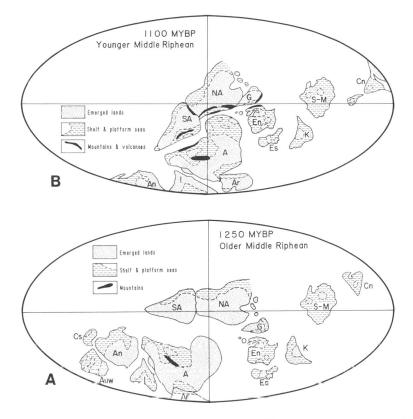

11.16 Highly provisional reconstruction of younger Proterozoic lands at 1,250 (*A*) and 1,100 (*B*) MYBP. Rough approximations from paleomagnetic data, combined with inference from stratigraphy. *Ar*, Arabia, *Aue*, eastern Australia, *Auw*, central and Western Australia, *Cn*, north China, *Cs*, south China, *En*, northern Europe; other abbreviations obvious.

ent-day China consists of once-separate miniplates that were not finally joined until later. The probably many separate pieces to this geological jigsaw puzzle have not yet been arranged in a fully plausible paleogeography.

Paleomagnetism to the Fore

Earth magnetism, official justification for an important spate of nineteenth-century global exploratory cruises, has undergone a new birth of interest with the extension of its records to remote times. The abundance of little-altered younger Proterozoic redbeds and associated volcanic and intrusive rocks provides a wealth of favorable subject matter for paleomagnetic analysis. Redbeds, volcanics, and intrusive dikes and sills all yield paleomagnetic data, while the igneous rocks provide for radiometric dating and bracketing. Advances in sampling practice and analytical technology have made it possible to strip away the so-called magnetic domains of weaker postdepositional magnetic orientations. In this way, one gets back to a primary orientation—that acquired at the time of

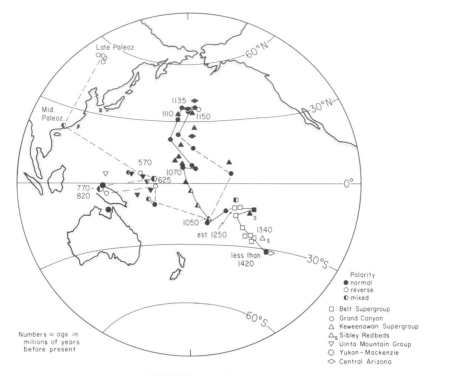

11.17 APPARENT POLAR WANDER PATH (APWP) FOR NORTH AMERICA BETWEEN ABOUT 1,400 AND 250 MILLION YEARS AGO. Solid lines are confident connections, others less so. *[Adapted from D. P. Elston and G. S. Gromme, 1984, Eos, v. 65(45), abstract GP 21-08. Copyright by the American Geophysical Union.]*

initial sedimentation or volcanism, or recording immediately postdepositional effects. A lot of history can also be extracted from postdepositional magnetization events—heating and cooling history, rotation of plates, and so on.

Employing such methods, the paleomagnetic signatures of now-isolated exposures of younger Proterozoic strata in the western half of North America (Fig. 11.2) and their positions in apparent polar wander paths (APWPs) have been measured and correlated by Donald Elston and associates of the U.S. Geological Survey. Such information

seems to resolve, to a degree not previously considered possible, the sequence and historical relations of younger Proterozoic strata deposited in widely separated basins and often concealed beneath overlying Phanerozoic deposits.

Figure 11.17 shows the Elston–Gromme APWP determined from paleomagnetic data on samples from different North American basins. That path, plus paleopolarity (whether normal or reverse), and concordant uranium–lead radiochronology, are the basis for the previously unlikely correlations indicated in Figure 11.18.

11.18 PALEOMAGNETIC EQUIVALENCES AND AGES OF YOUNGER PROTEROZOIC AND OLDER PHANEROZOIC STRATA IN WESTERN NORTH AMERICA. Thicknesses not to scale. Correlation based on APWP of Fig. 11.17 and polarity as in column to right. *[Adapted from work in progress by D. P. Elston and S. L. Bressler, U. S. Geological Survey, with permission.]*

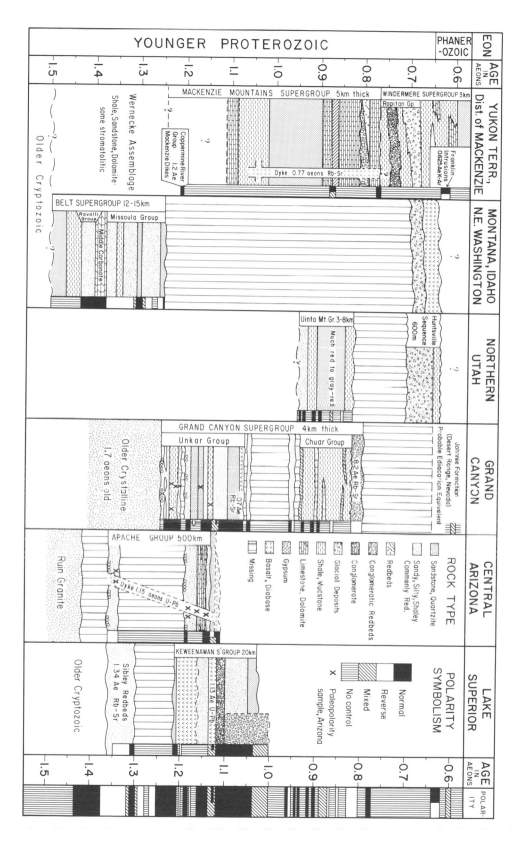

283

From detail concerning rock types there graphically summarized, one might infer much about source and mode of transport, deposition of sediments in the different basins or geoclines, their relation to land or sea, their depth if marine, and the paleoclimate as well. That is not the intent of such a diagram at this place, where apparent thickness is purposely distorted to fit a constant scale of radiometric time and paleomagnetic signatures. The purpose here is to illustrate how widely scattered sequences of rock with little in common besides their paleomagnetic signatures can be related to one another in an historical succession.

Also not obvious in Figure 11.18, is that, as in navigation before accurate chronometers, paleolatitude alone can be accurately determined from paleomagnetics, not paleolongitude. Approximate paleolongitudes are roughly estimated by dip and apparent direction to the pole where there has been no horizontal rotation at the site of sampling. Apparent polar positions in APWPs are plotted, after rotation to the horizontal, as if the plate that generated them was otherwise fixed in position, whereas it was in fact moving and may have been rotating. The longitudinal jogs are a product of the method.

The reader will recall that *apparent* polar wander paths really track the motions of plates, not poles. The rotational axis itself only wobbles and the magnetic poles move but slightly. It is merely simpler to communicate in terms of APWPs, as long as everyone understands that. If one wishes to find the position occupied by a given sampling site with reference to the hypothetical position of the rotational pole for a given position in a polar wander path, one must be guided by spherical geometry.

When that is done for the apparent polar positions indicated by an imaginary point at the Grand Canyon (Fig. 11.16), it turns out that this point, and presumably the whole continent, traveled a tortuous, almost dizzying course between about 1.4 aeons ago and the end of Proterozoic history. The present latitude of the canyon is roughly 36°N. At 1.4 aeons BP, its latitude was about 25°N. By 1.25 aeons BP it had drifted to 30°N. It then shifted to 15°N at 1.21 aeons BP before returning to roughly its present position at about 1.19 aeons BP. Following another slight south jog, it reached its farthest north at 1.13 aeons BP at a latitude of 39°N. From there it slid jerkily to 2°S at about 950 million years BP and finally to 16°S about 800 million years BP. By early Cambrian time it was back to slightly north of the equator again, thence to 15°S in mid-Paleozoic and eventually to 5°N in late Paleozoic time. Confusing? In fact, that simplifies the actual motions of our geologically spastic ancestral continent.

One might not guess all this from a simple inspection of Figure 11.17. Because North America was essentially intact and occupied a nearly east—west orientation during that time, what look like short east—west motions on the APWP can hide big latitudinal shifts. The upright-hairpin shape of the APWP near midway of its course, however, does approximate real latitudinal motion. The time of turnaround from northerly to southerly drift about 1.13 aeons ago coincided approximately with the opening of the mid-continent rift zone (Fig. 11.2), the extrusion of middle Keweenawan lavas (Fig. 11.4), and the abrupt deepening of the Lake Superior basin. The directions of stress generated by the paleomagnetically implied motion on the then east–west oriented continent would have paralleled these features. This coincidence of events suggests a connection with the Grenville orogeny, itself perhaps the product of a grand plate collision that seems to have affected the whole present southeastern margin of North America between about 1.25 and 0.85 aeons ago, apparently peaking around 1.1 aeons ago with the Grenville orogeny.

Most extensive by far of the sequences so drastically condensed in Figure 11.18 are those of the Belt basin and eastern Yukon boundary area (two columns at left), with partial equivalents from Montana to Alaska. The fault-bounded and nearly landlocked Belt basin of northwestern Montana and Idaho occupies a minimal area of 170,000 square kilometers. It ranged during its existence from emergent to deep and quiet, opening into the ancestral Cordilleran sea on the west. Between 1.45 to 1.25 (possibly as late as 0.85) aeons BP it was a subsiding reentrant of the sea, filled with sediments to a total thickness estimated to be locally in excess of 20 kilometers.

That sequence, the Belt Supergroup, now consists of siltstones, shales, and quartzites with a thick medial carbonate interval of commonly stromatolitic dolomite. It, its extensions northward into Canada (the Purcell Group), and the thick sequence of then passive-margin, miogeoclinal strata (the Windermere Supergroup) that overlies the Belt equivalents northward to the Arctic Ocean, are the objects of intense scientific and economic interest. Like other sectors of Riphean history, they parallel Phanerozoic sequences of comparable structural settings in every way except for metazoan fossils right up almost to the very top of the Windermere strata.

Riphean history represents the greater part of the younger Proterozoic and a continuation of historical trends already begun with the Wopmayan sequence. Together with the Wopmayan it comprises an age of stromatolites and of copper. It was the time of the origin of eucaryotes and the expansion of planktonic species. Its distribution was global.

Omitting that global history, most regrettably its great Southern Hemisphere extent, we see that Early Riphean (about 1.7 to 1.4 aeons BP) represents a time of marked continental emergence. Medial and Late Riphean, in turn, were times of marked tensional rifting, and of sedimentation that is confined mainly to broad platforms and rifted epicratonal or plate-margin basins. They were part of the long prelude to a great terminal Proterozoic glacial succession that emphasized the already general continental emergence and introduced the Metazoa and the Phanerozoic Eon about 670 to 650 million years ago.

In Summary

The unfolding younger Proterozoic Earth was a restless, changing place. Accumulated subcrustal stresses were beginning to express themselves in a succession of rifts and collisions in northwestern Canada (the Wopmay orogeny) just as the Sudbury asteroid thudded to Earth about 1.85 aeons ago (Fig. 11.1). Was it pure coincidence that it landed just where rifting of the eastern seaboard may have begun at about the same time (Fig. 11.3)? The Hudsonian or Penokean orogeny ensued—a shuffling of subcontinental plates, with great effusion of 1.8- to 1.7-aeon-old potassic granites that welded the pieces of proto–North America (Laurentia) together.

Beneath the growing continent's thermal blanket grew pervasive tensional stresses. Their eventual release probably accounts for much that is distinctive in the subsequent Proterozoic history of North America. Signs of that release include sets of extensive, deep-reaching, basalt-filled vertical cracks or dikes—dike swarms. Distinctive are the great Keweenawan flood basalts and many smaller, rifted, epicratonal sedimentary basins with their distinctive associations of redbeds and increasingly sialic volcanism.

A notably colorful assemblage of non-orogenic sialic intrusives from probably mantle sources is distinctive of younger Proterozoic plutonism. Such rocks seem to be concentrated at about 1.45 and 1.1 aeons BP, but they are reported at intervals from about 1.7 to 1 aeon ago.

Stromatolites play a lead role in younger Proterozoic history. They are important in environmental interpretation and approximate stratigraphic correlation. They are represented by a great variety of shapes of laminated, mostly calcareous buildups, mainly products of the photosynthetic and sediment-binding activities of proalgae. They underwent a striking expansion in numbers, diversity of form and microstructure, and reef-building activity in the extensive epicontinental seas of younger Proterozoic time.

Because microbial processes are the main influence in the precipitation, binding, and shaping of stromatolites, their presence is taken as presumptive evidence of a microbial presence. As a result, stromatolites have been prime targets in a successful search for new Cryptozoic biotas, especially where early replaced by, or originally consisting of, chalcedonic black chert—in effect hermetically sealed within it. This search, plus the increasingly fruitful search for planktonic, organic-walled cells, cell clusters, and filaments in the abundant dark shales of the time—symbols of local anoxia—has revealed changing microbial life-styles. Foremost among these is the strong implication that the eucaryotic cell had already evolved by or before about 1.4 aeons BP, providing the key biological breakthrough from which all advanced forms of life were to issue.

The initiating Wopmayan sequence began with a brief cycle of rifting and narrow separation of plates, in which volcanism, and rift-filling sedimentation on the west was matched by slope to platform sedimentation on the east, the region then being at subtropical latitudes. Following two unsuccessful tries for subduction—first toward the west, then eastward—opposing plates were joined together by granitic intrusion and volcanism. Elsewhere in the world this historical interval is represented most characteristically by the oldest extensive continental redbeds, commonly associated with mixed mafic to sialic volcanic rocks in rifted depositional basins. A prevalence of such records of atmospheric oxidation and continental sedimentation characterizes younger Proterozoic history.

Post-Wopmayan history from roughly 1.7 to perhaps 0.9 aeons BP is represented by the Riphean sequence, from an old name for the southern Urals, where rocks of that interval are well exposed. That history is most extensively represented by marine platform and miogeoclinal deposits that occur throughout the Soviet Union. Those strata are notable for the variety of stromatolites found in their carbonate intervals, for their biostratigraphic subdivision based on stromatolites, and for their parallelism to Phanerozoic platform and miogeoclinal deposits, *excluding* Metazoa and their works.

A vast extent of such otherwise unremarkable rocks is also widely preserved on most or all other continents. In North America, formerly extensive continental and marine sediments of this age appear in windows through blanketing Phanerozoic rocks across much of the western United States and in extensive outcrops northward from Montana to Alaska. The correlation of these Riphean sequences in central and western North American is made possible by the prevalence among them of magnetically susceptible continental redbeds and associated dateable volcanic and mafic intrusive rocks.

As history ascends toward the Phanerozoic Eon, signs of life become more abundant, more promising for biostratigraphy, and more eagerly sought. Here one can expect to find a wealth of yet-unharvested data, while remembering that overeagerness has, in the past, often brought delusion. All claims so far of metazoan animal life older than about 670 million years have been discredited, are suspect, or await verification, either as to supposed metazoan affinity or age. The believable Proterozoic fossils, including the stromatolite builders, are all microbial. Yet an embarrassing number, even of microscopic objects described as fossils, have proved to be spurious. It becomes necessary to discriminate among fossils, pseudofossils, dubiofossils, all-too-frequent garden-variety contaminants like fungal spores, and artifacts—products of laboratory procedures. Contaminant-free environments do not exist on our planet. A healthy and open-minded skepticism toward unverified data is the best safeguard against premature and misleading conclusions.

CHAPTER TWELVE

THE LONGEST WINTER

The terminal Proterozoic world, onward from somewhat less than an aeon ago, was a world of active plate motions and climatic change. Old plates broke up and new ones were cobbled together. Seas and marginal basins appeared where plates rifted open and folded mountain ranges where they collided. Ice blanketed much of the continental surface. Although extensive continental glaciation has challenged life on our planet during at least five long intervals of directly recorded history, none before or since recurred so persistently over so long a time as that of the terminal Proterozoic.

Rifting on the heels of the closing collisional event of the Grenville orogeny in North America opened the narrow Iapetus Sea, where now the Atlantic reigns. Pulses of the Pan-African and related orogenies between perhaps 900 to 550 million years ago reflect repeated separation and rejoining of plates. Restoration of sea level and widening of the continental shelves upon melting of terminal Proterozoic ice sheets foreshadowed the emergence of animal life and the onset of Phanerozoic history.

A challenging mystery, however, remains unsolved. What are the causes of continental glaciations and why do they recur? How, in particular, may one explain long, recurrent, at least locally sea level glaciations at times and places where, as in the terminal Proterozoic, so much of the available paleomagnetic evidence has been considered to imply low paleolatitudes?[1]

Paleoclimate:
Caloric Accountant, Atmospheric Purser

Our hominid ancestors began their perilous but serendipitous, knuckle-walking ascent from the plains of Africa to modern humankind barely 2 million years ago—just as the last great Northern Hemisphere ice sheets were reaching the vicinity of New York and Moscow. During the past 12,000 years, the remnants of that ice withdrew to Greenland and the tops of mountains. The sea, however, has never allowed it to travel far beyond Antarctica. As it

12.1 TERMINAL PROTEROZOIC TILLITE ON GLACIAL PAVEMENT AT VARANGER FJORD, NORTHERN NORWAY. Fragments of diverse rock types in a finer-grained matrix here rest on a striated and polished glacial pavement. Two directions of striation (*arrows*) indicate separate glacial incursions from the present east and southeast.

melted, sea level rose, great natural seaports came into being, and much of Earth's surface was cleared of ice as if for the occupation and support of humankind. At intervals during subsequent Pleistocene history, now known to have lasted only about 1.6 million years, ice covered as much as 30 percent of the present continental surface. Impressive though that is in terms of human history, however, it is but one of a sequence of low-altitude glaciations on a continental scale that have blanketed sectors of Earth's surface at intervals over geologic time. The longest-lasting succession of recurrent advances and retreats of ice on such a scale was that of the last 2 or 3 geocenturies of Proterozoic history.

Imagine a world where ice is so extensive that India, Siberia-Mongolia, and Antarctica alone among present continental regions lacked extensive ice sheets, where a third of the then-continental surface might at times have been ice-covered during scores of millions of years. Think of a planet where one might have skied year-round on almost any mountain more than a couple of thousand meters high, a world where coastal fog and icebergs would have been universal hazards to shipping. Imagine polar regions without penguins or polar bears—in fact, with no visible life more complicated than algal or other microbial associations, a world, moreover, that had been like that since time immemorial. Think of a world awakened, as it were, from a long sleep beneath its blanket of ice to find continents colliding and separating all over the place, leaving mountain ranges to mark the sutures that knit them together at the sites of collision and seas or oceans where they separated.

If one can imagine such a world, one

begins to comprehend what that of latest Proterozoic history was probably like compared to most earlier and later times. That world was not ready for us, but it portended advanced forms of eucaryotic life. Within the waning years of such a setting were to emerge the first animals, a variety of invertebrate Metazoa, our earliest multicellular forerunners. Perhaps stress does drive evolution.

The balance of variables that now moderate Earth's capricious surface temperature is not eternal. It has undergone and will undergo changes, some imperceptible, some dramatic. Without an atmosphere, surface temperatures would range from well below the freezing point of water to well above its boiling point. Given a blanketing atmosphere rich in carbon dioxide and water vapor, incoming solar radiation would be back-radiated in the infrared. It would become trapped beneath the cloud layer, raising temperatures (as on Venus) too high for comfort or even for life—an extreme result of the much-discussed "greenhouse effect." Recorded climatic history has varied within these two extremes. Liquid water has existed continuously somewhere on Earth since at least 3.8 aeons ago, while, over the same interval, Earth's seas have never boiled.

Without an atmosphere there would be no climate in the sense we know: dramatic day-to-night temperature extremes but no climatic variation, no glaciers, not even a hydrosphere in which to store and transport heat. Life could not exist. Planetary history following the loss of any primordial or outgassed atmosphere would have been entirely the product of internal processes, meteoritic bombardment, and stellar evolution. No breeze would lift the dust nor any sound be audible, for neither sound nor wind can exist in so nearly complete a vacuum.

Earth does have an atmosphere, however, and has had one at least as long as it has had water. Because of that atmosphere, the hydrosphere beneath, and Earth's spin, there is weather, there are climates, and there is a history of climatic change. Records of that history reflect variations in the CO_2 and O_2 content of the atmosphere, as well as the effects of other terrestrial and extraterrestrial processes and events.

The climate at any given place and historical interlude is a function of location with respect to poles and equator, of the size and height of land, and of position within that land. It responds to the effects of wind and ocean currents, to the blocking of Sun's rays by dust and gases of whatever origin, to the position of the solar system within the galaxy and Earth's orientation with respect to Sun, to rainfall, and to vegetation or the lack of it. It reflects human intervention—as in the oxidation of old carbon to CO_2 in industry, transportation, the regulation of interior temperatures and agriculture. It reacts to the heat- and sun-blocking dust of nuclear testing, warfare, volcanism, and meteoritic impact. It is affected by plate tectonism in many ways. It responds to the distribution of land and sea, to the height of the land and its position with respect to oceanic and air currents, and to changes in volume of the ocean basins. It is affected even by

the dissolution and precipitation of limestone, which are among the many variables that regulate the amount of CO_2 in atmosphere and hydrosphere.

Given all these and other variables, predicting climate is an uncertain venture. But if we cannot yet predict climate with much confidence, we can read the signs of past climatic variation and apply them to improving theoretical models and testing such models against the historical record. That involves paleoclimatology, of which we have seen examples in previous chapters. Terminal Proterozoic glacial deposits, although by no means the oldest, are extensive, unusually well preserved, and represent very much the longest span of time of any glacial age.

These deposits came on the heels of an older sedimentary record in which mainly dolomitic platformal carbonate rocks and perhaps the oldest extensive calcium sulfate deposits (gypsum) were conspicuous—indicative of mild climate and extensive aridity. Prominent also during preceding younger Proterozoic history were rift-basin redbeds and a variety of volcanic rocks. Those arid lands were too dry, too low, too extensive, and perhaps too close to the equator for sea-level ice sheets (Fig. 11.16B). At least no glaciation is authentically recorded during the more than a billion years that intervened between the Grenville event of an aeon or so ago and the 2.4-aeon-old early Proterozoic glaciation.

Evidence for Glaciation

Although everyone knows about the "Great Ice Ages" during which our

hunter-gatherer forerunners branched off from the australopithecines, some outside of geology may not be familiar with the evidence for that and other great glaciations. Ice melts. Glaciers vanish. How does one even know that ice sheets existed in the first place, and over what extent? What spoor betray their passing?

A classic terminal Proterozoic glacial deposit is illustrated as Figure 12.1—diverse angular rock fragments sus-pended in sands and silts on a striated rock surface. Most conspicuous to the untrained eye is the work of mountain glaciers (Fig. 12.2). Anyone who has hiked or even flown across a once exten-sively glaciated mountain range such as the Alps, the Andes, or the Rocky Moun-tains has probably noticed at their crests the bowl-like basins called *cirques* (Fig.

12.2 THE WORK OF MON-TANE GLACIERS: UINTA MOUNTAINS, EASTERN SIERRA NEVADA. Scene *A*, bowl-like cirques of for-mer mountain glaciers indent the Uinta Moun-tain crest, Utah, exposing a thick, late Proterozoic sequence of shales and sandstones. *B*, glacially shaped ancestral valley of Green Creek, east slope of Sierra Nevada, California, showing moraines left by a wasting tongue-shaped glacier along a classic U-shaped glacial valley; glacial cirques in back-ground. [*Courtesy of John S. Shelton.*]

12.2*A*). Here one sees the effects of prying, or plucking, by ice—of the freezing, melting, and refreezing of ice in rock crevices. Some such cirques are still occupied by remnants of the original glaciers. Downslope from them are U-shaped glacial valleys like that of the Merced River in California's Yosemite Valley or Scandinavia's fjords, contrast-ing with the characteristically V-shaped vertical profiles of purely stream-cut valleys. Where such valleys open to gentler, lower slopes, it is common for them to be lined with low ridges of glacially transported debris—the cumulative marginal records of glaciers now wasted away. An accumulation of such debris, consisting of commonly angular gravel

12.3 MANIFESTATIONS OF CONTINENTAL GLACIATION; (*A*) EASTERN WASHINGTON; (*B*) EASTERN CANADA. Scene *A*, terminal ground moraine of a wasting continental ice sheet, showing exotic boulders and depressions where blocks of ice within the moraine melted; Waterville Plateau, eastern Washington. *B*, glacially shaped, crystalline, Cryptozoic rocks of Quebec–Labrador boundary area; the winding light-colored line is an abandoned, gravel-filled, Pleistocene subglacial stream course, an *esker*. [A, *courtesy of John S. Shelton.*]

and sand, with dispersed and often scratched or striated pebbles, cobbles, and boulders, is called a *moraine*. Ridge-like *lateral moraines* line the sides of former montane glaciers and horseshoe-shaped *terminal moraines* mark their ends. Either or both may be composite, recording successive positions of glacial retreat as climate moderated and glaciers wasted away. Both are well displayed by former glacial valleys along the east slope of California's Sierra Nevada (Fig. 12.2*B*), as in Alpine settings worldwide.

Even more extensive are the manifestations of continental ice sheets (Fig. 12.3). In considering such phenomena, it is important to realize that although glaciers really do advance under the influence of gravity and internal flow as ice accumulates, ice sheets retreat only to a limited extent. Instead, they literally waste away with warming climate. Cutting tools of hard rock from various sources, frozen into the basal layer of advancing glacial ice, scratch or striate and shape underlying rock and are themselves scratched and shaped. Such striated surfaces and shaped rock demonstrate the advance of ice and record the varying directions of its flow. Other landforms record the shaping of deposits of gravel and sand by glacially related flowing or ponded water, by the dumping of glacial debris as ice melts, and by the later formation of ponds and other depressions upon further melting of ice within such debris.

In addition to the pleasantly undulating and often pond-dotted *ground moraines* of such now-vanished ice sheets—prize residential real estate in middle and higher latitudes of the temperate Northern Hemisphere—other distinctive landscapes record the works of recently vanished continental ice. Figure 12.3*A* displays the characteristic ground moraine of a now completely wasted ice sheet that covered a large part of western North America until about 12,000 years ago. The scattered large exotic boulders, the depressions left by melting of ice within the morainal debris, and the hummocky contours of the subdued landscape reveal at a glance the former presence of a great ice sheet at this place.

Equally revealing is the glacially denuded and shaped terrain pictured in Figure 12.3*B*, in the subarctic border region of Quebec and Labrador (where surface deposits have been mainly stripped away to be redeposited at places such as shown in Fig. 12.3*A*). Striking in such landscapes are long, narrow ridges of sand and gravel that wind across the rock-bound and sparsely wooded countryside as if so many river beds. Indeed that is what they were—the sands and gravels of fossil streams that once flowed beneath a glacial cover, building up their beds as the ice wasted away, undeformed by glacial motion. Rocky surfaces in such regions are everywhere grooved and striated by now-scattered stones that were once imbedded in the antecedent ice sheet. One sees them in town and country alike, at latitudes north of 40° or 50°, all around the lately glaciated Northern Hemisphere. More than one set of such striations may be found on the same surface (e.g., Fig. 12.1), recording successive advances of ice from different directions, following stillstands or interglacial intervals. The sequence is easily deciphered by means of Hutton's principle of crosscutting relationships.

Indirect evidences of once-frigid climates are of several sorts. Elevated strandlines, for instance, may record the isostatic rebound of regions formerly depressed under heavy loads of ice (Fig. 12.4*A*). Imprints of ice crystals, formed at the surface of freezing muds before burial are occasionally preserved (Fig. 12.4*B*).

Distinctive as such evidence is, records of pre-Pleistocene glaciation are not ordinarily extensive or well preserved. With few exceptions they must be carefully sought out and evaluated. Most common are striated and polished rocky surfaces, glacially shaped and scratched pebbles, and deposits similar to glacial *tills*—all to be seen in the reputedly 650-MY-old rocks of Figure 12.1. Tills (tillite when rock, as here) are conglomerates of glacial origin in which larger, commonly angular rock fragments are dispersed in a finer-grained matrix. Cobbles and pebbles in such tills may themselves be shaped and striated. Sequences of graded, cyclically laminated siltstones and mudstones similar to those of annually varved lake deposits (described in Ch. 4) imply lacustrine ice-margin sedimentation. That becomes a certainty for Proterozoic deposits if they also contain erratic pebbles whose imprints show them to have fallen from above (e.g., Fig. 10.5*A*). For floating ice was their only means of transport at that time in history, and the clotting mechanics of mud in the clay layers rule out a marine origin for the glacially varved lamination.

During any part of Earth history other than Mesozoic, the presence of polar lands was an invitation to glaciation—geography and atmospheric circulation permitting. In the absence of some form of "greenhouse" mechanism for trapping and retaining heat from the faint

12.4 Indirect evidence of frigid climate; emerged strandlines, moulds of ice crystals. Scene *A*, elevated strandlines at Cape Krusenstern, Arctic coast of Canada, a product of isostatic rebound of Earth's crust owing to melting of formerly thick Pleistocene ice. *B*, probable imprints of ice crystals in older Paleozoic sedimentary rocks, Tasmania.

early Sun, ice would very likely have been the commonest rock on the Archean Earth. That no signs of this are seen is one reason why an Archean "greenhouse" is inferred. It also seems improbable that glaciation would ever be

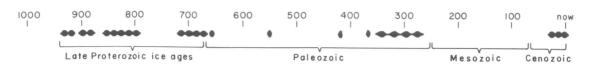

12.5 THE LAST BILLION YEARS OF GLACIAL HISTORY.

completely absent except at times of truly equable global climate (the Cretaceous) or of equatorial clustering of continents.

It is no cause for surprise, then, that ice sheets of large extent were already present as early as 2.4 to 2.3 aeons ago, and it is even possible that glaciers existed locally as much as 2.8 aeons ago during the deposition of the Witwatersrand Supergroup in southern Africa. The glacial history of the past aeon (Fig. 12.5),

of course, is much better documented.

It is, nonetheless, unclear what brought on these different ages of glaciation (Fig. 12.5) and what their ranges in time mean. To what extent are they connected with the timing of polar transit by drifting continents? To what degree are they related to global climatic change? Were extraterrestrial events involved? Has the record been distorted by vagaries of preservation?

The Perplexing Terminal Proterozoic Glaciations

Glaciated surfaces and deposits of former glacial regimes record an extended history of late Proterozoic glaciation on all great landmasses except Antarctica, India, and Siberia-Mongolia.

The best described local glacial sequence of this age consists of forty-seven paleoglacial intervals within the 750-meter-thick Port Askaig Tillite, on Scotland's rocky isle of Islay. The age of those tillites, however, is uncertain and may be little more than about 650 MYBP—the apparent (but presumably minimal) age of the terminal Proterozoic tillite called Varangerian or Varangian in Europe (from Varanger fjord). They represent, moreover, only the uppermost of two to four glacial intervals that comprise much of the late Proterozoic succession worldwide.

The Port Askaig Tillite is important not only for its classic display of late Proterozoic glacial deposits, but also because it demonstrates the value of a now-standard test for paleomagnetic significance. Early paleomagnetic studies giving its paleolatitude as 10°S have been shown to be in error by this Graham Conglomerate Test, in which paleomagnetic orientations of pebbles in an associated conglomerate are determined and compared. Because the pebbles came from different sources, their primary paleomagnetic orientations should be random and different from that of the surrounding matrix. If, however, all have the same orientation, that records a secondary, postdepositional magnetization. The Port Askaig Tillite, when so tested, showed a clustering of paleopoles. The supposedly primary 10°S paleolatitude had actually been imposed around 200 MY later, during Ordovician tectonism. New measurements, following the thermal removal of secondary magnetizations, show deposition at more believable middle to high latitudes.

A review of late Proterozoic glacial records worldwide leads to recognition of the Umberatana Group of South Australia's Flinders Ranges and vicinity as the best reference sequence for this sector of Earth history (Fig. 12.6). Up to 12 kilometers thick locally, the *Umberatanan sequence* includes repeated intervals of tillite with striated stones, varvite with lamination-deforming dropstones, and glacial-marine deposits separated by

intervals of marine, fluviatile, and lacustrine strata. As much as 30 percent of the total is reported to consist of demonstrably glacial deposits and related sediments, roughly bunched at three different stratigraphic intervals within the Umberatanan. Outcrops of this sequence range northward more than 500 kilometers from Adelaide on the Gulf of St. Vincent, reaching east–west widths of 400 kilometers. Glacial debris seems to have fed basinward from multiple ice centers.

The age is not well defined, but it is better defined than that of most late Proterozoic glacial phenomena—few of which are yet confidently dated within the range of a hundred million years or

12.6 VERTICAL EAST–WEST PROFILE OF THE PROTEROZOIC–PHANEROZOIC TRANSITION IN SOUTH AUSTRALIA, SHOWING THREE LEVELS OF TERMINAL PROTEROZOIC (UMBERATANAN) GLACIATION. *[Rearranged, simplified, and supplemented from R. P. Coates, in M. J. Hambrey and W. B. Harland, eds., 1981,* Earth's pre-Pleistocene Glacial Record, © *Cambridge University Press, p. 538, Fig. 2.]*

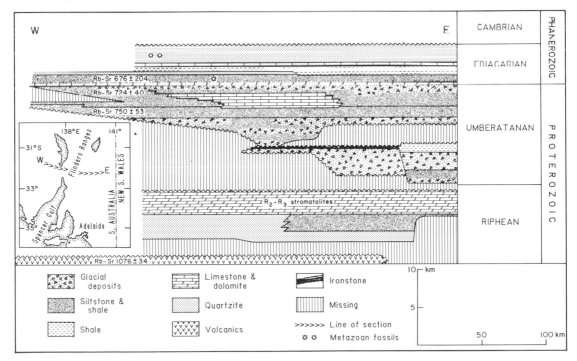

more. Rubidium–strontium whole-rock ages suggest rough limits within those indicated in Figure 12.6. These and other numbers, the relationships outlined, and other considerations permit an estimate of perhaps 720 to 670 MYBP for the upper glacials, more than 750 million years for the middle glacials, and per- haps 800 to 850 or 900 MYBP for the lower fluvio-glacial and tillitic deposits.

The grip of the long Umberatanan winter spread a wide swath of icy deso- lation across a large part of the western two-thirds of present-day Australia. It spread northwestward from now-humid Tasmania across the barrens of Broken

12.7 PLEISTOCENE VARVES (*A*), COMPARED WITH VARVITES (*B–C*) AND OTHER EVIDENCE FOR TER- MINAL PROTEROZOIC GLACIATION IN AUSTRALIA (*D–E*). View *A*, Pleistocene varves from Seattle area, seen edge on. *B*, varvite from drill core in Elatina Formation, upper glacials, east of Port Augusta, Australia; varves, 0.1 to 3 mm thick, record periods of 11 and 22 years, as do recent sunspot cycles. *C*, Sturtian glacial-like sediment, middle glacials, deformed by small dropstone, northeast of Port Augusta, South Australia. *D*, striae and arcuate chatter marks record movement on glacial pavement beneath 700 MYBP tillite, Margaret River, Western Australia. *E*, exotic cobble in tillite of uppermost Proterozoic Elatina Formation, Flinders Ranges, South Australia. [*A, courtesy of J. Hoover Mackin; B, courtesy of G. E. Williams.*]

Hill, to the rocky eucalyptus groves of the Flinders Ranges, to the wastelands of central Australia, and on to the Indian Ocean. Figure 12.7 illustrates some of the supportive evidence.

Glaciation somewhere within this long climatic interval is known almost worldwide and at surprisingly low reported paleolatitudes. A brisk survey of the main facts will outline the problem and consider causes.

The presently known minimal global distribution of late Proterozoic glacial deposits of the Umberatanan interval and equivalents worldwide, and the locations of their associated landmasses is provisionally indicated in Figure 12.8. This reconstruction is based on an attempt to reconcile available paleomagnetic and stratigraphic data with paleoclimatic probability. Figure 12.9 is an alternative, pole-centered paleogeography of the same land areas based on the simplifying assumption that all sites were glaciated at high latitudes and mainly around 750 to 680 MYBP, acquiring low-latitude paleomagnetic signatures postdepositionally. Considering that some key samples are being found, on clearing of secondary affects, to have primary remanent magnetizations indicative of high latitudes, one may reasonably hypothesize that much or all of this glaciation occurred during a long interval of geologically rapid drift of continents across a pole, most likely the South Pole. Its now-cosmopolitan dispersal is epitomized in Figure 12.10, showing samples of the terminal Proterozoic and early Paleozoic work of ice from Namibia, Yukon Territory, Australia, China, and Europe.

The most extensive of these generally late Proterozoic glacial deposits is found in present-day Africa. They occur intermittently in a wide belt that extends from now-northwest Africa through central and west Africa almost to its southwestern tip, in Namibia. Like most late Proterozoic glaciation, however, ages either are not well established or are not available. Deposits of glacial or marginal glacial (*periglacial*) origin include tillites, laminated sediments with dropstones, and polygonally segmented deposits whose depressed margins probably represent former ice wedges in permafrost. Limited evidence suggests that all these deposits could be very late Proterozoic, centering on a widely preferred but not established age of around 700 MYBP. Although paleolatitudes have not been confidently established in northwestern Africa, striated pavements and glacially shaped rocks indicate motion toward the present south-southeast.

Turning to central Africa, where apparent polar wander paths suggest low latitudes, two well-documented levels of glaciation are estimated to have ages of about 900 to 1,000 and 770 to 870 MYBP.

It is in Namibia and adjoining southwestern Africa that the most extensive likely records in time are found. Tillite-like deposits and other possible indicators of glaciation are found at seven or more stratigraphic levels extending from perhaps 1,000 to 550 MYBP. Although contemporary and later Pan-African tectonism in the region has complicated the problem of interpretation, where local gravity sliding is capable of producing striated boulders and diamictites, other evidence supports glaciation. Convincing dropstones and the faceting of striated stones show that at least some of these deposits are glacigenic. That sequence presents the most persuasive evidence

299

12.8 HYPOTHETICAL GLOBAL DISTRIBUTION OF LATE PROTEROZOIC LANDS AND INFERRED MINIMAL EXTENT OF UMBERATANAN GLACIATION ABOUT 700 MYBP. From limited paleomagnetic and geologic data; *Ar*, Arabia; *Aue*, eastern Australia; *Auw*, Western Australia; *Cn*, north China block; *Cs*, South China block; *En*, north Europe. Other abbreviations obvious. Compare Fig. 12.9.

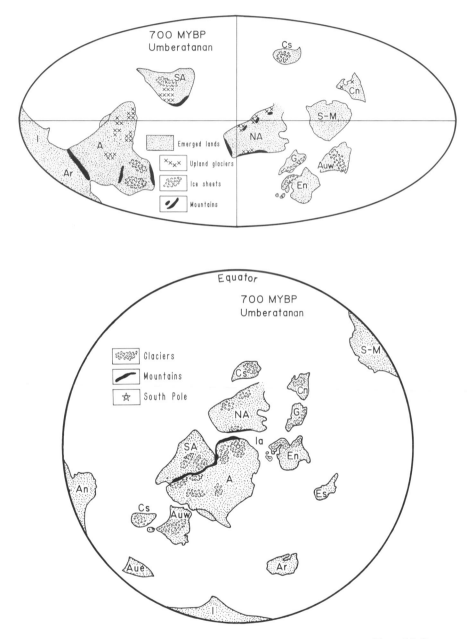

12.9 ALTERNATIVE PALEOGEOGRAPHY FOR SAME HISTORICAL INTERVAL AS FIG. 12.8, ON A POLAR PROJECTION, DISREGARDING PALEOMAGNETICS. Symbolism as in Fig. 12.8 plus *Es* for south Europe.

for long time-transgressive glaciation and apparent paleopolar migration.

The youngest of the Namibian frigid intervals is seen at a single locality where a Cambrian channel has exposed a grooved and polished surface on older Phanerozoic (Ediacarian) strata. An unusual kind of then soft-sediment deformation (Fig. 12.10*A*) there has been explained as the product of grounding sea ice. Yet if the work was that of ice, the particulate sediments on which it grounded must have been frozen as well, in order for them to deform so readily and coherently, accept such a polish, and preserve the deep grooves. Ambiguity remains, but not about the local frigidity of initial Paleozoic climate. Unless montane glaciation was involved, Africa would seem to have experienced a polar regime during much of late Proterozoic and early Phanerozoic history.

12.10 RECORDS OF EDIACARIAN OR CAMBRIAN FRIGIDITY IN NAMI-BIA AND OF LATE PROTEROZOIC GLACIATION ON FOUR OTHER CON-TINENTS. View *A*, gouged and polished surface, probably shaped by ice grounding on a frozen beach, Klein Karas Mountains, Namibia. *B*, striated stones (*arrows*) in 700-MY-old tillite-like rocks of the Rapitan Group, eastern Yukon Terri-tory, Canada. *C*, glacial surface beneath Eggan tillite, and *D*, ice-shaped and scratched quartzitic pebble (*arrow*) amidst ripped-up limestone, Eggan Formation, Kimberley district, Australia. *E*, Nantuo tillite, about 720 to 680 MY old, gorge of Yangtze River upstream from Yichang, China. *F*, detail of late Proterozoic tillite of Fig. 12.1 showing diversity of pebble types.

Late Proterozoic glaciations in Brazil extend over a region 2,000 kilometers square. Was that a single great ice sheet, more than one, or a congeries of montane glaciers? *Estimated* ages span a suspected range from around 600 to 1,370 MYBP. The only clue to paleolatitude is the evidence for glaciation itself and the likelihood that the glaciation may have been continuous with Africa.

Upper Umberatanan (Varangerian) glaciation occurred along the Scandinavian spine of northern Europe, out to Scotland and northern Ireland, and eastward into the Ural Mountains. Two intervals of very late Proterozoic glaciation are inferred, minimal (potassium–argon) ages being 650 MYBP and a younger one between about 650 and 610 MYBP.

China was glaciated on two then-separate miniplates in the present southeast and northwest during Umberatanan history (e.g., Fig. 12.10E). These glaciations are considered to be about 800 to 760 and 720 to 680 MYBP. Paleomagnetic data from a Ph.D. thesis by Lin Jinlu at the University of California, Santa Barbara, place the North China plate at a low paleolatitude (perhaps montane) in contrast to two high-latitude Proterozoic tillites from South China.

North America itself has a record of late Proterozoic glaciation inland from both present coasts. That record is fairly extensive on the Pacific side but limited on the eastern seaboard south of Newfoundland. As usual for strata of the Umberatanan sequence, ages and paleolatitudes are uncertain, but ages are widely inferred to be around 700 MYBP in Utah, Idaho, Montana, northeastern Washington, British Columbia, Yukon Territory, and Alaska.

An age of 700 MYBP for these deposits, *assuming all represent the same glacial interval,* has become so widely accepted on such uncertain ground that it should be challenged. In fact, a Pacific-coast sequence south of those mentioned may be substantially older. It is that of the Death Valley region of eastern California, within the Kingston Peak Formation. Tillite-like deposits are common among the mainly detrital rocks of that formation, which also includes lacustrine, fluvial, and avalanche deposits, striated and faceted stones, and probable dropstones. Because local avalanching of the same historical interval is recorded by large slide blocks and associated debris in the mainly eastern source areas, it also seems likely that local montane glaciation was involved. Debris, presumably including ice, was fed downslope along with slide blocks to accumulate or be reworked into lowland depressions, locally to impound saline lakes, and probably at places to reach the then-Cordilleran sea on the west.

In contrast to ages suggested for other late Proterozoic glacial deposits in the North American Cordillera, those of the Kingston Peak Formation may be significantly older. They are overlain by a shallow-water marine dolomite (Noonday Dolomite) that, for much of its extent, cuts far downward across the probably glacial deposits, and, in addition, displays a distinctive and unusual proalgal or algal fabric. It is strikingly similar to that of rocks found near Lake Baikal (Siberia) to have K–Ar ages of about 1,070 million years.

In the eastern United States, unequivocally late Proterozoic glaciation is found as an isolated occurrence on Mount Rogers, in the Blue Ridge Mountains, sunny Virginia's highest. The evidence for that glaciation consists of lenses of clearly tillitic and laminated pebbly mudstone that include a variety of sizes and types of larger fragments, some shaped and striated. These deposits, up to 100 meters thick, outcrop for about 25 kilometers along the strike of the beds. They occur at the top of the mainly vol-canic 3-kilometer-thick Mount Rogers Formation, once confidently dated at 810 MYBP until reexamination indicated the dated zircon crystals to be in part recycled from older rocks. The true age is now considered to be closer to the 700 MYBP number so widely accepted along the Pacific margin.

The Causes of Glaciation

When Earth was young and Sun was cool, all that would have been needed for global glaciation was sufficient H_2O and a low level of "greenhouse" gases. After present solar luminosities were reached, perhaps somewhat more than 2 aeons ago, glaciation was to be expected only in polar regions or in special circumstances. How, then, might one explain the recurrent glaciations of late Proterozoic history, especially if truly at low paleolatitudes?

Many hypotheses have been proposed. All call on some combination of reduced solar radiation at Earth's surface, the bootstrap effects of increased reflectivity resulting from more extensive and longer-lasting snow and ice, and moisture supply. These, in turn, are affected by volcanism, the positions of glaciated lands with respect to the rotational axis and its inclination to Sun, variations in the orbital path, and perhaps other extraterrestrial causes. Different theorists stress different factors or combinations in hypotheses whose variations are too numerous to review. Perhaps most involve some element of truth.

John Crowell of the University of California at Santa Barbara has persuasively reasoned that many of the factors sug-gested have probably played roles in the initiation and growth of continental ice sheets at one time or another—the essence of his Eclectic Hypothesis. Many processes interact. Other factors being equal, surfaces more obliquely inclined to the Sun's rays are more likely to be glaciated than others. Surfaces most inclined to Sun are now and probably always have been at high latitudes with respect to Earth's spin axis. Sea ice is likely to be more extensive the higher the latitude, thereby increasing short-wavelength reflectivity (albedo) and reducing temperature. The cooling of the planetary surface following great volcanic eruptions is well documented by tree-ring research and historical accounts, while asteroidal or cometary impact (and nuclear warfare) would presumably have similar effects. A succession of cold winters that led to more extensive year-round snow cover or sea ice could trigger glaciation by increasing reflectivity, assuming a solid continental surface on which glacial ice could accumulate.

Glaciers and grounded ice sheets don't form on the open sea, although parts of them may reach it, float there, and calve icebergs that drift to lower latitudes, scrape bottom on shoal areas, and release dropstones on melting (e.g., Fig. 12.10*D*). So continental surfaces at high-enough latitudes for large ice sheets to accumulate are prerequisite.

The climatic record of pre-Pleistocene history, moreover, shows clearly that the formation of continental ice sheets is episodic, not cyclic. Contrary to once-widespread belief, there is no law of nature that says ice sheets will expand over the continents every 250 MY or at any other approximately regular interval longer than those of orbital variation.

A mechanism does exist, however, for episodic glaciation of any part of the Earth or for persistent glaciation of polar lands. That mechanism is plate tectonics. As plates move over the poles, passenger continents are likely to become glaciated, assuming that there is open water close enough and in the right orientation with respect to winds to supply necessary moisture to a planet not much warmer than now. Such a mechanism can account both for continental glaciation and for its variant times of appearance on different continents. It is also consistent with latitudes suggested by Phanerozoic paleomagnetic data, but not with many reported late Proterozoic paleolatitudes.

How can this dilemma be resolved? A 1981 analysis of some eighty late Proterozoic tillites by Soviet paleoclimatologist Nicolai Chumakov takes issue with low latitudes. Chumakov finds, "using all data," that two-thirds of the positions considered useful were at middle to high latitudes at the time of glaciation; only one-third were dubiously at low latitudes. His conclusions appeal, for they would rationalize the longest and most extensive record of continental glacia-

tion in all of Earth history. Yet most students of paleomagnetism up to 1986 were still finding low paleolatitudes. Clearly more research is needed, both as to existing latitudes at times and sites of glaciation (determined after removal of any secondary paleomagnetic overprints) and as to how anomalies found might otherwise be explained.

Could supplies of polar ice simply have become so great that the ice flowed to low latitudes under its own weight? Could a clustering of sufficiently elevated continents at low paleolatitudes account for meteorological conditions that would lead to extensive equatorial and low-latitude glaciation? Were the terminal Proterozoic glaciations a product of general frigidity brought on by changes in Earth's orbit or obliquity or the position of the solar system in the galaxy, including passage though a galactic dust cloud? Might all be Alpine? Here might be a good project in which to invest some "nuclear winter" funds.

Insofar as currently available evidence goes, none of these simple alternatives seems to work without assistance from other processes. Many of the glacial records of Umberatanan age are near coastal regions that could have been elevated by preceding continental collisions to heights where moisture from newly opened seas could result in montane glaciation (Fig. 12.8). Glacial deposits, or the glaciers themselves, might then descend to lower levels, as they may have done in the Death Valley region. It is even conceivable that all of the North American late Proterozoic glacial deposits had montane origins. And contemporaneous African glaciation might even be the mainly montane product of recurrent Pan-African orogeny.

Other glaciations of this age range, however, were clearly those of lowland ice sheets, some of which evidently deposited glacio-marine sediments from floating marine extensions. If the paleolatitudes of Australian glaciation, for instance, were truly subtropical, there is no convincing way in which they can be attributed to anything other than sea-level ice. Such ice may have descended from remote elevated regions somewhere in the distance, but was clearly not the product of local montane glaciation.

To account for observed records at low paleolatitudes, either the last few geocenturies of Proterozoic history would need to be scenes of global refrigeration or else the records of abundant moisture from open seaways that survived as year-round snow and ice in some other unexplained way, on mountains and lowland surfaces alike.

The only easy answer is polar clustering with secondary paleomagnetic overprints on glacial deposits that drifted rapidly equatorward following high-latitude glaciation, and that seems almost too easy.

Still, where secondary paleomagnetic orientations have been removed from glacial deposits of Umberatanan age, it has not infrequently been found that glacial deposits previously believed to have formed at low paleolatitudes could owe that orientation to post-drift, post-depositional rather than primary paleomagnetism. If true, that also implies that glaciations recorded were time transgressive—that they occurred over different intervals on different landmasses as those lands moved across the polar regions.

The problem remains obdurate. No broadly acceptable solution has been found. Different apparent polar wander paths for the time involved, published by different students of paleomagnetics since 1970, do not agree. There remain late Proterozoic glacial deposits for which the so-far-available data imply origin near the equator. Until all have been put to the conglomerate test and restudied if they fail, agreement is unlikely.

Late Proterozoic Tectonism: Its Causes and Consequences

Active separation and rejoining of plates, openings of mainly narrow seas, and tectonism resulting from convergence and collision of lithospheric plates seem to have been features of late Proterozoic history. Was this the inevitable result of the long buildup of thermal and tensional stress beneath preexisting supercontinents, of rifting events that eventually split the supercontinent of which North America had been a part? Such rifting had been in evidence as far back as 1.2 or more aeons ago (Fig. 11.2). It preceded a plate convergence that peaked with the probably multiple collisional events and metamorphism known as the Grenville orogeny, ranging from around 1,100 to 850 MYBP. Evidence of related crustal mobility is seen in scattered exposures and recorded in the subsurface from Canada's Grenville Province all along the then probably southern and southwestern seaboards of North America, as far as Columbia in

South America and northward across the Atlantic to East Greenland, Northern Scotland, and Scandinavia.

That heralded a new and even more active breakup and shifting of plates that endured into Phanerozoic history. The probable distribution of collisional mountain ranges, exotic terranes, and rifting implied by internal zones of volcanism and fold belts suggests a kind of horizontal yo-yoing of plates. Such movements went on at different times and places right through the so-called Pan-African and related orogenies that affected some two-thirds of Africa and parts of other continents from perhaps 900 MYBP to as late as 550 or even 450 MYBP at some places.

These were times not only of mountain building but of extensive redbed sedimentation, associated with mafic to locally sialic volcanism such as seems to go with rifting and the breakup of plates. Similar redbeds and volcanism are widely associated in present Northern Hemisphere sequences of early Mesozoic, middle Paleozoic, and late Proterozoic age. The striking similarities of these associations imply similar histories of mantle-tapping processes that brought iron-rich mafic lavas to the surface at the same time that rift valleys were opening in arid continental interiors. At such places coarse, poorly sorted, oxidized rift-valley debris could accumulate during the opening of rifts which, in some instances, became new seaways. Modern examples are in progress in the rift valleys of eastern Africa, yet unsevered from the rest of that geologically ultrastable continent, despite repeated efforts at breakup over the past aeon.

A major historical consequence of Umberatanan unrest was the opening of the late Proterozoic to older Paleozoic waterway known as *Iapetus*, after the brother of Oceanus, sons of the mother goddess Gaea (Figs. 12.8, 12.9). The Iapetus sea, or ocean, was not the real proto-Atlantic because it closed again in middle Paleozoic time before that sea opened at nearly the same site during early Mesozoic history to become the Atlantic. The opening and closing of seaways between North America and lands to the east and south was to play a major role in Phanerozoic history.

Biology of the Umberatanan Earth: Transition to Higher Forms of Life

Convincing records of life on the younger Proterozoic Earth are still all microbial. Markings that have been interpreted as tracks, burrows, or body imprints of animal life have so far proven to be either not fossils, not animals, not older than about 670 MYBP, or dubious. And, whereas vertical burrow-like structures resulting from fluid evasion are seen throughout the geologic record, burrows of demonstrably biological origin are not known until Phanerozoic time. Advanced forms of plant life are also missing.

In fact, the transition from Proterozoic to Phanerozoic history was an interval of diminished biological variety. Even the relatively abundant and well preserved records of younger Proterozoic microbial life became impoverished in variety as that era drew to a close. Although pluri-cellular microbial assemblages (Fig. 12.11D–H), a few other mainly sphe-roidal microfossils, and crudely pre-served microbial fabrics (Fig. 12.11I) were locally abundant, diversity declined.

The first major extinction may have taken place then—brought on perhaps by the climatic frigidity, lowering of sea level, and reduction of shelf seas associ-ated with the growth of glacial ice. The oldest known clumping of cell-like structures in something other than a lin-ear chain or algal mat, however, seems to have begun then. Such clumping seems

to have arisen where individual mature cells divided into sporelike bodies within a membrane that did not rupture to release them, so that they grew together as a clonal colony (Fig. 12.11D–G).

Microscopic, floating (planktonic), pluricellular to unicellular bodies like these settled into marine muds all along the ancient Cordilleran seacoast inland from the present western margin of North America and elsewhere. They probably floated worldwide and are prospective time-markers.

A roving mind may hypothesize that one may be seeing, among these

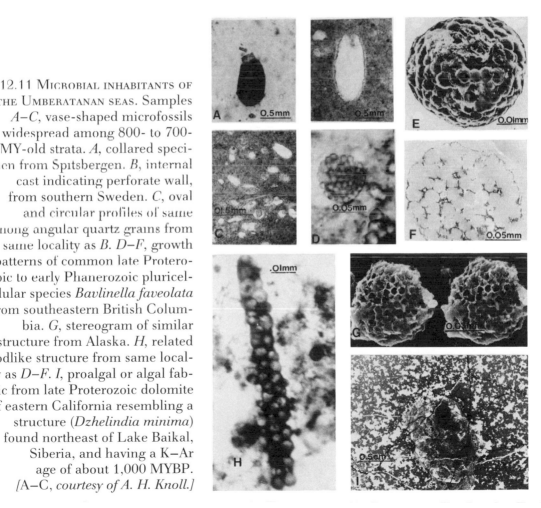

12.11 MICROBIAL INHABITANTS OF THE UMBERATANAN SEAS. Samples A–C, vase-shaped microfossils widespread among 800- to 700-MY-old strata. A, collared speci-men from Spitsbergen. B, internal cast indicating perforate wall, from southern Sweden. C, oval and circular profiles of same among angular quartz grains from same locality as B. D–F, growth patterns of common late Protero-zoic to early Phanerozoic pluricel-lular species Bavlinella faveolata from southeastern British Colum-bia. G, stereogram of similar structure from Alaska. H, related rodlike structure from same local-ity as D–F. I, proalgal or algal fab-ric from late Proterozoic dolomite of eastern California resembling a structure (Dzhelindia minima) found northeast of Lake Baikal, Siberia, and having a K–Ar age of about 1,000 MYBP. [A–C, courtesy of A. H. Knoll.]

microbes, something like the first rudimentary clumping from which differentiated, truly multicellular organisms
subsequently arose. Tenuous though such
a connection may seem, the pluricellular
Bavlinella spheroids illustrated in Figure 12.11 are reminiscent of some more
advanced forms or their early embryonic
stages. As far-fetched as that may seem,
we have no better candidates at this time.
There is only a temporal space in which,
like rabbits from a magician's hat, the
unequivocally metazoan faunas of the
Ediacarian System emerged about 670
MYBP.

But there is one hint that a protozoan
(protoanimal) sort of unicellular organism may already have emerged in the
icy seas of around 800 to 700 MYBP—
the vase-shaped microfossils of Figure
12.11*A–C*—perhaps tintinnid ciliates of
a protozoan suborder that sometimes dons
a resistant outer covering.

So ends the first 85 percent of Earth
history. Earth warmed. Ice sheets waned.
Their meltwaters increased the volume
of the global sea beyond the capacity of
contemporaneous ocean basins to contain it. The glacially beveled continental
margins were flooded with shallow,
warming shelf waters. Mother Earth
became impregnated with the seeds of a
novel and potentially exuberant form
of life—cellularly differentiated, multicellular animal life, the Metazoa, root
stock to *Homo sapiens*.

In Summary

Latest Proterozoic history from about
950 MYBP to the appearance of multicellular animal life around 670 MYBP
was a time of marked plate-tectonic
activity and climatic severity. New seaways opened. Mountain ranges arose
where plates collided. Glaciers flourished, storing water distilled from the
sea. Sea level fell and rose repeatedly
over a range of hundreds of meters,
keeping pace with the waxing and waning of glacial ice. This Umberatanan history witnessed the longest-lasting and
perhaps the most extensive succession of
continental glaciations known. Their
estimated ages differ from place to place,
however, and the episodes they represent are probably all time transgressive
in some degree, reflecting shifts in latitude of their mobile continental platforms.

Available data, viewed worldwide,
suggest two to four principal glacial
intervals within the Umberatanan
sequence. They approximate the following ranges, in order of downward
increasing age (in MYBP): (1) from about
670 or fewer to 720, (2) from 750 to 800
or as much as 870, (3) from 800 or 850
to 900, and *perhaps* (4) from 900 or
younger to about 950 or 1,000. Some
evidence implies local extensions of glaciation into early Phanerozoic history
(Ediacarian and even Cambrian). Before
the terminal Proterozoic there was a long
interval during which no older glaciation
is recorded until about 2.3 to 2.4 aeons
BP—the Gowganda and equivalent rocks
of Chapter 10.

Evidence for glaciation includes striated
and polished subglacial pavements, tillites with striated pebbles and good till

fabrics, and varvites with dropstones and splash-up structures. These are the most usual and convincing criteria observed in pre-Pleistocene glacial sequences. If all are authentically present, the case for glaciation is established. But criteria employed must be weighed critically. Rockslides, volcanic mass flow, faults, quarrying practices, and even old-fashioned steel-rimmed wagon wheels have scratched surfaces and pebbles in patterns that have been or could be misinterpreted as glacial. Avalanches, mudflows, and marine density currents can produce tillite-like deposits. Nonglacially laminated sedimentary rocks can be misinterpreted as varvite, and lonestones that glided into place may be taken for dropstones. It is estimated that no more than about one-fourth of all reported tillite-like rocks are demonstrably tillites.

Authentic glacial phenomena, for all that, are widely associated with late Proterozoic sedimentary rocks of the four intervals mentioned. Not all of these intervals, however, are represented in all glacial sequences of this age. Deposits of the apparently oldest suggested interval are rare and may not exist. Those of the third-oldest interval are only slightly more usual. The second and first intervals are well represented. The most frequently reported age of terminal Proterozoic glaciation is about 700 MYBP. Some hopeful stratigraphers even claim a near-synchronous universal glaciation roughly 700 MYBP that can be used for global correlation. Considering the limited and often imprecise age constraints, that is unlikely. But, synchronous or not, the termination of this glacial history does commonly approximate what is seen on other grounds as the transition from

Proterozoic to Phanerozoic history. It is here utilized as a convenient, and, to a degree also causal, nonsynchronous working approximation.

Late Proterozoic glacial history is most extensively recorded in the Umberatanan sequence of South Australia, proposed as a reference standard. Yet, even there, age control is skimpy and imprecise.

Plate tectonism of this general age range is manifested in the collisional elevation of mountains at former plate boundaries, in the opening and eventual closing of new seaways like Iapetus, and in the presence of ophiolite-like sequences of this age range at a number of places (back endpaper). The Grenville deformational events in North America preceded and introduced that history. Contemporaneous Pan-African orogenies recorded collisional events on the western and northeastern coasts of and within Africa.

The resultant elevations of lands and openings of seaways may have contributed to glacial history by providing sources of moisture and upland sites where it could accumulate as snow and ice. Later lowering of sea level would have destroyed shallow marine habitats where stromatolites had flourished, provoking evolutionary changes among them and related microbial floras.

It is probably no coincidence that there was a reduction in the variety of microbial life at this time, or that it was followed by a surge of evolution as climate ameliorated, ice melted, and rising warm water flooded the coastal lowlands. Protozoan-like fossils, perhaps ciliates, became widespread about 800 to 700 MYBP, presaging the onset of multicellular animal life.

PART III

METAZOA INHERIT THE EARTH: LIFE IN THE FAST LANE

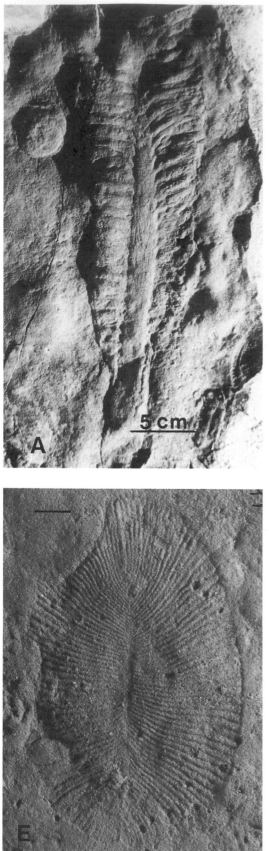

CHAPTER THIRTEEN

THE PALEOZOIC: PRELUDE TO THE MODERN WORLD

The radiant modern world of many-celled animals and plants reflects the synergistic evolution of form and function over the past 650 to 700 million years, of which almost the first 450 represent the Paleozoic. In contrast to the muted refrain of biochemical evolution among microbial forerunners, Phanerozoic evolution was more conspicuous. Among Metazoa, groups of cells cooperate as tissues and organs. Specialized systems assimilate and distribute O_2, fluids, and nutrients in many-celled bodies. Physical leverage in the form of body armor and bony skeletons followed. Earliest metazoan life diversified to occupy previously empty habitats and to create new ones. Interdependent communities evolved. Scavengers and predators arose. In good season, plants readied the way for animals to move ashore.

O_2 continued to build up, fitfully but inexorably, as carbon was sequestered, first reaching present levels around 100 MYBP. Marine sedimentary rocks approached modern proportions—replete with shelly Metazoa, including a variety of reef-building organisms. Igneous rocks continued to intrude, swell, and flood the continental crust, recording the geochemical and physical history of their deep sources. Oceans opened and closed. Old highlands wore down as plate convergence raised new ones. So went the Paleozoic movement of Earth's grand symphony, ascending to the crescendo, then decrescendo, of the waning Permian. In the end, the lands of Earth converged into the universal supercontinent of Pangaea. Glaciers that had come and gone over 50,000 millennia melted. More than half the families of animals then living vanished in history's greatest extinction.

13.1 THE OLDEST ANIMALS, FOSSILS OF THE EDIACARIAN SYSTEM. Imprints of soft-bodied animals from the roughly 620-million-year-old Ediacara Member of the Rawnsley Quartzite, South Australia. Bar scales 1 cm except as marked. Species names and closest resemblances among living organisms are: *A, Charniodiscus oppositus*, sea pen; *B, Spriggina floundersi*, annelid or arthropod; *C, Parvancorina minchami*, naked arthropod; *D, Tribrachidium heraldicum*, naked echinoderm (?); *E–F, Dickinsonia costata*, annelid; *G, Ediacaria flindersi*, medusa. *[Courtesy of Neville Pledge; the South Australian Museum.]*

Dawn of a New Era

The emergence of recognizably tissue- and organ-differentiated multicellular animals marks the beginning of the Paleozoic Era and Phanerozoic Eon. There were four prerequisites to that event: (1) life itself, (2) oxidative metabolism, (3) sexually reproducing eucaryotic cells, and (4) appropriate protozoan ancestors. Nearly four long aeons had elapsed before all were satisfied. The end of that long gestation, covering roughly 85 percent of Earth history so far, marks a great turning point. A new kind of evolutionary history began, leading eventually to the modern world.

The seas that surrounded the ice-gripped terminal Proterozoic continents were a stressful setting, even for their biochemically ingenious microbial populations. The extensive reduction of microbial variety at that time records the earliest probable mass extinction. Plate movement carried the lands of those times equatorward and their glaciers to oblivion.

As the great ice sheets melted, the expanding ocean spilled across the continental margins, flooding alike the ice-worn shelves and the veneering glacial deposits left behind by wasting ice. Extensive, warming, shelf seas flourished, conducive to marine life. Evidence for this is recorded in the veneers of carbonate rock that cap the underlying glacial sequences almost globally. The new types of life that then appeared included not only microbial species but also the oldest Metazoa so far confidently known, among them creatures such as those illustrated in Figure 13.1.

In contrast to Cryptozoic history, which concerns primarily the physical environment and its responses to microbial evolution, information about the flowering of the more familiar forms of life is far richer during the Phanerozoic. Data are so abundant during this phase of history, despite gaps, that a relatively high degree of historical resolution is practicable. The graphic summation of Phanerozoic highlights on the right front endpaper, therefore, is at a scale almost 10 times that of the facing Cryptozoic, and the discussion that follows is proportionally more extended.

Because of this enhanced resolution and the relatively short transitional intervals between Phanerozoic systems and periods, their "boundaries" plot, in that graph, as if their ages were precisely known and invariable worldwide. That is not the case. The numbers used to designate "boundaries" in the uneven flow of Phanerozoic history still vary by a few millions or tens of millions of years with author, place, sample, and method. In geologic terms, however, that variation is proportionally small enough that Phanerozoic "boundaries" can be graphed *as if* they were real time markers—keeping in mind the likelihood that they are, in practice, not only time transgressive to some degree but subject to revision when new data call for it.

Paleozoic rocks are well represented in North America by the Appalachian sequence (front endpaper) and elsewhere in similarly eroded old mountain belts that arose with extensive intercontinental collision toward the end of the Paleozoic history about 250 MYBP.

The big problem in this chapter, in

fact, is how to manage the abundance of information available. My solution will be to focus on the highlights and on the new, with emphasis on the biosphere. I shall try to fish the most interesting pools in the mainstream of history, without wandering too far up its many brushy tributaries.

Securing a Beachhead: Experiments Bizarre and Mundane

The beachhead from which animal life eventually radiated to Earth's far corners, transforming and being transformed as it progressed, was secured during Ediacarian history—first period of the Phanerozoic, just before the Cambrian (named for the fossiliferous Ediacara Member of the Rawnsley Quartzite, Ediacara hills, South Australia). It was taken by a feeble host of jellyfish-like floaters, flimsy sheetlike animals reminiscent of annelids and arthropods, elaborately wrinkled sea pen–like structures, creatures so far known only from sinuous wormlike tracks on sedimentary surfaces, and a submillimeter-sized microburrower from the basal Ediacarian of China. No larger burrows are known before Early Cambrian or very late Ediacarian time.

All of these animals were naked and without backbones, shielded only by their skins or flexible epidermis. They needed no shelly armor nor leathery integuments, for they were opposed by no enemy, subject to no predators, probably not even sought by scavengers larger than bacteria. Nor had they special requirements for rigid leverage of muscles or appendages. Indeed, before the origin of respiratory organs, it would have been to the advantage of the earliest Metazoa to have permeable body surfaces that were large in relation to the organism's volume, in order to acquire enough O_2 by epithelial diffusion into a thin layer of metabolizing tissue.

Consisting at this writing of but 64 named species, plus the tracks of a dozen or so equally or more delicate contemporaries, this Ediacarian fauna (Figs. 13.1, 13.2) presents a remarkable diversity. Some of its members resemble modern sea creatures closely enough to be assigned to still-living phyla. Others became evolutionary dead ends, even in those days of little competition. It has even been proposed that the prominent elements of the fauna represent an experiment that failed, a kind of alien population, perhaps distinct from the mainstream of metazoan evolution.

What is the substance of that provocative idea? Dead ends in evolution are well known. It is certain that some Ediacarian fossils represent extinct lines, while others are not seen again until much later, if at all. That is only to be expected among pioneers, especially unshielded ones that leave faint signs of their former presence or none at all. Indeed, the natural preservation and recovery of soft-bodied Metazoa of any age and number is so unusual that one has enough fingers and toes to count off the complete list of significant occurrences younger than Ediacarian. Why was this type of preservation so extensive just then? Could the Ediacarian fauna represent samples from what was to become a shifting undercurrent in the mainstream of evolution, one whose traces do not ordinarily persist after death but one which came and went in differing manifestations all the same?

Compare this fauna with a marvelously preserved and diverse soft-bodied metazoan fauna no more than perhaps 30 million years younger than Ediacarian, found in the Middle Cambrian Burgess Shale of southern British Columbia. Justly renowned for the insights it brings

concerning early life, this Burgess Shale fauna has recently been intensively studied (Figs. 13.3, 13.4). Of the 30 or so living phyla of animals, mainly consisting of species with few or no hard parts, only 8 are represented here, together with 10 previously unknown contemporaneous groups that, if living, would probably be considered of phylum rank, all soft-bodied. One of these extinct lines, abundantly represented by *Hallucigenia* (Fig. 13.3B), is so grotesque that it is hard to imagine what any descen-

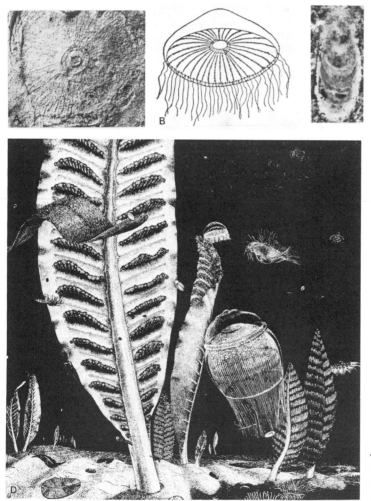

13.2 (*A–B*) Ediacarian and modern medusae; (*C*) feeding burrow of a basal Ediacarian micro-metazoan; (*D*) reconstruction of an Ediacarian habitat. Sample *A*, the fossil *Cyclomedusa*, about 4 cm across. *B*, medusoid phase of a modern aequorian hydroid for comparison. *C*, basal Ediacarian microburrow, lower Yangtze Gorges, China. *D*, components of the Ediacarian fauna in South Australia. [A, *after Reginald Sprigg, 1949,* Transactions of the Royal Society of South Australia, *v. 73(1),* Plate 15, Fig. 1, *by courtesy of the Royal Society of South Australia.* B, *after Libbie H. Hyman, 1940,* The Invertebrates, © *McGraw-Hill Book Company, p. 416, Fig. 121C. Reproduced with the permission of McGraw-Hill Book Company.* C, *courtesy of S. M. Awramik and David Pierce.* D, *drawn by Robert Allen from information supplied by Mary Wade, for Queensland Museum, Brisbane. Publication authorized by the Director.*]

13.3 BENCHMARKS IN
THE BURGESS SHALE:
(A) OLDEST DEMON-
STRABLE CHORDATE;
(B) MEMBER OF AN
EXTINCT PHYLUM.
Fossil *A* is an
Amphioxus-like fossil
called *Pikaia*, remote
ancestor to all verte-
brates, head to left.
B, *Hallucigenia*, rep-
resenting an appar-
ently extinct phylum,
head to right.
*[Courtesy of Simon
Conway Morris.]*

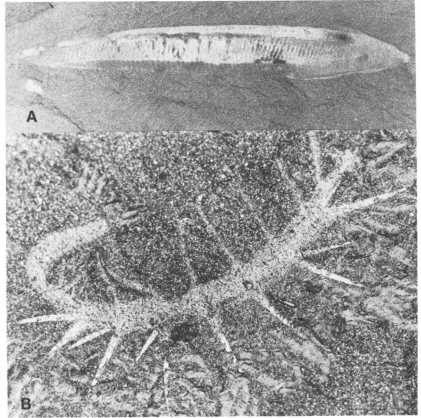

dants might have looked like. Trilobites, the dominant shelly Metazoa of the time, are represented by but 14 of the 120 Burgess Shale species known, yet few would deny that the Burgess Shale fauna samples the main flood of early metazoan evolution. It has even yielded 60 specimens of a species assigned to the Chordata (Fig. 13.3*A*), the phylum to which we belong.

So blind alleys, failed experiments, and unique fossils are normal vagaries of evolution—although on a larger scale during the early Phanerozoic burst of metazoan phyla. Even today some species are barely hanging on, assisted toward their ultimate fate by no less than *Homo sapiens* itself.

A remarkable thing about the Ediacarian fauna, considering its fragility, is its essentially global distribution. Representatives have been found at nearly thirty widely separated localities in different types of sedimentary rocks worldwide. Might such a distribution and the disappearance of several conspicuous elements almost concurrently with the onset of Cambria's armored hordes share the same explanation? Could it be that such wide dispersal, preceding an apparently abrupt extinction, signals the appearance then of something other than bacteria to take over the easy and essential job of scavenging? Did the Cambrian begin with garbage collectors?

Some of the Ediacarian soft-bodied

organisms left elongate imprints up to a meter long and other circular to elliptical ones up to half a meter in diameter (related to forms shown in Fig. 13.1*A* and *G*). Others were so flimsy that preserved specimens may be wrinkled and folded like discarded paper napkins (Fig. 13.1*E*). Still others were enigmatic in different ways (Figs. 13.1*B*, *C–D*, *G*). And some strange, elaborately branching forms from southeast Newfoundland resemble plants as much as animals, at least superficially.

Were these distinctive Ediacarian fossils related in any way to modern phyla? Or do they represent wholly anomalous creatures that left the scene without descendants? In the absence of relevant molecular evidence, response to that question rests, like so many matters of descent, wholly on morphology.

Indeed some of the Ediacarian imprints are quite similar to members of successor phyla (Fig. 13.2*A–B*). One of the more exotic forms (the genus *Dickinsonia* of Figs. 13.1*E–F*) survived into

13.4 BURGESS SHALE FAUNA. The mainly soft-bodied Middle Cambrian animals of the Burgess Shale lived on and in anaerobic black muds at the foot of a biogenic reef in now-southeastern British Columbia. Besides the oldest known chordate (*28*) and five uncertain phyla (*6, 17–19, 24*), the twenty-two other numerals represent eight living phyla. In order of abundance they are; priapulid "worms" (*4, 7–9*), polychaete annelids (*1, 2, 8*), arthropods (*3, 13–15, 20, 26*), sponges (*5, 12, 22, 25,* and perhaps *27*), a brachiopod (*11*), an echinoderm (*16*), a cnidarian colenterate (*21*), and a primitive slipper-like mollusk (*23*). Dark masses at base of cliff represent fallen rock. [*From Simon Conway Morris and H. B. Whittington, "The Animals of the Burgess Shale," Scientific American, July 1979, pp. 126–27. Copyright © 1979 by Scientific American, Inc. All rights reserved.*]

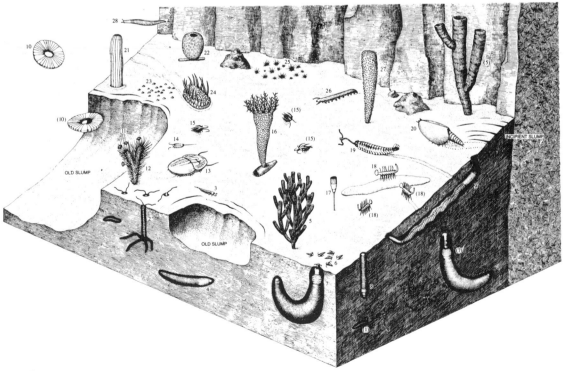

Cambrian time. Some critical observers interpret it as a marine annelid. It has also been considered plausibly ancestral to a shelly Early Cambrian mollusk called *Cambridium*, and some think it may even represent a jellyfish. Sinuous, wormlike tracks (Fig. 13.5*B*) resemble biological traces that imply a continuity of similar modes of life on up the geological column. And tiny, calcareous, tubular and conical shells that are known at least throughout the upper half of Ediacarian history have convincing, if insignificant, Cambrian descendants.

How could it be true, then, as was claimed in 1984 by a perceptive and respected student of trace fossils (Adolph Seilacher), that an unbridgeable biological gulf separates Ediacarian from Cambrian and younger Phanerozoic animal life? Most likely it isn't, despite much natural pruning of the tree of life. But all ideas now accepted were once minority views. The apparent is not always correct. Challenge, even when whimsical, is always good for science. It prompts reconsideration of the evidence, of its interpretation, and of its presentation.

What earlier students of Ediacarian strata had interpreted as jellyfish-like (medusoid) imprints, Seilacher interpreted to be the feeding traces of burrow-dwelling organisms. Yet comparison of an Ediacarian medusoid (*Cyclomedusa*) with a living hydroid medusa (Fig. 13.3*A–B*) shows striking similarities. A rare lobate species (*Mawsonites spriggi*), presented by Seilacher as "clearly such a burrow system," features a prominent concentric furrow and marginal lappets that strongly imply affinities to younger fossil and living scyphomedusan jellyfish of the order Coronatida. Still other Ediacarian imprints resemble modern siphonophores, familiar to many from the windrows of little ovoidal disks that pile up before the wind on today's seashores. And Adelaide University's R. F. J. Jenkins, after years of study, agrees that frondose creatures up to a meter long (e.g., Fig. 13.1*A*) are most likely some kind of colonial octocoral similar to living sea pens.

Suggestions of more advanced forms of life are also seen. The axial structures shown in Figures 13.1*B–C* and *E–F* hint at the presence of a gut (and a coelomate grade of life). Concentric wrinkles in the essentially two-dimensional *Dickinsonia* (Fig. 13.1*F*) betoken contractile body muscles. And even the body plan of Figure 13.1*D*, unassigned to phylum, is reminiscent of some peculiar later Paleozoic echinoderms, without their loosely articulated calcareous dermal skeleton.

It would be more than whimsical to argue that all of these morphological resemblances are purely superficial. More likely is that the things that so strongly resemble floating medusae, sea pens, and even conularids are, in fact, ancestral to younger fossil or living coelenterates, and that others are at least arguably related to annelid worms and arthropods. Probable Ediacarian ancestors exist for at least some of the Cambrian shelly Metazoa and medusae. Metazoan life did not have to start all over again at the beginning of the Cambrian. Indeed there were failed experiments. But enough succeeded to assure the continuity of metazoan evolution from Ediacarian into Cambrian history.

Plausible protozoan ancestors in latest Umberatanan deposits (Fig. 12.11*A–C*), U-shaped basal Ediacarian microburrows (Fig. 13.2*C*), and *perhaps* more recently discovered frondose structures of disputed origin from basal Ediacarian strata in South Australia are new signposts to the past. They are consistent with the expectation that obscure forms of metazoan life on this planet appeared with dawning Ediacarian or fading Umberatanan history and underwent a burst of phyletic diversification during the following 150 MY or so.

13.5 SOME DISTINCTIVE FEATURES OF LATE PROTEROZOIC (*A*), EDIACARIAN (*A–D*), AND CAMBRIAN (*E–F*) ROCKS. View *A*, Proterozoic–Phanerozoic contact; dolomite above midphoto caps an underlying glacial sequence; Flinders Ranges, South Australia. *B*, sandstone in Ediacara Hills records tracks of soft-bodied organisms as well as a medusoid imprint (*arrows*). C, short-spined microbial fossil from far-eastern Soviet Union. *D*, tiny coiled fossil alga and its cross-section from Saudi Arabia. *E* and *F*, vertical burrows of biologic origin in basal Cambrian quartzite of central Norway and Upper Cambrian sandstone of Minnesota; arrow in *E* marks base of U-shaped burrow. Annelids and arthropods both form comparable burrows in modern sediments.

The wide extent and relative continuity of unequivocally metazoan creatures during Ediacarian and later history, and their absence, rarity, or obscurity in older rocks, has other interesting consequences. It places the Ediacarian firmly in the Phanerozoic on any logical grounds and implies geologically rapid emergence of the metazoan level of organization and its principal phyla. The Ediacarian and Cambrian fossils together sample an unprecedented and unequaled "burst" of high-level evolutionary experimentation on a planet where, in effect, all potential metazoan habitats were then unoccupied. Indeed, all of the formal phyla of shell-bearing animals are first known from the earliest 150 MY or so of Phanerozoic history (as well as more than a dozen major kinds of animals that do not fit any now-living phyla). Diversification from structural, physiological, and biochemical themes then extant produced the current living world. Even chordates were already present in the Burgess Shale (Fig. 13.3*A*). Spineless though they then were, they were ancestral to all the familiar backboned animals of later times.

Integrating the Account: Fossils, Strata, Minerals, Oxygen

The hoary belief that a stream of metazoan diversity will be found to extend far back into Proterozoic history finds faint comfort in the evidence. Of the more than 200 examples of so-far reported Cryptozoic metazoans, all have been found to be either nonbiologic, nonmetazoan, misdated, of uncertain age, or postdepositional intrusions. At best they are *dubiofossils* or dubiometazoans—similar to known structures of nonbiological origin or of uncertain metazoan affinity.

Figure 13.6 shows a few of the many known examples of deceptive nonbiogenic markings (*A–F*), as well as others that illustrate problems of interpretation even of known biogenic imprints (*H*) and tracks (*G*). On the other hand, some competent paleontologists familiar with late Proterozoic strata still expect to find a yet-unknown extension of metazoan roots 200 to 300 MY or more into pre-Ediacarian history.

As research continues, that gap may narrow. The record of life is so incomplete that paleontologists can never know precisely the very first or last of anything. Recall the coelacanth fishes, thought to have gone extinct with the dinosaurs until fished up off eastern Africa in 1938. Think of the monoplacophoran mollusks, known as fossils only from Cambrian to Devonian, then dredged alive from deep-sea trenches in 1952. Their paired serial muscles and other organs bear witness to an early common ancestry and rapid adaptive radiation of annelids, arthropods, and mollusks. We may be glimpsing the roots of this radiation in Figure 13.1*B–C, E–F*. All that can now be said with reasonable confidence about a pre-Ediacarian metazoan record is that, if found, it is unlikely to be either of great duration or conspicuous. Such a discovery is not expected to shift the present base of the Phanerozoic far below the Ediacarian in any operational sense.

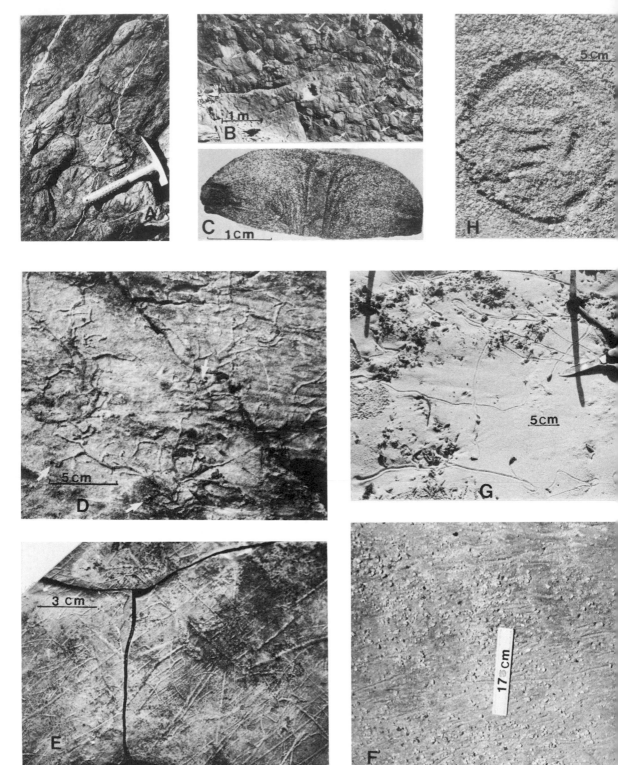

Here it is of more interest to follow the stronger spoor of ascending evolution.

Traces of that spoor, ascending from the basal Ediacarian dolomitic cap rock of the austerely beautiful Flinders Ranges of South Australia are illustrated in Figure 13.5*A–F*. Upward from what is so far a single occurrence in that thin but persistent cap rock (*A*) it reappears as meandering crawl tracks and imprints on fine-grained sandstones at Ediacara itself (*B*) and in equivalent strata worldwide. One picks up the trail under the microscope in the cherty dolomites of then equatorial Siberia and Saudi Arabia (*C–D*). And then, at the very base of the Cambrian in Norway and elsewhere, are the oldest vertically burrowed sandstones (Fig. 13.5*E–F*).

It is important also, in reconstructing the past, not to omit the stories in stone itself. Analysis of the rocks containing them swells the tide of historical testimony. Information from sequences like that of the northern Yukon, shown in Figure 13.7, joins the global paleonto-

13.7 Ediacarian–Cambrian transition in the Wernecke Mountains, east-central Yukon. Scene *A*, view west from headwaters of Bonnet Plume River, down section into strata of the upper Windermere Supergroup, terminal Proterozoic. *B*, reversing directions, the view is eastward, up section in the same sequence across the sparingly fossiliferous, Ediacarian–Cambrian transition to cliff-forming dolomites of Cambrian and Ordovician age. Arrows denote approximate Proterozoic–Phanerozoic transition (*lower right*) and Ediacarian–Cambrian contact (*center*).

13.6 Traps in the paleontological fairway. Views *A* and *B* show dewatering mounds with radial runoff streaks on surface of basal Proterozoic graywacke, south of Darwin, Australia. Diameter of mounds 8–45 cm. *C*, vertical section through similar, small dewatering mound from Upper Carboniferous graywacke in Cornwall, England. *D*, dewatering centers (*white arrows*) and distributaries on surface of 1.85-aeon-old Chelmsford Graywacke near Sudbury, Ontario. *E*, lineations on a lower Hamersleyan siltstone from Western Australia record a windy day on a 2.7-aeon-old mud flat; compare with *F*, made by wind-blown chips of sediment on present north shore of Great Salt Lake. *G*, tracks of small snails on mangrove flats, Bahamas; similar tracks of unknown origin are often interpreted as "worm tracks." *H*, imprint of a modern jellyfish on a sandy West Australian beach. [*C, after R. V. Burne, 1970, Sedimentology, v. 15, p. 217, Fig. 3. © Elsevier Science Publishers B.V.*]

logical record to provide grounds for climatic and historical generalizations such as those graphed in Figure 13.8. Such a graph, with all its approximations and uncertainties, results from the many linkages between plate tectonics, climate, mountain building, rises and falls of sea level, evolution, and mineral resources.

The capacity of oceans to hold water, for instance, increases, as we have seen, with early rifting and spreading, resulting in lowered sea level. It then decreases, and sea level rises again, as spreading ridges grow, warm up, absorb water, and expand. It may increase or decrease with convergence and collision of plates and mountain building. Sea level at any given time then is a product of the combined effects of the growth and decay of glaciers and of plate tectonics. They may either reinforce or oppose one another.

Sea-floor spreading during later Ediacarian history (Fig. 13.9A), by increasing ocean capacity, broadly lowered sea level which then rose again to flood the continental margins upon warming and melting of continental ice sheets and orogeny, reducing oceanic volume. High Late Ordovician epicontinental seas receded to low levels with terminal Ordovician to perhaps earliest Silurian glaciation (Fig. 13.8, *tiers 2–3*). The response of sea level to Carboniferous and Permian glaciation and orogeny, on the other hand, was ambiguous.

13.8 STORIES IN STONE AND FOSSILS. Evolving Phanerozoic marine habitats and biotal responses. *[Inspired by a sketch from A. J. Boucot, 1983,* Journal of Paleontology, *v. 57 (1), p. 8, Fig. 3.]*

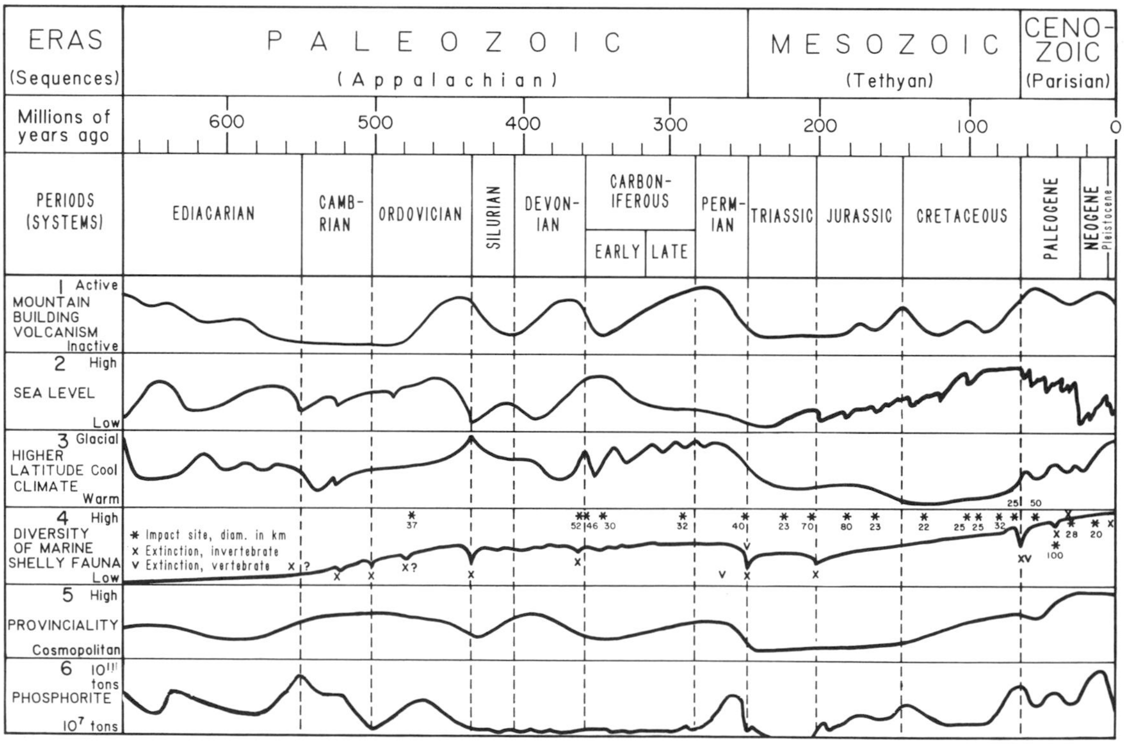

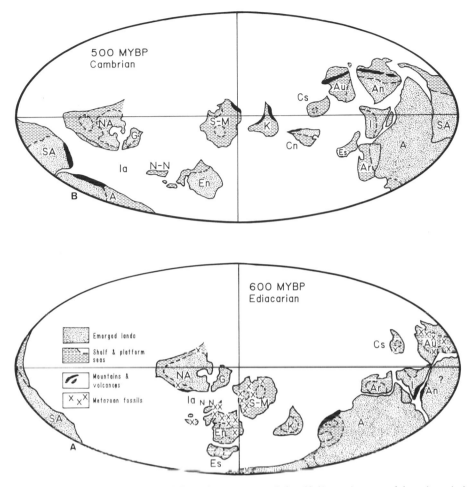

13.9 OPENING OF THE IAPETUS SEA. Map *A*, cartoon of the Ediacarian world as it *might* have looked from space about 600 MYBP. *B*, same for Late Cambrian–Early Ordovician paleogeography 100 million years later. Abbreviations obvious except for *Ia*, Iapetus sea; *N + N*, Nova Scotia + Newfoundland; *S-M*, Siberia-Mongolia; *En*, north Europe; *Es*, south Europe; *Cn*, North China block; *Cs*, South China block; *Ar*, Arabia; and *K*, Kazakstania, the southern wedge of present-day Soviet Union, between the Caspian Sea and Mongolia. *[Adapted from C. R. Scotese, 1984, American Geophysical Union, Geodynamics Series, v. 12, p. 2, Figs. 1–2. Copyright by the American Geophysical Union.]*

The tales rocks tell speak most clearly to those who attempt to integrate the whole decipherable physical and geochemical history with that of any associated fossils. In nature there are no completely independent variables.

Comparison of Figures 13.8, 13.10, and 13.11 suggest some of the complexities. The volumetric increase of carbonate rocks toward the present (Fig. 13.10) represents decreasing CO_2 and, indirectly, increasing O_2 (Fig. 13.11). Phos-

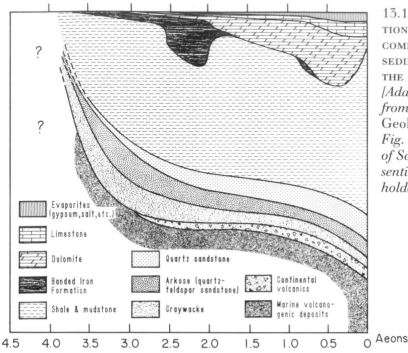

Legend:
- Evaporites (gypsum, salt, etc.)
- Limestone
- Dolomite
- Banded Iron Formation
- Shale & mudstone
- Quartz sandstone
- Arkose (quartz-feldspar sandstone)
- Graywacke
- Continental volcanics
- Marine volcanogenic deposits

4.5 4.0 3.5 3.0 2.5 2.0 1.5 1.0 0.5 0 Aeons

13.10 ESTIMATED VARIATION IN THE PERCENTAGE COMPOSITION OF EARTH'S SEDIMENTARY MASS OVER THE COURSE OF HISTORY. *[Adapted with modification from A. B. Ronov, 1964, Geokhimiya, no. 8, p. 716, Fig. 1, with the permission of Scripta Technica, representing the Soviet copyright holders.]*

phate, highly soluble in waters rich in CO_2 precipitates only as cooling and reduced pressure release dissolved CO_2. A similar response is characteristic of calcium carbonate, which is why it and phosphorites are so commonly associated. Phosphate deposits are rare in or absent from Archean and older Proterozoic sedimentary rocks (one of the reasons iron ores of that age are prized). Only a few deposits are known from rocks as old as 1.8 aeons in northwestern Australia and elsewhere, and in younger (middle Riphean) phosphatized stromatolitic rocks in India. No major phosphorites are older than about 800 to 700 MYBP, at which time the percentage of dolomite and limestone in the geologic column also increased conspicuously.

The uniquely marine and generally equatorial phosphorites are a primarily Phanerozoic phenomenon. They accompany reduction of CO_2 in solution, fluctuating with it as deep, cool waters well upward onto shallow shelves. There they warm up, experience reduced pressures, release CO_2, and thereby encourage not only the deposition of phosphatic and carbonate sediments but also algal growth (Fig. 13.8, *tier 6*). It seems that glacial melting, calving, and mixing cooled the later Proterozoic and earlier Phanerozoic seas enough to generate a vigorous vertical circulation. Dissolved phosphate, entrained in bottom currents, precipitated on warm, leeward, continental shelves when these deep, cool waters welled up to replace surface waters driven offshore by prevailing winds.

These latest Proterozoic to Ediacarian phosphorites are the oldest really large deposits of their kind. A similar coincidence between continental refrigeration and phosphorites is observed in the later

Permian (Fig. 13.8, *tier 6*) and Cenozoic. The Jurassic to Cretaceous phosphorites more likely reflect the reestablishment of vigorous oceanic circulation with increasing continental separation as Pangaea broke up during the last great episode of plate fragmentation and drift.

The link between phosphorites and lowered levels of CO_2 reinforces the biochemical evidence for the rise of O_2 to critical levels as unicellular and pluricellular modes of existence gave way to functioning multicellularity at the tissue level of complexity (right front endpaper). It also suggests insights to the origin of Metazoa as Ediacarian history commenced. The late Proterozoic existence of recognizable Protozoa, a critical level of O_2, warming shelf seas, and a plentiful supply of phosphate to nourish plant growth on which to feed would all have been conducive to metazoan origins then. The times were right. Evolution has a long tradition of opportunism. Would it fail to respond to such an opportunity?

Both phosphorite and carbonate rocks were deposited episodically, mainly in warm shelf or platform seas, throughout subsequent history. In a rough way they vary together, although phosphates are much less frequent and more erratic in abundance than carbonates. Both vary with CO_2 and therefore with climate and volcanism. One should expect significant variations in detail of the generalized curves of Figures 13.10 and 13.11 as data improve. Volcanism, for example,

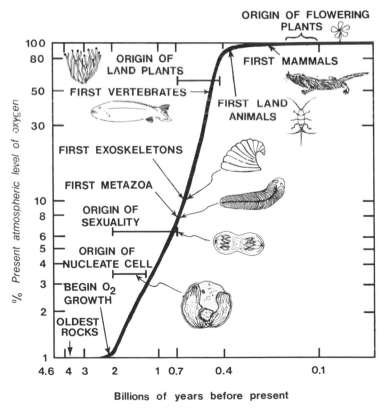

13.11 BUILDUP OF EARTH'S FREE OXYGEN. Approximation based on geochemistry and biologic O_2 requirements. Sketches not to scale. Not shown are departures from a mean level near the present over the past 400 MY. *[Modified from Preston Cloud, 1978, Cosmos, Earth, and Man, Yale University Press.]*

adds CO_2 to the atmosphere. Some of it is absorbed by Earth's waters, while the remainder contributes to climatic warming via the well-known greenhouse effect. More CO_2 in solution combines with the O_2 and H_2O to make carbonate ion, leading to precipitation of carbonate rocks, whereas the loss of dissolved CO_2 from upwelling waters facilitates the deposition of phosphorites.

In turn, local depletions of O_2 (anoxic episodes) have resulted in local extinction or emigration of metazoan and other O_2-dependent species throughout Phanerozoic history. Recall that Metazoa are not simply masses of cells. The metazoan state involves the organization of cells as tissues and organs to perform special functions. Oxygen must be brought to these groups of cells in some way to supply metabolic energy. It is also needed to make the connecting tissue called collagen that binds the cells together and later builds muscles and helps attach soft bodies to shielding external coverings.

The most direct way for an organism to get its O_2 is by epithelial diffusion—a kind of skin breathing. That supplements the O_2 supply of some animals with lungs and gills even today. Dependence on it, as earlier observed, may be why the conspicuous animals of the Ediacarian seas were either thin sheets, complex surfaces, or floaters in which inert interiors were covered with thin metabolic veneers. That may also be why external shelly coverings or impervious integuments did not appear until later Ediacarian and Cambrian time. Such coverings are first seen in much smaller, tubular to conical animals and only later in larger, more complex species.

Studies by Yale's Donald Rhoads of the Black Sea region, where bottom waters are anaerobic, suggest that now-living Metazoa first appear at sites where O_2 is about 6 to 7 percent of present surface levels, while shelly forms appear at about 10 percent. Here is a clue to early metazoan requirements. Of course our protozoan ancestors also needed O_2, but not as much as is needed to supply collagen to Metazoa. Combined with previous evidence, such data are used as approximate points on the curve of O_2 buildup suggested in Figure 13.11.

Following the major dumping of dissolved ferrous iron in the insoluble ferric state from older Proterozoic seas, decreasing abruptly about 2 aeons ago (left front endpaper), free O_2 began to accumulate in hydrosphere and atmosphere more rapidly than it could be neutralized by reduced substances. There were still brakes on that acceleration, however. O_2 can accumulate no faster than carbon is buried or otherwise sequestered (or than hydrogen escapes the grip of Earth's gravity). The record of microbial fossils supports the conclusion that levels sufficient for the operation of the nucleate cell and sexuality were first reached sometime between about 2 and 1.4 aeons BP. Roughly 670 MYBP saw the oldest yet-known metazoan fossils, showing that collagen was then in demand, and implying epithelial diffusion of O_2 at perhaps 6 to 7 percent of its present atmospheric level of 20 percent. The idea of so late a buildup of O_2 is now being strongly reinforced by geochemical studies and the evidence of late Proterozoic varves that reflect strong solar activity, consistent with the expected lower and thinner ozone layer. Somewhat before the end of Ediacarian history, perhaps by 600 MYBP, the presence of tiny conical and shelly organisms implies that gills or other organs for the collection of O_2 were functioning. And that suggests that O_2 had probably increased to around 10 percent of present levels (that is about 2 percent of the contemporary atmosphere).

It deserves mention here also that most mineral deposits convey interesting messages about the environments and history of our planet, including O_2 levels. They represent special settings where ordinarily dispersed things have been brought together in unusual and accessible concentrations for reasons that are intellectually challenging and economically profitable to understand. They tend to be concentrated in mineralographic provinces during specific intervals of history. Understanding the times, places, and circumstances of these concentrations improves the prospect of further discovery. Examples are iron at the margins of anoxic older Proterozoic shields, phosphate on marine shelves that once occupied low latitudes in zones of upwelling, coal in low-latitude Carboniferous sequences, and salt in Siluro-Devonian and Permian landlocked basins.

The organic precursors of oil are believed to have accumulated mainly from microbial sources in marine basins throughout at least post-Archean time, and probably from the first appearance of life onward. In older sedimentary rocks, however, because metamorphic processes have usually converted oil to natural gas or solid compounds, little has yet been recovered from deposits of pre-Phanerozoic age. It is found in rocks of all Phanerozoic systems, but especially of particular ages and places, in porous and permeable sedimentary rocks above suitable source beds, beneath impervious cap rocks, and in sufficiently well sealed sedimentary or structural traps. Paleozoic oil in North America is produced especially from the Ordovician and Permian of the southern mid-continent, the Devonian of the central mid-continent, and the Ordovician of the northern Great Plains. Most or all of this oil started out as once-flourishing microbial life at the bottom of the nutrient pyramid. It is found in sediments of seas and basins whose geologic history resulted in oil being trapped and sealed from leakage, but under mild-enough temperatures to prevent conversion to natural gas.

The Coupled Evolution of Paleozoic Geography and Ecosystems

Indeed, one sees many linkages between physical and biological elements of Earth history during the Paleozoic. These are illustrated in conventional accounts of historical geology by numerous vertical profiles or "cross-sections" of stratigraphic sequences, reconstructions of their paleogeography, and correlation charts.

Suffice it to say here that the numerous stratigraphic cross-sections portrayed in such accounts represent six main kinds of depositional settings. It is also important, in observing such cross-sections (e.g., Figs. 7.10, 10.4, 10.14, 11.15), to keep in mind that the deposits and structures represented are the end products of dynamically evolving features whose characteristics are best shown as grossly foreshortened caricatures at the scale of these pages. They display important geological relations but necessarily at great vertical exaggeration relative to the horizontal. Although in such portrayals they may look like deep primary basins, most of these basinlike features are, in fact, largely the product of continued subsidence of Earth's surface beneath an accumulating pile of sediments rather than the filling of once-open depressions.

The six main kinds, oriented with respect to continents, are:

1. *Simple interior basins:* roughly symmetrical, subsiding regions of continental interiors, with or without marine connections. The sediments that record their history characteristically thin and coarsen toward their margins and thicken and become finer grained basinward—unless the basin is so wide or sediment so limited as to result in basinward as well as marginal thinning. The sedimentary fillings of simple basins may be marine, lacustrine, evaporitic, fluvial, or some mixture of these.

2. *Fault basins:* occurring as pull-apart structures on rifting continental margins like those of the older Mesozoic in eastern North America (see Ch. 14), or in continental interiors such as the earlier-discussed 1.13-aeon-old mid-continent rift zone. Like some simple basins, their sedimentary fills may be roughly symmetrical from side to side, but they typically display thick marginal conglomerates. Association with volcanic rocks is common.

3. *Epicontinental platforms and shelves:* such as Hudson Bay and the continental shelves.

4. *Marine trenches and basins:* mainly bordering active continental margins and deriving their asymmetrical sedimentary successions principally from continental sources. They occur where subducting ocean floor descends, or in structural traps such as the offshore basins of southern California.

5–6. *Geoclines:* being composite, asymmetrical structures, recorded by continental slope and shelf debris landward—the *miogeocline*—and by pelagic sedimentary rocks, density flow deposits, volcanogenic graywackes, and basaltic volcanics seaward—the *eugeocline*. In an adjectival sense these words refer to deposits or structures of the nature described—for example, eugeoclinal deposits, miogeoclinal settings. With that, plus variations in thickness, the reader can visualize a typical cross-section of any specified sedimentary trap.

Of all these categories, paired miogeoclines and eugeoclines are the most important for global tectonics. Paired geoclines record convergence of plates. Together with linear, fault-bounded basins that indent the old continental margin at or near right angles (*aulacogens*) they identify triple junctions and suggest the former sites of hot spots (see Ch. 9).

Miogeoclinal and eugeoclinal sediments, moreover, are easily identified and separated from one another in Phanerozoic records, not only by their distinctive sedimentary characteristics, but by their distinctive kinds of fossils. The characteristic fossils of eugeoclinal sediments include both planktonic microorganisms and the strongly-patterned searching, feeding, and resting tracks of bottom-dwelling Metazoa of the open

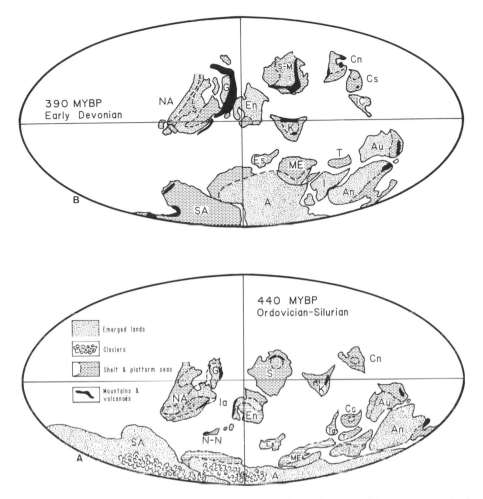

13.12 THE CLOSING OF IAPETUS. Map *A*, one version of how the Late Ordovician–Early Silurian world *might* have looked. *B*, same for the late Early Devonian. Symbolism and abbreviations as in Fig. 13.10, plus *ME* for Middle East, *T* for Tibet, *C* for China, and *IC* for Indochina. Extent of glaciation shown is minimal. [*A, adapted from C. R. Scotese, 1984, American Geophysical Union, Geodynamics Series, v. 12, p. 3, Fig. 4. Copyright by the American Geophysical Union. B, adapted from R. K. Bambach, C. R. Scotese, and A. M. Ziegler, 1980,* American Scientist, *v. 68(1), p. 33, Fig. 10.*]

sea. The intercalation of coarse sands containing shelly fossils in pelagic muds or oozes indicates downslope mass transportation of such sands from shelf or slope by seaward-flowing density currents. Similarly interbedded volcanogenic graywackes with few or no visible fossils record similar density currents from adjacent and commonly seaward volcanic arcs.

Figures 13.9 and 13.12, taken *cum grano salis*, reflect the now well established transience of early Paleozoic lands and seas and suggest their influences on physical and biological states and processes. Be advised, however, that such maps, of such ancient settings, are highly provisional and individualistic reconstructions. They are based on limited paleomagnetic and paleontologic data and are subject to revision. The point they are intended to emphasize is that the breakup of megacontinents during and following latest Proterozoic glaciation highlighted plate motions and created new and changing configurations of land and sea—perhaps like those suggested. The early Paleozoic ocean called Iapetus was of special significance for the geologic history of the present Northern Hemisphere. Gone now, it occupied the space between Europe and North America, where the Atlantic Ocean is today.

The extensive flooding and regional regularity of facies patterns during Cambrian history imposed a broad similarity on Cambrian biogeography. Seaward from central or marginal continental uplands were sandy deposits that graded offshore to muds (now mudstones). Still farther offshore were shallow carbonate banks and reefs, both with lots of conventional trilobites and high levels of provinciality. The contrasting faunas in mudstones and siltstones of the same age along seaward shelf margins and slopes, however, were highly cosmopolitan, being characterized by small, probably planktonic, trilobite-like arthropods called agnostids, as well as trilobites.

Continental flooding continued through Ordovician history, with declining provincialism, until latest Ordovician to possibly earliest Silurian glaciation. Sea level then fell again (with growth of continental glaciers) and provinciality began to rise to an older Devonian peak

before giving way to marked later Devonian and early Carboniferous cosmopolitanism (Fig. 13.8), all reflecting plate motions. Planktonic organisms called graptolites (similar to some colonial chordates) had short-lived global distributions throughout Ordovician and Silurian history, providing precise correlation and a high degree of historical resolution for strata of those systems.

What are the determining factors? Sea level, as we have seen, varies with the capacity of the global sea and how much of the hydrosphere it can accommodate. Recall that the capacity of the oceans, in contrast to the volume of water in them, varies as spreading ridges expand and contract. A newly initiated spreading ridge heats up, absorbs water, and expands, reducing oceanic capacity so that seawater spills across the continental margins. As new sea floor ages, cools and contracts, some or all of that water returns to the ocean basins. Similarly, we have seen that sea level falls as water becomes locked up in glacial ice, rising again when the ice melts. The global changes of sea level resulting from such processes are called *eustatic*,—standing at the same level worldwide. They are why it is often possible to correlate the same event by biologic or seismic criteria from one part of the world to another even more precisely than by means of radiometric dating. Local (noneustatic) changes of sea level result from faulting or from changes in crustal load (e.g., Fig. 12.4A) related to plate tectonism, glaciation, mountain building, impounding of waters in large lakes and reservoirs, and volcanism.

The likelihood of glaciation increases, of course, as continents drift across the poles. Volcanism and mountain building are also products of plate motion. All of these processes affect world climate as a result of temperature and sea-level changes, the addition of dust and gas to the atmosphere, and the effects of differences in elevation on atmospheric circulation.

Such processes shaped the older Phanerozoic world and its life, as they do our own. They contribute to many of the more sweeping biologic changes observed throughout history. Supplementing such effects are the weathering and erosional processes that transfer surface materials to the epicontinental seas and basins and to the miogeoclinal seas whose sedimentary deposits record so much of Earth history. And, finally, we may no longer overlook the record of asteroidal, cometary, and meteoritic impact, the potential effects of which have at last begun to be taken seriously into account (starred on Fig. 13.8 and on front endpaper).

Phanerozoic history then, is a kaleidoscope of changing patterns of land and sea, of rising and eroding highlands, of variable climate, of changing atmospheric and oceanic circulation, and of the interactions of all these things with biologic evolution. Plate tectonism is the primary forcing factor, climatic and biologic evolution are the products, rocks are the historical record, and asteroidal and cometary impact are the wild cards.

The Progressive Changes through Middle Paleozoic History

Cambrian followed Ediacarian, beginning some 550 MYBP—an estimated age that, like all important transitional age estimates, fluctuates with geochronologic method, nature and position in the stratigraphic sequence of the sample or samples dated, and other variables. The initial Cambrian sediments have long been recognized as characteristically time transgressive sandstones and quartzites, deposited from an onlapping sea that spread with time far into continental interiors. But, if glacial meltwater accounts for the initial transgression of the Ediacarian sea, what might explain the later retreat and then renewed advance of this same seawater across the Cambrian continents (Fig. 13.8, *tier 2*)? Plate tectonism perhaps? Rifting that enlarged the capacity of Ediacarian oceans could have counterbalanced earlier flooding from deglaciation. Then as spreading ridges again expanded, ocean capacity would have decreased once more, forcing the progressive Cambrian marine *overlap* (or *onlap*).

Preceding that transition, a temporary lowering of sea level or local elevation of shorelines is needed to explain the observed local erosion of Ediacarian deposits, resulting in unconformable relations between terminal Ediacarian and initial Cambrian strata at many or most places. Sequences exist, however, in which an Ediacarian–Cambrian transition might logically be sought: for example, that of the thick (up to 10 kilometers or more), detrital, continental-slope deposits of the Windermere Supergroup, extending all the way south from northern Yukon Territory (Fig. 13.7) to Montana and Idaho in the northern United States.

The earliest Cambrian seas brought a host of mostly small calcareous and phosphatic shelly Metazoa. They included a variety of tiny conical shells (some already in the Ediacarian), bivalved brachiopods, the exclusively Lower Cambrian spongelike reef-building archaeocyathids, the occasional snail, rare clams, a variety of rare and unusual echinoderms, and, with time, trilobites (Fig. 13.13). Those long-ranging arthropods were overwhelmingly first in number among preserved fossils, if not in order of appearance, during Cambrian history. Before that was over, all now-known shelly phyla except the twiglike, lacy, and encrusting colonial forms called bryozoans had joined the scene. So had many naked forms, known only from trails and burrows or preserved with great fidelity by carbonaceous films in the Burgess Shale (Fig. 13.4).

The most fundamental diversification of metazoan history had expanded to occupy a broad array of previously empty niches. Everything that followed was variation on themes then introduced. Biochemical universals confirm that.

A probably staggered progression to ever more exuberant metabolic activity responded to a broadly continuing increase in atmospheric O_2 up to present levels. Fossils observed and the requirements of metazoan metabolism imply that the evolution of breathing organs and the probably related acquisition of shelly coverings occurred at perhaps 10 percent present atmospheric level of O_2 (2 percent O_2) in late Ediacarian or early Cambrian time. Continued increase thereafter then finally approached existing levels about 400 MYBP (Fig. 13.11), to judge from the abundance then of large, active fish in the sea, and of insects,

arachnids, and advanced (tracheophytic) plant life on land.

Indeed, life already seems more familiar with the Cambrian—probably because generations of college students have come to think of "real" fossils as shells and bones and of the Cambrian as the beginning of the fossil record. Even children's coloring books include trilobites, those once-ubiquitous arthropods that account for two-thirds of the known Cambrian genera and many or most of its species.

Despite the relative familiarity of its fossils and the recognizable continuation of most Cambrian shelly phyla to the present (right front endpaper), however, its known soft-bodied fauna (Fig. 13.4) contains at least as many kinds of odd creatures and extinct phyla as did the Ediacarian. And neither Ediacarian nor Cambrian faunas would look all that strange in some modern tropical seascapes, replete as the latter often are with brightly colored naked snails, elaborately tentaculate and poisonous Portuguese men-of-war, stretchable, snakelike echinoderms (holothurians), and soft-bodied colonial members of our own chordate phylum.

Omitting longshore sands and conglomerates, the observer sees that shales and extensive layered bodies of magnesium-rich calcareous sediments (now mostly dolomite) dominated the late Cambrian to early Ordovician platform seas around Iapetus. The Paleozoic separation of Europe from North America and the equatorial orientation of continents were greatest at that time (Fig. 13.9B). A degree of plate jostling and mountain building went on locally, nevertheless, during Middle and Late Cambrian (e.g., in southeastern Australia).

Thereafter movement seems to have reversed. Throughout most of middle Paleozoic history (Ordovician, Silurian, and Devonian) Iapetus shrank as the previously passive margins of Europe and North America converged. During the same intervals, comparable movements were deforming Upper Ordovician and Upper Silurian strata in southeast Australia, middle Asia, Japan, and southeast China. Volcanic-island arcs appeared at the margins of Iapetus, and evolving faunas claimed an array of new or previously unoccupied habitats. Repeated convergence between North America, Greenland, and northern Europe elevated mountains during younger Ordovician and Devonian (the Taconic, Caledonian, Acadian, and equivalent orogenies), extending north beyond the present Arctic Circle. The growth of these mountains was extensively accompanied by volcanism and plutonism, including the intrusion of potassic granites from which, on weathering, red continental sands spread both east and west to become the well-known Old Red Sandstone, upon filling rift valleys during later Devonian history.

Elongate miniplates of other lands adhered to eastern North America when the colliding continents separated again.

Known middle Paleozoic fossils are overwhelmingly shelly marine species (Fig. 13.13). Brachiopods of great variety were the dominant component for the rest of Paleozoic history—bivalved and hinged, like clams, but with valves (shells) opposed bottom to top instead of side to side. After the Cambrian, trilobites were on the wane but still common. Clams, snails, and cephalopods were increasing and diversifying. Small, black, chainlike, floating colonial creatures, the earlier-mentioned graptolites (*g* on Fig. 13.13)—early colonial chordates perhaps, and important time indicators for Ordovician and Silurian—were gone early in Devonian history. The graptolites are known mainly from black shales but are also found in other fine-grained sediments into which they settled from overhead. That floating habit, combined with short stratigraphic ranges, makes them superior time markers, called *index fossils* because they make it possible to subdivide Ordovician and Silurian history into dozens of global marine biozones. Three-dimensionally preserved forms are known from chert and chemical limestones from which they are easily recovered by acid solution. Anaerobic basins, denoted by extensive pyritic black shales, also became important for the preservation of planktonic and pelagic organisms, particularly the rare soft-bodied forms.

The greening of the lands may have begun with simple creeping bryophytes during Middle Ordovician if not sooner, to judge from Libyan spore tetrads described by Jane Gray and associates. Air-breathing arachnids (scorpions) and wingless insects, arthropods both, followed plant life to the continents by late Silurian time, preceded by freshwater fishes. The early herbaceous plants were followed by lignification, and the oldest woody plants came in with the transition from Silurian to Devonian. Subsequent adaptive radiation of the advanced vascular plants initiated a cover of large, perennial, woody plants along established water courses. And from these, in times of drought, exploratory lobe-finned and air-gulping fish emerged to become amphibians at the very end of Devonian history.

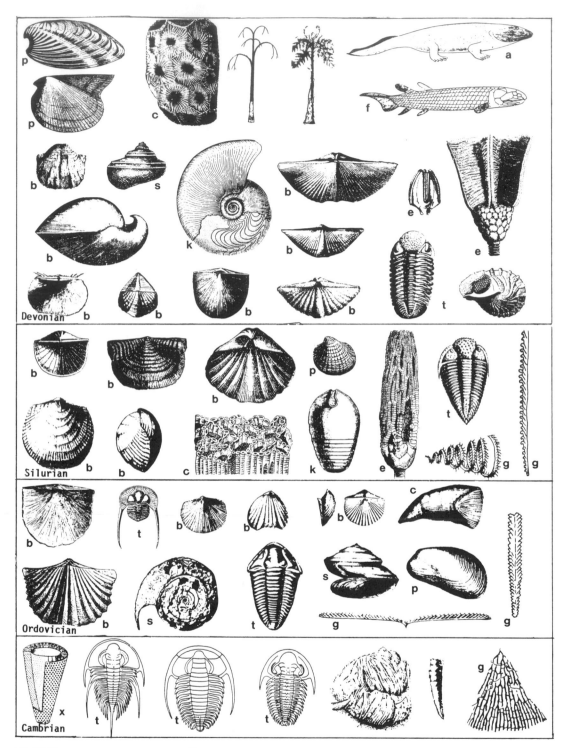

13.13 Representative older Paleozoic fossils, greatly reduced. Age increases from bottom to top. Key to fossils is *a*, amphibians; *b*, brachiopods; *c*, corals; *e*, echinoderms; *f*, fish; *g*, graptolites; *k*, cephalopods; *s*, snails; *t*, trilobites; *x*, archaeocyathids; *unmarked*, plants and miscellaneous.

Fishes were differentiating on opposite sides of the broad highlands that separated North America from Europe. They are found in the above-mentioned *Old Red Sandstone*, oxidized continental deposits famed for the fashionable brownstone mansions built from it at the turn of the century and from a popular nineteenth-century book of that name by the eminent Scot, Hugh Miller. This prolonged colonization of the lands and their fresh waters began as early as Middle Ordovician and extended to the Late Devonian. Evolutionary adaptations such as osmoregulation, air breathing, sensory adaptations, terrestrial reproduction, and (for animals) terrestrial locomotion, permitted plants, annelids, arthropods, and finally fish-become-amphibians to radiate into a growing range of new habitats.

Bony fishes found in the Old Red Sandstone of rift valleys in the Euro-American Caledonian collision zone include both *ray-finned* ancestors of the familiar bony fishes and the drought-adapted lobe-finned or *crossopterygian* fishes. It is an old and likely story that these lobe-finned fishes, in search of haven, became the first amphibians (the labyrinthodonts)—ancestral eventually to all higher vertebrates. The first clue to an amphibian presence in Devonian time was a single skull that was identified as either a very amphibianlike fish or a very fishlike amphibian, found in the uppermost Devonian strata of eastern Canada in the early 1940s. Since then, unequivocal amphibians of the most primitive order known, crocodile-shaped forms more than a meter long, have been found in highest Devonian and lowest Carboniferous (Mississippian) strata in Greenland and other regions.

Middle Paleozoic seas did repeatedly inundate and recede from the lands, just as all textbooks say and as happened earlier and later. That reflects varying combinations of expanding and subsiding spreading ridges, growth and decline of glaciers, and uplift and erosion of mountains. Accumulations of sediment and volcanic debris accumulated to thicknesses as great as 18 kilometers in deeply subsiding marginal basins, compared with a few hundred meters of sand, carbonate rock, and shales on flooded continental platforms.

The global ecosystem responded. Shelly marine faunas, provincial during Early and Middle Devonian history, became more cosmopolitan toward the end. New faunas and floras replaced the old. Kilometers of mostly reworked and some new sediment accumulated in basins and on subsiding and colliding shelves—thin and coarse shoreward, thickening and fining basinward or seaward. These sediments record changing current directions, paleoecology, and biotal composition. Extensive continental ice sheets covered a large part of the present Sahara and perhaps adjacent South America during latest Ordovician and perhaps earliest Silurian history. Iapetus narrowed, and its northern end closed, as North America and northern Europe collided during Devonian time, raising mountains and altering oceanic circulation. Plant and animal life expanded on land and continued to diversify everywhere. Amphibians radiated into available ecospaces during latest Devonian and early Carboniferous history. And the lands of the Southern Hemisphere completed their aggregation into the megacontinent of Gondwana—"the land of the Gonds."

Cyclic Sedimentation

Cyclicity is our earliest and most universal experience: sleeping and waking, feeding, day and night, the seasons, the tides, youth to old age. No wonder that Aristotle, pagan guru to the early Christian world, saw time and the universe itself as cyclic and eternal. The history of our planet, however, says clearly that even cyclic activities produce historically irreversible results. Over the long term, Earth and its components change systematically through a succession of distinctive modes, albeit the causes and processes of change repeat at varying intensities.

The frequency of cyclicity or rhythmicity in geology depends to some extent on the observer. Likely examples are the annually layered lake sediments associated with continental glaciation of all ages as well as variations in the extent and volume of Pleistocene ice that seem to correlate with regular orbital variations. Persuasive also is the wide prevalence in mid-Cretaceous deep-sea sediments of oxidation–reduction repetitions that have been estimated to reflect regular variations in orbital eccentricity and precession. But is such a rare event as asteroidal collision cyclic as well as catastrophic because it has happened, on a long average, once every hundred million years? Probably not. Solar physicists, on the other hand, with no better statistical data than are recorded in 700-MY-old strata (Fig. 12.7B) speak of sunspot cycles, the most impressive record of which is reflected in tree rings and varves. Geologic megathinker Alfred Fischer even visualizes all Phanerozoic history as comprising two great supercycles of alternating "icehouse" and "greenhouse" stages, with a third icehouse stage now well begun. Available evidence leaves room for differences of opinion.

A more complex (and therefore, to many, more convincing) illustration of geologic cyclicity is found in the Upper Carboniferous *cyclothems* that apparently record the repeated growth and decline of remote Gondwanan continental glaciations (Figs. 13.14, 13.15B). Ice flourished intermittently in different regions of Gondwana during much of Late Carboniferous and Permian history as they drifted across the South Pole. During such high-latitude glaciations, previously inundated near-equatorial lowlands emerged from the shrunken seas to become veneered with nonmarine deposits from eroding adjacent uplands. At times of extensive melting, expanding shallow seas inundated emerged lowlands to cap their nonmarine sedimentary veneers with marine carbonate rocks and shales, often richly fossiliferous.

Repeated over and over again, this rhythmic succession of nonmarine to marine sequences permits the correlation of equivalent sedimentary sequences, fingerprint-like, across large regions. The deposits of each sequence comprise a cyclothem. Upwards of a hundred of them, some including commercially significant coals, are found in the Illinois coal basin (Fig. 13.16A). This then-tropic region, like similar regions in Europe and elsewhere, records previously unimagined details of the growth and decline of polar ice.

13.14 SIGNS OF PERMIAN GLACIATION IN GONDWANA CONTINENTS. View *A*, varvite of the Parana basin, Brazil. *B*, Dwykka tillite near Durban, South Africa. *C*, groove in striated pavement with tillite west of Melbourne, Australia. *D*, glacial pavement at Halletts Cove, south of Adelaide, Australia. *[A–C, courtesy of John C. Crowell; Betty Crowell—"Faraway Places."]*

Southeastern North America offers the best known and perhaps most extensive cyclothems. They are found in the continental interior, rimmed by rising borderlands on the east and extending from the Great Lakes to Texas and from Pennsylvania to Nebraska. This vast inland platform, alternately flooded and drained a hundred or so times at the bidding of distant Gondwanan ice, is also a major coal producer. The coal and its distinctive root-bearing *underclays* (ancient soils or *paleosols*) are both the principal punctuating elements of the cyclothemic sequence and faithful recorders of biogeologic wonders.

At times of low sea level, streams meandered west and south across this mostly featureless surface. They spread terrestrial sands and muds across it before ponding locally, as today they do in the Okavango Swamps of northern Botswana or Virginia's Great Dismal Swamp. As glacial ice decreased, land gave way again to sea. The lacy greenery of the coal swamps (Fig. 13.17), buzzing with more than 500 species of insects, flourished and then drowned. It accumulated and aged to form peat and eventually coal. Its roots and stumps in the underclays and fallen trunks in the coals are mute recorders of truly primaeval forests

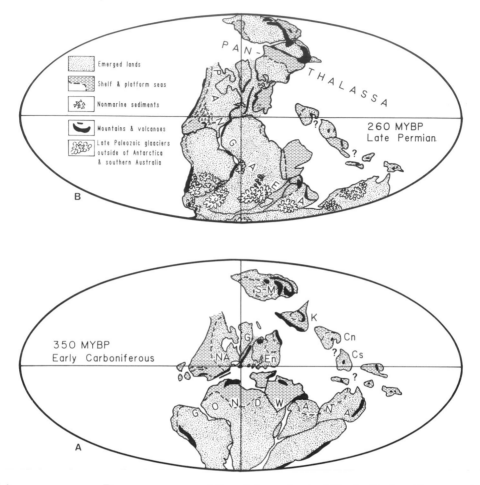

13.15 ASSEMBLING THE PANGAEAN LANDS. Map *A*, hypothetical Early Carboniferous paleo-geography; Gondwana is now assembled. *B*, 90 million years later, toward the end of the Permian Period and Paleozoic Era, Pangaea is complete. Symbolism as in Fig. 13.12. Indicated extent of glaciation is minimal. *[Adapted from C. R. Scotese, 1984, American Geophysical Union, Geodynamics Series, v. 12, p. 5, Figs. 8 and 10. Copyright by the American Geophysical Union.]*

where trees more than a meter in diameter attained heights of 35 meters or more—as big as a large Norway pine today.

As might be expected, many of the coal beds are thin and lenticular, testifying to local and short-lived ponding. But others are thick and extensive enough to be commercial coals. A few, Pennsylvania's great Pittsburgh coal for example, reach thicknesses up to 2 meters or more and extend for hundreds of square kilometers. In Illinois, where Pennsylvanian strata are only 150 to 600 meters thick, some two score coals are recognizable over large areas and a half-dozen are thick enough and pure enough to be commercially interesting.

Elsewhere a widely variant history is recorded. Nearer to detrital sources in Pennsylvania and southeastern Canada, sequences are thicker and contain more sandstone and little or no limestone. Upper Carboniferous rocks of Canada's Maritime Provinces are entirely non-marine and up to 4 kilometers thick at places. Upright stumps of trees found at twenty levels in an 800-meter-thick sequence in coastal New Brunswick reach heights of 6 meters or more. Toward the western and southern margins of the Kansan and Oklahoman lowlands marine strata prevail. Abundant marine limestone, for instance, is seen in the sequence of Figure 13.16C. Equivalent rocks are thick and almost entirely marine in northern and central Texas.

Cycles of one sort or another reflect distinctive circumstances in different environments. Cyclicity of a sort is manifested not only as varves and cyclothems, but by successions of upward-fining littoral sands and the episodic intercalation of coarse sediments into basinal shales by density currents, for example. Yet nowhere do we know a longer continuous record of cyclicity than that of the Upper Carboniferous cyclothems. Although coal deposits of Permian age are known from South America, Australia, northern South Africa, India, and Siberia, cyclothems and commercial coals of this age are so far reported only from South Africa. Coal is the key to conspicuous cyclothem development. Although a little coal is actually mined from Upper Devonian deposits in Svalbard (Spitsbergen Island) and more is known from the Lower Carboniferous, older coals are rare. Their abundant development had to await the radiation of woody plants.

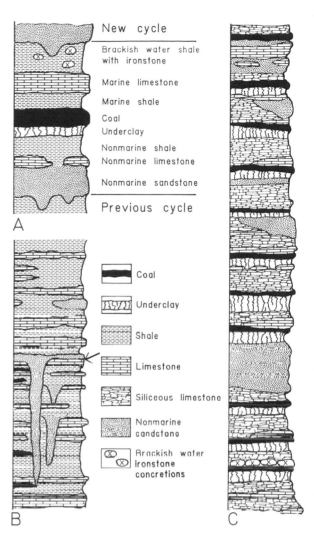

13.16 CYCLOTHEMS. Sketch *A*, idealized Illinois-type cyclothem, all members present. *B*, generalized eastern Kansas cyclothems at Carboniferous–Permian boundary (*arrow*). *C*, generalized middle Pennsylvanian cyclothems, southeastern Kansas; eleven cyclothems shown, each terminated by a coal bed. [*A, after J. M. Weller, L. G. Henbest, and C. O. Dunbar, 1942, Illinois Geological Survey, Bulletin 67, p. 10. B–C, U.S. Geological Survey, Professional Paper 853, part 1, Figs. 31 and 27.*]

13.17 Window on an Upper Carboniferous coal swamp. Diorama made to scale from actual fossils. [© *Field Museum of Natural History, Chicago.*]

The Making of Pangaea:
Converging Continents, Gondwanan Glaciation,
the Terrestrial Biosphere, a Tragic Finale

The central physical fact of later Paleozoic history was the continuing convergence of lands to form the ancient supercontinents of *Gondwana* in the Southern Hemisphere and *Laurasia* to the north, culminating in extensive mountain building. Eventually they collided to form the universal landmass *Pangaea* (Fig. 13.15). Iapetus vanished. The Appalachian (or Alleghenian) orogeny, creator of the so-named mountains, finds its extension in the Hercynian orogeny of Europe. Europe and northern Asia became fused along the Uralian orogen. Contemporaneous orogenesis elsewhere may include the initial movements of the Himalaya—even now still rising after widely separated collisions over the last 250 MY. Extensive glaciation left tracks on all Gondwanan conti-

nents as northern South America, southern Africa, and central Antarctica successively drifted across the South Pole.

Circulation in the universal ocean *Panthalassa*, blocked by the extensive north–south barrier of Pangaea, presumably centered on a pair of great hemispheric gyres—one northern, rotating clockwise, one southern, rotating counterclockwise under the influence of the Coriolis effect. A strong equatorial countercurrent and local variations around the minicontinents east from Pangaea would be parts of that system. As a result, marine faunal provincialism increased from beginning Carboniferous to latest Carboniferous and Permian, with a steady rise in diversity (Figs. 13.8, 13.18). The collapse of diversity during terminal Paleozoic extinction then made

Permian

Pennsylvanian

Mississippian

13.18 YOUNGER PALEOZOIC FOSSILS, ALL GREATLY REDUCED. Age increases upward. Key to fossils as in Fig. 13.13 plus *ä*, spider; *i*, insects; *r*, reptiles (plants unmarked).

way for the less varied but more cosmopolitan and longer-lasting shelly marine faunas of the older Mesozoic.

Later Paleozoic was a time of strong zonal climates. On land, at then low latitudes, were the flourishing coal swamps that give the Carboniferous its name (Fig. 13.17). In the mainly platform seas and basins, their marine equivalents peaked in diversity with recurrent Paleozoic reef-building episodes (Fig. 13.19). The larger setting of such a view, seen in Figure 13.20, represents the rim of a large, deep, land-locked basin in west Texas and southeastern New Mexico, the famous carbonate buildup known as the El Cap-

itan reef in the arid southern Guadalupe Mountains (Fig. 13.21).

At the other end of the marine ecologic spectrum, in the contemporaneous polar seas, there flourished cool to glacial-marine Gondwanan faunas in the Southern Hemisphere and only slightly less cold boreal faunas in the north. A distinctive high-southern-latitude Permian flora is characterized by the tongue-shaped leaves of the tree *Glossopteris*. At the upper right of Figure 13.18 is a fertile twig of this plant, adjacent to a bit from an early *Sequoia*-like conifer.

The humid swamps, windswept deserts, and grassless lowlands of the time saw an explosive diversification of terrestrial

13.19 RECONSTRUCTION OF A PERMIAN REEF AS IT WOULD HAVE APPEARED IN LIFE. Sponges and corals left and right. Brachiopods and ammonoids center. [© *Field Museum of Natural History, Chicago.*]

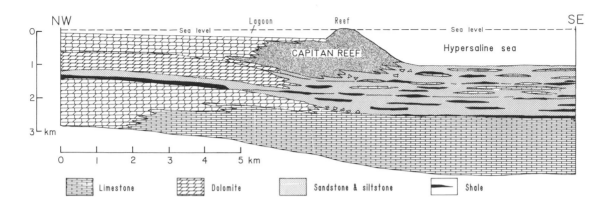

13.20 VERTICAL PROFILE THROUGH PERMIAN SEQUENCE AT EL CAPITAN, SOUTHERN GUADALUPE MOUNTAINS, WEST TEXAS. This classic buildup of carbonate rocks and its extensions ringed the deepening, oil-rich, hypersaline Delaware basin of West Texas and New Mexico as a late Permian wave-resisting structure or *reef*, in part of biogenic origin. *[Adapted from P. B. King, 1948, U.S. Geological Survey, Professional Paper 215, p. 98, Fig. 11.]*

arthropods, as well as an increasing diversity of amphibians and reptiles. The reptiles, later to include dinosaurs, were yet to undergo their great Mesozoic diversification, but the beginning of their ascent is recorded in younger Carboniferous swamp deposits of Nova Scotia where remains of the small, lumbering cotylosaurs are found. Permian amphibians ranged from animals a few centimeters long to fat crocodile-like creatures up to 3 meters long. They were outperformed, however, by their scaly cousins the reptiles, freed, at last, from bondage to a watery milieu by a new kind of egg that could survive and hatch on land— the *amniote* egg, a great leap forward in evolution.

In this new-style incubator (of which the oldest may be from the Lower Permian of Texas), the embryo lives within a shell that is tough enough to protect it but porous enough to let O_2 in and CO_2 out. Its own private embryonic compartment is connected with two other special compartments—one for nourishment, the yolk, the other for wastes. Possessed of such a clever cradle, as well as a desiccation-resistant scaly covering, reptiles were no longer tethered to a watery environment by their reproductive requirements. As soon as they hatched, they could wander freely over the dry land. Having achieved such a breakthrough, they underwent their first prominent diversification during later Permian history, when six of the sixteen orders ever to live originated (an *order*, being a group of *families*, is also the next-largest division of a kingdom of life, after phylum and class).

The early reptiles included the curious sail-backed pelycosaurs (Fig. 13.18, *top center*), and the mammal-like reptiles

13.21 AIRVIEW NORTH OVER THE NEARLY COMPLETE PERMIAN SEQUENCE OF THE SOUTHERN GUADALUPE MOUNTAINS IN AND AROUND EL CAPITAN PEAK (ELEV. 2486 M), WESTERN TEXAS. Dramatic changes in facies of strata and their biologic content are illustrated here, reflecting those graphed in Fig. 13.20. The massive Late Permian reef dolomites of El Capitan, with more than 250 reefal marine invertebrate species give way northwest (*left*) to laterally equivalent bedded limestone and dolomite with only 86 lagoonal species and then to evaporitic strata further west (shoreward) with but a handful of hypersaline species. To the east (*right*) the bluff-forming reefal dolomites descend and grade laterally into silts and sands of the Delaware basin at sea depths then of around 500 meters, later to be capped with a thick evaporite sequence rich in hypersaline salts. Underlying facies changes as sketched in Fig. 13.20. *[Courtesy of John S. Shelton.]*

(the therapsids)—most successful of all late Paleozoic reptilian orders.

Only four more important major categories of organisms remained to appear and evolve—the birds, the mammals themselves, the flowering plants, and the grasses. But first there was to be a devastating turnover of the families (groups of genera), genera (groups of species), and species of animals. From a distance of 250 MY it is hard to know exactly how devastating; but all estimates agree that it was both numerically and proportionally the most ravaging mass extinction of all Earth history so far. During the last few million years or less of Permian history, more than half the families and 90 percent of the species of then-living shallow-water marine animals (invertebrate and vertebrate alike) disappeared. Of four orders of amphibians present at the end of the Permian, only one survived, with fewer than half the genera then living. Among reptiles, all six then-existing orders survived, but there were heavy casualties at lesser systematic levels. Of the approximately 50 genera of mammal-like reptiles then present, only 1 (the genus *Dicynodon*) survived to give rise in Late Triassic time to the first true mammals. So close did we come to having our ancestry cut off at the roots!

Trilobites, already dwindling, are never seen again. Gone are most of the families of brachiopods—hallmark of the Paleozoic marine faunas. A whole class of the prickly echinoderms (the blastoids), previously waning, vanished, and the once highly diversified phylum never recovered its former variety. Paleozoic corals with calcareous skeletons and a fourfold radial symmetry (orders Tabulata and Rugosa) made way for Mesozoic corals (the Order Scleractinia) having less

stable calcareous skeletons (aragonite instead of calcite) and a sixfold radial symmetry. Modern reef builders are of the latter types. Nautiloid cephalopods—ancestors to the modern pearly nautilus—dwindled almost to extinction, as did the once-prominent coiled ammonoids with their complexly folded body chambers. Even the common large Protozoa of the time were affected. After 80 million years as conspicuous builders of biogenic limestone, these creatures, the fusulinids, were gone.

The collapse was awesome. Why did it happen? If it was a product of glacial severity, why was it delayed until the glaciers were in full retreat, or gone? Could warmth have been the culprit, doing in cold-adapted faunas? If it were caused by extensive upwelling of anoxic waters, what would have knocked out the quadrupeds? Were extinctions the product of reduced ecospace variety, resulting (in the sea) from shortened coastlines, narrowed continental shelves, and altered ocean currents, and (on land) from spreading desert conditions? Were key nutrient sources or prime producers eliminated, initiating catastrophic collapse or rearrangement of nutrient pyramids? Or could it have been some even more abrupt catastrophe—addition of a chemical poison instead of elimination of nutrients, perhaps? Or radiation from a nearby supernovation? Or collision with an asteroid or a comet?

All of these and so many other explanations have been proposed for this and lesser extinction events. The subject calls for a chapter of its own. It shall have one when we come to the most widely known and actively discussed extinction of all—the demise of the dinosaurs at the end of the Mesozoic.

In Summary

What are the most salient features of Paleozoic history?

The era began and ended with waning continental glaciation. At the beginning, the melting of late Proterozoic ice as a result of equatorward drifting of the continents or warming global climate, initiated conditions favorable to the proliferation of metazoan life from protozoan or simpler metazoan ancestors. In warming shallow seas, on flooding continental shelves, a limited variety of flimsy, soft-bodied, invertebrate animals found all potential metazoan habitats then unoccupied.

Products of what may be described as the first wave of experimental diversification into this inviting seascape comprise the unique fauna of earliest Paleozoic and Phanerozoic age—that of the Ediacarian System and Period. Some of these early experiments failed, while others survived as components of the initial metazoan radiation.

In contrast to the prevailing Ediacarian mode, the more diverse Cambrian undercurrent of soft-bodied creatures was largely expunged from the record by contemporaneous scavengers. A flowering of hydroid and scyphozoan coelenterates such as that of the Ediacarian was never again to play so visible a role in the expanding theater of metazoan evolution. The justly famous Burgess Shale of southern British Columbia preserves rare Middle Cambrian glimpses of the evolving cryptofauna only because of their anaerobic burial.

By the end of Cambrian history, nearly all of the now-known major phyla of organisms, including our own remote chordate ancestors, had evolved, as well as many that are different enough from previously known forms of life to be ranked as extinct phyla. The reigning trilobites then abdicated in favor of the already successful brachiopods, the best-known marine metazoan fossils during the rest of Paleozoic history.

Earth history is an amalgamation of all lines of evidence. When records are viewed concurrently, it is seen that the early Paleozoic record of increasing O_2 levels signalled by fossils and strata continued. Phosphate, as calcium-phosphate ($Ca\ P_2O_5$), precipitates from seawater only at sufficiently low CO_2 levels, resulting from expulsion of dissolved CO_2 on warming and decreasing pressure. That is observed nowadays mainly or wholly at low latitudes, and particularly on leeward coasts where cold bottom waters well up to replace surface waters driven seaward by offshore winds. Variation in phosporite abundance, therefore, in some complex sense, records variations in CO_2 and O_2 concentrations, oceanic circulation, and surface temperature at sites of origin, as well as volcanic emission of CO_2 and its conversion to carbonate ion and limestone. Even after O_2 first reached present levels, perhaps 400 MYBP, it probably fluctuated widely from them with variations in the rate of introduction of volcanic CO_2 and reduced gases.

Paleozoic geography and ecosystems were related products of plate tectonism and its interacting consequences. The opening of the Iapetus Ocean during Ediacarian and Cambrian time, and its closure during Ordovician to Carboniferous times, played major roles. Degrees of provinciality and selective evolutionary pressure were, to a large extent, products of such events.

The final closure of Iapetus during later Devonian and earlier Carboniferous time, and the mountain barriers that then arose, placed a premium on adaptation to a terrestrial mode of life. Quadrupeds arose and diversified—amphibians from lobe-finned crossopterygian fish in latest Devonian and reptiles from amphibians during later Carboniferous and Permian history. Descendants of simple, tetraspore-bearing, vascular land plants, apparently present by Middle Ordovician time, claimed a wider territory and broader range of diversification as time marched on. They luxuriated in lowland swamps and forests as Gondwana gathered in the Southern Hemisphere and then joined the Northern Hemisphere continents (Laurasia) to complete the jigsaw puzzle of Pangaea during Carboniferous and Permian history.

Coal forests flourished in low latitudes at the very height of the late Paleozoic glaciation that gripped different Gondwanan continents at intervals during 50 MY or more. A detailed account of the growth and decline of these forests, glaciers, and seas was being recorded at the same time by low-latitude witnesses—the cyclothems of Late Carboniferous to Early Permian history. Successive short cycles that began with continental deposits, often capped by coals, gave way upward with glacial melting to shallow marine sequences that completed the cyclothem. A succession of as many as a hundred or more such cyclothems records the glacially mediated retreats and advances of the North American midcontinent sea south from the Great Lakes. Similar cyclothems were recording similar events wherever tropical lowlands and polar glaciers flourished at the same time.

Like Tchaikovsky's 1812 Overture, Paleozoic history closed with fireworks. Landmasses collided all over the place, raising the Appalachian, Hercynian, Uralian, and other mountains and provoking volcanism. More than half of the families of then-living animals vanished, never to be seen again. No closing of a geologic era or period, no mass extinction—before or since—was quite so devastating.

JURASSIC

CARBONIFEROUS

PERMIAN

CRETACEOUS

TRIASSIC

CHAPTER FOURTEEN

THE MESOZOIC: REPTILIAN HEYDAY

Consolidation, then breakup, of Pangaea, with related tectonic, climatic, and biogeologic consequences, is the main theme of Mesozoic history. The mosaic of lesser plates that is now China was completed during later Permian and Triassic time. The western 12 percent of North America arrived piecemeal from distant sources. The opening of the continuous subequatorial seaway called Tethys dominated sedimentary and climatic history. Unfilled ecospaces became available. The Paleozoic dominance of amphibians, scale trees, and primitive conifers on land gave way to reptiles, cycads, true conifers, and flowering plants. Mammals and birds appeared, awaiting their turn in the limelight. New kinds of sea life flourished. Dinosaurs flowered and fell.

Distinctive sedimentary products bear witness to plate motions and their effects. The extensive Pangaean supercontinent, reduced coastline, and moisture-intercepting highlands fostered extensive older Mesozoic deserts. Later Cretaceous history saw the last great continental flooding, a time of flourishing biological productivity. Alas, this was laid waste by another devastating extinction—less drastic in total reduction of diversity than older ones but dramatized by the passing of the once-lordly dinosaurs. Birds alone survived to represent them. No flying or specialized swimming reptiles, nor any of the most distinctive marine shelly fossils, survived. Among plants there were curiously selective losses, with great reduction of the cycads. Early Cenozoic mammals were quick to claim their long-awaited priority.

The Concept of Mesozoic

The Mesozoic periods (and systems)—the Triassic, Jurassic, and Cretaceous—correspond roughly to the early diversification, flowering, and eventual decline of the dinosaurs and associated large reptiles (Fig. 14.1). Most who think *Mesozoic* think first of dinosaurs. Yet these fascinating creatures are

14.1 Panoramic view of the age of reptiles, from Carboniferous to Cretaceous. *[Peabody Museum of Natural History, Yale University; R. F. Zallinger, artist.]*

so much a product of historical processes and interacting physical conditions that their meaningful discussion best awaits a view of those shaping factors.

The Mesozoic and Cenozoic eras are known to geologists as the cradle of historical geology. They and their subdivisions illustrate the principles underlying sequential geochronology and the growth of the geological column. The basic concept was of a graduated succession of biotal components, punctuated by episodes of extinction and repopulation—going back to plain William Smith and aristocratic Baron Georges Cuvier early in the nineteenth century.

By 1840 most of the Phanerozoic systems now recognized, as well as the Paleozoic Era, had already been named. At that time John Phillips, a nephew of Smith, thought it such a good idea to have historical subdivisions larger than periods that he expanded the earlier-proposed Paleozoic Era and suggested others. Nineteen years before Darwin's *Origin of Species*, but on the heels of Smith and Cuvier, Phillips wrote (in the *Penny Cyclopaedia*, 1840) that general terms for stratified rocks should be "formed upon a consideration of their organic contents, which appear to follow a great law of succession." He proposed that Kainozoic (now usually Cenozoic among English- and French-speaking people) and Mesozoic be joined with the Paleozoic to denote the succession of global changes in the prevailing record of life. The validity of that judgment has since been emphatically validated by the universal recognition that Mesozoic history is indeed set apart from that of older and younger strata by two dramatic extinctions. Other extinctions (indicated by Xs on Fig. 13.8, *tier 4*, as well as on

the front endpaper) also coincide with or are close to then-recognized period and epoch boundaries.

As the Phanerozoic periods are grouped in three eras, so the Mesozoic Era is divided into three periods and systems, recognized by their fossils—a trilogy within a trilogy. There is a degree of mnemonic utility and charm to much early geologic nomenclature (e.g., Fig. 3.12, Table 3.1). Practical Jurassic helps us to picture its reference sequence, named from exposures in the scenic folded Jura Mountains of the Franco-Swiss borderlands. Cretaceous announces that chalk is a distinctive rock of that system.

Triassic, the initial Mesozoic system, is a punster's dream. An artistic geologist I know once cast a three-tailed fantasy reptile just so he could display it as a Tri-assic dinosaur. The name in fact refers to the threefold sequence of Triassic rocks in central Europe, as contrasted with England and western Europe where the Triassic (upper New Red Sandstone) comprises entirely nonmarine red sands and multihued shales and marls. Southeastward, on the continent, an otherwise similar sequence is split by a medial, fossiliferous, shallow-water limestone and dolomite from a subequatorial (Tethyan) source. Triassic commemorates this. Such a threefold division, however, is distinctive of only a small part of Europe. The much thicker (up to 3,000 meters) and more extensive Triassic limestone and dolomite of the Tethyan seaway itself (Fig. 14.2) reaches from the Pyrenees to the Himalaya. Although the name, in use since 1854, persists, the widely dispersed Tethyan stratigraphic sequence is taken as representative for the era (front endpaper).

14.2 Triassic dolomites of the Tethyan sea. Scene *A*, Middle and Upper Triassic, Dolomite Alps viewed northeastward over Cortina d'Ampezzo, Italy. *B*, upper Middle Triassic near Glerish summit, Karwendel Alps, Austria.

Pangaea Consolidated: Vast Lands, Shrunken Inland Seas, Accretional Complexities

The assembly of continents and microcontinents called Pangaea was not quite finished as the Mesozoic dawned. Parts still lay dispersed eastward from the supercontinental mainland and nestled against then-subantarctic India and Africa (Fig. 13.15*B*). They were assembled to become China and welded to the rest of the Asiatic landmass during Triassic and latest Permian time. The southern supercontinent of Gondwana simultaneously rotated counterclockwise, achieving a more symmetrical distribution of lands between Northern and Southern hemispheres, elevating new mountains, and completing the consolidation of Pangaea. It then seems to have remained intact during most of the remainder of Triassic history.

Such conclusions emerge from the data of paleomagnetics and observation of the red sediments and mainly basaltic volcanics that mark the sites of rift basins, formed when Pangaea began to break up again in latest Triassic and Early Jurassic time. This began a new episode of plate tectonism. Old spreading ridges had cooled and subsided or had been subducted. The capacity of the global ocean was at a maximum. Because of that, sea level, which should have risen a hundred or more meters with melting of the Permian glaciers, instead receded.

The result was an unprecedented and as yet unrepeated extent of continuous land—the Permo-Triassic maxicontinent Pangaea. Its high degree of continentality played perhaps the central role in the emergence of the dinosaurs. The seemingly endless north–south Pangaean landmass and its lofty mountains of convergence limited the flow of moisture-bearing maritime air inland, giving rise to extensive deserts and continental sedimentation during latest Permian, Triassic, and much of Jurassic history. The connectedness of the lands allowed open migration by even weakly thermoregulatory dinosaurs from one to another.

Sands and shales deposited in arid environments and those of marked seasonal rainfall commonly display the intense red colors that result from oxidation, then dehydration, of even fractions of a percent of dispersed iron, following initial deposition. Redbeds of this sort (Fig. 14.3*A*) and windblown sandstones (Fig. 14.4) are found worldwide among latest Permian, Triassic, and Jurassic rocks. They blanketed large sec-

tors of western interior North America from west Texas to Nevada and northward across the wide Colorado Plateau and now-soaring Rocky Mountains to Montana.

Extensive dune sands today are aptly called *sand seas*. The remnants of a sand sea that eventually covered the whole of the present Colorado plateau to a depth of 300 meters or more during Early Jurassic history (the Navajo Sandstone) loom as the colorful and spectacularly cross-bedded bluffs of Zion National Park in southwestern Utah (Fig. 14.4*A*). Outside North America, deserts of Triassic and Jurassic age were also widespread in northern and eastern South America, in eastern Africa from the Cape to Kenya, in North Africa and the Arabian Peninsula, in Europe and the Eurasian borderlands, in Mongolia and China, and in eastern Australia. In habitable parts of these warm-to-temperate regions the great reptiles flourished then as their mammalian successors do now in and adjacent to wooded seasonal wetlands like the Okavango swamps of northern Botswana. The oldest mammals, perhaps nocturnal, scurried beneath their feet—tiny rodentlike insect eaters in Late Triassic oases in areas now in Arizona, South Africa, and southwest England. Swamp deposits, now commercial coals, are found in the Jurassic of eastern Australia and South Africa. Figure 14.3*C*, of the Upper Triassic Chinle Shale, replete with petrified logs, reflects a similar scene. Such circumstances recurred worldwide during Mesozoic history, varying mainly in the extent of aridity and in the abundance and types of life preserved.

The disintegration of Mesozoic redbeds and aeolian sandstones is not uncommonly the source of sand in modern sand seas, as older sandstones were for Mesozoic ones—the grains becoming more rounded and frosted with each recycling. Study a handful of such sand with a magnifying glass and you will see the results.

Varicolored shales in redbed terranes like those of Figure 14.3C, signifying former oases or stream courses, have been the source of a number of dinosaur discoveries. Settings such as this have rewarded search in Triassic strata of west Texas, the Colorado Plateau, and parts of Africa. The best Late Jurassic finds are from similar strata in the Morrison beds and related strata of the Rocky

14.3 PALEOZOIC–MESOZOIC SEQUENCE (*A*) AND TRIASSIC STRATA (*B–C*) OF THE WESTERN INTERIOR UNITED STATES. Scene *A*, north slope of Owl Creek Mountains, Wyoming; marine Upper Carboniferous sandstones (*C*) and Permian phosphatic strata (*P*), are here overlain by Triassic redbeds (*T*), Jurassic marine to nonmarine gypsiferous shale (*J*), and basal Cretaceous conglomerates (*K*). Scene *B*, Middle and Upper Triassic marine dolomite and shale, Augusta Mountains, Nevada. Scene *C*, agatized Upper Triassic logs of ancient swamp deposit at now-arid Petrified Forest, Arizona.

Mountain region in North America and the Tendaguru deposits of Tanzania, in East Africa. In more humid Late Cretaceous settings, the Colorado Plateau and high plains west from the Missouri River in the United States vie with Alberta's high plains for variety and excellence of

preservation of the then thriving but soon to be extinct great reptiles.

This is not to imply that Mesozoic marine deposits are in any sense unusual, apart from their mostly restricted pre-Cretaceous extent and distinctive fossils. It is only that known nonmarine deposits

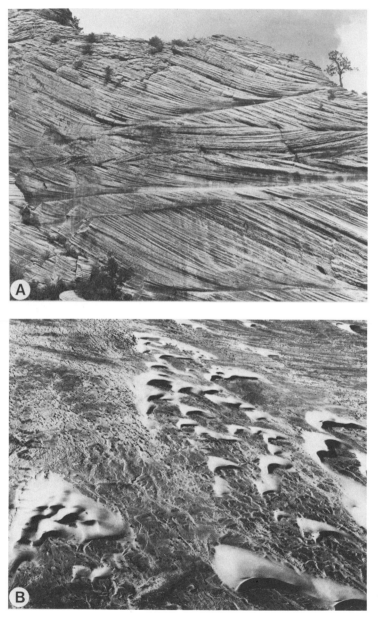

14.4 SAND DUNES OF EARLY JURASSIC AND MODERN TIMES. Scene A, large-scale "festoon" cross-bedding of the Navajo Sandstone, Zion National Park; vertical profile of ancient dune sands at varying angles. B, similar modern barkhan dunes advancing downwind toward the observer, west of Salton Sea, southern California. [A, *courtesy of U.S. Geological Survey film library*; B, *courtesy of John S. Shelton.*]

were usually extensive during Triassic and Jurassic history, as well as important for studies of the dinosaur clan. In fact, marine as well as nonmarine deposits of older Mesozoic age are found on most present continents and are locally extensive even within mainly nonmarine sequences (Fig. 14.5). In North America a nearly complete Lower Triassic marine sequence more than 2 kilometers thick extends northeastward from southern California across the Cordilleran region as far as western Wyoming. Shales with thin layers of marine limestone there

14.5 DISTINCTIVE TRIASSIC AND JURASSIC SANDSTONES AND REDBEDS (*DARKER STRATA*) OF THE COLORADO PLATEAU REGION. Scene *A*, Upper Triassic nonmarine strata near Moab, Utah. Thin white band at mid-height is the extensive and distinctive Shinarump Conglomerate; above it the Chinle red and green shale (see Fig. 14.3*C*). *B*, overlying strata near Uravan, Colorado. Thick band at base is the same massive Wingate Sandstone as at the top of *A*. Above it are more Triassic redbeds to fossiliferous marine limestone (*X*) and then Jurassic nonmarine strata, terminating with the dinosaur-bearing Morrison Formation (*M*). *C*, Upper Jurassic sandstones and red shales of the San Rafael Swell, Colorado. Uranium occurs intermittently throughout this sequence.

give way eastward to red continental shales and sandstones. A north-trending belt of upwarped land a few hundred kilometers wide separated the shallow seaway in which they were deposited from a more westerly boreal sea whose deposits contain marine fossils that descended from Yukon Territory to the international boundary. Linkage between Tethys and the Arctic sea did not occur until Late Cretaceous.

Westward from that Triassic inland sea and the present Rocky Mountain region, the western fringe of North America during all of Mesozoic time and even into early Cenozoic was complicated by a long history of accretional tectonics (Fig. 9.9). During that interval (Figs. 14.6, 14.7) the western 12 percent of the present continent (the western *North American Cordillera*) was being transformed by processes associated with

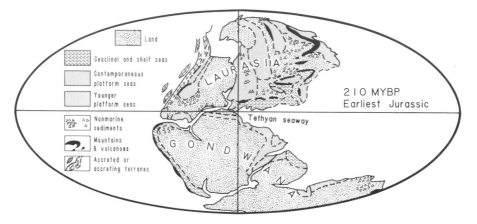

14.6 HYPOTHETICAL RESTORATION OF PANGAEA IN EARLIEST JURASSIC TIME; BREAKUP BEGINS. Continental positions about 210 MYBP.

14.7 PANGAEA DISPERSED; EARLY CRETACEOUS OCEANS OVERFLOW. Continental positions about 130 MYBP. Patterns as in Fig. 14.6.

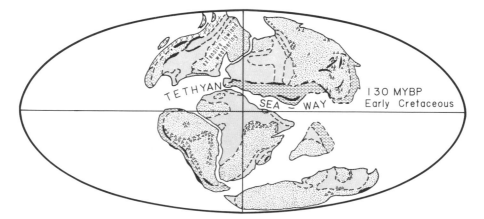

plate convergence and collison. It was growing by the addition to its western margin of exotic microcontinents that came from somewhere far to the southwest. A score or more of such detached terranes sidled obliquely into and against North America along Mesozoic transform faults. Convergence resulted in descent (subduction) and ascent (obduction) of exotic terranes. Associated rotational and shearing forces stressed the region of Nevada and adjacent western states, slicing them into linear north–south basins and ranges and accounting for much of the volcanism and mineralization of this region. Any map that indicates vertical relief will show why an eminent early geologist described the Basin and Range structure as reminiscent of an array of giant caterpillars crawling south toward Mexico.

The more easterly of the Lower and Middle Triassic strata of this complex North American Cordillera are not exotic but were actually deposited in the region. Underlying Permian phosphorites in southeastern Idaho and adjacent regions tell us that. They signify a subsiding continental shelf environment within which Lower and perhaps Middle Triassic marine rocks accumulated. Subduction beneath western North America was either wide and shallow or multiple—most likely both, to judge from recent seismic profiling.

After decades of study the region is again a great geological frontier. Westward from the North American Rocky Mountains all the way to Alaska, a rebirth of interest in accretion tectonics has resulted in an accelerating growth of knowledge that seems at each discovery to deepen the paleogeographic mystery. The region is a many-layered tectonic puzzle whose precise historical restoration challenges the best geological minds and sturdiest feet.

Breakup of Pangaea: Tethys Rules the Waves

The long transition from Triassic to Jurassic, virtually undetectable in continental redbed sequences that lack dinosaurs, is registered by a marked change in the dominant shelly marine faunas of that age. A sweeping marine extinction reduced family diversity by about 12 percent, all but wiping out the most characteristic Mesozoic shelly invertebrates, the coiled and complexly chambered cephalopods called *ammonoids* (from the temple of Jupiter Ammon in Libya). Only a few genera, representing no more than three families, survived to reestablish ammonoid preeminence in the sea. In fact there seem to have been two closely spaced extinctions, during which many nonmarine vertebrates, mostly dinosaurs, also became extinct.

Pangaea, meanwhile, showed signs of breaking up to spawn a new generation of oceans. The crystalline basement rocks of eastern North America and adjacent lands in Africa and Europe began to pull apart during Middle Triassic into Early Jurassic history from south-central North Carolina to Nova Scotia and along equivalent continental margins to the east. That opened up a set of rift valleys similar to those of present East Africa. Such pull-apart basins became filled with mainly red detrital sediments and associated basaltic flows and sills. They are now seen as wooded basalt ridges, encircled by fertile red plains as if stained by the bloody battles of the American Civil War, fought on such terrains at Gettysburg and Bull Run.

These rift valleys continued to open and accumulate sediments until some time during early Jurassic history, after which rifting became concentrated in more easterly sites and the Atlantic began to open. The northern supercontinent *Laurasia* (Fig. 14.6) then began to drift northward, away from the equator. Thus began the western marine reentrant of what was to become Tethys (the Panamic Seaway), the ancestral Gulf of Mexico, and the proto-Atlantic. By the end of Early Jurassic time (Fig. 14.6), Laurasia and Gondwana were completely separated. Mesozoic Tethys had split Pangaea from east to west, and inundation of its margins accounted for the warm, shallow, locally agitated waters in which extensive, calcareous, "fish roe–like" sediments called *oölites* accumulated, especially in the Middle Jurassic of Europe. This was also the time and place of origin of similar, economically important, oölitic ironstones, the industrially important minette ores of Europe.

Rifting of the North Atlantic accelerated. Thick deposits of common salt and other products of evaporation of paleo-Pacific source waters accumulated in the widening but still-narrow north arm of the proto-Atlantic and the incipient Gulf of Mexico. Thick Upper Jurassic sediments that grade from normal marine limestones and shales to shoreward redbeds loaded the lighter salts until, at places, they began to ascend gravitation-

ally. *Salt domes*, rising like bubbles from a diver's helmet or aqualung, deformed and even penetrated strata, creating important oil traps and adding salt and sulfur to the useful products of Mesozoic sedimentary history. Similar salt beds and domes beneath the Atlantic continental shelves of South America and Africa record similar events in the restricted Early Cretaceous proto-Atlantic as its southern arm began to open (Fig. 14.7). The then-opening Viking graben became a future site of North Sea oil. It would be no great surprise were salt domes and perhaps oil to be found in quantity beneath the North Atlantic continental shelves as well.

During some 30 MY or so of Cretaceous history following the continental positions suggested in Figure 14.7, Tethys attained its maximum width and influence. Its near-equatorial currents extended a third of the way around the world from what is now the Persian Gulf to the open Pacific, establishing a warm circumglobal circulation. Its principal remnants today are the Black and Caspian seas, and perhaps parts of the Mediterranean. Deflections poleward of eddies from Tethys as a result of Earth's rotation (Coriolis effect) carried moderating thermal currents into all oceans, resulting (with higher levels of CO_2) in a prolonged interval of unparalleled climatic amelioration and pluviality. No wonder reptiles flourished. No wonder palm trees grew at high latitudes. No wonder reefs and planktonic microflora flourished in the Cretaceous Tethyan seas.

Continental motions resulting from this most recent cycle of plate tectonism have persisted to the present. Post-Cretaceous India drifted northward still thousands of kilometers from its prior position against Antarctica toward its present location. The Atlantic Ocean continued to widen as the American plates rode westward over the subducting and narrowing eastern Pacific.

The opening of the Tethyan seaway was the central historical event of later Mesozoic time—the principal influence on climate, sedimentation, and paleoecology until the great dying at the end.

The Clear Limestone Seas of Late Mesozoic History

Redbeds, sand seas, and evaporites, then, are hallmarks of the prevailingly nonmarine depositional history of terminal Permian, Triassic, and earlier Jurassic. That contrasts with the prevalence of calcareous strata that delimit the warm seaways of later Mesozoic time, especially in the Northern Hemisphere. The most distinctive of these rocks is chalk—dead white, entirely biogenic, and consisting almost exclusively of the tiny calcareous plates of floating (*planktonic*) marine microalgae, known only as *coccoliths* (or coccolithophores). Coccoliths first appeared about 225 MYBP in Late Triassic time and proliferated as the modern sea floor began its history during the early Jurassic (no part of the present ocean floor is older than Jurassic). In late Cretaceous and younger seas, beginning about 100 MYBP, these tiny biogenic particles became extensive, open marine, calcareous oozes. They accumulated, however, only at places that were not so deep and so cold that weakly acidic solutions of CO_2 could redissolve the settling calcareous particles.

Symbolic of the Cretaceous and of Fortress England are the now-emerged White Cliffs of Dover—merrie Albion, welcome sight to returning natives, unbreached by surface assault since 1066 (Fig. 14.8*A*). Pre-Cretaceous chalks do not exist, nor are they as common in younger rocks. Mesozoic carbonate rocks of the Tethyan succession, moreover, are generally different (Fig. 14.8*B*). Shallow later Mesozoic seas of that region left behind an extensive variety of thick deposits of limestone and dolomite, and, during Late Jurassic and Early Cretaceous, biogenic reefs and banks. These reefs were built by corals and algae, or by thick-shelled, cup- and crock-like clams up to 1.5 meters tall (the massive *rudistids*).

The scene envisaged, even where limestone and dolomite is scarce, is one of extensive and expanding later Cretaceous platform and shelf seas, deepening from barely awash to nowhere more than a few hundred meters. These onlapping seas, products of expanding spreading ridges, inundated continents so leveled by weathering and erosion that detrital sediments are mainly shales except where shed directly from existing or rising mountains (Figs. 14.9, 14.10).

Around their shores and inland reigned the dinosaurs—the "terrible lizards." Only they were not lizards, but a diverse assembly of more active and apparently more sociable reptiles.

14.8 Cretaceous of English Channel and Spanish Pyrenees. Scene *A*,
White Cliffs of Dover; banding visible in upper third of this photograph
reflects a 40,000-year periodicity
similar to that of the tilting of Earth's
axis of rotation. *B*, Lower Cretaceous
marls overlain by Upper Cretaceous
limestones, Sierra de Rialp, Spain.

14.9 Vertical profile through prefolded Upper Cretaceous of west-central United
States. Restoration extends from present west-central Utah through Denver to northeastern
Kansas. Sandstone wedges on the west record rise and erosion of Wasatch Mountains. An
extensive basal sandstone, source of artesian water beneath the High Plains, transgresses
time upward in both directions from a midway late Early Cretaceous age. The terminating
Paleocene freshwater limestone on the west (*left*) overlies thin basal Paleocene sediments
that contain early mammalian fossils. Other limestones marine. See also Fig. 14.10.
[Adapted from P. B. King, 1952, Geological Society of America, Special Paper 62, p. 731.]

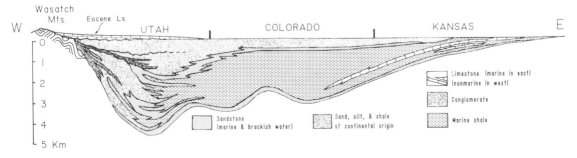

14.10 Views of Upper Cretaceous and adjoining strata equivalent to those profiled in Fig. 14.9. Scene *A*, view north from Mesa Verde, southwest Colorado, over Dolores River basin. The slope left from the bluff-forming, nonmarine, Mesa Verde sandstone is occupied by the marine Mancos Shale, shown by dashed lining at center of Fig. 14.9. *B*, basal artesian Dakota or Lakota sandstone, stippled in Fig. 14.9, near Beulah, Colorado. *C*, sequence similar to that of *A*, west of Green River, Utah. *D*, horizontally layered Paleocene freshwater limestone, the ridge-capping white band in the distance, and the upended Late Cretaceous strata beneath it record intervening orogeny and erosion of late Cretaceous age.

The Lives and Times of the "Terrible Lizards"

Large dinosaurs and other reptiles associated with them in the public mind have been a source of ever-renewed excitement, adventure, intrigue, confusion, and entertainment from the time of their first discovery. The first giant reptile of record (not a dinosaur) was found in a Dutch chalk mine in 1770 but had to await the maturing of the versatile Georges Cuvier (Fig. 3.3), then an infant, for its recognition as a lizard. It was only later that real dinosaurs were found and recognized as something different, but mankind's affair with large Mesozoic reptiles has gone on ever since, unbridled by niceties of paleontological terminology. Rivalries in the field of dinosaur paleontology have been as overscale as some of the great beasts themselves. They have led to unrelenting effort, lifetime feuds, bribery, hijacking of collections en route from the field, even telegraphy to the public press in order to be first into print with a new name, and other extreme procedures. Far from the honorable colorlessness often attributed to paleontology, no holds were barred.

Although Cuvier correctly identified the Dutch *mosasaur* as a swimming lizard, the size of similar reptiles estimated from comparative anatomy on the assumption that their bones were also those of lizards, proved to be unbelievably large. They could not possibly be typical lizards, said able but coldly pedantic Richard Owen in 1841, a London surgeon (and like Cuvier, an antievolutionist). So he called them *dinosaurs* (meaning "terrible lizards").

Dinosaur became a household word with the royally endorsed public display in 1854 of fanciful life-sized models made under Owen's direction (Fig. 14.11). That display was installed on the grounds of the famous Crystal Palace in Sydenham, formerly the main building of the Great Exhibition of 1851 at Hyde Park. It was opened by Queen Victoria and Prince Albert themselves, to the cheers of 40,000 onlookers. Hoop-la worked. Rarely since then has dinosaur collecting and research lacked public support.

The saga of dinosaur discovery continues with another perceptive Englishman, Harry G. Seeley, in 1887. Seeley realized that animals referred to as dinosaurs were of two distinct evolutionary lineages. Their association in a related group made no sense to him unless it were also to include the crocodiles, an ancestral group called *thecodonts*, and the birds. To this superficially motley but related assemblage we now add the flying reptiles (pterosaurs), often calling the whole related lineage (*clade* in the new evolutionary lingo) by the group term *archosaurs* (Fig. 14.12).

In North America, the oldest confirmed discovery of dinosaurs was of three-toed footprints on Early Jurassic red sandstones in the Connecticut River Valley, found by Pliny Moody in 1800. Moody may be forgiven for identifying them as the footprints of Noah's raven, made, presumably, as the flood subsided. Like real birds, they did display three toes, and, as biblical legends are often overscale, it was only natural for Noah's raven to have unusually big feet.

14.11 MODEL OF THE DINOSAUR IGUANODON (*CENTER*) UNDER CONSTRUCTION
FOR EXHIBITION AT THE CRYSTAL PALACE, LONDON, 1853. Reconstructions
as fancied by Richard Owen and artist B. W. Hawkins in *The Illustrated
London News*, V. 23, pp. 599–600. Tongue-lolling object in right fore-
ground represents what is now known to be a mammal-like reptile. At
lower left is Owen's idea of a Triassic labyrinthodont amphibian. *[From
A. J. Desmond, 1976,* The Hot-Blooded Dinosaurs, *London: Blond &
Briggs, 1975, p. 20, Fig. 4.]*

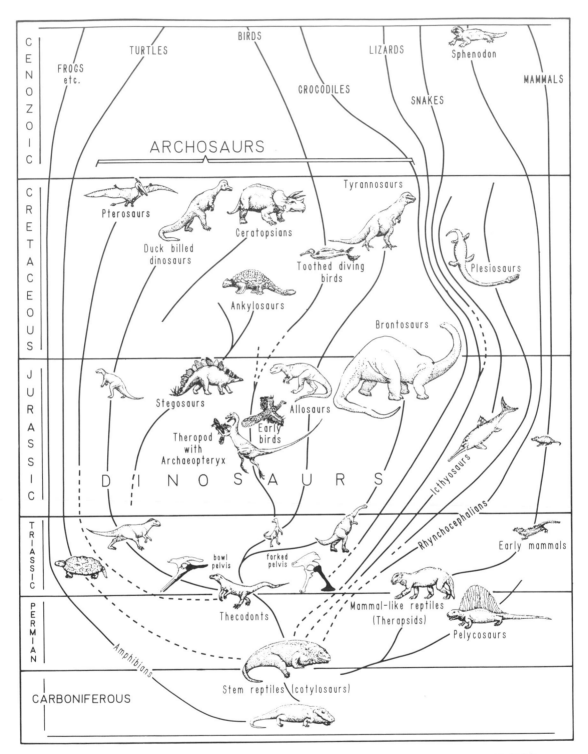

14.12 SCHEMATIC MAINSTREAM EVOLUTION OF THE DINOSAURS AND OTHER REPTILES. Solid lines indicate approximate ranges of suggested lineages, broken lines uncertain extensions. Nothing else is to scale, although relative differences in size are implied.

366

That was a good bit less far-fetched than the identification, in 1820, of the first real dinosaur bones from the same area as "possibly human"—a mistake neither Cuvier nor Owen would have made despite their then virtually mandatory creationist views.

The serious study of dinosaurs in North America began in 1835 with Edward Hitchcock's interest in the three-toed Connecticut Valley footprints. Like Moody, he understandably thought they were the tracks of birds. Hitchcock, noted divine and professor of natural history at Amherst College, began a collection that eventually occupied the whole lower floor of a museum there. After 13 years of study and 3 years into the presidency of Amherst, he wrote (in 1848): "I have seen, in scientific vision, an apterous [wingless] bird, some twelve or fifteen feet high . . . followed by many others of analogous character, but of smaller size. Next comes a biped animal, a bird, perhaps, with a foot and heel nearly two feet long. Then a host of lesser bipeds . . . and among them several quadrupeds." Mistaken, but by no means naive, Hitchcock made major contributions to the geology of his time while clinging to the end of his life to the belief that the three-toed, bipedal tracks were those of giant wingless birds (Fig. 14.13).

Although true dinosaurs had no wings and never flew or swam, some related archosaurs (the pterosaurs) did have wings on which they soared and which, some now believe, they may actually have flapped. At any rate, as birds, archosaur descendants eventually did become power flyers. The giant reptilian swimmers—the ichthyosaurs, plesiosaurs, and mosasaurs—are, in fact, not archosaurs, but more closely related to true lizards.

14.13 THREE-TOED TRACKS OF BIPEDAL DINOSAUR FROM TRIASSIC–JURASSIC REDBEDS OF CONNECTICUT VALLEY. Scale is a meter stick. *[Courtesy of Marshall Schalk and Allen Curran; Hitchcock collection, Pratt Museum, Amherst College.]*

Dinosaurs (loosely synonymous with archosaurs) dominated the Mesozoic scene. Following major Triassic setbacks, when 28 percent or more of all then-living reptilian families became extinct around and preceding the time of the great Manicouagan asteroid impact in Quebec, survivors radiated adaptively from thecodont ancestors. The next 140 MY or so saw them range into a diversity of surface environments—arid to humid. They populated oases in the deserts, wooded swamplands, and even waded about and fed in shallow waters of varying salinity. They included both carnivorous and vegetarian feeders. And they ranged from equatorial to polar latitudes. How did they manage where no reptiles manage today?

Viewed from the Permian (Figs. 14.1, 14.12) the ascent of the dinosaurs would not have been predicted. The mammal-like reptiles (therapsids) were the most diverse and most numerous of the six orders of reptiles then extant. On their way to becoming warm-blooded, any biologically informed gambler would have identified them as most likely to succeed. Yet, although representatives of all reptilian orders of Permian age survived the Permian extinctions, the then-ascendant mammal-like reptiles were the hardest hit, losing one of their three suborders and many genera. The thecodont reptiles, on the other hand, newcomers to the terminal Permian, not only survived but went on to establish the archosaur lineage that came to dominate Mesozoic history.

The emergent and extensive lands of Triassic Pangaea were all so interconnected that there were, at first, few barriers other than deserts to the migration of land-dwelling tetrapods. From latest Permian into earlier Jurassic history, for instance, a similar redbed terrain with similar distinctive plants (the *Glossopteris* flora) extended almost or quite continuously across Gondwana from South America to Africa and Peninsular India. Climates were mild, if mostly dry, permitting the contemporaneous reptiles and even certain amphibians (the labyrinthodonts) to range from Spitsbergen to Antarctica—then not all that far from their present locations. The mammal-like reptiles also recovered to expand over much of the same territory, along with the stem reptiles (cotylosaurs).

The vagaries of Mesozoic reptilian life were many. Middle Triassic seas in Europe became a haven for marine reptiles, but soon dried up, giving way again to terrestrial habitats and species. Triassic extinctions saw the dwindling or disappearance of once-dominant reptilian groups (including the mammal-like reptiles) and severe losses among the recently arrived dinosaurs—partly compensated by the entry of turtles, true crocodiles, pterosaurs, mammals, and, it is now claimed by some, *perhaps* birds. The spread of Early Jurassic sand seas followed, during Middle and Late Jurassic history, with a renewed diversification of habitat and the dinosaurs.

The generally more lush Jurassic vegetation, giving way locally to coal swamps, supported a variety of large forked-pelvis herbivores, including the monstrous sauropods *Brontosaurus, Diplodocus,* more recently *Supersaurus* and *Ultrasaurus,* and now, from the upper Jurassic of northern New Mexico, the Earth shaker *Seismosaurus.* These animals, up to 20 meters long for *Brontosaurus* and an estimated 37 meters for *Seismosaurus,* weighed as much as 80 to 100 tons. They were the largest quadrupeds ever to live—perhaps as large as could support their own weight in the open air, if that they did (Figs. 14.1, 14.12). Carnivorous predators, if any, probably also had to become very large in order to cope with such prey. Brute strength and small brains ruled the planet in those days. Aimless giants strode its surface. Small mammalian cousins, perhaps nocturnal, survived by keeping clear—to avoid being stepped on or becoming hors d'oeuvres.

Extensive sedimentary evidence records the presence of lowlands, floodplains, glades, swamps, and, locally, shallow marine embayments in the latest Jurassic. Such settings in the Morrison Formation of the Colorado Plateau and Rocky Mountains (source of all truly huge American dinosaurs), along with those of the Tendaguru beds of Tanzania, contain strikingly similar dinosaur assemblages—large and small. They share similarities with the latest Jurassic deposits of southern England which also contain marine reptiles. Even with the continents jammed together in Pangaea, these reptiles had to do a lot of traveling to attain such wide distributions—the ponderous ones have been estimated to move at perhaps 12 to 17 kilometers an hour, the agile ones 2 to 5 times as fast. Indeed, new discoveries of nesting, care and feeding of the young, and herding customs among dinosaurs (by Jack Horner of the Museum of the Rockies and others) are creating a new respect for these creatures as active social animals, far more birdlike than reptilian.

In contrast to the archosaurs, Mesozoic mammals and birds are rarities. The oldest true mammals were late Triassic. They were the size of small rats, presumably hairy, insectivorus creatures from the Cave Sandstone of South Africa, the Kayenta beds of Arizona, and fissure-filling deposits of southwest England. More than a century of quarrying the 150-MY-old Upper Jurassic lithographic limestones of Bavaria has produced a bare half-dozen specimens of the oldest confidently known bird (*Archaeopteryx*), one of them a single feather. Ironically this bird, once thought to be descended directly from root thecodonts with a birdlike pelvis, is now unequivocally identified as an offshoot of the theropod branch of the archosaurs, characterized by a reptilelike or forked pelvis.

Recent announcement of two fragmentary, birdlike skeletons from 225-MY-old upper Triassic sediments in the Texas panhandle would push that record back 75 MY and raise new questions about avian ancestry if confirmed to be birds. The relatively short forelimbs and lack of evidence for feathers, however, advises reservation until more conclusive evidence for affinity with birds shows up.

Cretaceous history was a reprise of later Jurassic, only more so and with different actors. Continental accretion was active and mountains were rising anew (Figs. 14.9, 14.10*D*). Flowering plants, probably insect pollinated, became established during Early Cretaceous history. In fact sycamore- and magnolia-like flowers are known from 100-MY-old strata in North America, Europe, and Japan; and studies of pollen take the record back to at least 115 MYBP. Leaves even suggest a pre-Cretaceous origin, perhaps as remote as early Triassic. In any case, Late Cretaceous flowering plants, including deciduous forest trees, were set for a burst of diversification. Shallow inland seas and extensive wooded swamps and floodplains veneered wide lowlands. Birds began their ascendancy, especially toothed diving birds. Eleven kinds are known by 90 MYBP, and sixteen kinds of water and shore birds, plus one land bird, are found in later Cretaceous strata, as if preparing for their big opportunity. The culmination of archosauran evolution is recorded in rich and wonderfully well preserved accumulations of Late Cretaceous dinosaur remains and those of other reptiles in southwestern Canada (Alberta) and the northern United States, in Patagonia and Brazil, and in outer Mongolia.

Then variety dwindled, and suddenly, as geologists keep time, they were gone. After almost 200 MY of varying success, and having recovered from later Triassic extinctions, the remaining dinosaurs, the pterosaurs, the swimming reptiles, and all the remaining archosaurs except the birds and crocodiles vanished over the course of a few million years. Why? How could puny mammals and birds survive and prosper for another 65 MY when dinosaurs couldn't? We shall return to that in the next chapter.

The next most difficult question related to dinosaur success and then failure is also an old one that has been the focus of much discussion and research. Were the dinosaurs warm-blooded? Some archosaur descendants, the birds, are. Others, the crocodiles, are not. The mammal-like reptiles were evidently on their way to some form of temperature regulation, if not already there. And, of course, their descendants, the true mammals, are warm-blooded, or, more accurately, *homeothermal*—able, that is, to maintain constant internal temperatures. Feathers and fur insulate modern warm-blooded animals, making it possible for them to operate over a wide range of temperatures—at a cost in metabolic energy. There is little evidence that any archosaurs (except perhaps some pterosaurs) or mammal-like reptiles had such insulation, but neither has that possibility been eliminated.

More persuasive evidence, however, has been advanced in favor of some form of temperature regulation by the dinosaurs. That is suggested by the relatively porous structure of their bones, by estimated ratios of predator to prey populations, and by the slender builds and probably agile gaits of some bipedal dinosaurs. It is consistent with their birdlike nesting practices and sociability. The small, carnivorous, birdlike theropods, for instance, may very well have been warm-blooded in the conventional sense (e.g., center of Fig. 14.12, with *Archaeopteryx*). Present consensus is that thermoregulation of some sort probably existed in dinosaurs, although it may have fallen short of constancy. In this we see a likely explanation for dinosaur success that also carried the seeds of failure. Their level of thermoregulation may have been too little and too late for the thermal bottleneck of falling terminal Mesozoic paleotemperatures.

Sea Life on Middle Earth: Focus on Ammonoids and Coral Reefs

Earliest Triassic marine life, still reeling from the halving of its Permian familial variety and drastic reductions of genera and species, was impoverished worldwide. Whole phyla or large groups are uncommon in or missing from the earliest Triassic record, although amply present in older and younger rocks. Corals, bryozoa, sponges, calcareous protozoans, echinoderms, snails, and brachiopods are rare or uncommon in strata of that age. Mainly mollusks, this unimpressive initial Triassic fauna consisted of a few cosmopolitan species of clams (*p* for pelecypods in Fig. 14.14) and a growing variety of *ammonoids*—coiled shelly mollusks related to the octopus. Fossils of these ammonoids (*k* for cephalopod in Fig. 14.14), the most distinctive shelly creatures of Mesozoic seas, consist of successive growth chambers, like those of the fabled nautilus only much more complicated. Partitions between such chambers appear on rocky internal fillings of fossils as a regular tracery of intricate lines, logically if incorrectly called *sutures* (e.g., lower right in the Jurassic box on Fig. 14.14).

In time, however, Triassic and later Mesozoic seas became populated with the diversity of marine life so beloved of paleontologists and so useful in relating and differentiating the deposits of ancient seas to and from one another. A great variety of fishes attracted hungry reptiles to take up a seagoing life. A horde of new marine invertebrates appeared, among which ammonoids were the leaders, with pelecypods runners-up. Some thin-shelled, predatory ammonoids attained the diameter of truck and tractor tires (up to 2.5 meters). They underwent a complex evolutionary history from tiny (diameter as little as a centimeter or so) and more symmetrical Triassic styles with relatively simple sutures to large Cretaceous genera with very complicated sutures and often bizarre shapes.

Ammonoid history included two nearly complete extinctions, followed by two remarkable returns to dominance before the terminal Mesozoic showdown. Although abundant and useful zone markers during later Paleozoic history, the group (Order Ammonoidea) first almost struck out with the Permian extinctions. It rebounded, however, to dominate Triassic marine faunas, diversifying to include a couple of hundred new genera and subgenera in a probably inflated three dozen families before later Triassic extinctions. Following them, only a few genera in no more than three surviving families of ammonoids were left to try again. These Jurassic survivors proved their resilience by outstripping even the Triassic rebound. Their irruptive radiation gave rise, it is estimated, to more than 1,000 new genera over the course of remaining Mesozoic history, before dwindling to fewer than half that number and then final collapse.

Ammonoids make good zone fossils because of their progressive succession of sutures and shapes (*k* on Fig. 14.14) over apparently short intervals of history. Evidence for a geologically rapid rate of evolutionary branching is seen in the thirty-five global *stages* into which Triassic strata are divided on the basis of ammonoids over the roughly 35 to 41 MY of Triassic history. That suggests an average lifetime of about a million years per genus or subgenus.

Although ammonoids ruled among the invertebrates, the rapid if less spectacular evolution of Mesozoic clams and snails bridged the way from Paleozoic forms to new and more modern families and lifestyles. The epicure may be grateful both for the appearance of the edible burrowing clams and the present absence of

14.14 SOME SHELLY MARINE INVERTEBRATES OF THE MESOZOIC SYSTEMS, GREATLY REDUCED. Letters represent *b*, brachiopods; *c*, corals; *e*, echinoderms; *k*, cephalopods; *p*, clams; *s*, snails.

ammonoids, whose predatory earlier presence may have contributed to the adoption of the protective burrowing habitat by clams.

Here too are the elegantly symmetrical echinoderms. Along with starfish and the now-dwindling lilylike crinoids (Fig. 14.14, letter *e*, at lower right, for "echinoderm"), one finds an increasing variety of the bun-shaped sea urchins or echinoids (also *e* on Fig. 14.14). Like starfish, they were and are mobile and adorned with spines that rotate about small knobs on separate calcitic plates. They are also decorated with feeding grooves and hydraulic "tube feet" that circulate and seize suspended nutrients and direct them to a mouth. They descended from Paleozoic ancestors, and have persisted until now as pretty much indiscriminate suspension feeders—illustrating by their longevity the advantage of being a generalist.

Extensive biogenic reefs are a distinctive feature of warm Mesozoic seas. Coral atolls and islands, home to the seafarer's nut-brown maidens of yesteryear, exist today only because descendant corals with a sixfold symmetry of aragonitic mineralogy replaced their extinct calcitic and tetrasymmetrical, reef-building Paleozoic ancestors. These *hexacorals* (the reef-building Scleractinia) and associated algae, however, were slow to displace the pinch-hitting sponges and algae that starred in Permian reefs between tetracoral and later hexacoral dominance. Modern types of corals and their algal associates have been active builders of biogenic reefs and banks only since the Late Triassic. In the Alpine Jurassic, they shared that role with both calcareous and siliceous sponges, and, during Late Jurassic and Cretaceous history, with the

earlier-mentioned heavy-shelled crocklike and smaller doubly-coiled clams called *rudistids*. Rudistids built massive reefs and shell banks in clear, offshore, shoal waters all along both sides of the Tethyan seaway before disappearing with the dinosaurs. Next to ammonoids and a related group with straight internal shells or stiffening rods (belemnites), rudistids were the most distinctive denizens of the Mesozoic seas.

In fact, biogenic reefs flourished in clear equatorial and subequatorial later Mesozoic seas in the Alps, the Jura, east Texas, and elsewhere—as they have at other places and times before and since. However, the essential elements of the basically tropical modern reef complex only came together in Late Triassic time. That was a major ecologic advance in the transition from Mesozoic to Cenozoic. For *reef*—in the biological and nautical sense of a biogenic, wave-resistant structure—refers to features of different compositions at different times. It is an ecological, rather than a biological, concept. A reef's structural properties are a product of an energy-diffusing or deflecting framework or mass, having little to do with biologic affinities but a great deal to do with ecology and Earth history.

Preponderantly calcareous biogenic buildups having the appearance and structure summarized by the term *reef* arise in favored nutritional circumstances that, in turn, offer sheltered and productive habitats to other organisms. They were constructed at different times and places by an ever-changing diversity of organisms over three extended Paleozoic intervals during which calcitic tetracorals played the lead role. Still earlier, during Early Cambrian history, the calcareous spongelike or algal archaeocyathids were the primary frame builders. Even in pre-Phanerozoic time, as far back as 1,800 MYBP and earlier, proalgae precipitated substantial to huge reeflike mounds and banks with waveresistant properties.

The only feature that all of these buildups have in common with one another and with the post-Paleozoic reefs, however, is that algae or proalgae were always there in important supplementary or major roles. In fact, the most wave resistant part of the modern "coral reef" is its toothed buttress of coralline algae—the algal ridge of reef ecologists. In many reefs of modern seas, corals are uncommon and "stony" hydrozoans locally play an important and even commanding role. Indeed, the bulk of a reef mass commonly consists of a variety of biogenic debris, trapped within the open frame, depending on what takes advantage of the sheltering and productive reef ecology.

Biogenic reefs, like rain forests on land and mangrove swamps where land meets sea, are thus among the biologically most diverse and ecologically most productive natural communities on the planet. They now play, and formerly played, the role of oases in marine deserts, surrounded by fine-grained muds or thinly stratified, fine-grained calcareous deposits, or they stand above and face into deeper basinal deposits. Although they flourish best in relatively quiet clear waters, they depend on nutrient-bearing currents, and reefmargin debris records episodes of violent turbulence. As today, ancient biogenic reefs were periodically battered by storms. In areas that are continuously calm, the tiny colonial coral animals and other reef builders are unable to keep their surfaces clear of suspended mud. They smother and die.

Reefs, then, tell as much or more of habitat and oceanic history as they do of time or biological affinities. Those of biogenic origin are absent or limited at times of extended glaciation as well as at sites of excessive turbulence and suspended sediment. Warm, clear, usually quiet waters are required, as well as a reliable, current-borne supply of essential dissolved mineral nutrients for their symbiotic algae and of suspended organic matter for the nourishment of their animal builders. They grow today primarily at low latitudes along lee shores where seaward displacement of surface waters by offshore winds makes way for nutrient-rich upwelling. And so it seems to have been at most places where there is an extensive record of fossil reefs. The subequatorial shores of Mesozoic Tethys had all the favored attributes, from late Triassic onward.

Tethyan Oil, Boreal Coal:
Patrimony from a Warm and Equable Age

Consider now the legacy to modern industrial society of coal from the lands and oil from the seas of this Mesozoic golden age of biological productivity. The effects of geography, oceanic circulation, and volcanism on climate interacted with biology to generate extensive coal-swamp forests on land and thick, organic-rich muds—potential source beds of petroleum—in anoxic marine basins.

Evidence for equability of climate greater than can be accounted for by the equatorial clustering of continents and the circulation of warm water from low to high latitudes calls for higher concentrations of atmospheric CO_2. Computer modeling by climatologists at the U.S. National Center for Atmospheric Research suggests the need for 2 to 10 times the present third of a percent of CO_2. As CO_2 increased and plant growth responded, atmospheric moisture and cloud cover also increased; and pretty soon, a lazy, humid greenhouse climate would be on its way—just the thing to bring joy to the heart of a reptile. How could this happen?

Volcanism, if it does not counter the greenhouse effect with sun-shielding dust, is the likeliest source of the needed excess CO_2 and plate tectonism the trigger. The Mesozoic was an age of volcanism as well as of dinosaurs. Eruption began with the earliest rifting and recurred intermittently right through to the end, with relatively dust-free flood basalts being widespread during the climatically equable later Jurassic and Cretaceous.

Large volumes of Triassic basalt are known from North Africa, central Siberia, and eastern North America. Huge outpourings of gas-rich Jurassic basalt were erupted in eastern Argentina and Uruguay, southern Africa, East Antarctica, and eastern Australia. Flood basalts of Cretaceous age covered a large part of central India and southern Brazil. Assuming volcanic CO_2 overbalanced O_2 buildup resulting from carbon burial, one could then account for the level of Cretaceous CO_2 called for by climate modelers.

So what? Such levels of CO_2 would favor the growth of coal vegetation, its preservation after death, and the generation of anoxic basins where circulation is weak. Such Late Cretaceous settings together account for more than half the world's reserves of coal and oil, with assistance from salt domes and rainfall patterns. Paleomagnetism and paleoclimatic criteria record that Northern Hemisphere coal fields cluster in areas of high rainfall at paleolatitudes of around 55°N in contrast to salt and related evaporites which occur in arid paleolatitudinal settings of about 25°N. Both formed at comparable latitudes south of as well as near the equator. Coal formed at intervals throughout Jurassic and Cretaceous history. Evaporites accumulated mainly during the early stages of rifting in narrow, older Jurassic proto-Atlantic and Tethyan embayments.

Coal, once thought to be more limited stratigraphically, is now known from Lower Jurassic strata of present southeastern Eurasia and the North American Cordillera as well as in the big Cretaceous coalfields. Volumetrically important deposits are found through much of the Upper Cretaceous of North America (Fig. 14.15) as well as in parts of northern South America, northwest Africa, and eastern Siberia, extending locally into the lower Cenozoic. These coals and associated plant remains are the principal evidence for the warm, humid climates that then supported a flourishing plant life over much of the planet.

The largest coal reserves of North America today are Late Cretaceous, a storehouse of history as well as fossil fuels. Associated flowering plants and, with them, social insects were then already present and ecologically important. The oldest yet-known blossoms certify to their advanced state at the dawn

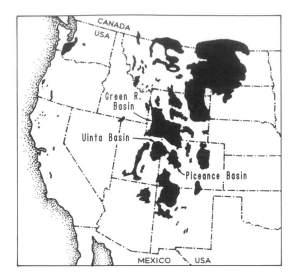

14.15 Late Cretaceous to earlier Cenozoic coal fields and oil-shale basins (*labeled*) of the western United States. [*U.S. Geological Survey Circular 94, Fig. 1, 1950.*]

of Late Cretaceous history. Pollen even tells us that the legume family, whose 14,000 living species comprise so significant a source of vegetable protein, had also made its debut before the curtain fell on the Cretaceous scene. Not only legumes and other flowering plants, but also deciduous trees, subpolar and even polar forests, and all the major elements of the modern flora except the grasses were then already present and flourishing worldwide. It was clearly a good time for the formation of major coals at the right places.

Evaporites, anoxic shales, and oil—on the other hand—may not sound as if they could have had much to do with one another, whereas, in fact, the ties are close.

Half of all known global oil reserves originated during only some 20 to 30 MY of medial Cretaceous history, fol-

lowing the ocean-to-ocean breaching of Pangaea by a continuous Tethyan sea. Some 70 percent of this oil is located in Saudi Arabia and other nations adjoining the Persian Gulf. The remainder is in basins marginal to ancient Tethys in Libya, Venezuela, and around and beneath the Gulf of Mexico.

The history of this Cretaceous oil began with deposition of the previously mentioned earlier Tethyan salts. It continued with the burial of uncountable trillions of Early to Late Cretaceous microorganisms in black muds that accumulated in anoxic marine basins during two intervals at around 120 and 95 MYBP. Light oil droplets from the defunct microbial hordes migrated upward into overlying biogenic reefs and other porous and permeable reservoir rocks. Sealed by overlying impervious shales and warped upward by rising salt domes and folding, they became oil traps. Shortly after World War II they were found by teams of Western oil geologists who soon recognized them as the world's principal oil reserves.

Source sediments, permeable reservoir rocks, impermeable sealing or cap rocks, confining structures into which oil can migrate and remain, are all essential preludes to filling your gas tank. So too is their discovery, production, and accessibility to refineries and markets. No other known petroleum province is so richly endowed, so conveniently located, or so vulnerable. It is easy to see why it looms so large in global politics—compared, for instance, to the remote and weatherbound polar regions.

Changing Scenes:
Sudden Death, Bigger Mountains, Squeaking By

Everybody knows that the Mesozoic ended with the abrupt extinction of the dinosaurs. Or did it? What if you were looking for the boundary in a sequence of nonmarine strata? If you were to find dinosaur bones limited to lower levels, with advanced mammals higher up, you would be in luck. You'd have bracketed the transition. But what if there were no dinosaurs? Suppose there were only plants and associated insects?

Late Cretaceous plants already show a preponderance of modern types—oaks, magnolias, sycamores, palms, pines, sassafras, along with gingkos, redwoods, and other archaic forms. Late Cretaceous insects all belong to living families. Only one knowledgeable about fossil plants or insects could find that boundary or bracket that transition. The plants and insects themselves were not conspicuously affected by it. The main elements of those often-coevolving groups were already present and merely expanded into greater diversity during Cenozoic time. So it was with the social insects, where termites, bees, and wasps were already poised for their great leap forward before the Cretaceous was over. What could have made so clean a sweep of large reptiles (excluding crocodiles) without more significantly affecting contemporaneous plants, insects, small reptiles, and mammals?

A similarly puzzling transition was taking place in the sea. Whereas some 80 percent of the species and genera of the microscopic floating plants (phytoplankton) went extinct during a geologically brief interval at the very end of Cretaceous history, the decline of the ammonoids was less marked. They went from a score or more families at the beginning of the Cretaceous Period to half that many at the end before vanishing. Some of their nautiloid cousins survived. Despite the extinction of many genera and species, the clams and snails retained a broadly modern appearance across the boundary.

Did the Cretaceous end with a bang or a whimper? The answer for shelly marine invertebrates seems to be that it ended with a bang for about 12 percent of the families but for a much larger proportion of the species. Ammonites received the coup de grâce perhaps when the base of their nutrient cycle (the phytoplankton) collapsed. The dinosaurs had long been fading. Of the 240 or more genera known to have lived, only about 30 are reported to have been extant 10 MY before the end of the Cretaceous (and Mesozoic), and that number was down to a bare dozen near the end. If the record is read in detail, many discontinuities are found in the inferred lineages of Figure 14.12, any or all of which might be interpreted as extinctions.

Meanwhile, the hypothesis that terminal Cretaceous extinctions were the product of an impacting asteroid or other cosmic body is stimulating much-needed and long-neglected research. And that research is enriching our understanding of the dinosaurs, of extinction, of asteroidal and cometary impact, and of the geochemistry of the platinum-group rare elements. The results are explored and evaluated in the next chapter.

Consider here some subtler, but also important, terminal Cretaceous events. Deep weathering and erosion were then taking place in the continental interiors. Rocks at the terminal Cretaceous surface of North America were being weathered to high-alumina clays as the Cretaceous gave way to Cenozoic, some of which, as bauxites, were to become aluminum ores. Where Cretaceous and early Cenozoic nonmarine sediments were stripped from buried Appalachian folds during later Cenozoic history, some of these clays, then preserved in valley-bottom sinkholes, became the first-mined North American bauxite ores. Those ores and associated high-alumina clays extended as local sheets and areally limited vertical pipes and bowls to depths as great as 120 meters beneath the valley surface in soluble carbonate rocks from Appalachian Georgia and Alabama to Vermont. At places they contain early Cenozoic or latest Cretaceous plant fossils and are associated with lignites that stoked colonial hearths and bog iron ores whose smelting provided rifle balls for the American Revolution. From Vermont north to Labrador and west to northeastern Minnesota and adjoining Ontario, the same weathering interval has also left its tracks. They also are high-alumina clays (and locally bauxite), leaves, and wood of earliest Cenozoic or latest Cretaceous trees (Fig. 14.16).

In eastern North America this represents the most intense, the deepest, and the most extensive weathering episode

14.16 TERMINAL CRETACEOUS OR EARLY CENOZOIC PLANTS FROM A PROTEROZOIC IRON MINE. Specimens *A*, *Sequoia*, and *B*, *Sassafras*, from clays associated with secondarily enriched Proterozoic iron formation in Redmond No. 2 mine, Schefferville, Quebec.

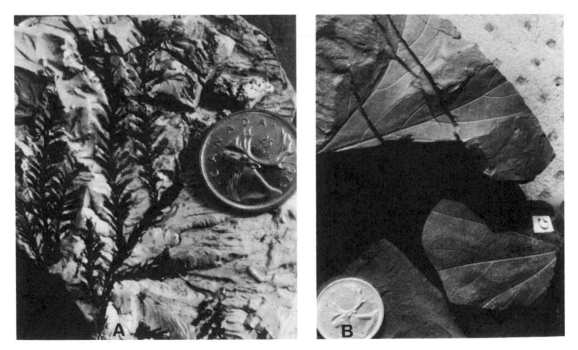

of record except for that of latest Proterozoic to early Phanerozoic time when the Proterozoic iron formations underwent their first deep oxidative enrichment. A decrease in the rate of delivery of nutrients to the terminal Cretaceous sea as a result of such weathering and the reduction of continental surfaces has been suggested as one possible cause for terminal Cretaceous extinctions of the marine phytoplankton.

Mesozoic tectonism in North America was a sometime thing, occurring in different times, places, and styles during intervals of active plate convergence. It increased intermittently from later Jurassic to late Cretaceous, with an early Cenozoic peak (the *Laramide orogeny*). A complexity of collisional features arose where Africa and India drifted toward and impinged on the sites of Alpine and Himalayan orogeny. Similar complexities involve the margins of the Pacific basin, overridden on the one side by Asia and on the other by the Americas—folding older sediments, generating new linear depositional troughs, and narrowing the Pacific Ocean itself. Sialic and intermediate magmas rose above subducting plates at zones of convergence and collision, giving rise to Andean-type plutonic mountains and effusive volcanism along the western coasts of the Americas. California's Sierra Nevada arose in this way during five or more episodes of Early Jurassic to latest Cretaceous intrusion, constituting the most significant North American metallogenic episode and source of precious metals since Archean time.

A dozen or more mainly north–south orogenic belts of this age range and type are recognized in the circum-Pacific fringe, all with different names. East–west folded mountain ranges like those that stretch intermittently from the Pyrenees eastward through the Alps and the Tethyan realm to the Himalaya began, at different and often-overlapping times, to grow in earnest. Details of the main Alpine folding of later Cretaceous time are still debated. Was it primarily the product of obduction overthrusting from the south, or subduction from the north, or both? Eugeoclinal pelagic muds interbedded with volcanic detritus (flysch) are found both north and south of the main ranges. Detached blocks from advancing thrust sheets slid downhill to foreland basins, breaking up en route to become the exotic components of blocky Alpine *wildflysch*. Some tabular, Switzerland-sized slabs, even where already deformed into recumbent folds, seem to have broken loose to glide downslope over previously emplaced thrust sheets (*nappes*) or later-deposited foreland sediments (the *molasse* of Alpine geologists). Features of such Alpine tectonism and stratigraphy, all of which continued into or recurred during Cenozoic history, are pictured on Figure 14.17 and interpreted in Figure 14.18.

Related processes resulted in similar but less dramatically displayed features during the roughly contemporaneous Laramide orogeny of western North America—an orogeny whose product was the Rocky Mountains and associated structures of the eastern North American Cordillera. As impinging exotic terranes moved in from the west and the Pacific sea floor continued to subduct, concurrent rotation of a few degrees clockwise by the Colorado Plateau (perhaps an independent microplate) may have accounted for much of the deformation observed. Arcuate Laramide folding to the east and north, and rifting to south and west, is consistent with such opposing rotations of the Pacific and Colorado Plateau plates. Alpine-Himalayan, Andean, and North American Cordilleran tectonism also shared in the endowment of most of the world's present copper reserves—the famous low-grade copper–porphyry deposits, found worldwide above subduction zones of latest Cretaceous to middle Cenozoic age.

14.17 THE ALPS AT LARGE AND SMALL SCALES. Scene *A* looks east across Wallensee, Swiss Alps, toward recumbent syncline and stacked nappes. See Fig. 14.18 for interpretation. *B*, shearing deformation in siliceous crystalline rocks, Cortina area, Italian Carnic Alps; diameter of lens cap 5 cm. Movement from right to left is indicated by structures visible in *A* and upper half of *B*. Rocks in the overriding nappe of *A* are lower Cretaceous limestone and dolomite at crest. [*B, courtesy of Richard Sibson.*]

14.18 SKETCH INTERPRETING GEOLOGY PICTURED IN FIG. 14.17A. The entire thick Tethyan
Lower Cretaceous in this region is shown here in vertical profile, resting on Jurassic strata at
the distant bottom edge. Upper strata within the northward (*left*) overturned crestal syncline
consist largely of a smallish, thick-shelled, rudisted clam, *Caprotina*. This prominent white,
late Early Cretaceous limestone extends the width of south Europe and into northern Africa.

In Summary

The Mesozoic shares honors with the
Cenozoic as the birthplace of stratigraphy, and of a geologic time scale reflecting the concept of a systematic
progression of life as the basis for
sequence in Earth history. It remains a
good example of that principle even as it
portrays a view of a world at once remote
from and directly antecedent to our own.

What were the main elements of life in
that world—dinosaurian and otherwise?
What were the physical processes, events,
and settings with which they interacted
to create Mesozoic history? How come
so prevalent an equability of climate?
Why did so many of its most distinctive
denizens fail to complete the course?
The middle kingdom speaks for itself in

381

the vocabulary of fossils, rocks, and geo-chemistry, and the syntax of plate tec-tonism, paleoclimatology, and evolution.

The movements of the Mesozoic tril-ogy reveal themselves alike in plate motions and dinosaurian evolution. The Triassic saw the completion of Pangaea and the onset of the tensional stresses and premonitory rifting that were to tear it apart. It was a time, above all, of relatively low sea level, of extensive con-tinentality, of aridity, and of deserts, redbeds, and generally limited epiconti-nental seas. It was the coming-out party and the proving ground of dinosaurian (and archosaurian) evolution (Fig. 14.12).

A variety of new types of life arose to fill ecospaces vacated during the termi-nal Permian annihilation, responding adaptively to physical changes. We see in the seas a restoration of the coiled, complexly chambered, marine ammo-noids, and on land the dinosaurs and the first mammals and birds. The ammo-noids, following their major resurgence during earlier Triassic history, again barely survived an oddly selective pair or set of extinctions of seemingly long duration which also eliminated a large percentage of the then-extant reptile families (including many dinosaurs). That happened before and around the time an asteroid impact announced the end of Triassic history at Manicouagan Lake, Quebec, around 210 to 213 MYBP.

The Jurassic brought recovery. With it came the soaring pterosaurs and with them a great diversification in shape and size of their dinosaurian cousins, includ-ing the largest animals ever to walk the Earth—some having the weight of a dozen large elephants. A second come-back of ammonoid diversity radiated from a small handful of surviving Triassic

genera to give rise to a thousand or so new ones during the course of ensuing Mesozoic history. Spreading rates increased and Pangaea began to break up. Tethys, the subequatorial seaway that was to play so large a role in later Meso-zoic history and climatology, began to open as Pangaea wedged apart from east and west. The North and South Atlantic oceans began to open, allowing succes-sive intrusions of paleo-Pacific waters from the west into an embayment now represented by the Gulf of Mexico. Thick deposits of salt from evaporation of these waters were deposited there, as in east-ern Tethys—breeding stuff of the salt domes whose later intrusion provided structural traps for the impounding of late Jurassic and Cretaceous oil.

The arid climates of Triassic and ear-liest Jurassic, a time of sand seas and deserts, gave way at last to increasing humidity and climatic equability. For-ests spread. Coal began to form at upper mid-latitudes. The Atlantic continued to open as the Americas overrode the Pacific basin. Subduction, volcanism, and the growth of Andean-type mountains were consequences. Microcontinents from southwestern sources sidled up to the Pacific coast of North America to dock as parts of a growing accretionary col-lage that now makes up the western margin of the continent.

Separation of Cretaceous from Juras-sic is more a matter of modes than abso-lutes. The Cretaceous was above all a time of benign climate, coupled with extensive marine inundation of the still actively separating plates. Dinosaurs reached their acme as vegetation spread. Deciduous trees and other flowering plants brought color to the biosphere. And coals achieved their greatest abundance. A variety of biogenic reefs grew in the shallow, warm, shelf waters of an expanding Tethys. Nearer shore, reefs of more turbid waters (mostly Jurassic) consisted mainly of sponges. Offshore banks and reefs were commonly the work of the distinctive Tethyan rudistids—improbably massive clams that appeared with the late Jurassic and faded with the ammonoids and dinosaurs. Others were essentially modern coral reefs. Embayments of the Tethyan sea became the sites of half the world's oil reserves.

Accretionary tectonism continued apace. Separation of plates continued to widen some oceans and narrow others. Mountains arose, responding to plate convergence and rotation. In North America they reflect the Laramide orogeny, enduring from about 80 to 40 MYBP (latest Cretaceous into early Cenozoic).

The Mesozoic was truly a time of transition from archaic to modern Earth history and biotas. The present continents, crunched together at the beginning, spread apart, providing more seacoast and improving the view. Despite the ever more modern appearance of the vegetation, there were still no meadows of grass to produce the forage and grains that keep modern herbivores going. In the end, the archaic reptilian faunas gave way to mammals, birds, and smaller, more tractable reptiles. Even the humble invertebrates changed for the Cenozoic party. All Mesozoic insects belong to orders living today. Social insects prospered with the flowering plants. All elements of the modern coral-reef complex originated and established their relationships as the dinosaurs waned and mammals took the spotlight.

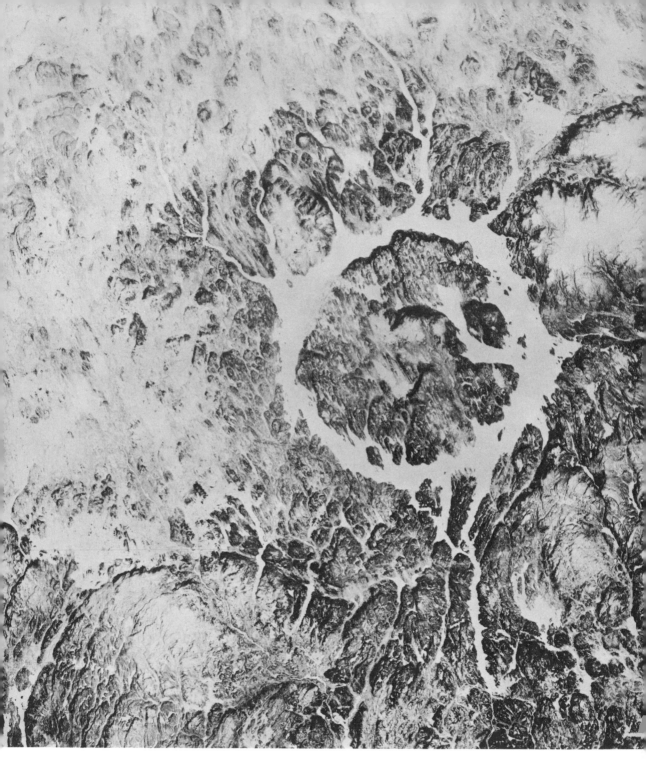

15.1 MANICOUAGAN IMPACT SITE, DIAMETER 70 KM, AGE 210 MYBP. Location 550 km NNE of Quebec City, Canada, at 50° 23′N by 68° 42′W; September 1974. The ice-ringed impact site maps geologically as an island of immediately postimpact volcanics surrounded by Cryptozoic crystalline rocks. Presumably asteroidal, this impact occurred within the time range of later Triassic extinctions. *[Courtesy of Paul Lowman; a NASA Landsat image.]*

CHAPTER FIFTEEN

BIOLOGICAL CHANGE
ON A MOBILE EARTH:
EVOLUTION AND EXTINCTION

Systematic change over time, especially where it exhibits a net trend toward increasing diversity and complexity, is called evolution. The universe and everything in it evolves— the galaxies, the stars, and the planets. Life too evolves. That is a predictable consequence of the persistence of a naturally variable system on an evolving planet. The proof is in the rocks, in the long record of paleontological change and its correlation with changes in climate, geography, and habitat—all of which reflect planetary evolution and particularly plate tectonism.

The issue is not did evolution happen, but why and how and how fast. The most widely prevalent view sees it as a product of intrinsic, but ultimately limited, genetic, physiologic, and morphological variability, channeled by selective pressures that arise from the kinds of occupiable spaces and ways of life available. Extinction is an important part of the evolutionary equation. It is the ultimate fate of all species, releasing ecospace for occupancy by qualified new or migrant forms. Natural selection, constrained by genetic potential, determines who is qualified. The particular causes and rates of evolution and extinction are perennial subjects of discussion.

Did Life Evolve?

Biological evolution, for many, is a troublesome, sometimes incendiary, subject. Constructive discussion of it begins with mutual understanding of words used. Dictionaries define evolution as an unfolding, a process of change over time, particularly if it seems to follow some trend. Evolution goes on all around us. It happens to the universe and all its parts. The everlasting hills wear down yet do not fill the sea. The rotating surface of our planet continues to receive and assimilate a thousand tons of new micrometeorites daily, with something the size of an asteroid or large comet about once every hundred million years. Earth's atmosphere, oceans, and solid crust interact with one another and the biosphere as all respond to internally driven plate motions. Why should life be exempt? It isn't. Extensive, independently verifiable evidence, presented earlier and to follow, shows that it is not. The composition of the biosphere changes. Lineages evolve and eventually become extinct. New species, genera, and families emerge to take their places. Biologic evolution has resulted in greater diversity and complexity with the passage of time. By definition, that is evolution—as is any directional change with time.

Hypotheses about biologic evolution abound, but its existence is neither an hypothesis nor a theory. It is, like gravity, an observable fact—calling for explanation, but deniable only by those who decline to observe. One can't *see* gravity, but it can be felt and measured. With biologic evolution (in this chapter simply evolution), although it cannot be felt, anyone who is willing to view the evidence can see that it has occurred. And, thanks to molecular biology, it can now be measured as well. The problem is not whether, but *how* evolution takes place. As is the case with gravity, there are hypotheses about evolution, and one very widely accepted general theory: Charles Darwin's theory of natural selection, with subsequent modification in detail.

Suppose you were little informed on biological matters but had read enough in the preceding pages or elsewhere to realize that there are firm grounds for the numerical ages and succession of Phanerozoic periods and systems shown on the right front endpaper. Then examine the array and succession of different creatures there sketched. All represent real organisms that lived at or near the indicated level in time. Even so limited a selection reveals that life on Earth has followed a broad succession of forms of generally increasing complexity. That records evolution.

Consider evidence that was intended only to represent some of the main recorded forms of life (principally marine) at different times in Phanerozoic history, with no intention to show anything about evolution. Figures 13.13, 13.18, and 14.14 do this for the Paleozoic and Mesozoic eras. They are literally random choices from illustrations of common or distinctive species available. But notice how different the representative forms are from one level to another. Or follow a set of images marked by the same symbolic letter from older to younger, paying attention to differences. The cephalopods (k), for example, from the Devonian through the Cretaceous, all belong to a group (ammonoids) whose successive living chambers meet along surfaces that appear on internal molds as lines (sutures) that twice became progressively more complex before final extinction. Following terminal Paleozoic and then a pair of Triassic extinctions, Mesozoic survivors repeat a similar but not identical trend (Fig. 14.14) before finally vanishing with the dinosaurs—perhaps in the wake of an event such as that pictured in Figure 15.1.

Other more striking examples of biologic evolution can be observed when apparent courses of development are followed through time—especially in the case of the larger, and more familiar, backboned animals (vertebrates). Such examples have been exaggerated in the past by oversimplification, and by conscious selection to emphasize prevailing tendencies (trends) without regard to deviations. Variations in the morphology of horse's feet and teeth and elephants trunks and tusks once taken to imply linear, highly directed evolution, do not occur in such regular order. The late George Gaylord Simpson (Fig. 3.3), the Darwin of his time, showed that a realistic diagram of horse evolution looked more like a bush than a ladder. All the same, the prevailing tendency of horse evolution over Cenozoic history *was* toward larger size, longer legs, fewer toes, and larger, more molarized grinding teeth. Small, early Cenozoic browsing animals with three or four toes and simple teeth evolved toward larger, swifter, hoofed grazers, having higher-crowned and more complex teeth. Key events affecting that progression were the appearance of grasses in mid-Oligocene time, followed by open prairies and large mammalian predators. Such elements favored the survival and propagation of those fleet of foot and better able to masticate siliceous grasses by reason of longer legs and infolding of the dentine layer.

Figure 14.12 schematically summarizes selected aspects of the well-known diversification of dinosaurs, other archosaurs, and related reptilians during Mesozoic history. All relationships there implied are supported by matching skeletal details (comparative osteology). Although the lineages and images shown are grossly generalized, they represent real animals, correctly located in time but only roughly proportionate in their highly variable size. The giant sauropods hold the record with lengths up to 25 or even 37 meters and weights of 80 to 100 tons, whereas some theropods were barely the size of barnyard fowl. The ranges in time of representative groups are approximated by solid lines. An evolutionary pattern emerges that, in broad terms, is seen repeatedly over time in different groups of animals, vertebrate and invertebrate alike.

Starting from the fact that Figure 14.12 is a highly simplified summary of the evidence, showing far more continuity than there is in reality, consider a likely interpretation of it. Biologically impoverished earliest Triassic terrestrial landscapes, with a diminished carry-over flora, some insects, a few amphibians, and a reduced reptilian fauna, presented many opportunities for recolonization by vertebrates. They were available to any survivors of the Permian extinction having the latent genetic potential to diversify into unoccupied ecospaces, survive there, and leave fertile descendants. The generalized thecodont reptiles filled the bill. They gave rise to a later Triassic expansion of "experimental" diversification. Successful survivors occupied the newly open ecospaces and continued to evolve along fewer successful lines. They were the archosaurs, comprising two main lines of dinosaur evolution that differed in pelvic structure, as well as the more reptilian crocodiles and the soaring pterosaurs. The sparse fossil record of later Triassic and earlier Jurassic terrestrial vertebrates indicates that this radiation was followed by the extinction of about a quarter of the then-existing lineages at intervals during the last two-thirds of Triassic and earlier Jurassic history.

Birds (then in effect small, presumably warm-blooded, theropod dinosaurs with feathers and wings of debated function) are a late Jurassic innovation—earlier *if* problematical late Triassic bones from the Texas panhandle prove to be those of true birds. Early descendants of *Archeopteryx* or contemporaneous forms reached the number of seventeen recorded species before the end of the

Cretaceous, after which they lost their teeth, became unequivocal power flyers, and, over time, gave rise to a diversity of descendants. These added features, along with feathers, set the birds apart as a distinct nonreptilian class of vertebrates.

An example of invertebrate evolution is suggested by the time-calibrated sketches of Figure 15.2, depicting hypothetical lineages of the sand dollars, sea urchins, and related marine invertebrates. These simple sketches represent all of the classes and orders of a long-lasting subphylum of echinoderms, the Echinozoa. The pattern, like that of the dinosaurs, is familiar. Beginning with an early irruption of diversification (in this case of uncertain ancestry), lineages evolve through a diversifying pattern of correlated form and life-style. Although prevailing tendencies (trends) can often be made out, other "patterns" present seemingly aimless diversifications of form.

The essence, as Darwin phrased it, is *descent with modification.* Whether or not there is always increase or decrease of diversity or complexity, the prevailing pattern over time so far has been increase of both, with briefer intervals of dramatic diversity decrease in the form of mass extinction. The observed pattern of total familial diversity changes over time as a consequence of the balance between extinction and branching (Fig. 13.8, *tier 4*). Nothing inherent in evolution, however, implies that it should always run in the direction of increasing complexity. Its only "goals" are survival, perpetuation of the species, and the colonization or reoccupation of any new ecospace.

Why There Are So Many Kinds of Life

Consider, then, the diversity of life. About 1.7 million species of organisms have been described and named—including 47,000 vertebrates and some 500,000 kinds of plants (including algae and fungi). Nearly half are insects, among which ants and beetles show the greatest diversity. New ones, mostly insects, are being discovered at the rate of about 10,000 yearly. Insects, in fact, are expected to add up to as many as 30 million, if the world's rain forests (already half gone) last long enough for a tally to be completed. How many kinds of life may have shared our planet before becoming extinct over the nearly 700 million years of Phanerozoic history is anyone's guess, but, were a count possible, their numbers might well tally in the billions, of which no more than a small percentage are known as fossils. The species we know are grouped in hundreds of thousands of genera, many thousands of families, several thousands of orders, hundreds of classes, and about thirty-five phyla (the highest taxonomic level below kingdom).

Why are there so many kinds of life? Given the biochemical and molecular evidence that all are interrelated in some way, what are these relations and in what historical sequence do they fall (e.g., Fig. 15.3)? How did the descendants of common ancestors become different? Inquiring people have been studying and thinking about such questions for the last two centuries, with the result that there is now a high level of consensus

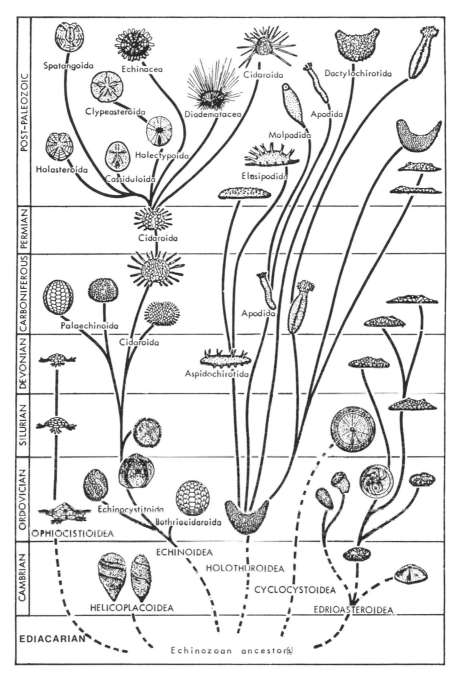

15.2 Hypothetical succession of related marine invertebrates spanning the whole of Phanerozoic history. Echinoderms of the Subphylum Echinozoa, their six classes and twenty-six orders are represented by the forty-five genera here linked in inferred ancestral-descendant order. The two lineages at the right lived attached to hard objects, others were free-living bottom dwellers. A sparse but well-preserved fossil record is known from earliest Cambrian onward. A possible Ediacarian ancestor is the soft-bodied *Tribrachidium* (see Fig. 13.1D). *[As hypothesized by H. B. Fell, 1966, Geological Society of America*, Treatise on Invertebrate Paleontology, Part V, Echinodermata 3, *p. U116, Fig. 96.]*

about the major features of evolution and a lot of healthy disagreement about details.

To the first question one could respond, without much fear of contradiction, that there are so many kinds of life because there are so many places and ways for things to live—each with different requirements and opportunities, and the whole modified by life itself and otherwise changing with time. As new ways and places to live (niches) arise or old ones are vacated by extinction, creatures that qualify to fill them tend to emerge—with some lag. It seems that nature really does abhor a vacuum.

Other questions arise, however, and will continue to arise. Why weren't niches or ecospaces where different things might live all filled up long ago? Why did they fill up in the order they did? What happens if and when all should become filled? Answers to such questions depend on what one thinks about how and how fast evolution works, an actively (and often hotly) disputed problem of modern evolutionary science.

As for the survival of any living component in a fine-tuned world, it must be able to read the music and play a role. It must have qualifications and be able to find a place where those qualifications ensure its functional continuation. The characteristics of the species must be consistent with and able to change with the demands of its environment. Yet the ability of any species to become adapted is limited by ancestry and by individual genetic (and other) qualifications and experience, as well as by its range of environmental variation and tolerance. Even the human species (and the leaf-cutter ant), with about 100,000 genes and a billion different combinations of its nucleotides, is limited in what it can evolve into. How is the successful mix decided upon?

When Darwin, then 22, set out on his voyage of exploration as naturalist on HMS *Beagle* in late 1831, he, like evolution, had no explicit goals. To him it was little more than a great opportunity to observe God's creative handiwork first hand, a youthful voyage of vaguely defined romance and adventure that would have had little chance of NSF support. Moreover, little credence was given in those days to Lamarck's ideas that species evolved as a result of natural processes. The only question, if any, in the contemporary Judaeo-Christian world was whether a divine being really made all species now living in a single unrepeated set of creative acts during the six biblical workdays of Genesis or continued to make new ones. (For the sequences of fossils observed by Cuvier and Smith implied repeated and perhaps continuing creative acts.)

When Darwin returned to England he was five years wiser. He had seen, especially among the giant tortoises and tiny

15.3 THE MAIN KINDS (PHYLA) OF ANIMAL LIFE AND THEIR INFERRED DESCENT, BASED ON FORM AND STRUCTURE. Sketches of most adult animals greatly reduced, of larvae enlarged. Underscored lettering denotes common names for phyla; scientific names in parentheses beneath. Capital lettering without underscoring indicates major body types. Subordinate body types are indicated by broken underscoring. Starred phyla known as fossils. No sequence in time implied. [After Preston Cloud, 1978, Cosmos, Earth, and Man, *Yale University Press.*]

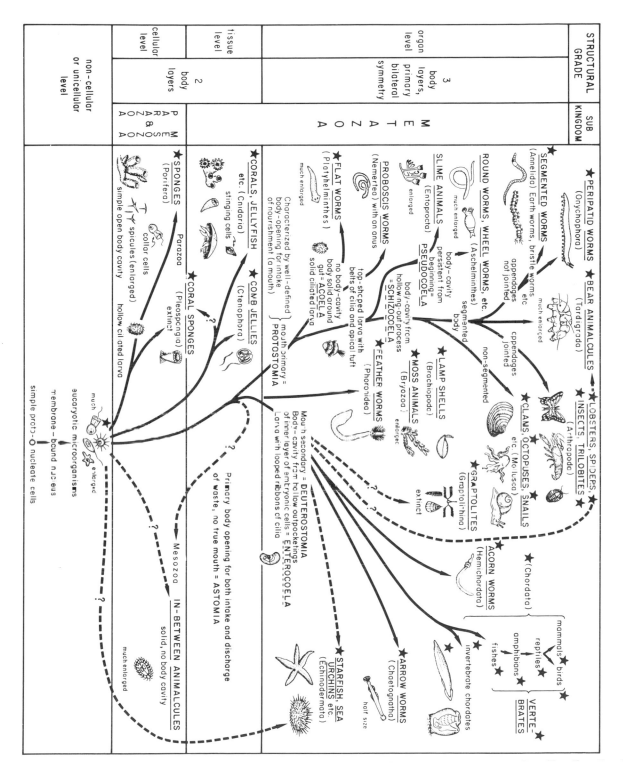

finches of the sixteen main Galapagos islands, that species differed from place to place and that, where similar species lived side by side, they had different feeding preferences or other living habits and did not interbreed. Each plied its own niche. So much duplication of effort did not seem worthy of an omnipotent creator. The thoughts of the musing Darwin paralleled words published by poet and hymnist William Cowper in 1785 in his great lyric poem "The Task." Could it be, Cowper asks, incredulously, that natural law, once set in motion, could operate to bring about the changes observed without further divine intervention?

That, at any rate, is what Darwin finally concluded, when, after another quarter-century of reflection and documentation, he published his epochal work: *On the Origin of Species by Means of Natural Selection*—a book that shook his Victorian world to its creationist roots. In what may be the most widely known and quoted (but least read) book in an Indo-European language, after the Bible, the Koran, and perhaps *Das Kapital*, he established the base from which all subsequent studies of evolution have sprung and to which most return. Even in the unlikely event that Darwin were eventually to be proved mistaken in other than trivial detail, this work would still be honored for scrupulous documentation, deep insights, high standards of objectivity, and its lasting heuristic effects.

What did Darwin conclude, with which so many still agree? He concluded that natural isolation of small populations, no more distinct from parent species than stock in the paddocks of animal breeders, could, in time, become new species by a kind of natural selection compara-

ble to the selective breeding of stock. Isolating mechanisms that could substitute for the breeders' fences include geographical separation—as on islands (e.g., Galapagos tortoises) and in deep valleys (Hawaiian land snails)—and sexual exclusion (some Galapagos finches). Others might be behavioral distinctions and preferences in food and habitat. Such mechanisms, acting on small populations of limited variability, were seen as the agents of natural selection and the *branching* or splitting of species. For, although simple variation within a species over time may eventually lead to recognizable differences between initial and later populations, no multiplication of species occurs without branching.

Otherwise, how could the observed variation in plants and animals have arisen and been parceled into distinct, noninterbreeding species in the time available—estimated by Darwin to be little more than 400 MY (not a bad guess for Phanerozoic time)? How could that happen when all the efforts of 10,000 years of later human history had failed to produce a dog or cat that, given opportunity, would not mate with others of the species *Canis familiarus* or *Felis catus* to produce fertile offspring? What was the cause of variation, anyhow? How could there be intrinsic differences that were inheritable from generation to generation when, as August Weismann was to show, the "acquired characteristics" favored by Lamarck were not a valid mechanism? And how to explain the observed differences within populations? Indeed Darwin never quite gave up the idea that Lamarck might be right about acquired characteristics. After all, genetics was then unknown and generations of stretching might seem as good an explanation as any for the giraffe's long neck.

The problem of time has been answered by geologists (Ch. 4). Time is ample. The first steps toward resolution of the question of inheritance were being taken by the Austrian monk, Gregor Mendel, while the *Origin* was on its way to publication. And the triple rediscovery (in 1900) of Mendel's work by the first generation of geneticists following him was to contribute to an advancement of the field beyond anything a priest of those times might have imagined, or perhaps approved of.

Inherited differences are programmed by means of genes and chromosomes that pass from one generation to another. The genetic potential of an individual defines its *genotype*. The characteristic physical and behavioral expression resulting from interaction between genotype and environment is the *phenotype*. The traditional object of study has been the phenotype; but behind every phenotype is a genotype, and both are targets of selection. Each species and population has a distinctive but variable gene pool from which the genetic composition allotted to each individual at birth specifies its intrinsic characteristics within limits set by environment and experience. The characteristics of the gene pool vary between generations because of mutations (changes) of genes and chromosomes and recombinations of their sequences. The recombinations result from genetic "mistakes," such as overlap or offsetting of the strands of DNA on which the genes are situated or the insertion of transferable genetic elements.

All of these variations and departures from Darwinian natural selection collectively have been called neo-Darwinism.

But *how*, exactly, does natural selection work? Turn-of-the-century geneticists, notably the Dutch botanist Hugo de Vries, saw evolution as directed by mutations. But that became an Achilles' heel. Mutation as a source of variation was one thing, but how could random variation direct? It can't, said students of form and function: it is the response of populations, of phenotypes, to the pressures of habitat variation and their relative reproductive success that channels genetic evolution. Previously futile discussion came to a boil in the decades of the 1930s and 1940s among a couple of dozen European and American evolutionists, some of whom finally convened a meeting at Princeton University in 1947, with gratifying results. The meeting of minds and evolution of outlooks that there occurred led to modification of previous views of neo-Darwinism that resulted in four mostly harmonious decades of cooperation and communication among evolutionists of different orientations. The consensus then reached has been accurately, if awkwardly, called the *synthetic theory* of evolution.

Despite the ring of phoniness that accompanies the word "synthetic" in many minds, there was nothing phony about the synthetic theory. It was, in fact, a kind of neo-neo-Darwinism that arose from a broad effort by evolutionists of different persuasions to assimilate the evidence for and understand the views of others. Paleontologists and systematists learned some genetics, and geneticists conceded the point that heredity does not work in an ecological or historical vacuum. That enlightening accommodation has, to a large degree, continued, while researchers at the new molecular front in particular have opened new avenues of understanding. And our still-incomplete but ever more refined understanding of how evolution works has now reached a level that might be called *trans-Darwinian*. Still Darwinian in the sense that natural selection remains the focus of evolutionary theory, trans-Darwinian thinking presents a more comprehensive perception of how evolution works than could have been visualized before mid-century.

Viewed through this ever-widening window of the mind, we begin to see how apparently random mutation events can accumulate in preadaptive directions while existing adaptations continue to function. The large organic molecules called nucleotides (Ch. 10) are the elementary particles of genetics, the components of DNA and RNA. They function in pairs to make up the two strands of the information-carrying double helix. There may be as few as about 4 million pairs in a bacterial cell (or only about 1,300 in a virus) to as many as a billion in the cells of some eucaryotes. Explicit sequences of these nucleotides a thousand or so long—like letters of a sentence in which the words are run together—carry the information that is coded as genes along the strands of DNA.

Not all of these sequences, however, carry information. Within the sequence of thousands of nucleotide pairs comprising a gene, all information is carried on repeated sequences of multiples of nine nucleotide pairs (known as *exons*) separated by noncoding sequences (*introns*)—the spaces between genetic "phrases." There being thousands of coding genes on a single molecule of DNA, ample opportunity exists for variation as a result of structural shifts during the reductive division of sexual reproduction—for example, shuffling of the genes and overlapping of the DNA strands.

As geneticists now tell us, the storage and transfer of genetic information is also highly redundant. Thousands of copies of the same gene are observed in some organisms. Duplicate and non-functional genes are common. A single complex gene may split up into separate genes. Separate genes may join to form a different one, as we may change the meaning of a sentence, a phrase, or a word by rearranging words or letters. Replicate genes may perform the same function; or one may continue to do the work while others drift, unused, in different directions until accumulated variation prepares them to function in unanticipated ways—as in the poison-injection mechanism of a viper or the directed scent gland of a skunk.

It is often asked why apparently useless or even harmful things evolve, or why, if selection is so effective, anything ever becomes extinct. What use are the tusks of an elephant? Why sickle-cell anemia? What good is an appendix? Most such features, on examination, have proved, or are expected to prove, functional in obscure ways. They may be genetically linked with advantageous features, or success in mating and propagation; or be residual from a time when they were functional. The gene for sickle-cell anemia, for instance, as for all survivable lethal genes, is recessive, functioning as a killer only when inherited from both parents. It survives because its carriers are rewarded by immunity from malaria.

Genetic isolation is important for the branching events that account for the multiplication of species. The interbreeding that goes on in large populations keeps the gene pool so diverse that local differences that otherwise might become fixed are soon swamped. Where sectors of a gene pool, however, are prevented from backbreeding by reason of geographic, behavioral (including sexual), or other genetic isolation long enough for the accumulation of differences incompatible with the parent species, speciation has occurred. The split is final. If and when contact with the parent species is restored, no or only sterile hybrid interbreeding is the rule. Small, isolated populations such as this, *the founder populations* of Harvard's Ernst Mayr, are widely considered to be the active scenes of natural selection among sexually reproducing organisms. They contrast with populous and widely distributed species which may continue virtually unchanged throughout long species lifetimes.

The central core of trans-Darwinism remains natural selection, but a perception of natural selection that is strongly linked with genetics, molecular biology, and ecological opportunism. Selection, like a downhill skier, can choose only from what presents itself. Despite the enormous potentiality of organisms for variation, variability in all species is limited by their ancestry and the composition of their gene pool. Despite the obvious advantages of wheels, for instance, a gene for wheels has apparently never been a part of the gene pool of any organism. Continuing attention now focuses on rates and mechanisms of evolution, on the origin of life and the eucaryotic cell, on the origins and early history of the metazoan phyla, on host–parasite relations, and on other lively issues. Future students of evolution will find no dearth of problems.

On the Kinds of Evolution

The kind of evolution considered so far in this chapter is that of creatures like ourselves—bisexual animal species that live, take in food and moisture, seek shelter, reproduce from fertilized ova, die, and eventually become extinct (on an average after about 1 to 10 MY). In addition to the plants that surround, sustain, and sometimes even poison and entrap them, animals share their world with a huge biomass of mainly wee creatures of different or no sexual persuasion and therefore different reproductive habits and genetic characteristics.

Many plants and some animals undergo an alternation of reproductive style between sexual and asexual or unisexual generations, or successions of generations. Many plants and coelenterate animals (e.g., hydroids) undergo a regular or seasonal alternation between sexual and asexual generations. In some insects and crustaceans (e.g., aphids, water fleas), the alternation is environmentally programmed between bisexual and unisexual. Under favorable conditions, as when groups of genetically identical female aphids (clones) are ravaging your roses or devouring your broccoli, they continue to replicate themselves. Only when conditions become unfavorable, commonly with seasonal change, does the aphid's fancy turn to sex. Then some of their eggs hatch as males, and the Oktoberfest is on.

Among organisms that normally reproduce by simple cell division (including bacteria and many single-celled eucaryotes, or protists), each cell simply replicates itself repeatedly until, for reasons still not understood, a curious thing happens. Individuals fuse in pairs to produce a combined cell in which chromosomal restructuring takes place. Even though it is possible or characteristic for neuters or females to propagate and maintain populations without sex, it becomes clear that genetic recombinations do, somehow, play a vital role, even if only episodically.

What could that role be? If so much of life is able to flourish without or with only occasional genetic recombination, what good is sex beyond entertainment?

Until the discovery of genes, chromosomes, and mutations, such questions, like that of life in other galaxies, would have been unanswerable. With the genetic code and mutations, the problems involved enter the realm of fruitful inquiry. Indeed it is apparent from experience with antibiotics that primarily asexual organisms can respond very rapidly when survival of the species is at stake. Why not, where a single surviving mutant could, in theory, generate by simple cell division a clone of identical resistant individuals that are, in effect, a new population—conceivably even a new species?

Bisexuality serves other purposes. There are advantages to the regular recombination of DNA, chromosomes, genes, and the ninefold sets of nucleotide pairs that occur when egg and sperm unite to share their genetic signatures. That gives the gene pool of the local population and the species both a diversity and a stability that asexual species lack. The diversity in all respects of a large interbreeding gene pool combines flexibility in response to environmental change with the potential for diversification that occurs with isolation of peripheral or otherwise distinctive sectors.

One sees, finally, in many plants and rare animals, unusual but real examples of, in effect, instantaneous speciation. That happens where the standard (*diploid*) number of chromosomes is multiplied from one generation to the next because pairing of male and female sets of chromosomes is not followed by the usual reduction-division (meiosis) that ordinarily precedes the joining of egg and sperm. Such multiplication of chromosomes, known as *polyploidy*, is reported to account for 47 percent of all species of flowering plants, including common agricultural types such as wheat and potatoes.

The point that evolution does not begin and end with sexuality in animals must be made but not dwelt upon. Instead, consider the processes of evolution.

On Selective Pressure

In order for natural selection to serve as a channeling process, countering the randomness of mutation, selective pressures must be at work. For, as most informed adults know by now, evolution is *not* random. The course of evolution moves in directions that to some have appeared programmed or even predestined, if not always beneficent. No one needs to be advised, as an Arkansas court was during recent hearings on an anti–First Amendment bill, that a gale in a junkyard would never assemble a B-29.

So, what channels natural selection? It is channeled, like the course of skiers on a hillslope, by environmental pressures, subtle or lethal. Like opportunistic skiers looking for short lift lines, organisms move into unoccupied potential ecospaces (*niches*), for which their attributes qualify them. They become extinct when they fail to adapt to changing conditions. Because a niche in the ecologic sense is by definition occupied, and by a single species, there can be, by definition, no empty niches. That expression, where used in this book, is figurative, including any potential vacancy within an existing ecospace. And any ecospace may include overlapping niches, depending on size and specific living requirements—like the bacterium in the hindgut of the termite in the forest. Interaction of potentially available ecospace with ancestral gene pools is the central channeling factor in evolution.

Oxygen was an early forcing factor. Considering the primordial anoxic Earth, after its waters became otherwise receptive, the biologist reconstructs a likely scenario in which the first minute anaerobic beings were something like bacteria. When one of them came to produce and then to tolerate O_2—and after the many early O_2 sinks were sufficiently neutralized by the combined forces of photolytic and photosynthetic O_2—the nucleate cell arose, sexuality began, and eucaryotes flowered. As O_2 rose to levels high enough to meet metazoan requirements, cellularly diversified, many-celled animals appeared, probably from different protozan ancestors, introducing the Phanerozoic Eon.

During an interval of perhaps 150 MY following the onset of the metazoan adaptive mode, a geologically rapid exploratory diversification of new lifestyles took place. It gave rise to ancestors of all of the now-living phyla of shelly marine invertebrates and most or all soft-bodied ones, plus a dozen or more extinct groups of phyletic rank (Fig. 15.4). During all of the more than 500 MY since Middle Cambrian, not a single new phylum (except perhaps Bryozoa) can be shown to have appeared, although classes and orders have. Even the chordates were then present and soon after acquired bony armor and then skeletons. If one backs off far enough to focus only on the record of near-shore shelly marine invertebrates, as James Valentine of the University of California has observed, it appears that classes and orders also underwent a slightly delayed early Paleozoic burst of diversity, then a jerky decline through the Paleozoic, followed by later stabilization. That was paralleled over the same interval by familial diversification, a useful if subjective measure. The big Permian decline was followed in due course by complete recovery of marine familial diversity and then a near doubling to a present peak of more than 400 recognized living families of shelly marine organisms alone.

Why was there such a proliferation of metazoan variety in the beginning? That so stretched an older view of stately unfolding that many in the past (and some still) have preferred to hypothesize a long pre-Ediacarian history of metazoan evolution, supposedly represented by rare markings of dubiously metazoan and even dubiously biogenic origin. Although much research is now focused on the problem of metazoan beginnings,

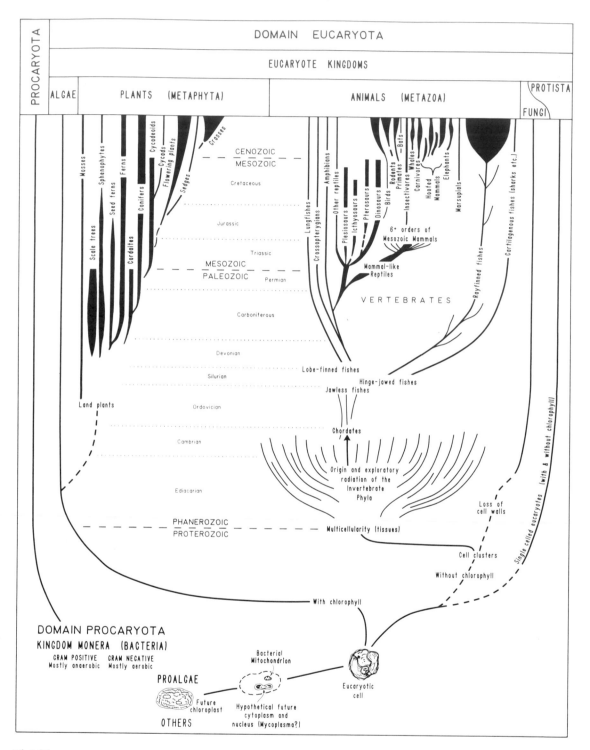

15.4 NEW ADAPTIVE MODES AND THE DIVERSITY OF AVAILABLE ECOSPACE CHANNEL EVOLUTION IN GENETICALLY ALLOWABLE DIRECTIONS. Featuring major groups (phyla, classes, and orders) of backboned animals and plants, where width of line designating approximate vertical range is *roughly* proportional to diversity.

however, unequivocal Metazoa are yet to be found in rocks of undoubted pre-Ediacarian age.

In any case, the first Metazoa would have found all ecospaces ever to be occupiable by metazoans to be then unoccupied. A great opportunistic diversification of metazoan form and function into such potential ecospaces seem likely, including a number of "experiments" that would prove to be nonadaptive. It has long seemed to me that this is what the admittedly skimpy preserved record of the Proterozoic–Phanerozoic transition is trying to tell us.

This early Phanerozoic irruption of diversification, impressive as it was, was not the last. There were others, following on the heels of, overlapping, or at some distance from, every significant extinction event and every emergence in time of important new adaptive modes—as when fishes with hinged jaws emerged during late Silurian history, the reptilian amniote egg in the Carboniferous, and grasses in the Oligocene (Fig. 15.4).

In a very real but nondeterministic sense, evolution is opportunistic. The observed historical pattern is that it takes new directions whenever unoccupied ecospaces become available, whether by the appearance of preadapted evolutionary novelty or by extinction of prior occupants, by geographic or climatic change, or by some combination of these things.

Ever since the beginning of life on this planet, the biosphere, atmosphere, oceans, climate, physical geography, and the crust itself have interacted to shape one another—Earth as *Gaea*, in neovitalistic imagery. Much foreclosure of old and opening of new ecospace results from the pervasive and ever-changing effects of plate tectonism, a major forcing factor in evolution. Landmasses join, separate, and change geography and latitude. Seaways open and close and ocean currents become redirected. Folded, volcanic, and plutonic mountains arise, affecting atmospheric circulation and rainfall patterns. Continents become glaciated as they drift across the poles and then are deglaciated as the conveyor belt of plate tectonism moves them equatorward again. Coast lines become longer and shorter. Migration routes and barriers to migration change. Deserts wax and wane. The circulation of oceans and atmosphere is affected. Climate changes. And the biosphere reacts. The list is as extensive as your imagination.

On considering external sources of selective pressure, one sees that many are related in some way to plate tectonism or to other dramatic natural processes (e.g., meteoritic impact). The record in the rocks implies that evolutionary opportunism responds, with structural novelty and higher rates of speciation, to major events in lithospheric and climatic evolution.

Rates of Evolution: Lethargic, Spastic, or Variable?

A perennial question in evolution, as in all geological processes, is that of rates. Nowhere is it better stated than in the earlier-quoted words of the nineteenth-century Cantabrigian divine and polymath, William Whewell, in his 1832 review of the second volume of Lyell's *Principles*. There Whewell asks, editorially, whether the changes we see have been "on a long average" uniform or whether they are paroxysmal, punctuating a generally "tranquil" state or even flow of events. He correctly predicts that the matter will long divide the geological world. It has and it does.

How does one best measure and describe rates of evolution or other geological processes, and in what units? When we estimate the rates of sea-floor spreading, for instance, by dividing the distance between equivalent magnetic reversal stripes by the time elapsed since their originally joined emergence at a common spreading center, what does this tell us about the mechanics and rates of spreading? Did the plates creep steadily along at the apparent average rate of, say, a fingernail's growth? Or did they move spasmodically, perhaps a meter every 50 years, or at irregular intervals and times? It would take continuous or successive, closely spaced measurements over long intervals to work this out, and perhaps longer still to decide whether the results of such movement were best considered uniform on a long average, episodic, or catastrophic.

Consider rates of biologic diversification, often discussed under the term *speciation*, but in fact including all hierarchical levels. To paleontologists who have attempted to view evolutionary rates over the long term, evolution has often seemed episodic—undergoing high levels of diversification over geologically brief intervals between longer intervals when nothing much seems to be happening. If 93 percent of the Permian shelly marine species became extinct, as has been guesstimated, it is clear that the follow-

ing marine replacements were, in effect, a new fauna. Evolution responds not only to extinction beyond background levels and to the geographical and environmental changes brought on by plate tectonism, but also to the challenges of new adaptive modes, some of which are summarized in Figure 15.4.

There we see that, perhaps 1,400 MYBP, following the appearance of the eucaryotic cell, a major expansion in numbers and kinds of microbial fossils took place. Following the Ediacarian onset of the tissue level of cellular organization (the oldest Metazoa), and extending into the Cambrian we have seen an interval, long in actual years but short by geologic standards, during which there occurred an exploratory radiation of metazoan phyla—the only one of its kind. Trilobites underwent a seemingly rapid expansion of diversity during Cambrian history, as brachiopods were to do during later Cambrian and Ordovician, contemporaneously with the bourgeoning and then reduction of invertebrate classes and orders. The appearance of invertebrate chordates in the Cambrian preceded a flourishing of jawless Ordovician and Silurian fishes and then a renewed diversity of fishes with the onset of hinge-jawed forms in the Devonian.

Meanwhile, late in Devonian time, came the first quadrupeds—creatures with fishlike tails and backbones, but with legs instead of fins. They were, of course, amphibians, because they could now manage on the land but remained tied to the water by the need to breed and hatch their eggs there. They were, of themselves, authentic transitional forms from fish to reptile. A diversity of amphibians followed, only to be outdone by the Upper Carboniferous and Permian reptilian diversity that began when some slippery amphibian became a scaly reptile, made the basic skeletal changes, and laid an amniote egg. The amniote egg, which could be hatched on land, and the scaly covering that protected its hatchlings against desiccation, freed the reptiles to explore the land and began a new adaptive radiation.

The adaptation that triggered the Mesozoic dinosaur radiation is not clear, but it is a fair guess that it was some form or forms of thermal regulation that allowed them to expand into and migrate across the deserts of their time. This is suggested, among other things, by the perhaps heat-venting "sails" on the backs of late Paleozoic and Triassic pelycosaurs. Dinosaur bone structure, proportions, and prey–predator ratios are also consistent with their functioning at higher metabolic levels than conventional reptiles. Taken en masse the evidence supports some form of internal thermal control. Some slender, presumably agile, dinosaurs might even have been fully thermoregulatory.

Although mammals had already split into placental and marsupial lineages early in Late Cretaceous history, they represent an adaptive mode that remained largely suppressed until the decline, then passing, of the dinosaurs made way for them to flourish. Modern bony fishes record a similar pattern. They underwent an enormous Cenozoic radiation that actually began while the great reptiles still ruled the lands. Why? Perhaps because, as in the case of mammals later, their predators and competitors became extinct. Mesozoic fishes were preyed upon by swimming reptiles (mososaurs, ichthyosaurs, and plesiosaurs) and probably competed with ammonoids for prey. One of the two major groups of ichthyosaurs dropped out at the end of the Jurassic, while ammonoids were beginning to wane. That may have been enough to start the expansion of piscine variety that persisted through the boundary event and is, or was, still going on.

Plants illustrate the same principles in their own way. The appearance of land plants (if not yet tracheophytes) by Middle Ordovician time is now convincingly supported, on grounds of spore morphology, by Jane Gray of the University of Oregon and associates. After some delay, while the simpler land plants became established, five main branches of advanced plant life appeared during latest Silurian and Devonian, covering an ever more extensive terrain with ever-larger plants during most of the remainder of Paleozoic history. Following some extinctions, a product of late Paleozoic and Permo-Triassic aridity, plant life continued along much the same lines until and after flowers finally brought color and social insects to the scene. That interesting preadaptive breakthrough occurred no later than late Early Cretaceous (100 MYBP) and perhaps as early as Jurassic or even Triassic time, but it became prevalent only after responses to continuing plate tectonism further ameliorated Cretaceous climates. The later Cenozoic expansion of the grasses and grains, and the response to it on the part of herbiverous mammals, represents the last of the great adaptive radiations so far clearly visible—one near the end of which the homonids themselves arose and spread worldwide.

The tranquillity of the long intermissions between scenes in the evolutionary theater is, nevertheless, far from absolute. Natural selection continues to function, albeit less dramatically. Gene flow in populous and widespread species results, as a rule, in high levels of stability over long intervals, whereas geographically limited species with small populations and shorter lifetimes come and go. Body counts imply statistically steady levels of background extinction between mass extinctions, while, intermittently, new species or groups of species arise or drift in from some peripheral site of emergence to fill resultant niches or create their own—as the placental mammals only recently (about 2 MYBP) were to do in South America. Although it has been suggested that as much as 90 percent of all extinctions and speciations are of the background type, that is clearly a very rough approximation, especially when one considers similar estimates and claims of more frequent extinctions. Whatever may be the correct ratio, it is clear that mass extinctions and their evolutionary replacements have been a major source of biological novelty, particularly at higher taxonomic levels (macroevolution). They provide the primary basis for major divisions of Phanerozoic history. Such an evolutionary pattern, again whatever the correct ratios, might best be described as *episodic* or *variable-rate evolution.*

That can be defined explicitly as a pattern in which longer intervals of dynamic evolutionary balance are interrupted by episodes of more rapid diversification in the wake of events that result in a diversity of unoccupied ecospaces.

Had I been asked, then, in 1970 to state briefly the thinking paleontologist's perception of biologic evolution, I might have phrased it as follows: Biologic evolution is a product of the response of existing adaptive modes and genetic potential to the changing nature and extent of available ecospace at rates that vary according to population sizes, opportunities, and genetic principles, and which are commonly episodic. Seeing evolution as a contingent process, as I do, I would have added that, under appropriate circumstances, it happens rapidly in a geologic sense and that once-prevalent outlooks leaned toward greater uniformity of process than my own.

The lively discussion about rates that has filled so many pages of scientific and other journals since 1970, however, calls for rethinking the matter. The most widely discussed view, termed *punctuated equilibrium,* seemed at first to be saying something consistent with but narrower and more categorical than the above. It was phrased in flat opposition to what its authors perceived to be the prevailing opinion—a view they characterized as *phyletic gradualism.* The term *punctuated equilibrium* (and the idea) began with the American Museum's Niles Eldredge and came to wide attention as a product of the enthusiasm and eloquence of Harvard's Stephen Jay Gould. That, and its stated challenge to conventional views, have given paleontology and evolution a new visibility, and stimulated or provoked renewed attention to the perennial problems of tempo and mode in evolution. It did for them what Bobby Fisher did for chess and Martina Navratilova for women's tennis.

The heuristic value of this renewed discussion, however, has been flawed by its distorting effects on the public view of evolution and evolutionists and by incidents of personal animus. The impression has arisen on the one hand that it stands Darwin and the architects of the "synthetic theory" on their heads or replaces them with Marx. Others have cheered to see a supposedly entrenched view of exclusively "phyletic gradualism" get its comeuppance. Some see more charisma than content in Gouldian punctuationalism. Still others have seen themselves falsely attacked—an outlook fanned by media accounts of speciation symposia that have been characterized by one eminent scholar as "thoroughly awash in unfounded and often contradictory speculation." How could there be so much equivocation if the evidence weren't equivocal? Why, suddenly, does it seem as if rate were the only interesting problem in evolution?

A good row is as American as apple pie and probably better for your coronary arteries, but what is this row all about? What does punctuated equilibrium really mean? And what is the entrenched fallacy it is meant to displace? In their first joint paper in 1972, Eldredge and Gould define their concept of evolution as one of long intervals of *stasis*, "punctuated here and there by rapid events of speciation in isolated subpopulations." The reason intermediates are uncommon, they stated, is not because the geologic record is so incomplete but because the intervals of speciation are so brief.

There is surely an important element of truth there on a geologic scale of time, but just as surely other equally important elements in the whole truth, including demonstrable gaps in the record. Speciation events "here and there" do not read like the episodic new appearances that either the gifted Cuvier or practicing paleontologists since his time have sought to explain. "Here and there" and "stasis," taken together, are not the same as steady

state with evolutionary interjections. Rather they convey the impression of dead calm zapped by the occasional speciation event. Nature presents few either/or choices. Rates of speciation are certainly related as much to accessibility of niches as to inherent qualities of organisms.

Given that rapid means geologically rapid and not ordinarily one or a very few generations and that *stasis* refers to modest rates of progression, I would have no problem accommodating punctuated equilibrium as a phase of variable-rate evolution. Either might apply to a classic 1966 paper by U.S. Geological Survey paleontologist Bill Cobban. In it he records a succession of twenty-two swift, predatory, and perhaps seasonally migrant species of cephalopods (in this case squidlike creatures called belemnites) at intervals through 1,000 meters of marine sediments that were deposited in eastern Wyoming between about 80 and 70 MYBP. No evidence there, however, suggests a lineage or any recognizably branching events, only a succession in which one species is replaced by another at intervals that average 500,000 years. In the absence of intermediates and the presence of evidence for sea-level fluctuations, peripheral origins with later introductions seem likely.

Considering more broadly what fossils have to tell of evolutionary rates, they seem to tell of a *statistically steady state* of extinctions and new appearances, interrupted by episodic mass extinction and then irruptive or more prolonged diversification. New appearances between such events are probably the products both of geologically brief intervals of speciation and of the small size of the isolated or peripheral populations and gene pools from which new species characteristically arise. Such populations are likely to escape detection or preservation in the fossil record until they become successful, more numerous, and more widely dispersed.

Being, perhaps, the first to claim (in 1948) that appeal to the incompleteness of the record is too facile a way out of troublesome anomalies, I must also agree that incompleteness is repeatedly certified by evidence for recycling and bypassing of sediments, subduction, nondeposition, and erosion. Complicating that are the collecting habits of paleontologists, which determine what is to be found in the museums whose collections so commonly are the data base for paleontological studies. Students of fossils characteristically prefer to concentrate on strata already known to be richly fossiliferous and to omit the usually much thicker and less fossiliferous intervals between. Relative abundances of preserved fossils in different kinds of organisms also bias the data. Least complete are the records of the vertebrates and most complete the various groups of microfossils.

Cesare Emiliani of the University of Miami, who has studied continuous sequences of shelly Protozoa in drill cores, sees evolutionary rates as "a continuous spectrum ranging from lethargic to spastic"—a good description of variable-rate evolution. Could that be what punctuationalists are trying to say?

The discreteness of species in time and the common absence of intermediates are other debated subjects. Species that comply with Ernst Mayr's tidy *biological species concept* are both discrete and contemporary. The *morphological species* of the paleontologist, however, has an extension in time—commonly millions to tens of millions of years. There are commonly so many intermediates that statistical analysis of variation is required for consistent species recognition.

Where intermediates are apt to be lacking is at higher hierarchical levels, although there are good intermediates between all vertebrate classes. The phyla and the invertebrate classes and orders, however, evolved so rapidly in a geologic sense and so far back in a record whose completeness varies inversely with age that the chances of intermediates being preserved are slim. That is where the biggest problems of macroevolution are. Although we should seek explanations for the known record *as if* it were all there is, to deny the likelihood of preservational gaps within a known sequence of species is to retard their recognition.

Is there only a tempest in this pot, or is there some good, nourishing stuff abrewing? Is the action all reaction, the research all bookwork, or are studies of rates in real fossils from seemingly uninterrupted sequences (apart from Lake Turkhana) being undertaken? How static is stasis? How gradual is gradualism? How rapid is geologically rapid? Do punctuated equilibrium and stasis really mean that nothing at all is happening between speciation events that appear like scattered rifle fire during a truce? Or is it more like the episodic or variable-rate, lethargic to spastic evolution described by Emiliani and here?

Better evidence and more explicit definitions are the missing ingredients. Until the evolutionary cinema can be run at the same speed for all viewers, it will be hard to resolve the matter of rates with either precision or accuracy.

When evolutionists have enough

information to make fair estimates of the number of generations required per speciation event in different lines of evolution, what will they find? Will they find a widely variable spectrum of rates, depending on ancestry and genetic potential, mutation rates, availability of survivable ecospace, and identifiable selective pressures? Or will rates be found to cluster about some general mean? Molecular evolution offers a promising approach to that problem.

As punctuationalists have now conceded, the idea of *geologically rapid* shifts of pace in biological evolution and other processes is neither new nor revolutionary to practicing paleontologists. Such a pattern is, in fact, the basis of stratigraphy and the classical sequence of Phanerozoic history, but it is not the whole story. The enemy is a straw man.

Where the punctuationalist view deserves strong support is in its position that concepts of *macroevolution,* the origin of supra-specific levels, and even very rapid morphological speciation, need to be taken more seriously and investigated more rigorously than has been usual. Much discussion over the years, including many entertaining pages by Gould, has been directed to what significance for macroevolution might be found in processes similar to the premature reproduction of a well-known Mexican salamander, the axolotl. Known broadly as *paedomorphosis,* such processes, in

some circumstances, could lead in theory to large changes over a few generations. Instances are known where such an interpretation would be consistent with the paleontological record (e.g., the genus *Homo*; the scleractinian corals).

To contribute more than anecdotal generalities to that endeavor, paleontologists will need to find and study places where fossils are well and abundantly preserved over geologically long and demonstrably continuous intervals. They will also need to deal with the now-convincing concept of *mosaic evolution,* wherein different characteristics of the same species evolve under different controls and at different rates. The oft-mentioned late Jurassic *Archaeopteryx* retained a fully reptilian jaw and skeleton after it had already acquired completely birdlike feathers (including a downy undercoat) and presumably efficient temperature regulation. The earliest amphibians were, in effect, fish with legs. *Australopithecus* retained the cranial capacity of an ape long after its skeletal structure had already adapted to upright walking.

Recently (*Natural History,* February 1984) Gould has made interesting and welcome concessions to variable-rate evolution. He no longer believes that punctuated equilibrium can solve the problem of the fluctuating directions and rates of life's history all by itself.

Causes of Extinction:
Triggers and Means

To borrow some appropriate words from Karl Flessa: "extinction is here to stay." It is clearly a major source of open ecospace and therefore a primary chan-

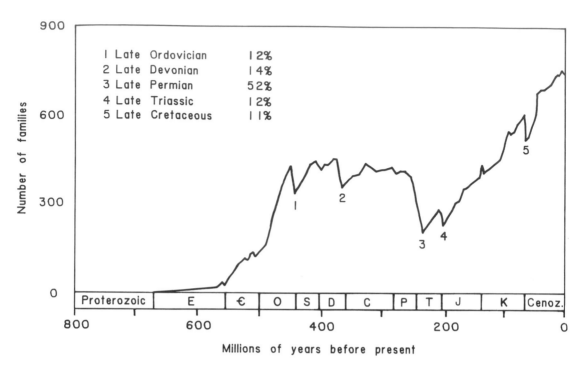

15.5 DIVERSITY LEVELS DURING PHANEROZOIC TIME FOR PRESERVED FAMILIES OF SHALLOW-WATER MARINE ORGANISMS, SHOWING TIMES AND PERCENTAGES OF MASS EXTINCTIONS. *[After D. M. Raup and J. J. Sepkoski, Jr.,* Science, *v. 215, p. 1501. Copyright 1982 by the AAAS.]*

neling influence on evolution. Two main sorts of extinction are recognized—*background extinction* and *mass extinction.* The focus here is on mass extinction, observed at intervals throughout Phanerozoic history (Figs. 15.5, 15.6). So many different and plausible hypotheses have been suggested for different episodes of extinction that a general cause threatens to remain forever beyond reach of all but the most eager believers. We can think of so many alternatives for any given extinction, and there is so much seemingly selective variation in the kinds of organisms affected, that common cause evades the searching mind. Beautiful new ideas are forever getting lost in forests of ugly old facts. It is also important, in

thinking about this problem, to keep in mind the possible distinctions between triggering events and the actual or proximal cause or causes of extinction, and between particular mass-extinction events and general hypotheses of mass extinction.

That is well illustrated by the discussion that swirls around the attractive and plausible hypothesis of cosmic (asteroidal or cometary) collision.

The cosmic collision (or Siva or Shiva) hypothesis was first provisionally proposed in 1970 by Canada's Digby McLaren to explain an apparently near-instantaneous late Devonian extinction event (*2* on Fig. 15.5) that wiped out nearly 15 percent of the then-existing families and perhaps 80 percent of the species of shallow-water marine invertebrates. It came prominently to focus and caught the public eye with the discovery of unusual trace-element concentrations suggesting a cosmic source and cause at the time of a later extinction event of about the same magnitude (Figs. 15.5, 15.6), the late Cretaceous extinction whose victims included the last of the dinosaurs. That discovery, announced in 1979 by Luis and Walter Alvarez and colleagues of the University of California at Berkeley, was in a thin but widespread clay layer (or succession of layers) that coincides or nearly coincides with the biological change from Cretaceous below to Cenozoic above. This or a presumably equivalent interval is now known as the *K/T boundary* from geologists' use of *K* for the German *Kreide* designating the Cretaceous and *T* for the now-anachronistic Tertiary (including all but the last 1.6 MY of the Cenozoic Era). A devas-

15.6 DIVERSITY LEVELS DURING PHANEROZOIC TIME FOR PRESERVED FAMILIES OF NONMARINE TETRAPODS, SHOWING TIMES AND PERCENTAGES OF MASS EXTINCTIONS. *A–C* represent dominant tetrapod faunas of Paleozoic, Mesozoic, and Cenozoic eras respectively. *[Adapted by permission from M. J. Benton,* Nature, *Vol. 316, p. 811, Copyright © 1985 Macmillan Journals Limited.]*

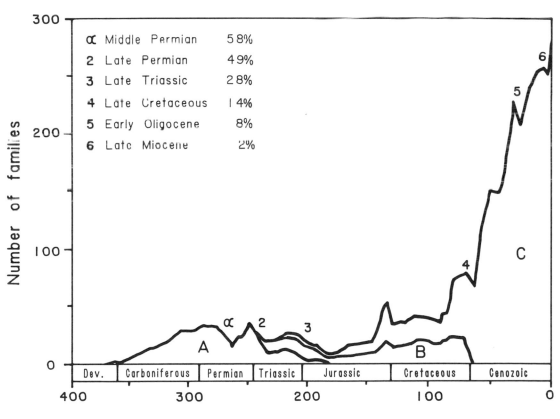

tating reduction of then-living floating marine microalgae (*phytoplankton*) and Protozoa was succeeded by an influx of functionally comparable but distinct Cenozoic forms. The once-flourishing Cretaceous dinosaurs are not to be seen again.

Most distinctive among the anomalously abundant minor elements of the discovery clay is the platinum-group element *iridium*, whose concentrations, along with those of osmium, gold, platinum, and others, here resemble meteoritic rather than crustal abundances. Comparable concentrations of 100 or more times crustal averages were soon found in shale at or near the K/T boundary at other places. The Berkeley group, therefore, quite reasonably concluded that this evidence favored collision with a large cosmic body, perhaps an asteroid, as the event responsible for the extinction of the dinosaurs and many lesser creatures.

Literally a smashing idea! What a pairing of spectaculars—falling stars and doomsday. Everybody loved it and justifiably so. As with punctuated equilibrium, the media rose to the occasion and the level of interest, both in meteoritics and in extinction, rose to unprecedented heights among all sectors of the public and the geologic profession alike. That flush of excitement led to a global search that soon revealed iridium anomalies at or near K/T extinction levels or intervals at now upwards of seventy localities worldwide from Denmark to New Zealand. The appealing simplicity of the connection, the statistics, and the will to believe were all strongly supportive. How could it be otherwise? In that revivalist atmosphere, the only questions seemed to involve explicit killing mechanisms.

Great ideas, however, are not always great because they are right. They are still great if they stimulate and focus research that may ultimately lead to their invalidation or replacement by a better idea. Both support and critical evaluation are essential to their healthy evolution. Front-runners attract challenges, and the Siva hypothesis has enjoyed its share. That is the way of science.

Devil's advocates were quick to think of problems. An impacting body, a bolide, of the size estimated (10 kilometers in diameter), striking Earth at the hypothetical speed of 25 kilometers a second, should open a crater 66 kilometers in diameter—about the size of the Manicouagan crater and volcanic rebound dome pictured in Figure 15.1. Although the Manicouagan object fell during a late Triassic extinction during which a fair number of dinosaurs became extinct, that extinction had two widely separated peaks. And no iridium anomaly has yet been found. If impact of large bolides were a major extinction mechanism, why didn't the one that opened a crater 100 kilometers in diameter at Popigai in Siberia about 40 MYBP bring about a more impressive extinction than is recorded for the Eocene–Oligocene event (front endpaper, Fig. 13.8)? Why isn't the 183-MY-old 80-kilometer impact crater at Puchezh-Katunki (also in the Soviet Union) reflected by a mass extinction nearer to the base of the Middle Jurassic than any yet recorded? Was the K/T bolide an even-larger asteroid or cometary nucleus that landed in the ocean and whose site of impact has since been filled or subducted, or which correlates, as has been suggested, with the origin of Iceland or the great Deccan flood basalts of central India? Why haven't iridium anomalies comparable to those at the K/T boundary been reported for the impact sites of Manicouagan, Popigai, and Puchezh-Katunki?

The estimated fallout time of dust injected into the stratosphere by the hypothesized Alvarez bolide was originally suggested as about 3 years. That was supposed to suspend photosynthesis, starve the dinosaurs (but not insects, mammals, birds, and most marine invertebrates other than ammonoids), and then give way to resurrection of plant life from preserved seeds and restoration of animal life from survivors. New estimates of fallout time have been revised to only 3 to 6 months, following global suspension during the few hours of its initial ballistic orbit.

Additional proposals for killing mechanisms that might have been triggered by such an event include: (1) an instantaneous heat wave; (2) prolonged darkness and cold; (3) a greenhouse effect, elevating atmospheric temperature by an average of perhaps 10°C and killing by heat death; (4) an excess of nitrous oxides and acid rain; and (5) poisoning by cometary gases. One variation on proposal 3 calls on sex-determining effects of temperature during the hatching period, as seen in modern crocodiles, to bring on a generation of unisex dinosaurs unable to reproduce. Another has reclusive mammals feasting on dinosaur eggs. One can but admire the ingenuity of the mind freed from the tyranny of evidence. As most of these and other theoretical killing mechanisms can also result from changes unrelated to impact, it is not easy to be explicit either about proximal causes or triggering mechanisms.

The K/T boundary, however, is no longer the only one from which above-background iridium concentrations have been reported. Additional records include four localities in the basal Cambrian and two at the Permo-Triassic boundary in China, one in the Late Devonian of Western Australia, one associated with Late Eocene radiolarian extinctions on Barbados, and one in a 2.3-MY-old layer from an Antarctic deep-sea core. Could high iridium concentration then truly be an extinction-related phenomenon? Or is the discovery at these levels an artifact of selective preservation or research? What besides a cosmic source could account for its enrichment? The beauty of the Siva hypothesis is in the way it provokes and focuses research by the questions it raises and the fascinating observations that have been and are being made as a result.

The iridium in the Devonian of Western Australia is concentrated in a fossil microbial structure. Algae and bacteria are well-known concentrators of certain rare elements up to thousands of times above background. Could that be the case for platinum-group elements? Deep-mantle sources of volcanism as at some sites above hot spots along spreading ridges and in plateau basalts bring excess iridium and associated elements to the surface, particularly in the gas phase. The Deccan Basalts, comprising perhaps a million cubic kilometers in east central India and Madagascar, have recorded modest iridium anomalies during some 10,000 to 100,000 years right around the time of the K/T extinctions, and Virginia Polytech's Dewey McLean has been suggesting them as the source of the global anomaly since 1981. These basalts have been related to the opening of the Arabian Sea when India separated from Madagascar at the time of a conspicuous magnetic reversal, perhaps reflecting a major event at the core—mantle boundary. Others ask whether an impacting bolide may itself have triggered such events.

In this provocative atmosphere, the most telling support for the inferred K / T bolide may be the low ratios of osmium isotopes and the apparently contemporaneous shocked quartz and probable impact glass at several K / T boundary localities. Because it appears that all crustal osmium started with high rhenium—osmium ratios, its initial ratios of $^{187}Os / {}^{186}Os$ are 25 to 35 times those of meteorites, whereas the K / T boundary ratios are almost the same.

The logic of a cosmic source for the platinum-element anomalies and an impact event or events at or near the K / T boundary is thus both strongly supported and persuasively challenged. Taking all evidence into consideration, however, the case for impact grows ever more compelling.

Given an asteroidal or cometary collision, then, what might that have had to do with the demise of the dinosaurs and other victims of the K / T extinction event? Was a collision directly or indirectly causal, contributory, or merely coincidental? Several plausible scenarios and a vast amount of discussion have not answered those questions. Surely the dust and gases raised by a collision of such magnitude would have had temporary climatic effects perhaps even exceeding those visualized for the much discussed nuclear winter. Had their internal temperature regulation been still primitive, surface-living dinosaurs could have succumbed to a level of climatic cooling that spared mammals, birds, insects, and most of the plants in low and middle latitudes, as well as the aquatic crocodiles and burrowing snakes, lizards, and turtles. Whatever the cause of their demise, the roughly simultaneous and abrupt death of the phytoplankton, devastating the marine nutrient pyramid, seems more than enough to account for other marine extinctions.

Still another awkward circumstance, however, mars the simple picture of instantaneous massacre of the dinosaurs by cosmic collision. The data are limited and circumstantial. To collect and study these animals is a costly business. Stratigraphic ranges are not well known. Of the roughly 240 genera of authentic dinosaurs reported to have lived, all but perhaps a dozen to 15, represented by no more than a score of species, were apparently already extinct by the middle of the last division of Cretaceous history, the Maestrichtian. Those remaining dinosaurs lived mainly around the then-drying western interior seaway of North America. In fact it seems that they were already in steep decline when the cosmic bell tolled.

None of the foregoing invalidates the importance of the K / T extinction. It was a major event, and abrupt in geological perspective. But it is now clear that it was neither instantaneous nor an immediate product of cosmic collision.

The principal current rival of the collision hypothesis calls on climatic cooling. That has been championed particularly by Steven Stanley of the Johns Hopkins University. Such a cooling, of course, could have been impact-triggered. It is as well or better accounted for, however, as a product of plate tectonism. Deep-ocean coring shows an abrupt shift upward at the K / T boundary of the level below which calcium carbonate dissolves in the sea—indicative of cooling oceans. Brown University's John Imbrie sees this cooling as a product of the separation of Australia from Antarctica, opening the Southern Ocean, initiating a circulation similar to now, and deflecting cold deep Southern Ocean waters far northward into seas previously warmed by Tethyan currents.

Such events would affect aspects of ocean chemistry, marine biology, and climate worldwide, as did the widely publicized effects of the El Niño currents of 1983–1985. Pleistocene temperature fluctuations of some 9°C in a temporal sequence that coincides with equivalent fluctuations on land is recorded by North Atlantic deep-sea cores. Supporters of impact have called on a similar range in temperature to account for selective K/T extinctions. Would they object to plate tectonism as a triggering cause? Or, as in the case of Deccan volcanism, is it possible that cosmic collision may have been causally linked with plate tectonism, simultaneously accounting for the absence of an identified impact site by having the impact itself initiate the separation of these Gondwana continents?

It would be tedious here to review and compare all of the hypotheses and variations on them that purport to explain different extinctions and their selective peculiarities. Some are rendered plausible by evidence and thoughtful analysis, others are amusing and perhaps meant to be, and still others are yet-unresearched flashes of imagination or insight. With so many hypotheses to choose from and such a variety of selective patterns to be explained, an even partially open mind must concede it possible that different extinctions had different causes. Whether or not a general cause or trig-

gering mechanism is ever established, both cosmic collision and plate tectonism have attributes that make them prime candidates. Climatic change, disruption of food chains, and geographic reduction of habitat are among the most likely proximal causes.

For closing thoughts, consider Earth-crossing asteroid orbits (Fig. 1.5) and the cratering record on Earth and its moon. Geologists George Wetherill and Eugene Shoemaker have estimated from crater counts that asteroids around 10 kilometers in diameter (that estimated for the suggested K/T killer) should collide with Earth *on an average* of once in 40 to 50 million years—enough to account not only for all conspicuous extinctions but for many of those perceived to emerge from statistical analysis. Why, then, aren't there more iridium anomalies? Why, indeed, don't we know more unequivocal mass extinctions—more events that can be seen without statistical manipulation? Why, in particular, haven't more been found to coincide in time with earlier tabulated major Phanerozoic impacts? Clearly it is high time both to pay more attention to the recalcitrant problems of extinction and meteoritics, and to open our minds to the still vigorously disputed and equivocal arguments for cyclicity in extinction. Such issues will long challenge the most rigorous geological and geophysical minds.

It is easier to ask questions than to answer them, but important to do both. Questions asked above must remain unanswered, perhaps until some reader of these pages finds the answers. But I observe that, in the current climate of eagerness for new "breakthroughs," where "outrageous hypotheses" are often published and almost as often vanish after their day in the limelight, a healthy skepticism is in order. The White Queen's ability to believe impossible things is all very well before breakfast, but the test is how well the impossible things hold up under the cold light of evidence in the late afternoon.

In Summary

Change with time is a universal phenomenon. Nothing in our universe remains forever the same or returns to exactly its former state except, perhaps, some elementary particles. That is evolution. Of course life evolves. Of course biological evolution is a fact. It goes on all around and in us, at times faster, at times slower. It is a problem with no final answers, no omniscient judges; a problem in which the central questions are how, why, when, and how fast. The only scientifically valid responses to such questions are ones with verifiable consequences that can be sought, observed, and validated or disproved.

Living things evolve from one state or level of complexity to another. Changing times and conditions set new demands. Charles Darwin described this as natural selection in 1859, when the idea of natural biologic progression was already accepted (if clandestinely) by a number of biologists and had been employed for more than half a century as the basis of historical sequence in geology.

Genetics was then taking its first steps in the cloister gardens of Gregor Mendel, eventually setting off a long and at times acrimonious discussion between geneticists and students of morphology and environment.

A turning point came during the 1930s and 1940s when paleontologists, geneticists, and organismal biologists (students of the whole animal) finally got together to learn one another's languages and try to identify common ground and the sources of differences. Natural selection emerged as the unanimously agreed upon primary external driving and steering force, linked firmly with the vast but ultimately limited potential for variation of all individual gene pools. A plant cannot become a tetrapod. A mammal may return to the sea, but not as a fish or an octopus.

The consensus of this discussion group, referred to by its organizers as "the synthetic theory of evolution," has continued to evolve.

Thoughts about evolution conventionally focus on sexually reproducing eucaryotic organisms in which a reductive division (meiosis) of the separately paired male and female chromosomes occurs before egg and sperm join to activate an embryo with the same number of paired chromosomes. Although the basics of what may now be called trans-Darwinism (natural selection of form and function from gene pools of large but limited potential variability) are considered to be the same for all, common modes of reproduction depart widely from the "standard" pattern. Aphids and some crustaceans alternate between unisexuality and bisexuality, depending on external circumstances. Many plants (including common garden plants) and even a few animals have discovered an ingenious evolutionary trick. They simply skip the reductive chromosomal step, thus doubling the number of nuclear chromosomes in the next generation—a feat called *polyploidy*. About 42 percent of all plant species have arisen in this way.

Selective pressures arise from the availability or scarcity of unoccupied ecospaces, limited by genetic potential and competition. Ecospaces and niches are opened for occupancy with the extinction of former occupants, the generation of new ecospaces as a product of plate tectonism, or the emergence of new adaptive modes. Evidence for this is seen throughout geologic history and especially its last 670 MY. Accelerated biological diversification has followed or accompanied all extinctions. The appearance and extinction of families of shallow-water marine invertebrates, for instance, has been closely linked throughout Phanerozoic history.

Evolutionary rates are a problem that has existed as long as geology and will continue. Nearly everyone involved agrees that rates of biologic evolution are variable, but they do not agree about the patterns of variability. The pattern here favored is one of variable or episodic evolution. It is seen as bursts of opportunistic diversification into new or recently vacated ecospaces, or following new adaptive modes, with occasional immigrants or immigrant groups from founder populations in speciation refuges—interrupting the more stately flow of background extinction and speciation. Although this suggests the phraseology of punctuated equilibrium, it is a good deal more inclusive, less categorical, and seems to be denied by definitions of that term.

All organisms eventually become extinct. Extinction simultaneously empties ecologic niches and invites and challenges competitors for the new ecospaces. Extinction—meaning that something disappears from the planet forever—includes the more or less continuing background loss of species and mass extinction, where large sectors of the biota disappear and general diversity is temporarily reduced. Although background extinction may well be the larger component, the focus here is on mass extinction, particularly at the Cretaceous–Tertiary (K / T) boundary, and on cosmic collision as a possible triggering mechanism.

The now-persuasive evidence for cosmic collision consists of anomalously high iridium and related platinum elements at or near the K / T boundary at many places and meteoritic osmium-isotope ratios. The actual (or proximal) selective killing mechanism or mechanisms remains uncertain, but cosmic collision at that time may be only indirectly causal. It is estimated that a global temperature decrease of perhaps 8° to 10°C would suffice for extinctions observed. That could have been the result of dust from massive volcanism in central India or of the plate-tectonic opening or significant widening of the Southern Ocean, either perhaps triggered by impact. Global temperature variation, for whatever causes, may be a more general proximal mechanism, working through its effects on the nutrient cycle and ecological balance. Of all the dinosaurs known to have lived, only a small percentage were still extant as the Cretaceous curtain fell. The abruptness and magnitude of their extinction, impressive as it was, may be *partly* an illusion of retrospect and size.

16.1 THE GOLDEN AGE OF MAMMALS. Typical mammals and plants of the western Nebraska short-grass plains on a fine early Neogene day 22 MYBP. Three-toed horses in center foreground, giant hog on right, and the large-clawed mammal *Moropus* on the left. Also wolflike dogs, gophers, small camels, pigs, small rhinos, and others. The modern-looking vegetation was typical of the times. [*Smithsonian Institution Photo No. MNH-933; Jay H. Matternes, artist.*]

CHAPTER SIXTEEN

THE PENULTIMATE SCENE: THE CENOZOIC ERA

The physical evolution of our planet during the past 65 million years, interacting with the biosphere, oversaw the adaptations that were to give rise to us. Stresses attending the northward flight of plates away from Antarctica resulted in regional deformations of contrasting orientation and variable consequences, roughing out the present physical geography. Collisional mountain systems of mainly east–west orientation stretched from the Himalaya to the Alps, simultaneous with closing of the Tethyan seaway, opening of the Southern Ocean, and amplification of the climate-shaping circum-Antarctic current.

Mid-Atlantic spreading, linked with convergence all around the Pacific margins, resulted in the mainly north–south, collisional fold mountains, plutonic ranges, and volcanic chains that ring the Pacific. The surging growth and recession of Antarctic ice from about 35 MYBP onward resulted in variable sea level, realignment of oceanic and air currents, and other climate-shaping effects. A related diversity of potential ecologic niches, at sea as well as on land, led to the greatest yet-known familial diversity, especially among mammals, birds, insects, flowering plants, fishes, and snails (both marine and nonmarine).

The most recent 15 MY saw marked cooling, with episodic warming. The interval from about 6.5 to 5.2 MYBP was notable for repeated desiccation and reflooding of the Mediterranean basin—prelude to the well-known Pleistocene ice ages, temporal cradle of humankind. As Cenozoic history progressed, an ever more detailed geochronologic resolution of events facilitated correlation globally and between marine and nonmarine history.

The Fretful Earth:
The Dimensions of Cenozoic History

The rain catchers of the world, the sky-scraping ridges and peaks that rim the continents or define their collisions in all regions of Earth, tell of former or present plate motions. With few exceptions, the prominent and familiar ones are post-Paleozoic, mainly post-Mesozoic, brought to their present heights during the last 65 MY. Many are still rising.

Three and a half centuries ago the populace of the Western world thought they lived on an immobile surface whose only changes were in response to the daily and seasonal behavior of the Sun. We know better now, thanks to the likes of Copernicus, Galileo, and Kepler. Indeed, our planet moves in ways they had no way of knowing. Earth's jigsaw pattern of crustal plates has been shifting like a slow-motion kaleidoscope for the past 2 or 3 billion years without ever reaching a permanent configuration. Those motions interact with any extraterrestrial influences, with geography, and with contemporary living systems to generate the succession of local ecosystems observed and to register in the sedimentary record the physical and biospheric evolution of the planet. Despite the rapidity of such changes in geologic retrospect, they are not so rapid on a scale of thousands or even millions of years. The younger Cenozoic mammals and plants (Fig. 16.1) are different, but not so very different, from those of modern times.

In the language of geology it is recorded that we and our fellow occupants of this planet live at the end of the Cenozoic Era. Perhaps it will seem more like middle Cenozoic from the vantage of another 65 MY. Or maybe the ever-hoped-for and promised new era will have dawned by then.

Few certainties exist beyond change itself. The idealized dimensions of our planet remain constant only by definition—the axis and poles of rotation, the equator, the gridwork of latitudes and longitudes, the mass of the planet. The real planet, the not-so-solid Earth within that geodetic framework, changes. Plates migrate, moving the continents across

that imaginary grid. Earth's spin axis wobbles relative to its solid crust in 14-month cycles. The planetary mass grows daily by a megaton or so of gravitationally captured micrometeorites. Lands move both up and down relative to sea level, and sea level changes with the growth and melting of glaciers and the size of ocean basins. Things simply don't stay in place. They never did; they never will. Rates vary. Stability is a temporary statistical illusion.

When tracing Earth history from ancient toward modern times, the geological rates of everything also seem to accelerate. This too is illusory, a product of better preservation and less erosion as one approaches modern times. More likely, rates were faster in the past when the flow of heat from Earth's interior was greater.

The resolving power of our geologic clocks, however, improves. Radiometric methods with shorter half-lives become applicable. Because of reduced opportunity for argon loss, the potassium–argon (K–Ar) method, with a K half-life of 125 MY, applies within ever-smaller limits of error. A detailed record of polar reversals related to radiometrically dated tie points brings that calibration widely into play, both on land and beneath the sea (Fig. 16.2). A comparable sequence of forty-one similarly calibrated planktonic microfossil zones is recognized in Cenozoic deep-sea cores and elevated marine sediments, with an average duration per zone of 1.5 MY. New methods in stable-isotope geochemistry now allow oxygen-isotope ratios of shelly marine microorganisms to be correlated with variations of Earth's orbital motion during the past 800,000 years within limits of error of only 3,000 to 5,000 years, and thorium-

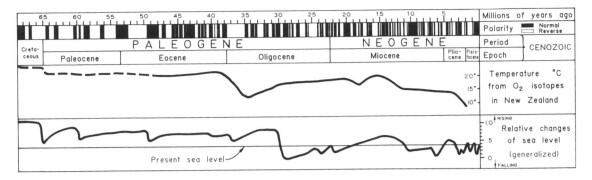

16.2 RADIOMETRICALLY AND PALEOTOLOGICALLY CALIBRATED PALEOPOLARITY, PALEOBATHYMETRY, AND PALEOTEMPERATURE HISTORY FOR THE CENOZOIC ERA.

230 techniques promise extension to 500,000 years with a 1,000-year resolution. In the range of carbon-14, applicable with special techniques to the last 70,000 to perhaps as much as 200,000 years, time resolution is within decades. And by Recent, or "post-glacial," time (the last 10,000 to 12,000 years) resolution is annual, seasonal, or briefer, as tree rings, glacial varves, and records of human activity become applicable.

Our knowledge of Cenozoic history thus becomes increasingly detailed with time and growing relevance to humankind, and, with that, the complexity of historical terminology has increased. Only the main elements of that nomenclature appear in Figure 16.2, and we shall mostly do with fewer. Paleogene serves as a collective term for the three older Cenozoic epochs and Neogene for the younger. Although rocks called Quaternary are often excluded from Neogene, they are a natural continuation of it. The much used terms Pleistocene and Quaternary (Pleistocene plus Recent) have been useful anachronisms in an anthropocentric world, but they represent only the last 1.6 MY, and their blending is overdue. Pleistocene is the preferred inclusive term.

The succession of strata in the Paris basin, the Parisian sequence, famed from the studies of Cuvier and Brongniart (Ch. 3), is considered representative for the era.

Plate Jams and Tectonic Complexes: The Himalaya and Other Young Eurasian Mountains

Despite the billions of years that Earth has had to wear down toward sea level, today's lands stand high. Wherever the current cycle of continental motion has caused plates to converge and collide, mountains have arisen, affecting prevailing winds and temperatures, precipitation, and life assemblages. Wherever once-adjacent lands have separated, new seaways have appeared and ocean currents have changed. Both in the assembly and deformation of Pangaea and in

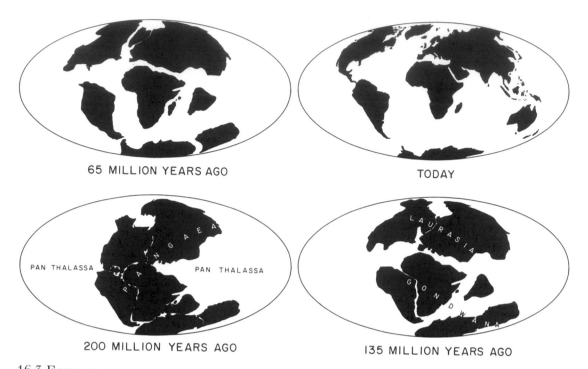

65 MILLION YEARS AGO

TODAY

PAN THALASSA

PAN THALASSA

200 MILLION YEARS AGO

135 MILLION YEARS AGO

16.3 FOUR STAGES IN THE EVOLUTION OF PRESENT GLOBAL GEOGRAPHY. *[Adapted and simplified from R. S. Dietz and J. C. Holden, "The Breakup of Pangaea," Scientific American, v. 223(4), pp. 34, 36, 37. Copyright © 1970, by Scientific American, Inc. All rights reserved.]*

its dispersal, plate tectonics has been the master architect and chief weather maker (Figs. 16.3, 16.4). Cenozoic history saw the opening and enlargement of seaways between Australia, Antarctica, and South America, with establishment of the cold circum-Antarctic current. It witnessed the closing of Tethys and the Panamic seaway, the opening and expansion of the Southern Ocean, the isolation of Greenland and Antarctica, and an unremitting growth of mountains.

From Triassic to late Paleogene, as Tethys opened and then closed, the warm waters of its shelf margins favored the deposition of calcareous sediments now seen as often richly fossiliferous and reefal limestones and dolomites. As Africa and Eurasia converged, however, Tethys was obliterated, except for isolated remnant seas. Bordering plate margins were broken up into microcontinents and separated from one another by slivers of ocean now recorded only as bits of obducted ophiolitic crust (back endpaper) within the Alpine ranges north from and within the Mediterranean Sea (e.g., Cyprus).

The Mediterranean, however, is not a simple remnant of Tethys, as once thought, although it did connect with it through the Black Sea. More realistically, it was a separate miniocean that opened up behind microcontinents as they became detached from the northern edge of Africa during Paleogene time and converged to make up Italy, the

Balkan states, and parts of the Near East. Alpine deformation across southern Europe and northwestern Africa occurred principally during late Cretaceous and Paleogene history. Nearly flat crustal slices (nappes) of huge extent were thrust relatively northward scores of kilometers, riding over and repeating themselves wherever grounded by friction. Erosion followed, with deposition of nonmarine gravels, sands, shales, and marls (the *molasse* facies) in adjacent forelands during late Eogene and early Neogene time. The present Alpine scene, however, owes its snow-dusted beauty primarily to subduction and the resultant vertical elevation during late Cenozoic history.

As the Atlantic widened, the American continents rode westward over a narrowing once-universal sea, reduced now to the present Pacific basin and rimmed by growing collisional mountains, intrusive sialic chains, and volcanic arcs. The prominent post-Paleozoic mountain ranges of the modern world, many of which began during Cretaceous or earlier Mesozoic history, were brought to their present states by Cenozoic collisions, sustained compression, folding and uplift. They cluster in two main sets. One trends broadly west to east, consisting mainly of folded mountain ranges that rim southern Eurasia all along the once-Tethyan seaway. The second—a more heterogeneous set of crystalline intrusives, volcanic belts, and folded sedimentary rocks—trends north to south along the western margin of the Americas. All are products of convergence and collision, but they vary in historical detail.

Most awesome among these often awesome mountains is the great Tibetan plate jam. A product, as now known, of 250 MY or more of episodic plate motions, this soaring snow-capped jumble of ridges and plateaus began its ascent about 50 MYBP, when India first encountered Asia. Radiometric evidence for age is supported by the appearance

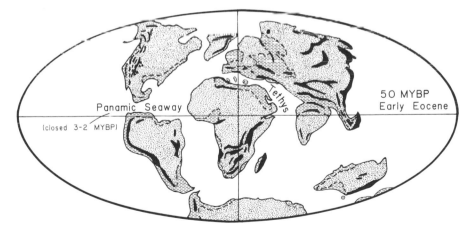

16.4 THE MODERN WORLD TAKES SHAPE. Continental positions at maximum marine flooding of the Cenozoic continents in early Eocene time. Shallow seas, *lined*; land, *stippled*; new mountains, *black*; North American nonmarine deposits, *black triangles*. Not shown here are scattered upland basins and other nonmarine deposits, including coal.

in India then, for the first time, of a number of Asiatic land mammals. As convergence continued, sediments that had accumulated and accreted south of Tibet, and Paleozoic and Mesozoic deposits that peeled away from the northern flank of advancing India, were squeezed together like curds in a cheese press. During the past 38 MY strata were deformed and normal crustal thickness increased to about 50 kilometers. Deformation culminated with vertical uplift in relatively recent times, responding to continued compression between the two colliding plates and the increased thickness of light crust beneath the Himalaya. The south edge of Asia (Tibet), on the other hand, is buoyed up by a broad, deep, sialic root twice the thickness of normal crust to a depth of 70 kilometers.

The dimensions alone are staggering. Within a mountainous region almost the size of Australia, an area larger than Greenland has been elevated to heights generally above 4 kilometers. From valley bottoms at altitudes higher than any of the North American Rocky Mountains, the land ascends another few thousand meters to the perpetually ice-crusted Permian marine limestones at the very top of Mount Everest (8,848 meters).

Microcontinents and island arcs within the narrowing space northward from India were in fact assembled into now-southern Tibet at intervals over the past 150 MY. When the leading (oceanic) edge of India itself first docked along the present northern margin of the Himalaya, it generated a 2,400-kilometer-long suture or zone of joining (Fig. 16.5), tracked now by the sacred Indus and Brahmaputra rivers. Seismic reflections from the crust—mantle boundary in this

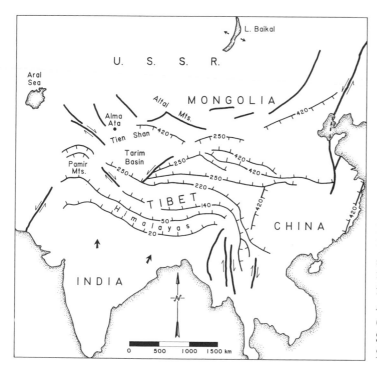

16.5 THE GREAT TIBETAN PLATE JAM. Long lines with ticks on the overriding side denote former subduction zones where exotic microcontinental plates or slivers have converged and joined. The numbers are approximate ages of collision in MYBP. Half arrows elsewhere show direction of strike-slip motion on transform faults. The Himalayan chain itself has arisen over the past 50 MY from compression between Tibet and northward-drifting India, with great increases of crustal thickness beneath Nepal and adjoining territories. [Adapted from Zh. M. Zhang, J. G. Liou, and R. G. Coleman, 1984, Geological Society of America Bulletin, v. 95, p. 296, Fig. 1, with additions.]

16.6 DISTINCTIVE ALPINE ROCK TYPES. Scene A, *wild-flysch* with large exotic blocks from an actively advancing mountain front; Kura River, Tbilisi, Georgia, Soviet Union. B, *molasse;* weathered "nailhead" conglomerates (*behind bicycles*) and nodular limestones in the foundation of a Munich church.

region indicate that India thereafter tucked its leading edge beneath that boundary and was overridden from the north in the main Himalayan mountain-building event, terminating perhaps 10 MYBP.

Altogether some 1,500 kilometers of north–south shortening is believed to have taken place in this region, where India is still wedging northward beneath its own once-leading edge and into Asia at about 5 centimeters per year. Will crustal compression continue? Assuming it does, will the Himalaya grow still higher, or will they at some point flow to lower levels under their own weight?

The Himalaya and associated mountains of southeast Asia were not the only products of the northward safari of Gondwanan continents and their collisions with one another and Eurasia.

So arose, at different Cenozoic times, the Atlas Mountains of northwestern Africa, the Pyrenees, the Alps, and their eastern European extensions and, with them, the distinctive flysch deposits of Alpine deformation (Fig. 16.6A). The orogenies responsible for that picture-book scenery then gave way to the vigorous erosion whose distinctive post-tectonic sedimentary product is the Alpine molasse (Fig. 16.6B). Coarser deposits were dumped in front of the eroding mountains as streamborne gravels and sands while muds and dissolved minerals came to rest in the foreland plains, meadows, and lakes of Switzerland and contiguous areas. The Alps, like the Himalaya, stand high because of continued compression, but not as high as the latter because of thinner crust beneath.

The most recent phase of Alpine movement saw broad uparching of the whole region during the last 5 MY. That drained all remaining shallow seas from the European continent, including the classic Paris basin, as Alpine and later lowland glaciers assumed the task of shaping the historic landscape.

Other Motions, Other Mountains: The North American Cordillera

Beyond the broad east-to-west Tethyan realm, most other great mountain ranges of Mesozoic and Cenozoic origin trend north to south. They ring the Pacific basin as it subducts beneath the overriding plate margins on all sides, reflecting motion away from the Mid-Atlantic spreading ridge and toward the Pacific margins.

Prominent in this tectonic scene is eruptive volcanism. It links plate tectonism with mountains of classic beauty, soils of fabulous productivity, and the grotesque imagery of fresh volcanic surfaces to be seen all around the Pacific rim and beyond. It is seen above the East Pacific rise, in mid-Pacific island chains, and, during the past 15 MY, throughout Indonesia. Episodic volcanism within the North American Cordillera left its products as cinder cones, lofty shield volcanoes, and sets of dikes, flows, and sills at different intervals along the Rocky Mountains and west to the Pacific Coast. In ocean islands such as Hawaii volcanic rocks are exclusively basaltic. The seaward-facing island arcs and continental-rim volcanics, however, are often of intermediate composition. Inland they range from basaltic to sialic.

The history of the North American Cordillera back to Triassic times encompasses volcanic, intrusive, and folded mountains, structural basins, and constructional plateaus. It comprises nearly the whole western third of the continent, extending from the Rocky Mountains westward to the sea (Fig. 16.7). Between the Rockies and the coastal ranges on the west is a complex of volcanic and sedimentary uplands and extensive interior-draining basins laced with north–south trending fault-block mountains. Largest of these is the westward sloping Sierra Nevada, faulted remnant of the extensive older Andean-type batholiths that rim the Pacific margin of North America. In the United States the most westerly ranges are a pair. The relatively low Coast Ranges are separated from the towering Cascade Range and Sierra Nevada to their east by the Puget Trough in the north, and, in the south, the Great Valley of California.

The terrain adjacent to the rising Rocky Mountains was close to sea level as Cenozoic history began. Paleogene drainage from Rocky Mountain uplifts was intercepted and impounded in large inland basins, some of which became the sites of chemical treasure-houses of saline lake deposits and oil shales. The mainly Eocene Green River basin, for example, occupied much of the southwest quarter of Wyoming. And the Green River oil shales continue on the south side of the Uinta Mountains of Utah and Colorado across the Uinta and Piceance basins (Fig. 14.15). Altogether they blanket some 100,000 square kilometers of present-day semidesert with thicknesses of up to

16.7 THE NORTH AMERICAN CORDILLERA TODAY. *[Compiled from U.S. Geological Survey Geologic Map of the United States and Geological Society of America Geologic Map of North America, with additions.]*

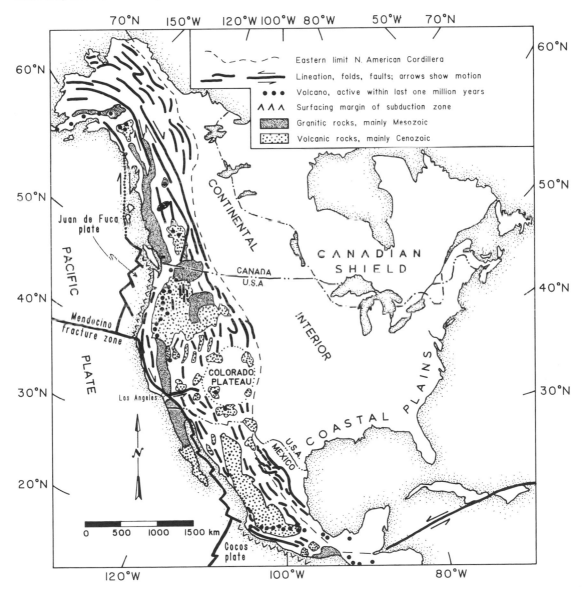

a kilometer of organic rich sediments that yield up to 280 liters (75 gallons) of oil per ton of shale. The smaller of these three basins alone (the Piceance basin) is estimated to contain about 900 billion barrels of extractable oil, about twice the known Arabian oil reserves. The Green River oil, however, is not "reserves," even at present high prices. Should it become economically extractable, it is probably not feasible to do so in terms of water needed, nor worth the environmental costs.

The scenic west coast of North America from Alaska to Mexico results from the erosional shaping and vegetational adornment of tectonic and volcanic landforms. It is everywhere the scene of mostly still-rising mountains and sinking or recently sunken coastal basins, resulting from collision with and gaps between linear microplates of the eastern Pacific rim—rich in now-dwindling oil for the lamps of the temporarily affluent. Lateral and rotational motions have shifted the Pacific sliver of California about 630 kilometers northward along the west side of the San Andreas fault system—some 300 kilometers in the Paleogene and another 330 kilometers during later Cenozoic history up to the present.

Early in the Neogene, perhaps 25 to 20 MYBP, as convergence yielded to transform motion, rotation, and tensional stress, mainly Jurassic granites of the Sierra Nevada batholith were rotated upward and tilted westward, hoisting a giant block into the reach of the master sculptor—erosion. Ice and water, with the help of wind to bring seeds, made nature's cathedrals there, including one called Yosemite.

Only the main elements of a stagger-ingly complex scene are summarized in Figure 16.7. Apart from unroofed granitic intrusives, a grossly simplified pattern of effusive volcanics, and explosively eruptive volcanoes, it shows only generalized trends of faults, folds, and lineations. Notice that active and dormant volcanoes of explosive sialic and intermediate composition are absent from this region between about 20° and 40° N. Their absence coincides with a similar absence of subduction. As most of the parent Farallon plate disappeared beneath North America (the Juan de Fuca and Cocos plates are remnants), it gave way to relative southward slip along the east side of the San Andreas fault. From Mount Lassen, at the south of the Cascade Range and opposite the southern end of the still-subducting Juan de Fuca plate, one must now go all the way to southern Mexico and the subducting eastern margins of the Cocos plate to find another such eruptive center.

Mount Lassen, then, is the southern member of a chain of thirteen major, intermittently explosive, volcanoes of chemically intermediate composition that, with associated volcanic rocks, comprise the Cascade Mountains of the northwestern Pacific states. The Cascade volcanic peaks are big mountains by North American standards: eight exceed 3,075 meters (10,000 feet) and two exceed 4,300 meters (14,000 feet). One must go to Mount McKinley in the Alaska Range (6,194 meters) or Mount Whitney in the Sierra Nevada (4,415 meters) to find higher peaks, and they are not volcanoes.

Cascadian eruptive history started in early Cenozoic time, peaked during the Pliocene, and culminated with the Pleistocene growth of its present eruptive centers. As Mount St. Helens showed in

1980, they can still be violent. Indeed, signs of volcanism are seen throughout the Cordilleran region—old and young volcanoes, patches or fields of lava, erosionally denuded volcanic necks and dikes, and a circle of volcanic fields and peaks around the Colorado Plateau.

Eastward from the Cascade chain are the famed *flood basalts* of the Columbia River Plateau (Fig. 16.8), part of the same northwestern volcanic province. Like older flood basalts already mentioned, they are the product of highly fluid mafic lavas that well up from probable mantle sources through glowing rifts of great extent to race across expansive surfaces as steaming sheets of lava. They poured out mainly between 16.9 and 13.4 MYBP as some 120 to 150 individual flows having a cumulative thickness

of 2.5 kilometers. And they cover a region the size of Nebraska, along and away from the Columbia and Snake rivers in Washington, Oregon, and Idaho. One flow alone (the Roza flow), 30 meters thick, spilled 1,500 cubic kilometers of liquid basalt across a fifth of the region in the space of a few days. Its erosional remnants line the scenic Dalles of the Columbia River, upstream from Portland, Oregon (Fig. 16.9*A*).

The Columbia River flood basalts, however, cover only about a quarter of the seven-state northwestern volcanic province outlined in Figure 16.8. That includes mainly other middle and late Cenozoic basalts and related volcanics in the Pacific Northwest. Eastward are the similar flood basalts of the Snake River Plains and other volcanic features

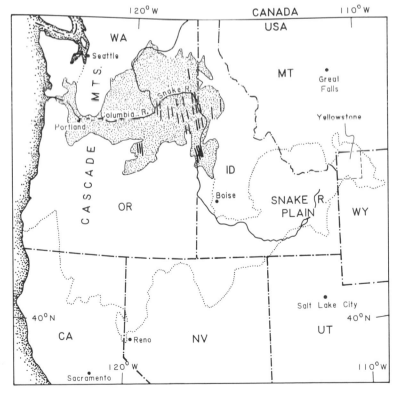

16.8 THE MIOCENE COLUMBIA RIVER BASALTS (*STIPPLED*) COVER 200,000 SQUARE KM OF THE NORTHWESTERN UNITED STATES TO A DEPTH GREATER THAN 1,000 M. The dotted line outlines the remainder of the northwestern Cenozoic volcanic province. The dark parallel lines represent known dikes, former lava conduits.

425

of generally decreasing age as far as Yellowstone Park. There, 700,000 years ago, a great eruption poured out debris of a range of chemical composition 2,000 times the volume of the recent Mount St. Helens eruption. That resulted from the collapse of a great caldera (Fig. 16.9*B*). The presence of helium-3 in Yellowstone fumarole gases indicates mixing of mantle sources and sialic crust as the North American plate tracked westward over the still-hot Yosemite Hot Spot (resulting in apparent eastward motion of the latter). It would have been an eerie, suffocating scene around Yellowstone, had anyone been there, as the fine

siliceous dust and ash came sifting down repeatedly over the millennia.

As the Cenozoic Era began, the spreading center now within the Gulf of California, the East Pacific rise (Fig. 16.7 and back endpaper), lay well to the west of North America, probably coming ashore somewhere near the southern edge of Alaska. It separated the Pacific plate to the west from what was then the Farallon plate on the east, where the latter was being overridden by the North American plate. Where continuing subduction of the Farallon plate resulted in the abutment of the North American and Pacific plates, subduction ceased and

16.9 Scenes from the northwestern Cenozoic volcanic province. View *A*, succession of basaltic Miocene lava flows at The Dalles of the Columbia River. *B*, 700,000-year-old sialic eruptives from the rim of Yellowstone River, Yellowstone Park, Wyoming.

sideslip motion on the San Andreas fault took up the slack. Some 1,500 kilometers of mainly sideslip motion now separate the northern and southern remnants of the former Farallon plate between the Cocos plate off southern Mexico and the Juan de Fuca plate to the west of Oregon.

Plate tectonics has simultaneously illuminated Cordilleran geology, raised new problems, and focused great exploratory activity on the region. Seismic profiling, with computerized reduction of data, is the exciting new source of information about the geometry of subduction and low-angle pull-apart faulting and rifting beneath the Great Basin of Nevada and adjacent states. One infers from the presence of the vast low-grade copper ores found characteristically above subduction zones that subduction extended inland as far as the Colorado Plateau and the Rocky Mountains between 70 and 30 MYBP. Indeed a number of peculiarities of Cordilleran geology might be explained had subduction of the Farallon and predecessor plates been both at a relatively low angle and repeated, as seismic profiling now implies. That could account for uplift of the Rocky

Mountain mass following K/T folding as well as some 1,500 meters of vertical motion that elevated the Colorado Plateau and challenged the Colorado River to excavate its canyon. Bit by bit the mysterious Cordillera gives up its secrets.

The physical attributes of North America from the Rocky Mountains to the Pacific coast thus reflect its convergence history over the past 160 MY—especially the past 65 MY. Later Mesozoic microplate accretion was followed by still-continuing subduction, thrust faulting, and folding. Then offset along the San Andreas transform eased compressive forces on the region of the present-day Great Basin, giving way to tensional rifting, and volcanism, enhanced by the opposing rotations of the Pacific basin and Colorado Plateau. In this way the tensional basins and intervening ranges east of the San Andreas fault, the mantle-tapping flood basalts of the Columbia lava plateau and Snake River plains, and the late sialic and intermediate volcanism of Yellowstone and the Cascades record the final tectonic and volcanic shaping of the Pacific borderlands.

Some Words on Sedimentary, Erosional, and Biospheric History

As for the region east of the Rocky Mountains, the Cenozoic story is a more passive one, revealed in two main parts. One involves the erosional shaping and extensive fluviatile blanketing of the High Plains eastward from the Rockies. The second records the blanketing of the distant Atlantic and Gulf coastal plains with yet-undeformed sedimentary rocks con-

taining marine fossils, deposits typical of trailing (passive) plate margins.

For most of Paleogene history, the eastern Cordillera, despite extensive early folding and thrust faulting, was a region of low general altitude and relief, and of warm, humid climate. How did it acquire the characteristics we associate with it today—high altitudes and relief, deep canyons, and contrasting local microclimates? Paleogene sedimentation in the region was largely confined to local, often large, lake basins and confluent stream termini. How did these lake basins become drained while accordant ridge and summit levels in the same region record an extensive, now-dissected, gently east-sloping upland at some time during that history?

Answers to those questions are suggested by the volcanism that climaxed around the Colorado Plateau in early Neogene time when lavas, ash, and coarse volcanic detritus blanketed parts of the region to thicknesses of several kilome-

ters. That perhaps reflected the 1,500 or so meters of uplift that elevated the whole Rocky Mountain region about then, initiating the deep excavation of the Colorado River and tributary drainage so characteristic of that terrain. The same uplift presumably also drained the older Rocky Mountain lake basins, some eastward to the high plains, some westward to the arid depressions of the landlocked Great Basin.

We see then that older Cenozoic sedimentary rocks were impounded mainly in those once-landlocked inland basins of or marginal to the eastern Cordillera (Figs. 14.15, 16.10). Latest Paleogene

16.10 CENOZOIC NONMARINE STRATA OF THE ROCKY MOUNTAINS. Scene *A*, basal Eocene sandy-silty rocks in the western rim of the Colorado Plateau, Richfield, Utah; dark strata at center and beneath are redbeds. *B*, lower Eocene along Green River near Kemmerer, Wyoming; high in bluff is Green River Shale, known for fossil fish. *C*, Eocene shales and siltstones in white bluffs, Oligocene in upper wooded bluff, Miocene in upper grassy slope, at Beaver Divide, near Lander, Wyoming; rich in mammalian remains.

16.11 MAINLY MARINE CENOZOIC
STRATA OF COASTAL CALIFORNIA.
Scene *A*, mainly marine shales
and sandstones of Eocene to
Miocene age in the Coalinga
anticline, an oil-producing
structure; thin white band at
midphoto is a marker diatomite.
B, fluvio-marine sandstone of
mainly Oligocene age in the
Coast Ranges south of Monterey, California. *C*, view west
across Paleogene sequence of
Santa Ana Mountains, Los
Angeles basin, California: Cretaceous marine to nonmarine
silts and sands at lower right
(*K*), are overlain by Paleocene
clay, silts, and lignite (*P*), then
Eocene sandstone (*E*), and Oligocene red-gray nonmarine
conglomerates (*O*).

and Neogene deposits, however, are erosion remnants of originally extensive
blankets of streamborne debris, happy
hunting grounds for mammalian paleontologists. Such deposits extended far
across the high plains of eastern Colorado, western Kansas and Nebraska,
and regions to the north and south. The
downstream profile of the eroded and
debris-blanketed surface flattens from
steep upper levels to about a 1 percent
grade on the plains—just enough, in that
semiarid region, to transport the deposits
observed after a good rain. This is about
when or not long after grasses appeared
in the record of plant fossils, quickly
generating the high prairies—to become
the empire of the bison and later of the
Cheyenne, the Arapahoe, and the Sioux
(the oldest fully modern grasses appeared
only about 7 MYBP, late Miocene). Here,
before their local extinction, *native* North

American horses and other grazing animals achieved their greatest extent and abundance, evolving adaptations of teeth for cropping and chewing siliceous grasses and long, hoofed legs for fleeing predators.

The Cenozoic coastal history of the Americas varied greatly with the nature of the coast. In plate-tectonic terms the eastern coasts were passive and the western active. The active western margins were scenes of mountain building, diverse marine to nonmarine basinal sedimentation (Fig. 16.11), and deformation, leading to sites of petroleum accumulation.

The history of the passive eastern margins was like that of passive margins everywhere—uneventful in geologic terms. The repeated incursion and retreat of fossiliferous marine sediments recorded rises and falls of sea level over a few tens to a couple hundreds of meters, attributable to melting and accumulation of glacial ice and variable plate motions. Baltimore and Miami, had they been present, would have been alternately high and dry and submerged. Allowing for such excursions, and including Gulf Coast sequences, a nearly continuous succession of Cenozoic sedimentary strata up to perhaps 1,000 meters thick can be cobbled together for the eastern coast of the United States. It is found above an erosion surface that overlaps Cretaceous and Jurassic marine sediments, truncates the mainly nonmarine deposits of Jurassic rift basins, and, at places, cuts across deformed and metamorphosed older crystalline rocks.

The region is seismically inactive but not completely so. In late 1958 a strongly felt earthquake near St. Louis reminded us that strong shocks destroyed much of Charleston, South Carolina, in 1886 and severely shook the thinly populated southern Mississippi Valley in late 1811 and 1812. Hidden forces slumber below. Faults from olden times, like mythical demons, may strike from beneath concealing sedimentary cover. Subterranean stress may yet succeed in ripping the continent open again along some ancient line of weakness.

All lands are fringed by coastal marine sediments of Cenozoic age (see Fig. 16.12) and blanketed internally here and there by discontinuous nonmarine sequences or basin fills of some sort. They are correlated from one sector of the planet to another and their history is amplified by means of distinctive fossils, some of which are illustrated in Figures 16.1 and 16.13. Floating calcareous microfossils, pollen, and paleomagnetics are especially important for global correlation, particularly between marine and nonmarine sequences and history.

16.12 DISTINCTIVE PALEOGENE MARINE STRATA OF EGYPT (*A*) AND NEW ZEALAND (*B*). Scene *A*, the great sphinx Harmakhis near Cairo, carved in fossiliferous Eocene limestone; behind it the pyramid of Cheops, built from blocks of the same rock. These legendary strata are called the Nummulitic limestone for their abundant coin-shaped fossils (Fig. 16.13*n*). *B*, right-dipping fossiliferous greensands and marly limestones of Paleogene age reach upward from terraces of New Zealand's Waipara River (*lower left*) to ridge-forming Oligocene limestone at right.

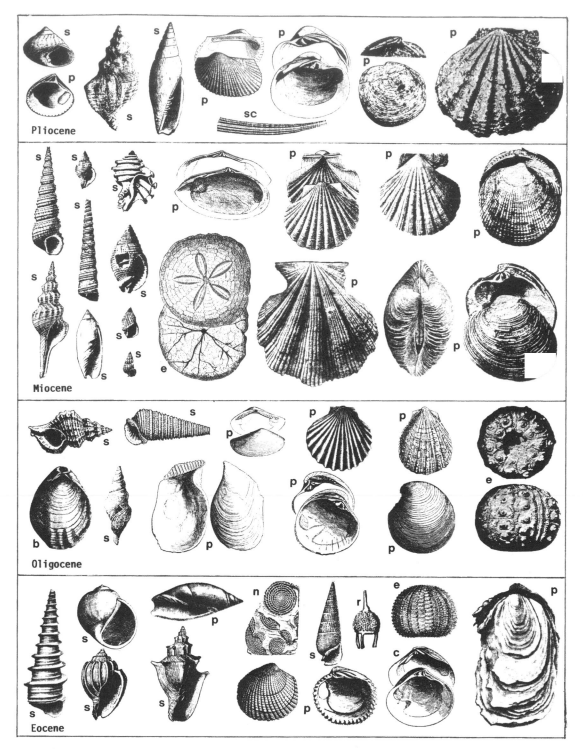

16.13 SOME CENOZOIC SHELLY MARINE FOSSILS. Lettering has some significance as in Fig. 14.14 plus *sc*, scaphopod; *r*, radiolarian; *n*, *Nummulites*, the Eocene rock-building protozoan of Fig. 16.12*A*. All greatly reduced except *r* (enlarged) and *h* (slightly reduced).

432

Cooling Polar Oceans and the Cenozoic Climates

Volcanoes, glaciers, oceans, and climate have been linked from mythical times. Juno, consort of mighty Jupiter, embarrassed by her lame son Vulcan, threw him into the ocean, where he remained for 9 years, dwelling ever after in different palaces underground, spouting lavas and shaking the Earth. Humankind, having watched and studied volcanism at first hand from ancient times, has long had a fair understanding of its history. Of the sea, however, we have known mainly about its present state, and of climate, mainly that it changes. Although evidence for the study of oceans older than about 200 MYBP remains severely limited, that for the last 65 MY and especially the last 50 to 35 MY is now increasing rapidly. Our understanding of Cenozoic oceans and climates has grown with recent advances in the study of stable isotopes, micropaleontology, analytical precision and instrumentation, and methods of dating and correlation applied to deep-sea sediments.

To begin with climate, paleontological records of temperate forests in the Eocene of Greenland and of tropical floras and marine shellfish of the same age in southern England (then already near their present latitudes) imply that still-earlier Paleogene chilling had by then given way to renewed warming. Concurrently, the earlier-mentioned Green River oil shales of the central Rocky Mountains (Fig. 14.15) were recording mild climate there. Famed alike for their annually layered lake sediments and exquisitely preserved fossil fishes and microfossils, these strata reveal a history of vegetation and climate about 50 MYBP similar to that of the southeastern states today. Global realignment of plates and ocean circulations then in progress, however, was soon to erect new mountain barriers to atmospheric movements, close equatorial and open Antarctic seaways, and trigger volcanism whose atmospheric dust would have extensive cooling effects.

A detailed record of those effects is found in deep-sea cores and surface samples now being returned from a multinational deep-sea drilling program (DSDP), and from these has grown the new subscience of paleoceanography. Cores that penetrate sediments deposited over the last 35 to 50 MY on upwarped sea floor between Australia and New Zealand tell of Paleogene volcanism and shifting plate orientation in and around the Pacific, concurrently with widening of the Southern Ocean around Antarctica. Shifts toward the heavier oxygen isotope (oxygen-18) in the composition of calcareous microfossil assemblages at about 42 and 38 MYBP imply cooling attributable to contemporaneous growth of an Antarctic ice cap. And that is linked with changes in oceanic circulation and thick volcanic ashfalls. Such phenomena may also be related to the impact event responsible for the 100-kilometer crater near Popigai, Siberia, about 40 MYBP. Likely products of that event would have been massive clouds of stratospheric dust, impact-generated volcanism, micrometeorite showers such as are recorded in oceanic sediments of that age, and, of course, climatic cooling nearly concurrent with late Eocene microbiotal extinctions.

Similar measurements of deep-sea sediments and estimates of their thickness and depth from seismic stratigraphy imply later sea-level and temperature fluctuations toward a peak of mild climate around 18 to 14 MYBP (medial Miocene), followed by fluctuating but mostly declining temperatures toward the present. So long have the most recent polar icecaps endured.

The thermal history of the last 25 MY, the Neogene (here including "Quaternary"), is recorded in ever-increasing detail, alike on land and in a growing abundance of submarine cores. Methods of age determination and correlation have multiplied, reaching new peaks of precision and dependability. Discrimination between events as few as 100,000 years apart is now made regularly, even in rocks beyond the range of radiocarbon, while radiocarbon ages discriminate down to centuries or even decades. Such precision promises a more consistent paleoclimatology. Indeed so much is now emerging from recent studies of cores and pollen profiles in oceans, lakes, and bogs that Pleistocene paleoclimatologists are now venturing seasonal "weather" reports for particular times and places in the past.

Whereas Cretaceous oceans were warm, temperature gradients weak, and deep circulation sluggish, Cenozoic marine oozes record a contrasting state. Polar glaciation, and the equatorward flow of deep, cold, circumpolar waters resulted in large latitudinal temperature gradients—a zonal climate. This thermal transport, together with the pattern of global air masses, constitutes Earth's thermal regulatory and nutrient-balancing system. A fluctuating but marked cooling toward 30 MYBP is followed by early Neogene warming and then final cooling toward ice-age lows (Fig. 16.2). Facilitating that was the growing circum-Antarctic current and the northward deflection of its ever more voluminous cooling waters.

There had, indeed, long been a trickle of Antarctic waters between that continent and South America (the Drake Passage), and that flow was reinforced with the widening Antarctic separation, first from Australia and then from South America. Neogene history was responsive to that. Significant enlargements of the Drake Passage are implied at about 23 MYBP and again about 18 MYBP, increasing the flow and cooling effects, and bringing on shifts in the composition and patterns of marine planktonic communities. Fluctuations between warmer and cooler climates set in as late Neogene refrigeration ensued, with swings from glacial to interglacial and back. They have been plausibly attributed to cycles in Earth's orbital patterns. Cold alone, however, is not enough to bring on or maintain glaciation. It must snow enough on the gathering ice to exceed summer loss of moisture from melting and evaporation or ablation. An ample source of moisture is needed, as well as an appropriate temperature regime to move it to sites where glaciers can accumulate.

Much more might be said of ancient oceans, deep-sea sediments, and their implications for climatic history. Here I note only that deep-sea sedimentation may no longer be imagined as a continuous rain of the finest-grained sediments into quiet deep-sea basins, disturbed only by the occasional surge of coarser debris that settles from overhead *density currents*. No longer does one think of the oceans as repositories of a continuous record, if it could but be recovered. Active deep-sea currents have episodically eroded or prevented the deposition of bottom sediments over large areas of the present ocean bottom. That leaves regional gaps in the sedimentary history that can be matched from one place to another and mapped, recording their own history and perhaps a pattern of eroding sea-floor currents—responsible for ripple marks in some deep-sea sediments.

Even the long-term stability of ocean chemistry is being questioned by some because of the recognition that ocean waters are still actively flushing through spreading ridges of basically mantle chemistry. From these they issue, mineral laden, at rates that would recycle the entire ocean every 8 to 10 MY. Among the important minerals recycled by that deep circulation are phosphorus and calcium. Their extensive Cenozoic and particularly Neogene (including Pleistocene) coprecipitation on low- and mid-latitude shelves and banks is a product of deep circulation and upwelling.

Biologic Diversity in the Age of Mammals

From an anthropocentric viewpoint, the end result of the traumatic Mesozoic–Cenozoic transition could hardly have been more favorable. Gone from the early Cenozoic Earth were the oppressive humidity and ruling "terrible lizards" of our planet's middle ages. The dinosaurian extinctions and others, supplemented by geographic partitioning and increasing climatic diversity, resulted in a diversity of empty ecospaces. Into such niches radiated a flood of biotic diversity at all surviving levels. Most significant for us was a burst of metabolic opportunism on the part of the furry, lactating, warm-blooded mammals, including our own remote primate ancestors.

Mammals, in waiting since the Triassic, had the metabolic equipment to claim the dinosaurian estate and expand their own. They yielded dominance only in the air, where all but six of the twenty-seven living orders of similarly warm-blooded birds had already evolved before the end of the Paleocene Epoch. All of the main plant groups save grasses were present from the beginning, as were all of the orders and 65 percent of the important families of Cenozoic invertebrate life, including insects. The root stocks of modern bony fishes were well started on an irruptive diversification, as were, if less dramatically, the surviving amphibians, reptiles, birds, and, most important for us, insectivores (moles, and their relatives)—ancestral to all placental mammals (Fig. 16.14).

Among marine invertebrates, mollusks remained dominant, but different groups prevailed. Gone were all cephalopods except the squid, cuttlefish, octopus, and the pearly nautilus. Snails (gastropods) came sharply to the fore with some 40,000 genera and more than twice that many species, commonly of bizarre shape and elaborate ornamentation. Their great marine speciation spree was matched on land and in fresh water by another 40,000 still-living genera. Second only to the fecund insects, snails were a great success among terrestrial invertebrates. Clams (pelecypods) were also numerous, among them many scallop-shaped forms of a great range of sizes, and other epicurean varieties—none as large or massive, however, as their extinct barrel- and trapdoor-sized cousins of Cretaceous time. Sea urchins and sand dollars (echinoderms both) are also conspicuous for the first time among the index fossils of the Cenozoic seas.

A wide diversity of small to large calcareous to siliceous Protozoa, both floaters and bottom dwellers, became important in paleoceanography, paleoclimatology, and oil-field stratigraphy. And biogenic reefs of corals, hydrozoans, calcareous algae, and sink-sized "man-eating" clams (*Tridacna*) were abundant at intervals over vast areas of the tropical realm, as they are today (commonly called coral reefs no matter how few corals may be present). Figure 16.13 illustrates a tiny selection of the richly varied Cenozoic marine invertebrates.

Most successful of all animals that have ever lived in the open water, however, in terms of numbers and activity, are the modern bony fishes (the teleosts). Their diversification began during Late Cretaceous history and shows no sign of diminishing. It now accounts for some 200 families, a few thousand genera, and many thousand species. Yet even the numerical success of fishes in the Cenozoic aquatic realm (literally a second age of fishes) pales before human eyes in comparison with the metabolic success and adaptability of mammals.

Members of our industrial society can well appreciate the key to mammalian success. It lies in the capacity to acquire and use energy to do work—to achieve higher levels of sustained and focused activity than competitors. Mammals realized high levels of metabolic success because their ability to maintain a *constant* internal temperature ("warm-bloodedness," or *homeothermy*) allows them to function effectively across a wide range of external temperatures. Because of this advantage, given opportunity, they could and did expand four- to sixfold from 3 to 4 surviving Cretaceous orders and fewer than 20 families to the 18 living orders, 120 families, and 4,500 species known today (Fig. 16.14).

The traits that made mammals successful where dinosaurs failed with changing times were present or potential from late Triassic onward. In addition to constant internal temperature regulation, they included hairy insulation, live birth with parental care among most, generally good night vision, advanced central nervous systems, and efficient metabolism. Why did it take so long for them to flourish? It can only be because the niches into which they might earlier have diversified and expanded were occupied or otherwise unavailable. The flowering of the mammalian adaptive mode had to bide its opportunity. Indeed those creatures waited twice as long as they have flowered, for the Cenozoic "age of mammals" represents but a third of their time thus far on this planet.

16.14 MAMMALIAN FAMILY TRELLIS. Historical ranges of all eighteen living mammalian orders except aardvarks, scaly pangolins, gliding colugos, and sea cows are shown here by solid lines. Placement of images has no time significance except very roughly for primates and hoofed mammals.

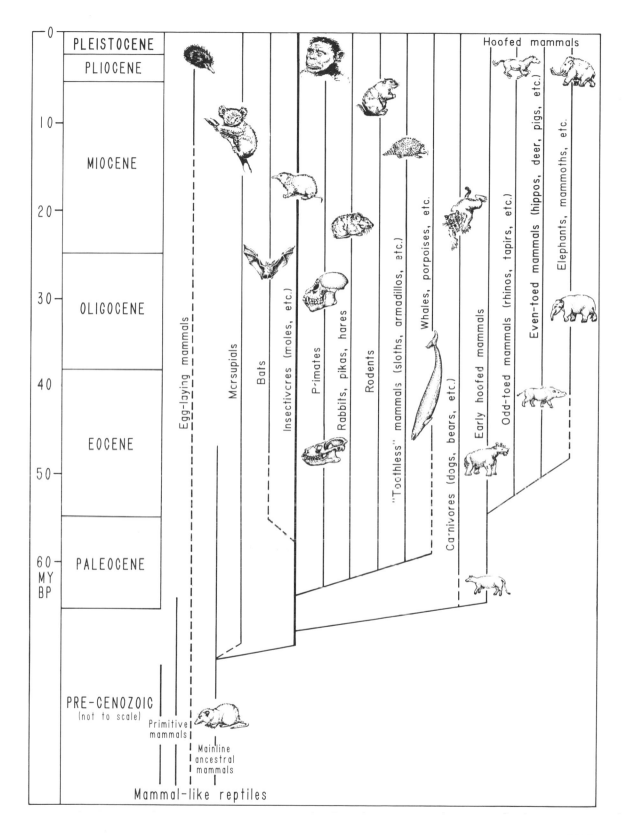

The Cenozoic, then, was above all, and in fact, an age of mammals—especially of placental mammals, for 16 of the 18 living mammalian orders are placental, as are 30 of the 35 orders that ever have lived. The pouch-bearing marsupials, although once widespread, are today native only to Australia and the Americas and prosper only in Australia. The teatless egg-laying mammals—known as fossils only from one Miocene, one Cretaceous, and one Late Jurassic record—today cling tenuously to existence only in Australia.

The seemingly capricious nature of evolutionary success is emphasized by the fact that all of the great succession of Cenozoic placental mammals, up to and including humankind, derive from an initial, late Cretaceous line of tiny, secretive, "bug"-eating insectivores. Within this unity is great diversity. Not only have the placental mammals radiated into a great variety of surface habitats, but their evolution of radar and sonar-like tracking systems has allowed them successfully to invade the air as bats and the seas as porpoises and whales—not to mention other uniquely adaptive mammals. In addition, their increasingly sophisticated nervous systems have given rise to stereoscopic vision, manipulative hands, conceptualization, and other prerequisites for an industrial society, including a degree of control of the planet. The mammalian world has pioneered the exploitation of energy—metabolic energy at first, then fire, mechanical energy, hydraulic energy, wind power, the fossil fuels, and finally nuclear energy. Cenozoic mammalian evolution can be seen as involving a small number of major developmental modes. First was an older Paleogene radiation, during which most of the surviving orders and many of the families arose, as well as then-dominant groups that failed to persist into later times. That gave way to a decline about 40 to 35 MYBP in which background extinction temporarily exceeded origination rate (Fig. 15.6, *event 5*), and archaic families were replaced by modern ones. Grasses evolved about then, eventually exceeding in one diverse family the number of species in the entire mammalian order, multiplying the number of potential grazing niches, and contributing to the appearance and survival of increasingly modern grazers in succeeding Neogene history (Fig. 16.1). Only during Pleistocene history, however, did the mammalian fauna achieve its acme—persisting until the wave of extinctions that came with the recession of mid-latitude continental ice and the appearance of modern humankind.

In a more ecological vein, the descendants of inconspicuous Cretaceous insectivores quickly evolved to fill roles formerly played by dinosaurs. Archaic hoofed mammals became the large plant eaters and early carnivores their predators. With the coming of grasses, an estimated 30 percent or so of the productive land surface of the globe became dominated by some variety of them. A diversity of grazers evolved or became adapted to crop and masticate grass and to flee predators (horses, cattle, etc.). This led to ever-fleeter grazers (horses, gazelles) and ever-swifter, stealthier, or better-organized predators—the big cats and pack-hunting dogs. In short, save for a few (to us) odd-looking extinct creatures, the global biota from about 25 MYBP onward already broadly resembled that now extant.

Where placental mammals of the usual

types were not present, their ecologic niches were likely to be filled by the parallel evolution of other kinds. South America, for example, was cut off from the rest of the world until the late Pliocene emergence of the Panamanian land bridge. There a unique order of one-toed placentals (the *Litopterna*) occupied the grazing niche of the horse. In the absence of placental carnivores, the litopterns and two exclusively South American marsupial orders became the prey of local marsupial flesh-eaters. That persisted until the litopterns and marsupial carnivores were wiped out by predation and competition after the Panamanian land bridge gave passage south to more efficient North American predators about 2 to 3 MYBP.

A much larger variety of parallel adaptations is seen among the marsupial mammals of Australia, which became an island continent and a haven for them before placental mammals could evolve or migrate there. Australia features marsupial "mice"; other small, rodentlike creatures; wolflike carnivores called thylacines; creatures known as dasyures that resemble small cats and martens; large woodchuck-like wombats; and everybody's favorite, the slothlike koalas. There are also kangaroos that function like deer and antelope and even have doelike faces. Could such parallelisms exist if there were not fundamental elements in the mammalian adaptive mode, gene pool, and adaptive physiology that cause different lineages to respond in similar ways to similar stimuli? One might reflect on what may eventually occupy our own ecologic niche on planet Earth.

The Withering of Tethys, the Mediterranean Desert, the Antarctic Connection

Travelers marvel at sunken Carthaginian suburbs along the Mediterranean coast of Africa and other evidence of recent sea-level shifts around that great island sea, ripples in the flood of change this region has experienced over the past 15 MY. Consider a more Wagnerian movement in Earth's unfinished symphony. Imagine a bone-dry desert basin larger than the whole of Greenland, having a relief 3 times that of the Grand Canyon, and descending to depths as great as 5,000 meters below sea level. Imagine that basin filling to the brim during the following millennium over a colossal briny waterfall, evaporating dry once more, and then repeating the same cycle again and again.

Such a place needs no imagining, for it already exists or did exist—not on some remote planet but right here on our own. It is the Mediterranean basin, keeper of Homer's "wine-dark sea" since its last refilling about 5.1 MYBP. That history, moreover, is only part of a chain of bizarre related events. How could there be such an outrageous place? Why should anyone believe it? No one did, and few even suspected it, until a detailed documentation of just such a multicycle sequence was found in deep borings through late Neogene sequences beneath three separate Mediterranean basins. A similar sequence was found in emerged strata on Sicily, and reflected by seismic profiling of the region's abyssal plains.

That discovery, a product of the earlier noted Deep Sea Drilling Project (DSDP) since 1970, relates Mediterranean climatic and geographic history to plate tectonics and the fluctuating Neogene growth and decline of polar and especially Antarctic ice. The repeated desiccation and flooding observed took place between about 6.5 and 5.1 MYBP. The convergence of Eurasia and Africa, with Eurasian consolidation of intervening microplates, had by then reshaped the older Tethyan link between the Atlantic and Indian oceans. By about 6.5 to 6 MYBP, when continuing convergence finally reduced the Strait of Gibralter to about its present dimensions, the growth of Antarctic ice had also lowered the sea beneath that shallow connection. And that initiated the sequence of desiccations and floodings described. It is all written in the sediments at the bottom of the shrunken and disconnected fragments of the former Tethyan ocean— water bodies of varying salinity that include the Persian Gulf, the Aral, Caspian, and Black seas, and parts of the Mediterranean.

Given such a geography, climate, and tectonic history, events described by Ken Hsu of the Swiss Technische Hochschule and others no longer seem surprising. The annual loss of water by evaporation from the Mediterranean surface today exceeds by 3 times that added by all its tributary sources, plus rainfall. Only the inflow of Atlantic water maintains the present Mediterranean level. When that was shut off sometime between about 6.5 to 6 MYBP, net evaporative loss set in at the rate of around 3,300 cubic kilometers yearly. At that rate, the 3.7 million cubic kilometers of water in the basin would dry up in scarcely

more than a thousand years, leaving an extensive layer of salt some tens of meters thick and raising global sea level about 12 meters.

The surface of just such a salt layer transmits a strong seismic reflection, and seismic profiling implies total thickness to be around 2,000 to 3,000 meters locally—the equivalent in salt of the evaporation of 50 to 100 or more basinfulls of seawater. The desiccation–refill cycle is also recorded by micropaleontology, geochemistry, and sedimentary structure within the salt sequence. Mineralogy and structural details establish chemical precipitation from shallow brines, with intervals of total desiccation and emergence. Fossils in thin layers of pelagic sediments within the salt sequences indicate intervals of deep water and locally of shallow freshwater pools. Such repeated associations mean either that the floor of the Mediterranean was bobbing up and down like a yo-yo, or that the basin episodically dried up and was refilled. The former defies even vivid imagination. Mechanisms to explain the latter exist and are, in fact, deceptively simple. Allowing for evaporative losses from refilling waters in transit, local concentration of salts of all types in deeper pools, and other variables, eight to ten such cycles are enough to explain the record in the drill cores.

Everything about this sequence of cycles, in fact, illustrates well how simple physical processes, operating in lawful ways at normal rates, can produce apparently catastrophic results—a classic example of Lyellian "uniformity." A time span as short as a thousand years, either to refill or to dry up the Mediterranean basin *is,* in fact, instantaneous on a geologic time scale. Its simple "uniformitarian" message is that where evaporative loss of water from a closed basin exceeds precipitation plus inflow, the basin dries up as a result of *causes now in operation*. It has happened many times at many places and is happening now—by design, and to public dismay—at Mono Lake in California. Evaporation and refilling of aquatic basins offers an illuminating display of the power of natural laws and the hindsight compression of time to dramatize events.

Inflow from the Atlantic is the major source of Mediterranean water. If, after complete desiccation of that sea, such a connection were to have been reestablished through these ancient Gates of Hercules, influx would still have had to exceed the rate of evaporation if the only slightly hypersaline Mediterranean fauna were to survive. That rate of restoring inflow calculates to be about 40,000 cubic kilometers a year, which is 10,000 times the rate of Niagara Falls, thundering over a 2-kilometer-high cascade. That we know nothing on such a scale today deserves the attention of neo-catastrophists.

It is hard to visualize either the enormity or the normality of such events. Did any pre-DSDP evidence favor a greatly reduced Mediterranean at some time in the past? It did. It was discovered a century ago that sub-sea-level river sediments underlay the valley of the lower Rhône. Other buried channels were later found where rivers entered the Mediterranean basin at levels far below that of the Herculean Gates. Few were bold enough, however, to suggest a dry basin instead of a lake where these channels debouched. Seismic recognition of buried Mediterranean salt domes in the early 1960s sharpened interest but was still inconclusive. Interbedded salt and deep-sea sediments on uplifted Sicily were early interpreted as evidence for some still-unknown form of chemical separation from heavy deep-sea brines. The present evidence for a history of deep canyon-cutting erosion all around the Mediterranean coast, however, powerfully reinforces other evidence for once-huge reductions of Mediterranean water-level. The original channel at the mouth of the Rhône is now a kilometer below sea level and an even-deeper channel was cut by the 3-kilometer descent of the ancient Nile to the Levantine basin.

Records of this later Neogene history, moreover, are not limited to the Mediterranean. The evidence is extended by drill cores from the bottom of the Black Sea basin. When the Black Sea and Mediteranean basins were separate, influx from the Danube, Dnestr, Dneper, Don, and other central and south European rivers transformed the Black Sea to a freshwater lake. It remained so until the first desiccation of the Mediterranean, when reorganization of European drainage systems diverted much of the Black Sea tributary water to the Mediterranean, accounting for local eastern Mediterranean sediments with distinctive nonmarine microfossils. The final movements or rise of sea level that opened the Herculean floodgate then inundated these basins and their connecting straits with water close to normal marine salinity all the way from Gibraltar to the Bosporus and beyond.

With the beginning of mid-latitude Northern Hemisphere glaciation, the then temporarily isolated Black Sea once more freshened and deepened. Rich in climate-sensitive pollen, its sediments became keepers of the European glacial record. Abundant pollens of shrubs and grasses, indicative of a steppe environment, set three major glacial intervals apart from intervening interglacial episodes when the pollen of hardwoods and conifers prevailed. These records are dated by reference to polarity episodes (Fig. 16.2). They extend almost to the present.

The most recent event in the Black Sea saga was about 10,000 years ago, when the Bosporus connection to the Sea of Marmara and the Mediterranean was restored. Thereafter, fresh or brackish waters from the surface of a stratified Black Sea have flowed outward above heavier, saltier waters, compensating the bottom inflow of hypersaline water from the Mediterranean side of the shallow sill that still separates the two water bodies.

The Modern Zonal Climate

Although the Mediterranean region is a large fraction of Earth's surface, it is a fraction still. Events premonitory of the present strong zonal temperature gradient from cold polar regions to warm equatorial and mid-latitudes were much more widespread. They tightened their grip about 15 MYBP with a steep cooling trend. That cooling trend is registered by a sharp increase in the heavy isotopes of oxygen and carbon in the calcareous shells of planktonic microfossils then being deposited. Cooling is also signalled by worldwide coastal deposits (especially along lee coasts) of siliceous microorganisms—for example, the Monterey diatomite of the central California Coast Ranges and correlative deposits elsewhere (Fig. 16.15). Such deposits imply strong zonal winds and upwelling of cold currents rich in silica from extensive contemporary volcanism.

Progressive cooling in later Neogene time resulted in further growth of the already sizeable Antarctic ice cap. That also may have played a key role in the late Neogene cooling that eventually led to Pleistocene and older mid-latitude glaciation, following by barely 3 million years a marked advance of Antarctic shelf ice. Moreover, by about 3 MYBP, records of Arctic sea-level glaciation are seen in the form of dropstones and grounding shore ice in northern waters, reinforcing earlier Miocene evidence for Alaskan glaciation.

Thus the once widely held idea that the historical interval generally known as Pleistocene began with the onset of Northern Hemisphere glaciation is at odds with its "official" paleontological beginning, now dated as only about 1.6 MYBP. Instead, sea-level glaciation has been going on somewhere since about 30 MYBP or earlier. The best candidate for the "real" beginning of the terminal Cenozoic "Great Ice Age" is probably the onset of steppe vegetation in central Europe and other evidence of northern mid-latitude frigidity following the last refilling of the Mediterranean basin. The Pleistocene global sea level and climatic curve records a varying succession of glacial and interglacial intervals with sharp temperature decreases at about 1.5, 1.2, 0.9, and 0.5 MYBP. It was during and between these chilly intervals that humankind emerged. The major features of the modern planet were then already in place. It remained only for erosion and volcanism to apply the fin-ishing touches and for humans to establish a relationship with their surroundings and the existing biosphere.

16.15 SCANNING ELECTRON MICROSCOPE IMAGE OF THE EARLY MIOCENE MARKER DIATOM *ACTINOPTYCHUS HELIPELTA* (ABOUT 0.02 MM IN DIAMETER). Pinpoint-sized bits of opaline silica comrise the diatomite at the base of the Calvert Formation near Dunkirk, Maryland coastal plain. *[Courtesy of George W. Andrews, U.S.G.S.]*

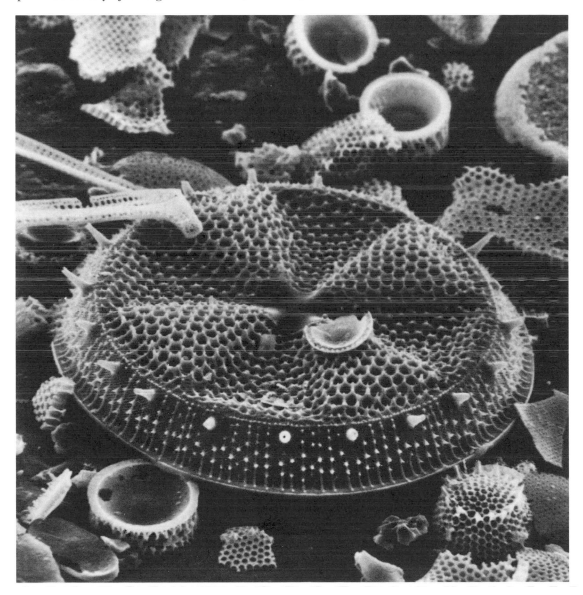

In Summary

In historical perspective, the physical and biological evolution of our planet seems to have intensified with approach to the present, as larger fractions of the record are preserved and its detail and resolution in time improves. Geologists have learned so much about the last tenth of Phanerozoic history (the Cenozoic, or era of "recent life") that the brief telling of it requires ruthless condensation. Here, then, I focus on topics distinctive of that history or in other respects well suited to convey the flavor of the Cenozoic scene.

Crustal activity symbolizes its physical history: all of the world's presently well known mountain systems, for example, were products of or importantly modified by plate motions during Cenozoic history. They are arrayed in two main sets (Fig. 16.4): (1) east–west, parallel to the collisional deformation of the Tethyan region, and (2) north–south, parallel to the Pacific belts of convergence, accretion, and volcanism. Examples of the first are the Alpine-Himalayan belt of deformation, of the second the western American Cordillera.

The highest and perhaps most complicated terrain on Earth is the awesome Tibetan plate jam, under construction since Paleozoic time (Fig. 16.5). The final collisional history, however—the events that deformed, locally granitized, compressed, and hoisted the region toward twice the normal thickness of sialic crust—began only about 50 MYBP. It peaked about 20 MYBP and is still going on—consuming Asian real estate at a linear scale of about 50 meters per millennium. The Alpine fold belts from Spain to the Caucasus—the Mecca and training ground of structural geologists worldwide since geology began—approach the complexity of the Himalaya, are perhaps even more beautiful, and are more accessible. They consist of structurally detached, recumbently folded, and stacked sheets of layered rocks of Mesozoic or younger age, piled northward over one another like platefuls of gigantic rumpled pancakes between continuing compressive forces that, as in the Himalaya, sustain their present elevations. These are classic folded mountains where compressive forces and crustal shortening are the principal tectonic agents.

Other motions make other kinds of mountains, exemplified by variations seen around the Pacific rim. The sialic intrusions at the core and the explosive intermediate volcanics in the glittering icy crest and western slopes of the Andes ascended from deep melts where Pacific floor is subducting into the asthenosphere. Crumpling and eastward thrusting of sedimentary rocks in the adjoining fold belts on the east is a joint product of plate convergence. The great elevations observed are sustained isostatically by a deep, broad sialic root. Relaxation of compression results in the current foundering of the Andean western slopes.

Similar processes and products are parts of an older and even more complex North American Cordillera, where Sierra Nevadan and British Columbian batholithic granites along the Pacific rim corresponding to the core and sialic crest of the Andes. They are, however, flanked by far-traveled accretional collages and, on the east, by folded mountain ranges, reminiscent of the Andes but more advanced. The Basin and Range Province or Great Basin, eastward from the Sierra Nevada, is a tensional or pull-apart terrain, reflecting the relaxation of compressive forces eastward from the San Andreas transform and perhaps the opposing rotations of the Pacific basin and Colorado plateau. Active intraplate research focuses on the complexities of this western third of North America, especially on what came from elsewhere and when.

Evolutionary complexity, paleoceanography, and paleoclimatology wrote large in Cenozoic history. The recorded familial diversity of invertebrates and vertebrates alike increased severalfold during this era (Figs. 15.5, 15.6). Most notable to the human eye was the emergence of thirty new orders of mammals from insectivore ancestors, of which slightly more than half survive. In sheer numbers and nomenclatural variety, however, the mammalian count is far outstripped by fishes, snails, and insects.

Ocean floor, in distinction to shelf and platform seas on the continents, has been regularly recycled down subduction zones about once every 180 MY or less, but only that of about the last 50 MY is extensive enough to provide much useful information as yet. A growing subscience of paleoceanography, therefore, has now emerged for this part of Earth history, utilizing paleomicrobiology, sedimentology, isotopic geochemistry, paleomagnetism, and seismology to decipher the paleogeography of oceans and current systems and to reconstruct global climatic history for that interval.

The continuing convergence of Africa and Eurasia disrupted Tethys, wedging the Mediterranean Sea from northern Africa and tenuously connecting it with a chain of remnant mini-Tethyan seas to the northeast. Closure of the ancestral Strait of Gibraltar about 6.5 MYBP finally isolated the Mediterranean basin, initiating a succession of cycles of desiccation and refilling at roughly thousand-year intervals until its last refilling 5.1 MYBP. That is revealed by an array of evidence, including the precipitation of great thicknesses of salt as that sea repeatedly evaporated and refilled.

Indeed, the amount of salt now at the bottom of the Mediterranean demands the evaporation of at least ten or a dozen basinfulls of Atlantic water. So many times and more did the Gates of Hercules open and close over a million years of Earth history, to admit or bar the inflow of Atlantic waters, either by faulting or by glacial control of sea level or both. Rising sea level or faulting, perhaps along the great transform suggested in the back endpaper, opened the spigot for the last time about 5.2 MYBP, to begin the Pliocene Epoch, harbinger of mid-latitude glaciation, of marked zonal climate, and of modern times.

CHAPTER SEVENTEEN

THE HUMAN HABITAT

Thanks to radiocarbon, the polarity reversal time scale, paleoanthropology, and Pleis-tocene stratigraphy, we now know more than ever about the emergence of humankind and its habitats over the past 1.6 million years or more. Water in its several states shapes the surface, aided by gravity, Earth's rotation, and solar energy. Plate tectonism, rough-hewer of geography, builder and destroyer of migration routes, is also the source of volcanism and seismicity. Such processes account for climate and the prevailing forms of life.

Recurrent, mid-latitude, Northern Hemisphere glaciations, alternating with warm interglacials and reflected by varying sea-level and biotal migrations, dominate Pleis-tocene history and with it the circumstances of later hominid evolution. Although lesser primates existed long before the earliest hominid, the oldest record of the genus Homo *itself is from the deposits of sparsely wooded, equatorial African grasslands near the beginning of mid-latitude glaciation.* Homo sapiens *emerged later, to be followed per-haps 50 millennia ago by anatomically modern mankind. Whether early post-glacial mammalian extinctions were attributable primarily to human or climatic causes remains unsettled—but there is no doubt about the causes of present habitat reduction and extinction. The impact on* H. sapiens *itself of natural hazards and limitations also now increases with growing population and expanding urbanization. Felicitous solutions call for an informed public, statesmanship, and appropriate technology, tempered by prudence and compassion.*

Shaping the Human Habitat:
The Roles of Water

It may seem to self-centered humans as if the long evolution of the planet we call "ours" was all in preparation for the arrival of mankind. Not so. Yet the present and the recent past deserve spe-cial attention because the processes that shaped them are the same processes as presided at the emergence and during the evolution of the species to which we belong.

The misty cyclopean cascades that transformed the dry late Neogene chasm

17.1 SOUTHERN HEMISPHERE VIEW OF EARTH FROM SPACE, FEATURING THE ROLE OF WATER. From Apollo 17. [*NASA, courtesy of Paul Lowman.*]

then separating Africa from Europe into Homer's "wine-dark sea" around 5 MYBP went unobserved by eyes then sensible to their significance. Emergence of humankind from apekind was only approaching or had barely reached the australopithecine level. Another 3 million years or more would pass before the arrival of unequivocal humans in the form of *Homo erectus,* coinciding with the maturing of the age of mid-latitude glaciation.

The Pleistocene of specialists officially began with a biotal and paleomagnetic boundary at 1.6 MYBP—near but not precisely at the time of arrival of *H. erectus.* Whether or not the roughly simultaneous arrival of mankind, mid-latitude glaciation, and the flourishing of open steppes and woodland-dotted tropical savannas was purely coincidental, important feedbacks were involved or established.

Before the onset of mid-latitude glaciation some 2 MYBP, a similarly oriented view of our planet from space would have resembled the present one (Fig. 17.1). There were, however, finishing touches yet to be added before that imagery might so faithfully repeat. These final shaping touches were primarily the work of water—water in all its inconstant states, as universal solvent, moderator of climates, agent of weathering and erosion, lubricator of Earth movements, foundation of the biosphere. Water, moved by the forces of planetary rotation, gravity, and changing temperature, shapes Earth's physiography, directs its surface processes, and modifies its climates.

Figure 17.1, a Southern Hemisphere view of the planet, epitomizes the role of water as a surface-shaping agent. It directs the eye to the chaste beauty of ice-bound Antarctica, 30 million years in the deep freeze since final isolation from its Gondwanan neighbors. Positioned over the pole by climate-shaping plate-tectonic motions and encircled by the prevailingly clockwise West Wind Drift of the Southern Ocean, Antarctica has become the central forcing factor in present global climate. The continuing motions that bestow such influence can be seen high in this global scene, at the triple-junction where the currently spreading Red Sea and Gulf of Aden meet eastern Africa's chain of rift valleys at Djibouti (see also back endpaper). The symphony remains unfinished. Indeed the Pleistocene remains unfinished. Although the last 10,000 to 12,000 years are often dignified as Recent or Holocene, the same agents are still at work. The only question is whether the current interglacial will continue to warm, as predicted, with growth of industrial CO_2, or whether it will decline with intensity of solar incidence, giving way to renewed glaciation.

Consider how Antarctica influences and directs the central agencies of change. Chilled by its glaciers and sea ice, the heavy, cold waters of the encircling Southern Ocean sink where they bend left to converge with warmer, lighter, peripheral surface waters at about 50° to 60°S (the Antarctic convergence). They become nutrient-rich bottom waters that flow northward even as far into the North Atlantic as Nova Scotia, as well as into the Pacific and Indian oceans. Where surface waters are blown or drift seaward off lee coasts, such bottom waters well up to support fisheries, deposit phosphorites on banks and shelves, and cool the neighboring lands. Cold air settling over icy Antarctica becomes entrained by the moisture-carrying westerly winds that ring it as the Roaring Forties. Influenced by the Coriolis effect of Earth's rotation, counterclockwise gyres spin away northward to drive the southeast trade winds and fade into the equatorial doldrums, dragging surface waters of the ocean with them. A similar but much weaker system affects the Northern Hemisphere—weaker mainly because of the presence of a continent-encircled polar ocean in place of an ocean-encircled polar continent.

These systems, and the plate motions that bring them about and raise mountains to complicate atmospheric circulation, are the main weather makers of the planet. Variations in global temperature seem to have coincided with the appearance and regression of continental ice at mid-latitudes. The question is how they are related. Glaciation requires only that more snow accumulate over winter than melts in summer, over a long-enough interval. That calls for a source of moisture for snow, for land on which it can accumulate, and for cold winters and cool summers. Once begun, the expanding areas of snow-covered land and iced-over sea increase reflectivity, thereby lowering regional temperature and enhancing the glacial process.

The process may be reversed by failure of moisture supply, or rising temperature, or both. Orbital variations in received solar radiation are widely thought to be the triggering event, but others are possible. In a Pleistocene north polar view of the planet (Fig. 17.2A), for instance, where glaciated land rings a polar ocean, one can visualize how that ocean might serve as the regulator of a delicately balanced system related to reflectivity and the extent and continuity of sea ice. Being situated, as the Arctic Ocean is, where it serves as the principal source of moisture to peripheral glaciers, the extent of sea ice can function as a control valve. It can turn glaciation on and off as the area covered by sea ice expands and contracts relative to some threshold extent—by increasing and decreasing reflectivity (or albedo), with reverse effects on temperature. Is that why the patterns of glaciation have been so different (despite some similarities in timing) between Northern and Southern hemispheres?

Isotope geochemists have found that an average global summer temperature difference of barely 5° to 7°C (9° to 12.6°F) separates the weather of the last mid-latitude ice sheets from that of the present clement mid-latitudes. This implies that a decline of average summer temperature by that amount, could, in a few thousand years, cause the global distribution of ice to return to that shown in Figure 17.2B instead of the south polar dominance seen in Figure 17.1. Northern Hemisphere ice sheets would then extend as far south as central Germany and southern Ohio. Most mountains more than about 3 kilometers high would be glaciated. The Arctic Ocean and peripheral North Atlantic and Pacific waters would be sealed with sea ice. And fertile blankets of silt, winnowed from the outwash debris of periglacial plains and similar sources, would collect downwind as the fertile silt deposits called *loess* whose productivity feeds the populous Chinese and sustains the Northern Hemisphere's main wheat belts (Fig. 17.2).

Thus, during the Pleistocene, water in the solid state repeatedly spread an erosive icy blanket over some 30 percent of

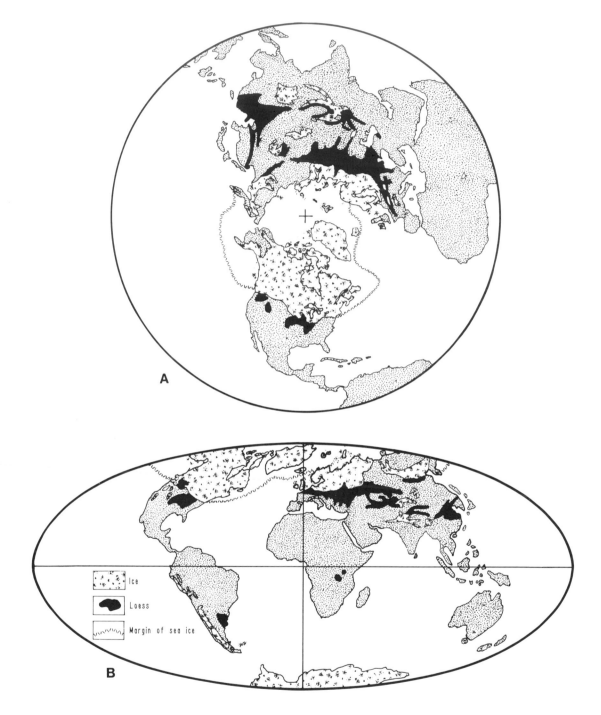

17.2 POLAR (A) AND EQUATORIAL (B) aspects of the global distribution of ice and windblown dust during the last glacial maximum. [Adapted, with additions, from R. F. Flint, Glacial and Quaternary Geology, Fig. 4.9, p. 75, copyright © 1971 John Wiley & Sons, Inc. with permission.]

Earth's land surface and just as often melted away (Fig. 17.3), roughly in harmony with cyclic variations in Earth's orbit, obliquity, and distance from the Sun. At its maximal extent, about a third of this ice was in Antarctica and the rest mainly in the polar sector of the Northern Hemisphere. The total volume was enormous—some 56 million cubic kilometers of ice blanketing some 45 million square kilometers of land. The waxing and waning of that ice could swing sea level through a range of about 140 meters (460 feet).

Flowing in all directions away from well-known centers, this ice scoured and shaped the underlying terrain. The directions of its motion are recorded by scratches on and shaping of the underlying rock, until, reaching water deep enough to float, it began to release dropstones—most distinctive of all criteria

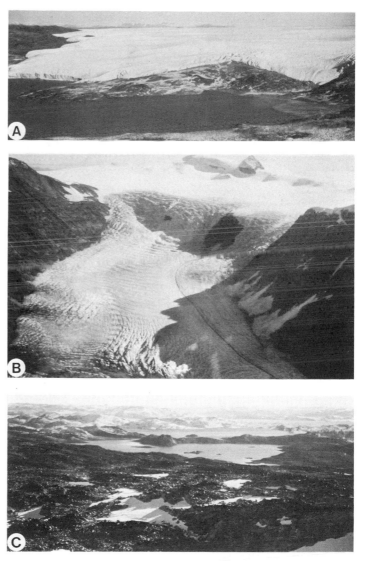

17.3 SCENES AT THE SOUTHWEST-ERN MARGIN OF THE GREENLAND ICE SHEET. Scene *A*, ice margin northeast of Godthåb. *B*, valley glacier draining ice sheet south of Söndre Strömfjord. *C*, recently deglaciated surface northeast of Godthåb.

17.4 ICE MARGIN FEATURES. View *A*, morainal surface showing flowage structures and mud volcanoes at Lake Tekapo, New Zealand. *B*, recently deglaciated U-shaped valley and alluvial cone at southwest margin of Greenland ice sheet.

(when authentic) for the older glacial epochs. Glacial motions are also recorded by other distinctive landforms and deposits of glacial debris (Figs. 17.3, 17.4, 12.1, 12.2, 12.3, 13.14). Glacial deposits commonly consist of angular rocky fragments suspended in silt and clay, the distinctive glacial *till* of geologists, called *tillite* when lithified. Often accompanying such deposits are a variety of stream and lake sediments that tell tales of nearly contemporaneous processes beneath, peripheral to, or beyond the active glacial fronts.

Conspicuous among the work that ice can do with imbedded rocks for cutting tools are the striated rocky surfaces and piles of till or other debris left where glaciers waste away. At the margins of former montane glaciers horseshoe-shaped terminal and ridgelike lateral *moraines* (Fig. 12.2*B*) rest on glacially scratched and polished rock surfaces. Where once were continental glaciers there now are ground moraines on similar surfaces (Fig. 12.3*A*). Such landscapes are often picturesquely rolling terrains with numerous ponds and lakes—

often indifferent farmland until deeply weathered, but preferred residential areas in once-glaciated regions north of about 40°N in North America and 50°N in northwestern Eurasia. Regions poleward of such terrains have usually been scraped so bare (Fig. 12.3*B*) that, austerely beautiful though they be, they are usually populated only by miners and seasonal fishermen, artists (e.g., Canada's famed Group of Seven), and tourist resorts. The exquisite Yosemite Valley and surroundings in California's High Sierra—a region of parklike U-shaped valleys, waterfalls, and imposing exfoliation domes (Fig. 17.5*A*)—is classic among examples of

surface modification by glaciation and related freeze–thaw cycles.

Even more remarkable are the flood-shaped *channeled scablands* of eastern Washington State, comprising some 30,000 square kilometers of gently sloping but grotesquely scarred basaltic terrain drained by the Columbia and Snake rivers (Fig. 16.8). Long a mystery, it is now well established that these scablands are the work of the abrupt and repeated release of much or all of the water in a many-lobed late Pleistocene lake the size of Lake Ontario. That was glacial Lake Missoula, then occupying the intermontane valleys of western

17.5 Shaping of the land by glacial agents. View *A*, glacial scouring and hanging valleys in Yosemite Valley. *B*, downstream scouring by glacial floodwaters, Dry Falls of Grand Coulee, Washington. [B, *courtesy of John S. Shelton.*]

Montana (see Fig. 17.6). The calculated rate of discharge from this lake upon rupture, undermining, or floating of the impounding ice dam in northern Idaho dwarfs that of the world's great rivers and waterfalls (Fig. 17.7). It is estimated to have flowed at a couple of thousand times the normal rate of the Colorado River through the Grand Canyon and more than 100 times that of the lower Mississippi at flood.

Enduring for only brief intervals until the ice dam restored itself or the lake

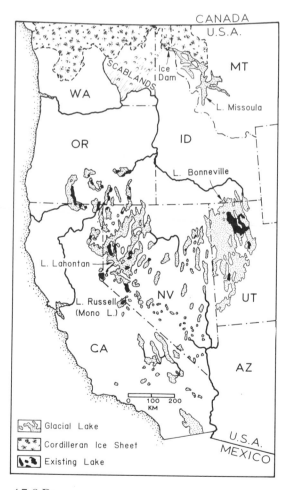

17.6 PLEISTOCENE COMPARED WITH PRESENT LAKES OF THE WESTERN UNITED STATES.

drained, such colossal floods raced at express-train speeds many meters deep across the nearly flat surfaces of Miocene lavas. In transit they detached and removed successive layers of basalts on a scale so far beyond prior experience as to be widely dismissed as a "wild idea" by literal (or "substantive") uniformitarians when first proposed by J Harlen Bretz in 1923. But evidence always takes precedence over "common sense," and eventually everyone came to accept that these floods surged, unconfined, over the whole eastern third of Washington, scouring surfaces with dislodged, water-driven blocks up to 6 meters across. They rolled, dragged, and bounced these and smaller blocks through the now dry Grand Coulee and over its so-called Dry Falls, gouging and deeply scouring the rocky surfaces beneath (Fig. 17.5B). They left huge potholes, scour basins, and upstream apexing erosional remnants reminiscent of those on Mars—scribing in large-scale arabesque and deep relief the entire scabland region.

Erosion of the Channeled Scablands is seen, then, as the product of repeated draining and refilling of glacial Lake Missoula during the last glacial interval, beginning about 18 to 16 millennia ago and lasting to perhaps 10,000 to 12,000 years BP, when expansive freshwater lakes temporarily filled large and often interconnected basins in the Cordilleran region. The estimated refill time for Lake Missoula after collapse and resealing was about 125 years. There may have been a score or more such floods during the 6,000 years or so of its duration, each quarrying new scars and erasing or defacing traces of older ones.

Meanwhile, current examples of the process envisaged have been found and

17.7 WATER AT WORK. Scene *A*, Victoria Falls of Zambezi River. *B*, inner gorge, Grand Canyon of the Colorado River.

described in Alaska and Iceland where the undermining and failure of ice dams and abrupt discharge of lakes have been observed. In Iceland, for example, water normally leaves the icebound Lake Grimsvötn through a tunnel at the base of the ice. Enlargement of that space increases exponentially when melting starts, until the structure is undermined or collapses, rupturing the dam and releasing the colossal floods called *jökul-hlaups* in Iceland. Similar, if less conspicuous, effects resulted from the abrupt spillway failure of Utah's glacial Lake Bonneville (ancestral to Great Salt Lake), lowering it by 115 meters, and scarring the downstream Snake River Plains. Considering how unusual the failure of great natural dams and the generation of scablands seems to be, it is interesting to find that the phenomenon was anticipated by Charles Lyell in Volume 1 of his *Principles*. He offered it, giving seismic release of Great Lakes water, as an illustration of how catastrophic results might be foreseen as products of yet-unrecorded natural causes where consistent with the operation of natural law. Of course no laws of nature prohibit the abrupt rupture of large man-made impoundments from having similar effects.

Glaciation, we see, is a key factor in the partitioning of Earth's hydrosphere between land and ocean. In addition, fluctuations of sea level regulate the opening and closing of migration routes and the composition and areal ranges of plant and animal communities.

About 98 percent of the modern hydrosphere resides in the ocean. Most of the remaining 2 percent presently covers about 10 percent of the continental lands in the form of ice, and the balance is in lakes, rivers, and ground water. The melting of present glacial ice alone would raise the level of the global ocean by about 40 meters, drowning most of the world's important seaport cities and disrupting existing agricultural patterns. Restoration of all Pleistocene ice, resulting in a lowering of ocean surfaces by 100 meters, would have opposite but equally dramatic effects. It would not go so far as to return the Mediterranean to a salt pan without tectonic assistance, but it would leave the world's seaports high and dry and shift agricultural and population patterns. Sedentary, densely concentrated modern humankind is not well prepared to cope with such changes in either direction.

The work that water in the fluid state is able to perform, and its value to humans, depends on topography, vegetation, and the weather. The best trout pools are to be found along hard-working mountain streams. In Arctic or arid to semiarid regions with little vegetation, given piedmont plains or broad valleys and lots of coarse-grained sandy sediment to transport, stream flow is commonly braided (Fig. 17.8*A*). In humid, well-vegetated lowlands and broad-bottomed alluvial valleys cut by larger antecedent rivers the familiar meandering pattern prevails (Fig. 17.8*B*) and erosion occurs mainly at times of flood. As if following the "Peter Principle" or the first law of thermodynamics, streams adjust to perform the work that needs to be done during the time that is available to do it in. They conserve energy, and they declare that to the observer by the flow regime displayed.

The mere fact that there is a lot of surface water around does not mean that a lot of work is being done, nor does its

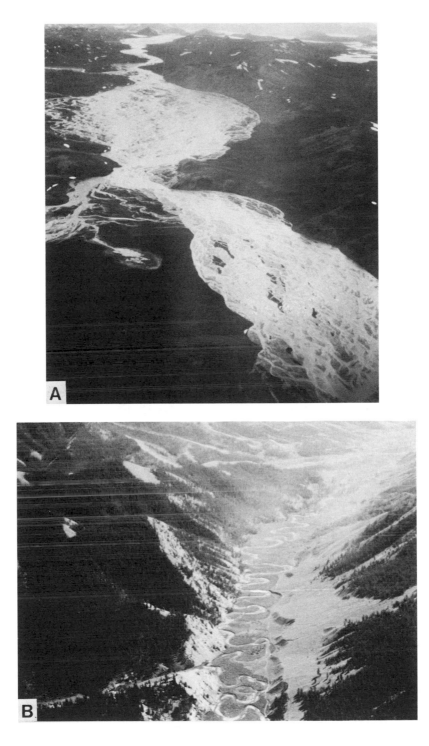

17.8 DISTINCTIVE STREAM PATTERNS. Scene *A*, braided stream in southeast Iceland. *B*, meandering stream in northwest Wyoming.

scarcity imply that nothing is happening. Although Minnesota boasts 10,809 lakes larger than 10 hectares (25 acres), a watershed that drains to three major seas, and the headwaters of the Mississippi River, the state is not noted for its high rates of erosion (Fig. 17.9*A*). More erosion is carried out by ephemeral streams in deserts where sediment binding and flood-resistant plants are scarce. In central Australia, where surface water is almost nonexistent most of the time, subsurface water is still present and vigorous seasonal stream erosion is not

unknown. Eucalypti in dry stream beds and rare perennial waters such as at Alice Springs (Fig. 17.9*B*) assure us that, despite the sad fates of some early explorers, drinkable water can still be found and life sustained by resourceful people who have come to terms with the desert.

In humid regions where little or no water is to be seen at the surface, it is still at work below ground. Limestone terrains, as in the Ozark hills of Arkansas and Missouri and the Causses region of central France, may show no surface

17.9 Sustaining waters of wetlands and deserts. View *A*, headwaters of the Mississippi River, outlet of Lake Itasca, Minnesota. *B*, Alice Springs, Australia.

water until an unexpected stream gushes from a hillside or a spring-filled chasm appears at one's feet. In the limestone valleys of Appalachia new sinkholes appear overnight from the collapse of cavern roofs. The universal solvent has been at work all along, opening sheltered spaces such as those in which, after dispossessing cave bears, humankind long frequented, left pictorial records of creatures hunted, and, more recently, have raised mushrooms and stored wine and contraband. Caverns, sinkholes, and the ragged solution-etched surfaces of lime-

stone outcrops are all comprehended by the term *karst*, from their conspicuous development around Kras, along the Dalmatian coast of Yugoslavia.

Although karst has developed to some degree in all humid regions that are also underlain by limestone, its most extreme development is the tower karst of southern China and adjacent Vietnam. This baroque region of crowded conical peaks and caverns, looking on a large scale like nothing so much as the bottom of an empty egg carton (Fig. 17.10), has inspired and sheltered artists, poets, and

17.10 THE ULTIMATE IN KARST TOPOGRAPHY. Scene *A*, tower karst in south China, along the Li River, south of Guilin; *B*, airview between Guilin and Guangzhou (Canton).

17.11 TROPICAL PACIFIC COASTS IN STORM AND CALM. View *A*, surf from a distant typhoon pounds normally leeward western Guam. *B*, windward reef edge, eastern Guam. *C*, grooves and buttresses of windward reef, Onotoa Atoll, Kiribati (formerly the Gilbert Islands). *D*, reef, near low tide, Arno Atoll, Marshall Islands.

460

revolutionaries from the legendary Monkey King to Jou En-lai. Renowned for its classic topography, much featured in Chinese artistry, it is also known for the variety of microhabitats that provide such a wide assortment of mushrooms and medicinal forms of life. It now also boasts one of the world's rare organized karst institutes (the Yan Rong Janjuesuo), at Guilin, in Guandong Province.

Last of the many roles of water to be touched on here is the work of salt water, mainly in the oceans but also in chemical corrosion and the precipitation of salts on land. Everyone knows about the destructive work of coastal erosion and the flooding of lowlands. Fewer realize the extent to which the same processes are also capable of extending the land,

with the aid of sediment trapping and binding salt-marsh grasses and mangrove swamps. Among the 10,000 or so fabled and often beautiful islands of Oceania, uninhabited or absent 30,000 years ago, we now find nearly 3 million islanders whose lives and possessions often depend on a tropical sea that builds and destroys impartially. Its typhoons and lesser storms (Fig. 17.11A–B) attack alike the impregnable high volcanic and limestone islands of this region, and the very coral reefs and low islands the sea itself has so patiently nurtured (Fig. 17.11C–D) and which the next shift of sea level could destroy. Among better-known regions, the low coasts of England and mainland Europe have been similarly beset, now gaining, now losing ground. And so it is wherever land meets sea.

Other Shaping Agents

Dominant though the agencies of water are in shaping the Pleistocene surface, and awesome though they can be in a great storm, they do not work alone. They are usually linked in some way with broader agents of change such as plate tectonics and gravity, which locally manifest their powers in volcanism and seismicity, involving water-lubricated shaking, rupture, and avalanching. Such processes are usually more abrupt, more infrequent, more conspicuous, and on a larger scale than the everyday work of water (soil creep, slope wash, and runoff).

Volcanism has affected the human habitat throughout its evolution. As the back endpaper shows, eruptive volcanoes and earthquakes are everywhere concentrated along and landward from

convergent plate boundaries. More than 60 percent of them cluster around the rim of the Pacific Ocean. Volcanism at such sites is likely to erupt lavas more sialic than basalt and to do so explosively and on a large scale. The more quietly effusive basaltic volcanoes and lava fields of oceanic regimes and mantle-tapping rifts on the continents from time to time pour forth impressive volumes of lava (as in the Columbia Lava Plateau), but are notably destructive only in the regions flooded.

Volcanic ash ejected from explosive craters of the Pacific rim may be ejected with such force, however, that it reaches the stratosphere. There it may orbit the globe, affecting global weather and rainfall. Figure 3.9 charts the impressive extent of a few identifiable ashfalls in the United States and Figure 17.12 compares the estimated volumes of some notable ones. Despite its toll of lives and property, the widely known 1980 eruption of still-rumbling Mount St. Helens, near Seattle, does not rank with the great eruptions of history. The gigantic, prehistoric, nearby eruption whose only conspicuous present remnant is Crater Lake (Oregon), the Mount Mazama of

17.12 CONSERVATIVE ESTIMATES OF VOLUMES OF EJECTA IN CUBIC KILOMETERS DISCHARGED BY SOME WELL-KNOWN ERUPTIONS OF THE PACIFIC RIM. Other estimates are 2 to 6 times as large as those here presented. [After USGS Professional Paper 1249, 1982, p. 118, Fig. 62.]

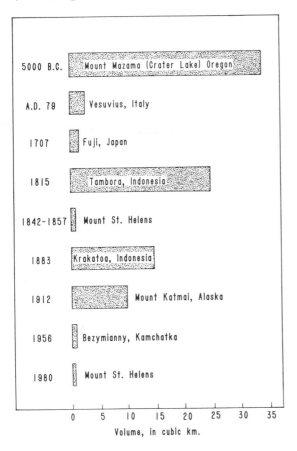

geologists, discharged a volume of ejecta at least two orders of magnitude greater than did Mount St. Helens in 1980— about 34 as compared to 0.34 cubic kilometers (0.1 cubic mile). And there were more ancient eruptions of unimaginable magnitude (e.g., Yellowstone, estimated to have expelled 2,500 cubic kilometers of eruptive debris).

In brief, volcanism transports ejecta and rock-making fluids from Earth's deep interior to surface and atmosphere, adding load, inducing instability, enriching future soils, affecting the weather, killing people, and destroying the works of humankind (Fig. 17.13A).

The peaceful scene at Mount Ruapehu's crater lake in 1939 (Fig. 17.13B) belies its hazardous potential, revealed in 1953 when a wet debris flow from it caused the wreck of a Christmas Eve excursion train in which 151 New Zealanders died. Eruptions of volcanoes like Ruapehu, with crater lakes or extensive summit snow and ice, generate destructive, fast-moving, wet avalanches ("mudflows") of the type called *lahars* in Indonesia.

A record of deaths from volcanism and its effects is available from A.D. 1500 until now. About 220,000 have been killed during that time, in addition to the unrecorded prehistoric deaths at Thera (Santorini), Crete, Pompeii, and in early Indonesia. Among eruptions for which records are available, about 92,000 people lost their lives in the great Tambora eruption of 1815 (Fig. 17.12). That eruption put enough ash and aerosol into the air to obscure the Sun and bring on the famous "year without a summer" (1816). The eruption of Krakatoa in 1883 generated a tidal wave that killed another 36,000 and dust enough to cause brilliant sunsets worldwide for the next 3 years.

More than 25,000 were buried alive at Armero, Colombia (90 percent of the town's residents), in November 1985 when lahars raced down the slopes of the ice-capped Andean volcano called Nevado del Ruiz at speeds up to 150 kilometers an hour, inundating the ancient lahar track of 1595 and later eruptions. The April 1982 eruption of El Chichon, in Mexico, put so much dust into the stratosphere that global weather conditions were significantly affected for the next 2 years. The effects of comparable events on the human habitat and its occupants over time are a large element of Pleistocene history. Footprints of *Australopithecus* and perhaps *Homo erectus* in wet ashfalls suggest the terror, probable loss of life, and awe imposed by volcanism in the African rift valleys as humankind began to ascend the evolutionary ladder.

17.13 VOLCANOES IN OUR LIVES. View *A*, cathedral engulfed by lava, Paricutin, Mexico, 1945. *B*, Mount Ruapehu, New Zealand.

Seismicity, expending most of its great energy to move huge volumes of rock, wreaks its destruction on the works of man mainly by surface-shaking at long wavelengths. These long waves are amplified in weak, watersoaked ground in populated areas like the old lake beds beneath part of Mexico City and the mudflats around southern San Francisco Bay. The magnitude 8.1 double earthquake that collapsed about 300 large buildings (some of supposedly earthquake-resistant design) and damaged more than 3,000 others in Mexico City in September 1985 also killed some 12,000 people, injured another 30,000, and left around 50,000 homeless. Its initial point (*focus*) was about 15 kilometers beneath the town of Lázaro Cardenas, along the actively subducting Cocos Plate (Fig. 16.7, back endpaper), 360 kilometers southwest of Mexico City. Little damage was reported between Lázaro Cardenas and Mexico City, but it is reported that resonance in the underlying lake clays beneath the latter amplified ground motion from 2 seconds to 2 minutes. That was responsible, it seems, for the terrifying and destructive toppling of six- to fifteen-story buildings.

The stated importance of seismic activity to people, like that of volcanism, is usually based on numbers killed—a statistic more related in the case of seismicity to subsurface conditions and construction than to magnitude. Mortality records from sixty of the world's most lethal earthquakes between the time of Charlemagne and 1925 add up to more than 3.7 million deaths from various earthquake-related causes, half from five great Chinese earthquakes. That number is more than 5 times the number of U.S. battle deaths in all of its wars from the American Revolution through the Vietnam War.

Earthquakes near plate boundaries and volcanic regions (where most of them occur) tell us that moving plates and subterranean magmas can shake and disrupt the surface. About 50,000 seismic shocks worldwide each year are strong enough to be felt unaided by instruments, and about 100 of them are strong enough to affect the human habitat adversely. Destructive effects and instruments include rockslides and avalanches, mudflows, seismic sea waves, foundation failure, and rupture of service facilities. A typical damaging 7.7-magnitude tremor that struck New Zealand in 1929 triggered some 7,000 surface breaks and slides larger than an acre each over a mountainous region smaller than the state of Delaware.

Landslides and avalanches regularly accompany earthquakes, but even in seismically quiet hilly areas, gravity, moisture, and geologically unstable slopes can start them. Approximately 2 million debris flows or landslides have been counted in the Appalachian Mountains. Large-scale geologic maps in much of California are sprinkled with landslide patterns. It has been calculated that a 9.2-magnitude earthquake would induce surface-altering motion over a region twice the size of Oregon.

Although thirty-nine U.S. states have experienced damaging earthquakes, North America's most seismically active region is the Pacific Coast. The locations of a few of California's 600 or so more prominent faults are shown on Figure 17.14, emphasizing those where damaging motion is most likely to be brewing. Damage can be minimized by sound engineering construction and earthquake forecasting. Studies now in progress worldwide have shown that the best prospects for reliable forecast are repeating sources that occur at *characteristic intervals* and particular depths over limited lengths of frictional hang-up. Several such repeating sources are found along the San Andreas fault and one, near Parkfield, California (*P* in Fig. 17.14), is now instrumented for observation. Five magnitude-6 earthquakes having a characteristic interval near 22 years occurred there between 1857 and 1966. The next is due in 1988, give or take a couple of years. Perhaps you will be (or were) there for it.

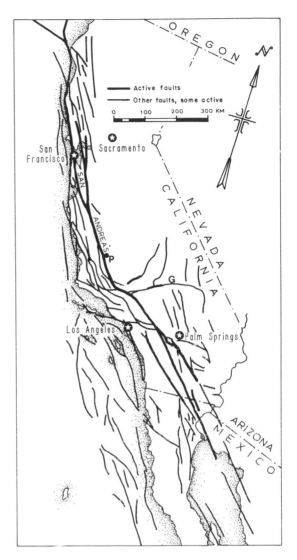

17.14 MAJOR FAULTS OF COASTAL CALIFORNIA AND NORTHWESTERN MEXICO (*P* IS PARKFIELD; *G* IS GARLOCK FAULT). [*Adapted by permission of the Geological Society from J. C. Crowell*, Journal of the Geological Society of London, *v. 136, Plate 295, Fig. 2.*]

Some Consequences

Plate tectonics, geography, orbital variation, and climate have shaped the human habitat in a variety of ways. Mid-latitude glaciation in the Northern Hemisphere has played a leading role, not only in the glaciated regions but far beyond. At one time or another during the Pleistocene, glacial lowering of sea level has connected England with western Europe, Japan with Korea, Ceylon with India, the Indonesia islands with Southeast Asia, and Alaska to Siberia.

Our bipedal forebears and many of their quadrupedal fellow travelers used all of these connections to achieve their present distributions. The Bering Sea bridge between Siberia and Alaska was not simply a narrow causeway but at times was nearly as broad as Alaska itself. It was opened and closed repeatedly by crustal movement and glacial fluctuations of sea level. Mammals, including Paleoindians, Aleuts, and Inuit, crossed the region as late as latest Pleistocene, and conceivably as early as 33 millennia ago.

Falling sea level compensated to some extent for the loss of habitable area as a result of continental glaciation. The Atlantic coastal plain of the United States, for example, was about 70 kilometers wider in late glacial time than now, offering a partial haven for qualified refugees from the north.

A striking feature of Pleistocene history is the evidence for stronger periglacial winds and increased rainfall at lower latitudes. Dune sands and loess (Fig. 17.2), although by no means uniquely glacial, are common periglacial features, reflecting the sinking cold air that formerly blew from glacial fronts across silty and sandy outwash plains and still does from modern ice sheets. The shapes of dunes register paleo–wind directions (Fig. 14.4) and the loess provides much of the world's most fertile farmland.

Evidence for thriving contemporaneous lakes and vigorous perennial streams records increased rainfall or reduced evaporation because of cooler temperatures at lower latitudes during times of mid-latitude glaciation. About half of the 200 mapped ''glacial'' lakes of the Great Basin and adjacent regions are indicated in Figure 17.6. They included Lake Missoula of scabland fame and the huge Lake Bonneville, once 300 meters deep, predecessor to Utah's Great Salt Lake. The formerly inundated flight of recessional terraces that now rings Salt Lake basin records the postglacial

evaporation of Bonneville's sweet water to the present bitter brines. Although such lakes were common only to about 13°S beyond the Cordilleran ice sheet (Fig. 17.6), they reached as far as Mexico City. These lakes were new-world havens for migrants from Beringia. Their waters and game sustained flourishing American Indian encampments for 12,000 years or more as the lakes evaporated—almost until the arrival of Lewis and Clark, the fur trade, and the European discovery of gold.

Similar conditions in Eurasia affected chiefly the Mediterranean region, central Asia, and northern China. The effects of this pluviality were evident right down to the equator itself. The climatic qualities of equatorial East Africa, famed for the discoveries of hominid evolution there, were well suited to primitive human comfort. As the only large subtropical landmass that lies mainly north of the equator, Africa received winter-season outbreaks of cooling air from circumpolar sources, air masses that even today reach within 15° of the equator. During the Pleistocene, when Northern Hemisphere climatic zones were displaced southward, such winter air masses, interacting with moist tropical air, would favor year-round precipitation and water supply, luxurious vegetation, grazing herds, and moderate equatorial temperatures.

Results of the alternation of glacial and interglacial climates on different scales of time and temperature affected the entire surface of the Earth—land and sea, tropics and high latitude alike. In North America most of Canada was under ice most of the time, with extensions across the northern United States reaching south as far as southern Ohio.

Glacial advances alternated with retreats of long duration, during which interglacial weathering converted loess and calcium-rich glacial tills into the fertile midcontinent soils. West of the hundredth meridian we see the effects of montane glaciation, continuing regional uplift, and the interior drainage that converted so many closed basins from ephemeral into perennial lakes.

Accelerated stream erosion within and eastward from the Rocky Mountains as a result of continuing Pleistocene uplift and mountain meltwaters transformed later Neogene landscapes into the present ones. Downward cutting and removal of large volumes of older Cenozoic basinal deposits brought these streams into erosional contact with underlying older sedimentary and granitic rocks into which they cut deep gorges. Streams that once drained northeastward from the northern Rocky Mountains into Hudson Bay were deflected southward by the north-

ern ice sheets to become the Missouri River and join the Mississippi at St. Louis.

In the western Cordillera, nearly continuous ice thrived along the still-rising and eroding Sierra Nevada crest, suckling its litter of glittering distributary glaciers in their moraine-ringed, U-shaped valleys below (Figs. 12.2*B*, 17.5*A*). It looked eastward over many a lonely summit glacier within the Great Basin, while, far to the southeast, the oft-raging Colorado worked its way to the sea.

The Pleistocene history of North America is recorded by the transfer of Canada's surface materials to the northern United States, the excavation of finger lakes elongated in the direction of ice movement, other glacial landforms, and the records of now dry or drying glacial lakes. The Great Basin has returned to its former warmth and aridity, where its improbable urban centers survive on imported or mined water.

Fellow Travelers, Inconstant Climates, and the Latest Extinctions

Although numerous similarities exist between plants and animals now living and their older Pleistocene ancestors, important differences are also evident. Many creatures that shared the uncertain journey through the Pleistocene ice ages with our hominid ancestors have not survived. Some were unable to withstand the rigors of climatic change or compete successfully against rivals. Others were eliminated by the actions of *Homo sapiens*—by hunting, territorial exclusion, or alteration of habitat.

Many Pleistocene animals, vertebrate and invertebrate, marine and nonmarine

alike, have migrated south or north, and to lower and higher altitudes with changing global and regional climate. Similar but usually more sluggish migrations are seen in plant life, recorded by the shifting composition of pollen profiles from cores of bog and lake sediments. Shells of marine creatures once found only south of Los Angeles now range northward as far as San Francisco. Gone are the once-abundant dire wolves, the saber-toothed cats, the giant American lion, the European cave bear, the mammoths, the giant condor, the native American horses and camels, and others. The remains of tundra muskoxen, today indigenous only to Arctic regions, are evidence for an older Pleistocene age when found in Nebraska.

Although adaptations to cold are seen— wool in elephants, rhinos, and muskoxen, blubber in polar bears—most evolutionary changes in mammals are not clearly related to temperature. At the same time there was plenty of evolutionary change for so brief a span of years, most likely related to relative fitness for ever-changing livable space. In that "competition," plants were conservative and animals often inventive. Some 80 percent of modern plants were already extant by the beginning of Pleistocene time, but the reverse is true of mammals. Of the 119 species of mammals now found in Europe and its Asian borderlands, only a handful are *known* to have been present as the Pleistocene began. The rest either remain to be found in older deposits or else evolved or migrated there during the last 2 to 1.6 million years.

Mammals are the focus of interest. Their history is a checkered one. During earliest Pleistocene, additions to Pliocene carryovers included true elephants, zebrine horses, modern camels, and the now-domestic cattle. Further additions beginning half a million years ago included periglacial adaptations such as reindeer, muskoxen, woolly mammoths and rhinos, moose, and lemmings. During later middle Pleistocene came the taiga antelope from the middle Asian steppes, with its inflated nose (for warming cold air before inhaling it, 'tis said). A succession of migrations across the Bering land bridge brought mammoths, modern horses, woodland muskoxen, antelopes, and many rodents to North America.

Finally, during the last major interglacial and its preceding glaciation, there came to North America bison, skunks, rodents, and other creatures now beautifully preserved in the Rancho La Brea tar pits and on eye-catching display at the handsome Page Museum in central Los Angeles. Besides 150 species of birds, the La Brea fossils include another 400 species, representing many known living families of mammals as well as reptiles, amphibians, invertebrates, plants, several extinct genera and species, and one young female human from 9,000-year-old tar. The collapse or extermination of that fauna at the end of the last (*Wisconsin*) glaciation is generally thought to have coincided with the establishment of humans and the beginning of desiccation in the Great Basin (although, as we shall see, new Brazilian data imply an earlier arrival).

Late Pleistocene extinctions are, in fact, no less impressive and almost as puzzling as older ones, with no record of asteroid or cometary impact to blame. The existing fauna of large land mammals can only be called impoverished compared with that at the height of Wisconsin glaciation about 20,000 years ago. Precise radiocarbon dating shows that most extinctions occurred between about 15 and 9 millennia ago, clustering around 11,000 years BP—except in New Zealand, where extinctions came later, and Australia, where they were earlier. The large mammals of the Americas and Australia were hardest hit, perhaps because humans and other animals of other regions had longer to adapt to one another. Some 75 percent of North American mammals (33 genera) having an adult weight of more than 44 kilograms (100 pounds) became extinct.

South America lost 46 genera, including two whole orders of mammals. And Australia, perhaps 30,000 years earlier than North American extinctions, lost 49 marsupial species, plus a lizard, a tortoise, three birds, and a giant snake.

Lesser losses were experienced elsewhere. During the same interval, Europe lost only 3 species by extinction, although 10 others migrated to Africa and Asia. Africa lost some 30 genera by extinction over the course of Pleistocene history up to about 130,000 years ago, but only a few species became extinct during latest Pleistocene climatic change. Asia too seems to have logged few late Pleistocene extinctions, although relevant data are yet scanty. Large flightless birds, the moa in New Zealand and the roc in Madagascar, became extinct, along with a number of small animals—all potential prey for humans and the rats and dogs that have accompanied their migrations. Although marine extinctions were few, the loss among mammals was similar in scale to that of the terminal Cretaceous reptilian extinctions.

Were these creatures all done in by the upright, big-headed, global wander-

ers and hunters they had not yet learned to fear and avoid? Or was climatic and environmental change the culprit? It is well established that the passenger pigeon and the great auk were victims of human hunters during recent times. The much earlier extinction of New Zealand's giant moa at the hands of the Maoris seems equally well established. And estimates of the human population of North America, if believable, indicate a marked increase at the very time that evidence for the abundance of American mammals shows a marked decline. That was between about 13,000 to 11,000 years ago, roughly synchronous with the final extinctions.

The seemingly well-adapted hairy mammoth, once known across the whole of northern Eurasia and south to China and the Mediterranean, also became extinct then. Carcasses in the permafrost of eastern Siberia have yielded radiocarbon ages from 45,000 to 12,000 years BP. A partly restored 45,000-year-old body with skin and hair can be seen in the Mining Museum at Leningrad (Fig. 17.15). Well-preserved skeletal remains are so abundant in Siberia that dwellings

17.15 *MAMMUTHUS PRIMIGENIUS* FROM A 45,000-YEAR-OLD BURIAL IN YAKUTIA, SOVIET UNION, AT MUSEUM OF THE MINING INSTITUTE, LENINGRAD.

were built of them, while ample evidence shows that the mammoth was hunted and eaten by nomadic tribes.

Mammoth remains in North America are associated with the well-known Clovis hunting culture. Other large mammals became extinct or rare with the climatic changes that favored the spread of *Homo sapiens*. Beneath a cliff at the Olsen–Chubbuck site in Colorado are the remains of some 200 giant bison, recording their destruction at the hands of pre-Columbian Indians.

Were these late Pleistocene extinctions wholly the result of human hunting, climatic change, or both? Why were the results and timing so different in different regions? Granting extinction and the arrival of potential human hunters to have been concurrent, what is the evidence that it was also causal? Why are there so few butchering sites for large mammals compared to those for the moa? In light of the evidence for rapid climatic and ecologic change, how could that fail to have been a factor? Arguments linking climatic change with this extinction, however, are also challenged.

Every challenge draws a different logical response. Those attributing extinction to overkill by early humans explain its absence in Africa by the argument that African mammals had more time to learn fear and develop evasive techniques as human weapons and hunting techniques evolved. They attribute earlier extinctions in Australia to the earlier arrival there of *Homo sapiens*, perhaps 40,000 years ago. They see the rarity of butchering sites as a probable result of rapid selective hunting.

It seems likely, in fact, that both climatic change and human intervention will be found to have played important roles in those Pleistocene extinctions. By no means, however, can the current rapid and tragic decline of biologic diversity, particularly in the world's rain forests, be attributed to anything but mankind's continuing destruction or preemption of natural habitats worldwide. It behooves us to reverse that trend as a condition of our stewardship of planetary resources, for posterity's sake, if not our own.

The Precocious Primate

The Pleistocene ice ages, last of many glacial regimes to grip our planet in the course of Earth's long history, provided the theater in which the last act of the evolutionary drama was played. It began, very nearly, with the appearance of *Homo erectus* on the stage—a tall, robust, upright hominid, unquestionably human in its stance, cranial capacity (Table 17.1), and general physical characteristics. First recorded about 1.64 MYBP in Kenya, *H. erectus* eventually spread across the whole of Eurasia to China and even Java. Where did it come from? What was its ancestry and its fate?

The steps to humankind were many and precarious; and with each came the opportunity to stumble. The Pleistocene emergence of the human state was but one of many events in billions of years of evolution in which the chain to our species could have broken at any of many branching points or times of mass extinctions. Without the eucaryotic cell and the metazoan level of evolution there could have been no vertebrates and therefore no mammal-like reptiles. The mammal-like reptiles barely survived terminal Permian extinction. Had they perished there would be no mammals. Had not the ancestral mammals carried through to late Mesozoic times, there would be no placental mammals. Without placentals there would be no insectivores, no primates, no hominids, no us (Figs. 16.14, 17.16).

Humans probably originated on only one of the present great landmasses; most likely Africa. Only later did they wander off to others now occupied. They were already in Papua New Guinea and Australia about 40,000 years ago, but there is disagreement about time of arrival in the Americas. The oldest reliable records in North America south of the former ice sheets do not exceed 13,000 years, but significantly older occurrences are reported in South America. The best documented of these is at Boqueirao de Pedra Furdada, Brazil, where French investigators have reported a radiocarbon-dated sequence of human occupation from 6,160 back to 32,160 years BP. Because ages found are of charcoals from hearthsites associated with worked stone tools, their authenticity seems undeniable. But could the ancestors of these people have traversed North America without leaving traces until 20,000 years later? Might they instead have come by sea? Or is some masterful hoax abroad?

Regardless of how they got about, several named varieties of the genus *Homo* have known our planet, concluding with the anatomically modern subspecies *H. sapiens sapiens*, which apparently first appeared about 50,000 years ago (Fig. 17.16), but which has been estimated on molecular grounds to date from about 80 millennia BP. They, plus ancestral

TABLE 17.1 *Cranial Capacities of Hominids Compared with the Chimpanzee*

	Age Range (Before Present)	Cranial Capacity (in Cubic Cm)	
		Average	Range
Homo sapiens sapiens	50,000 yr– present	1,350	1,100–1,600
H. sapiens neanderthalensis	350,000 yr?–32,000 yr	1,450	1,100–1,620
H. erectus	1.64 MY—300,000 yr	950	770–1,300
H. habilis	2.5 MY–1 MY	690	510–800
Australopithecus, robust species	2.5MY–1 MY	520	—
Australopithecus, gracile species	3.7 MY–0.75 MY	450	390–500
Australopithecus	5 MY?–1 MY	450	390–520
Chimpanzee	5 MY?–present	about 350	—

species, comprise the immediate family of humankind, the hominids—not to be confused with hominoids, including hominids and apes. The distinctive anatomical characteristics of the genus *Homo* can be recognized from skeletal remains alone. These are stereoscopic vision, grasping hands, bipedal gait, erect posture, opposable thumbs, and a large cranial capacity. By inference we include speech and a conceptual brain.

Consider the improbable progression from a likely ancestral mammal to that level. Crucial steps were to become pla-

17.16 THE ROOTS OF HUMANKIND.

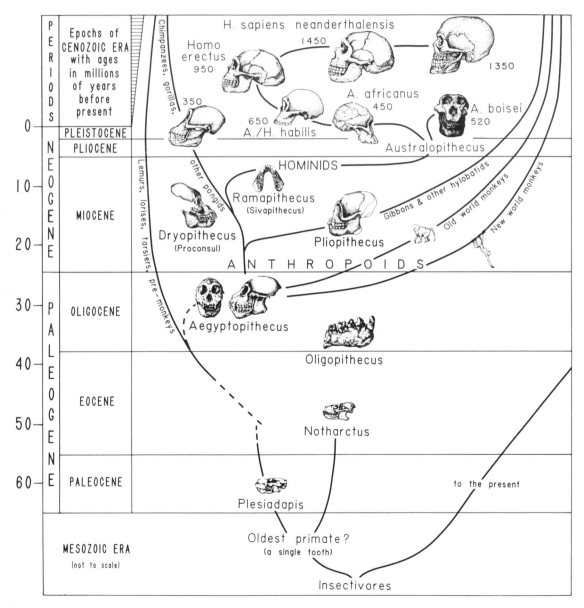

cental and primate—for our scurrying, four-footed, shrewlike forerunners to differentiate from their own insectiverous ancestors about the time the last of the dinosaurs were known. Next, and equally crucial, was the branching of the anthropoids (monkeys, apes, and hominids) from the mixed group called premonkeys on Figure 17.16, somewhere between about 30 to 50 MYBP. Only long after that did the hominids emerge and differentiate, eventually to become modern humankind.

Thanks to radiocarbon, the polarity reversal time scale, and the recent discoveries of an informal international consortium of paleoanthropologists, Pleistocene stratigraphers, and geochemists, we now have more and more accurately dated information on human evolution and its circumstances than ever before. The evidence, nonetheless, is still meager and fragmentary. Until the arrival of *Homo erectus*, and even for some time after that, the number of individual hominids and supposed hominids from any given site can usually be counted on the fingers of one hand. Skulls complete enough to allow measurement of cranial capacity are even fewer, and the most complete skeleton known is only 40 percent complete and lacks a skull. Thus, qualified specialists disagree about important details. Despite that, there is also strong convergence of opinion about the main lineage and the directions it may have taken.

Progression toward modern humanity was not an even course, with the whole assembly of human characteristics appearing at the same or coordinated rates. What occurred is aptly called *mosaic evolution*. It is like arranging bits of a jigsaw puzzle until something like a complete picture begins to emerge. Bipedal gait and erect stance came first, then the opposable thumb, dextrous hands, and enlargement of the brain, all before convincing signs of cognitive function and cultural evolution. The implied rate of that evolution suggests that employment of an already large brain in advanced conceptual thought may have had to await palatal or other changes favoring articulate speech and communication.

Bones and teeth claimed to be those of the first anthropoid are reported from two different regions. One set is from 40- to 45-MY-old deposits of Burma (*Amphipithecus*). The other and more likely is from 30-MY-old river sediments of the Faiyum desert in Egypt (*Aegyptopithecus* and *Propliopithecus*). The scanty Burmese material suggests a gibbon-sized creature with probably overlapping fields of view that would imply an early stage of stereoscopic vision and perhaps arboreal life. But how can such an ancestry be reconciled with evidence from the younger, more numerous and better preserved, unequivocally apelike Egyptian fossils? Unless the descendants of *Amphipithecus* somehow managed to get from Burma to Africa across the still-wide Tethyan ocean (Fig. 16.4) in time to be ancestors to the African apes, only one of these claimants is likely to be right. Because the oldest reasonably abundant remains of apes are found at Faiyum before the appearance of anthropoids is known anywhere between there and Burma, it seems likely that *Amphipithecus* was an independent, dead-end offshoot of the pre-monkey line. The Faiyum apes may well be found to have older and more monkeylike ancestors somewhere in Africa.

We know, with some confidence, only that these Faiyum fossils, fragmentary and rare as they are, were already apes. The Faiyum habitat, revealed by its sediments and the bones of birds related to living tropical families, resembled modern East African swamplands, bordered by forest and open woods or savanna, as in present Uganda. Expanses of open water with mats of floating vegetation were part of the scene. Following the collision of Africa with Eurasia about 17 to 15 MYBP, connections were established that would have allowed its itchy-footed anthropoids to seek more agreeable surroundings. They made their way across the crumpled ancient Arabian plate to and beyond the rising Himalayan foothills—shortly, as geologists count, to become known from Spain to middle Europe, the Near East, Pakistan, India, and China.

Until recently it was thought that the step toward hominids and away from apes may have been taken as long as 17 MYBP. Creatures called *Ramapithecus* or *Sivapithecus* were widely favored as the earliest hominid. The relationships of these creatures, however, are now questioned. Their faces and thickly enameled molar teeth suggest affinity with hard-shelled fruit- and nut-eating orangutans rather than *Australopithecus*, which, chimpanzee-like, is considered to be a forager and eater of foliage. In addition, DNA hybridization and gene-sequencing experiments imply close relations among humankind, the chimpanzee, and the gorilla. Our australopithicine ancestors, it now seems, probably separated from the line of descent of the apes no more than perhaps 5 million years ago.

A small species of *Australopithecus* was already present in northern Kenya not long after the Mediterranean basin last filled and warm, earlier Neogene climates gave way to a cooling Pliocene and Pleistocene history. The continuing convergence of Africa and India on Eurasia created new geographic arrangements, elevated lands, and modified atmospheric and oceanic circulation—all affecting climate. The still-continuing elevation of the Himalaya, inferred from fossil plants and pollen, had bowed up its southern foothills to an altitude of about 3,100 meters by then, and they went up another 400 meters by the end of the Pleistocene. Similarly broad regions of vertical uplift in Africa were being generated by elevation associated with rifting. Climates of the region became mildly seasonal, with variable rainfall, forest-dotted grasslands, and limited areas of rain forest.

In the East African rift valleys, where hominid evolution seems to have centered, present climate is largely a product of opposing oceanic air masses. Winds from the Atlantic dump their moisture on the Congo, whereas the easterly trade winds carry little moisture and that only seasonally. As a result, East African vegetation is also seasonal and drought-adapted. Our early ancestors arose under similar but moister conditions that rewarded movement, endurance, and ingenuity. Survivors had to be good at fleeing danger, finding edible plant products or carrion, tracking game, and frequenting safe watering places. As long as they were arboreal, the scattered woodlands among which they lived required crossing open grasslands from one to another. That favored individuals having an upright stance, for better visibility in pursuit and escape.

There remain disagreements as to the circumstances of human origins, legitimate differences of opinion that will take more evidence and further research to resolve. It is remarkable, nevertheless, to find so nearly unanimous an agreement that the study of *Australopithecus* is, in fact, the study of emerging humankind. Although that has proved to be a turning point in the study of human evolution, it had little support from the anthropological cognoscenti until after the discoveries of Louis and Mary Leakey in the Olduvai Gorge of northern Tanzania in the 1930s, beginning a family tradition in the field. But it has been a hot subject in both science and the mass media since Mary Leakey's discovery there, in 1959, of a well-preserved adult *Australopithecus* skull. The resulting and continuing accumulation of new East African materials has illuminated human ancestry in quite unforseen ways.

It is clear now that there were two main kinds of *Australopithecus*, robust and vegetarian, contrasting with gracile and omnivorous. Neither attained heights above about 150 centimeters (5 feet). In Figure 17.16 the robust kind is represented by *A. bosei* and the gracile kind by *A. africanus*. Variations within those clans are called by different names with varying degrees of validity, but here, and in Table 17.1, *gracile* and *robust* will suffice. Oldest of all was an early member of the gracile clan, perhaps as old as 4 to 5 MYBP. Its first good records were reported by Mary Leakey from 3.7-MY-old strata at Laetoli, near Olduvai (Fig. 17.17A–C). It has been called *A. afarensis*, and its best-known member is the late, belatedly famous Lucy.

Lucy's fame is well deserved. Although barely the size of a pygmy chimpanzee,

her discovery by Donald Johanson in 1977 in approximately 3.3-MY-old sediments from a rift valley setting in southern Ethiopia, confirmed suggestions first made 22 years earlier by Robert Broom and Wilford Le Gros Clark that *Australopithecus* stood and walked upright like a human and unlike an ape. Lucy's 40 percent complete set of bones clarified that. Despite her lack of a skull, analysis of the areas of muscle attachment showed an upright, head-high, bipedal primate that was basically human in stance, gait, and skeletal morphology except for slightly apelike legs that would have shortened her stride. The step from knuckle-walking apekind to upright humankind had clearly been taken 3 million years or more ago. Further evidence of a bipedal stance and gait for contemporary australopithecines was added only two years later by Mary Leakey's discovery of humanoid footprints in 3.7-MY-old strata at Laetoli.

The importance of these discoveries is that an upright stance and gait frees the hands to carry things, to use and perhaps to make simple tools. It may have stimulated progress toward opposable thumbs. And these advances may have stimulated cerebral development by creating challenges and opportunities—to throw a stone or a club, to wield a digging stick effectively, to design and build a better rainy-night shelter or a camp. Hand and wrist bones from the same dig as the rest of Lucy permit the reconstruction of a very human-looking hand, the proportions of which suggest that Lucy and friends may already have acquired opposable thumbs. Her well-preserved pelvis, 12 percent smaller than that of a comparable chimpanzee, and similar to that of a more recent gracile *Australo-*

17.17 Discovery sites of ancient hominids. Scenes *A–C*, Olduvai Gorge, Louis Leakey at 1961 discovery site for *Homo habilis*, other bones. *D*, Zhoukoudian, where the bones of some forty individuals of Peking Man were excavated from cave-filling rubble at and below *8–9*.

pithecus from Sterkfontein, South Africa, was already evolving in the direction of female *Homo* and the birth of infants requiring extended maternal care.

To analyze when and by what evolutionary progression those things came about is to encounter once more the red herring of nomenclature. Lucy and kin seem quite clearly to have been on the direct line to later forms of gracile *Australopithecus*, such as *A. africanus*. The gracile and robust forms, moreover, were already distinct from one another at the time of the first known appearance of the latter about 2.5 MYBP. On an average the robust form was taller and heavier boned. It also had a larger cranial capacity, consistent with its larger size. Gracile and robust forms might have been considered female and male except that they seem not to be found at the same places. They also seem to have had different temporal ranges and dietary preferences—the robust forms with massive, crushing molars were evidently vegetarians whereas the gracile species was probably omnivorous. The robust species seems to have been a dead end. Quite likely it branched off from the gracile species at about the same time as a truly different species of gracile proportions began to show the growth of cranial capacity that bridges the gap from gracile *Australopithecus* to *Homo erectus* (Table 17.1).

That animal, whose few available skulls imply an average cranial capacity of about 690 cubic centimeters, is generally called *Homo habilis*, first known from 2.5-MY-old strata at Olduvai. It might as well be called *Australopithecus habilis*, for it has all the attributes one would expect in a species transitional from *Australopithecus* to *Homo*. It lived in the same African rift valleys as *A. africanus* until both became extinct about a million years ago, probably separated only by dietary preference and social habits. They must have seen one another and perhaps waved or gave warning on passing, but there is no evidence that they either mingled, shared, or warred.

The average cerebral capacity of this creature is midway between that of *A. africanus* and *H. erectus*, and its extremes of range touch both. The enlargement of Broca's and Wernicke's areas of the brain, shown by a well-preserved skull, are associated with language in the modern human brain. That permits the interpretation that *Homo habilis* may have been able to speak—to communicate, to explain to companions how to shape the already known stone tools, perhaps to build a fire.

The line blurs. If a conventionally attired *H. erectus* attempted to board your bus, the driver might think he was heavy browed and a bit flat-headed, but would probably let him on. In the case of *Homo habilis*, he might call the zoo.

H. erectus came onstage from the wings of a north Kenyan rift valley about 1.64 MYBP, and it ranged from there to eastern China before last seen about 300,000 years ago. Comprising a nearly complete skeleton, its oldest remains are those of a tall (168 centimeters or 5.5 feet) heavy-boned boy of about 12. Had he lived, he would have reached an adult height of about 1.8 meters (6 feet), half again the size of an average *Australopithecus*. *H. erectus*' range of cranial capacity overlapped that of *H. sapiens*. Its range in time and geography also overlapped that of *H. habilis*, and no other qualified immediate ancestors have been found in that heavily searched region. The evidence that the genus *Homo* evolved from *Australopithecus-like* ancestors in North Africa around 1.7 to 1.6 MYBP is now convincing.

H. erectus, like "Neanderthal Man," was as big as modern humans, had an ample cranial capacity, and had been making stone tools (Acheulian culture) for some time when its oldest discovered bones became fossils. There can be little doubt that it was using its brain in a conceptual way to direct agile limbs and opposable thumbs, whether in the use of digging sticks, at toolmaking, or seeking the fresh kills of predators to diversify its diet. Whether or not it also spoke during the course of such endeavors may have depended more on the development of its vocal cords than its brain, for the necessary neural level may already have been achieved by *H. habilis.* It has also been suggested, however, that even early *H. sapiens,* the subspecies *H. sapiens neanderthalensis,* was unable to employ articulate speech. Were that true, it might also explain why cultural development seems to have first become conspicuous following the arrival of anatomically modern *H. sapiens sapiens.*

During its long sojourn on this planet (about 1.3 MY) *H. erectus* left a more representative record than any of its predecessors. Scattered localities across northern Africa and southern Eurasia, eastward all the way to Java, have yielded the remains of some seventy individuals apart from those of eastern China, where eight discovery sites may double that number. The most important *H. erectus* discovery site, in fact, is in China, at Zhoukoudian (Choukoutien), or Dragon Bone Hill (Fig. 17.17D). The bones of forty individuals of both sexes and all ages, including six nearly complete skulls,

have been found there, where they have been called "Peking Man." These people probably used fire, and their temporal overlap with some variety of *H. sapiens* supports the claim that China could be the place where *Homo sapiens* originated. During some 230,000 years of continuous residence there, the Chinese *H. erectus* is reported to have registered a progressive increase with time of skull size and refinement of stone culture.

The apparently oldest record of *H. sapiens* so far, however, was presumably a Neanderthal, with a cranial capacity of about 1,400 cubic centimeters. That was at Vertesszollos, near Budapest, in 1965, in terrace deposits of the Danube River. The reported radiometric age of 350,000 years overlaps the recorded range of *H. erectus,* doubtfully represented in central Europe by the Mauer jaw, from near Vienna. The wide distribution of *H. erectus* would make it a suitable ancestor for either of the subspecies of *H. sapiens,* but another dilemma arises. Although Neanderthal Man has been regarded as a dead end, there is a big gap in time between the last *H. erectus* and the first unequivocal *H. sapiens sapiens—unless* the Vertesszollos bones had already reached that level.

No one knows whether modern humankind, as *Homo sapiens sapiens,* evolved directly from *H. erectus* or by way of the neanderthals, but the time ranges of this very incomplete record, shown in Table 17.1, support the latter. Within *H. sapiens,* the most consistent differences between subspecies are the elongate flat-topped skull, large protruding face, heavy brow ridges, and receding jaw of *neanderthalensis.* If one were to board your bus wearing blue jeans or a suit you would probably notice little difference from the other passengers. Slightly heavy-browed the boarding passenger might seem, and a bit bigheaded, but scarcely deserving the unflattering image so long evoked by early reconstructions.

Humans have been around, it now seems, from very early Pleistocene onward, while the habitat was evolving under the influence of the shaping processes described. They lived at first mainly in the East African rift valleys, later in Asia and Europe, and only during the last 32,000 to 40,000 years in the Americas and Australia. We see signs of their level of acculturation everywhere, but it is only since about 30 millennia ago that this reaches levels that can confidently be attributed to creative thought and effective communication.

The first unequivocal signs of conceptually as well as anatomically modern humans are those of the celebrated caves, rock shelters, and living sites of southern Europe, especially in Spain and France. They date from the early part of what archaeologists call late Paleolithic, ranging in Europe from about 28,000 to 10,000 years ago, just before and during the last glacial interval. Records from a good score of caves and rock shelters and some 70 open dwelling and burial sites include a diversity of culturally advanced portable implements, charms, personal ornaments, clay figurines (mainly exaggerated female forms), ornamented surfaces, and even simple flute-like and percussion instruments.

Some 600 colorful and lifelike images on the walls and roof of Lascaux cave in the Petits Causses region of France testify to the talent and sophistication of the people who lived there between about 27 to 22 millennia ago. Although these images do not portray depth, they and similar works at Altamira (northern Spain) and elsewhere are thematic, artistic works of high order. Their composition and commonly obscure locations suggest that they served symbolic and ritualistic purposes connected with tribal life, magic, fertility, and the hunt. Animals depicted, often pierced by weapons, imply that these people were probably seasonally migratory, hunting-gathering tribes, whose game included large periglacial mammals such as muskoxen and mammoths. Illustrations of traps and pitfalls suggest that such large animals were immobilized before killing. Dwelling clusters were located at favored lookout sites only seasonally colonized by their hunter-gatherer occupants.

This impressive artistry was that of Cro-Magnon Man, the oldest known records of high conceptualization. It was not equalled in imagination and refinement by anything so far known to have followed it up to the beginnings of agriculture, organized warfare, and city-states. Here one sees the earliest signs of modern culture, community life, and perhaps religion.

Coming to Terms with Nature: Some Reflections on the Future of Humankind

What bearing does the story of Earth's history have on present realities? Most basically that ignorance of the main elements of that history is unbecoming to cognitive humankind. A more practical and longer response might deal with the bearing of this history on the discovery and extraction of metals and other minerals products—with energy sources, construction materials, building sites, transportation, avoidance or control of natural hazards, and the waging of diplomacy, war, and peace. Central among these challenges, and steadily worsening with the continuing growth and congestion of human populations, are matters of natural hazard, the quality of life, and limitations to the quality of our existence.

17.18 FLOE ICE OF THE GREENLAND SEA SURROUNDS THE ICEBREAKER POLARBJORN. *[Courtesy of V. A. Squire; Scott Polar Research Institute.]*

Like the Cro-Magnons, we too must come to terms with nature, and the terms must suit the times. The dilemma of modern humankind in an already over-crowded world is symbolized by the little ship in Figure 17.18. Beset on all sides by great floes of ice, it looks hopelessly entrapped. Similarly beset in Antarctic pack ice in 1915, Sir Ernest Shackleton's ship *Endurance* drifted helplessly for 2,430 kilometers before being crushed and sunk in the Weddell Sea. Yet, although the tiny vessel here pictured is only 50 meters (160 feet) long, it has but to rev up its powerful engines, nudge the ice aside with its shielded hull, and go wherever its skipper cares to take it. For it was designed and built to cruise the ice pack, and it was there on purpose in mid-1983 to study the marginal ice zone above a seemingly "permanent" eddy in the northern Greenland Sea.

We see that there are or can be tech-nological fixes for many problems, depending on the technology of the times, and that former problems can be turned to advantages—as here the ice floe to the left of the vessel becomes a platform for measuring the movement of currents beneath the ice. The problems do not really go away; they are overcome, less-ened, or adapted to.

Impressive though the level to which the powers of *H. sapiens* have now risen, putting it in a class with the forces of nature, they pale before the leading powers of nature. Nothing mankind could have done would have stopped the destructive 1980 eruption of Mount St. Helens (Fig. 17.19) or will prevent a future eruption of Seattle's Mount Ranier, in the same volcanic chain. Only natural processes could stop the continued upward doming of Los Angeles' Palos

Verdes Hills. Nothing else could have prevented the sea from cutting the thir-teen terraces that track its succession of emergences and stillstands, or perhaps even the landsliding shown in the right foreground of Figure 17.20*A*. Nothing we yet do well would have prevented the slide (right side of Fig. 17.20*B*) that dumped 50 million cubic meters of rocky debris into the path of western Wyo-ming's Gros Ventre River in 1925, creat-ing an instant lake. And nothing we yet know how to do could have arrested the fall of the object that produced Arizona's 1.2-kilometer-wide Meteor Crater (Fig. 17.21) as recently, perhaps, as the arrival of the Paleoindians.

Nor could anything except plate tec-tonism or changes in Earth's orbit or solar intensity have stopped the advance of the Northern Hemisphere ice, whose limits are shown on Figure 17.2, or the filling and then evaporation of the exten-sive late-glacial lakes of Figure 17.6. Despite humankind's stunning suc-cesses in coping with its everyday physi-cal world, the powers of technology are limited.

Although the possibility of such events may seem remote in terms of everyday life, all have occurred here on Earth in the past and will repeat in the future—even asteroid impact. Before the popu-lations of the world became so large, so concentrated, and so established, how-ever, few were affected by all but the greatest of such events. Survivors sim-ply took note of casualties, perhaps migrated, and became footnotes to his-tory. The most powerful North American seismic activity of record, the 1811–1812 earthquakes at New Madrid in southeast Missouri, toppled a few log cabins, swamped or overturned boats on the

Mississippi, and rang church bells as far away as Boston. But few, if any, lives were lost. A shock of similar magnitude in the same region today would wreak havoc and death over a large part of the mid-continent.

Even in the once more sparsely populated world, nearly 4 million have died from the effects of recorded large earthquakes over the last millennium. And, lest we forget, new earthquakes remind us periodically of seismic hazards. The 1976 Tangshan (China) quake, magnitude 8.2 on the Richter scale, killed 240,000 by official count and is reported to have snuffed out thrice that number. The 1985 Mexico City earthquake killed and injured thousands, and caused 5 billion dollars in damage. A big earthquake along California's San Andreas fault is long overdue. It has been estimated that, even with modern building

17.19 BEFORE (*A*) AND AFTER (*B*) VIEWS OF 1980 ERUPTION OF MOUNT ST. HELENS FROM USGS COLDWATER II OBSERVATION SITE, 9 KM NORTH OF SUMMIT CRATER. White arrows indicate identical locations. [*Courtesy of Harry Glicken; United States Department of the Interior, U.S. Geological Survey, David A. Johnston Cascades Volcano Observatory, Vancouver, Washington.*]

codes in effect, an 8- to 8.5-magnitude quake along the southern end might kill as many as 14,000, injure another 55,000, and inflict 17 billion dollars worth of destruction.

The long-expected and deadly 1980 eruption of Mount St. Helens, although it cost some 60 lives in a sparsely settled area (including that of U.S. Geological Survey volcano-watcher David Johnston), was an inconspicuous event in the history of its kind. There are, in fact, at least thirteen "dormant" volcanoes in the Cascade Range that could be the sites of similar or greater eruptions, including Mount St. Helens itself, and, of course, Mount Ranier.

One is reminded by sobering accounts of recent eruptions near settled places that "dormant" and "extinct" are relative. A quarter of a million or more people have succumbed to processes

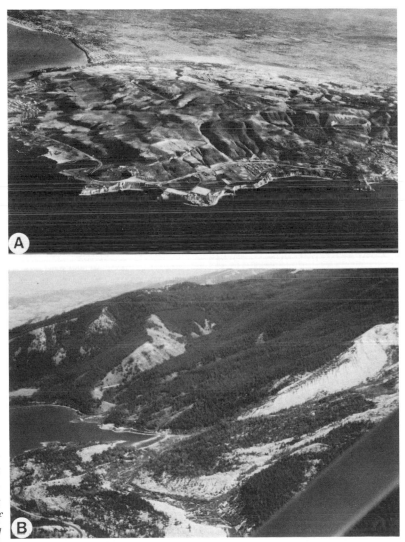

17.20 CONTEMPORARY SUR-
FACE MOVEMENTS IN
COASTAL SOUTHERN CALI-
FORNIA (A) AND WESTERN
WYOMING (B). [Courtesy of
John S. Shelton.]

associated with volcanic activity over the last 500 years (including volcanically generated mudflows and tsunamis). The only available preventive is care in the siting of human settlement. Broader ameliorative and monitoring measures that apply alike to volcanoes, seismic hazards, landsliding, foundation failure, outbursts of poisonous gases from lakes (and the sea), and floods involve appropriate land use and the enactment of legal implements that encourage it.

One of the potentially most disastrous hazards—after nuclear war, overpopulation, and perhaps wildfire—is a large cometary or asteroidal impact. Although statistically unlikely during any given human lifetime, such unwelcome visitors have left their calling cards repeatedly over geologic time (Fig. 13.8, *tier 4*; front endpaper). They arrive unpredictably, indiscriminately, inevitably, and unpreventably. Few are aware of the quantity of debris that reaches Earth's surface annually from outer space, most of it, happily, small and harmless. Fewer still might think of it as a source of damage to persons or property. Yet at least one person in living memory was struck by a 3.9-kilogram chunk of meteorite, fortunately after its velocity had been diminished by falling through the roof of a house and bouncing from an adjacent surface (in Sylacauga, Alabama, 29 November 1954). Computer modeling, based on statistical records for the United States, suggests that, worldwide, some sixteen buildings a year are damaged by meteorites.

But what might happen if a meteorite only the size of the small one that made Arizona's Meteor Crater (Fig. 17.21) were to land on the Kremlin or the Pentagon? Might that provoke a panic atomic response? How much more damage might a Manicougan-sized bolide (Fig. 15.1) inflict, not only in immediate panic-response and casualties, but through a meteoritic winter with consequences (save radiation) similar to and probably much greater than those predicted for a nuclear winter?

On a comparable but probably longer-lasting scale, dense sedentary populations that depend on complex urban support systems, a mechanized temperate agriculture, and efficient distribution would be severely stressed, if not disabled, by significant climatic change. Increases in global temperature attributable to the greenhouse effect resulting from the continuing emission of industrial carbon dioxide and other gases are predicted to amount to 5° to 7°C worldwide in another 50 years. Such a warming could disrupt current agricultural patterns and production levels. Were all the water now locked in glacial ice returned to the oceans, sea level would rise, drowning coastal cities. Although a disaster of that magnitude would presumably come on gradually and therefore give some warning, surges of warming polar ice into the oceans could speed the process. Evidence for such a surge and rapid rise of coastal sea level about 11,600 years ago, for instance, has been called upon to explain the widespread tradition of an ancient great flood, as in the biblical book of Genesis.

Present societies would find it traumatic to move "away" when hazard threatens. A northward shift of deserts and wheat belts would require agonizing policy reappraisals on the part of nations affected. Food shortages, including famines, could occur where there is now plenty, and now-unproductive northern lands could become grain belts. Whether or not it seems nifty to have the global bread basket in Siberia and deserts in Middle America depends on where one lives. And, of course, natural processes can reverse the order. The present interglacial climate could harshen, glaciers could again descend from the polar regions and the mountains, and the Northern Hemisphere could return to a more recognizably Pleistocene state within which the habitable regions shrink. Will industrial CO_2 remove that option? Or will it be balanced by a fading Sun or the buffering effects of the global sea and its capacity to absorb CO_2?

Geologists, meanwhile, are concerned that the material foundations of society must be maintained and supplemented from ever-leaner grades of ore, whose mining and processing requires ever-higher levels of energy and generates ever more serious environmental impacts.

I conclude that industrial societies need to become more cognizant of these and other natural threats to existence, security, and quality of life on this tiny planet. The pressures of still exponentially

17.21 METEOR CRATER, ARIZONA, 1.2 KM IN DIAMETER, 60 M DEEP. Note lobate and hummocky fallout debris blanket surrounding the crater and widening toward observer. To the rear is a deeply entrenched valley that acquired its meandering course on the flat surface of the Colorado Plateau before downcutting began. [From Geology Illustrated by John S. Shelton. Copyright © 1966 W. H. Freeman and Company. Used by permission.]

growing populations exacerbate inequity, generate unrest, create territorial disputes, and ultimately threaten war. Stabilization, followed by reasoned and democratically moderated reduction of populations generally and dispersal of excessively clustered populations particularly, would go far toward reducing hazard and improving the conditions of life. Sensible land-use planning would discourage, limit, or prohibit concentrations of people in the paths of potential floods, volcanic action, landslides, and other forseeable natural hazards, as well as construction on ground that is vulnerable to seismic shaking, liquefaction, and collapse. Good engineering-geology planning maps show where such places are. They should be utilized where available and made where not. Hazardous areas should be reserved for agriculture, parks, golf courses, greenbelts, and wilderness, and excluded from dense residential and industrial development.

Humanity has progressed far enough by now to appreciate the problems, to recognize limits to its mastery of nature, and to perceive where it must adapt.

Earth scientists and ecologists can recognize, define, and delimit many or most of the problems. Technology can solve some but not all. Students of human behavior, political science, and law must deal with the larger behavioral and regulatory problems. Politicians will become statesmen as they implement solutions that reach beyond their districts and terms of office. They might start by establishing qualified, full-time, national, state, and city *land use and natural hazards planning commissions* to study the problems, propose solutions, and recommend priorities. It is not difficult to identify cities that could have averted costly and injurious problems with the advice of such commissions, or which today are, or are rapidly becoming, disasters waiting to happen.

Humankind has come a long way over a geologically brief time from the rift valleys of Africa and ancestral species who, even earlier, may have seen the Mediterranean desert basin refill for the last time. Its path has been generally upward, its migrations far and resolute, its achievements heady, its follies distressing and sometimes appalling. It can do better. We know many of the problems and they are manageable. People alive today stand at the threshold of what could be, on average, a continuing ascent toward better lives for all or a descent into a new dark age.

Foresight and ameliorative action are needed. Although the potential is great, populations continue to increase at an alarming rate, and resources are as unevenly distributed and disbursed as ever. "Winning" is widely valued above honor and even personal and public welfare. And pride, prejudice, bigotry, and narrow nationalism thwart resolution of the threatening problems before us. We must do better.

How can our species most effectively come to terms with nature and itself? What positive steps may be taken? Our actions and inactions will set the tone for the yet-unfinished symphony. If the qualities we know as human are to prevail and reach all corners of this planet, concessions must be made—by people toward nature and one another. Boldness in technology and human affairs is to be encouraged but monitored. Boldness must be informed. And, being informed, it should also be balanced with prudence and compassion.

In Summary

The last 1,600 millennia, comprising the Pleistocene Epoch, saw the genus *Homo* emerge from protohuman *Australopithecus*, evolve into the anatomically modern human animal, spread over most of the world, and become culturally and technologically modern as well. That progression was, as all life is, a product of heredity responding to its environment. The human habitat is the work of external processes that reshaped the late Neogene surface into the one on which we live.

Water as liquid, solid, and vapor was the main reshaping agent and ice its most dramatic form. The many roles of water are seen most clearly in the deep Southern Hemisphere. That ubiquitous solvent, chilled by Antarctic ice, sinks at its convergence with warmer surface waters from the north, there to flow northward to and beyond the equator, moderating surface temperatures. En route it contributes to nutrient-rich upwelling currents that account for major fisheries and phosphate deposition, particularly along leeward coasts. The circum-Antarctic current and its branches deliver moisture to the southeast trade winds, which carry it equatorward. Thus do moisture-bearing winds join with cold bottom currents, glaciation, and geography in governing weather and climate. Similar but more complicated processes go on in the northern polar regions.

Water in its liquid state is the main agent of weathering, erosion, and surface failure. It supplements and intensifies landsliding, explosive volcanism, and seismic shaking.

Volcanism and seismicity, however, have more evident and immediate effects. They strike, often with little warning, mainly along plate boundaries, and commonly after long, supposedly inactive intervals. They destroy property, kill, trigger landslides, initiate glacial surges, and change geography.

Geography, solar intensity and angle of incidence, and oceanic and atmospheric circulation account for the physical state and distribution of water, while climate, varying with latitude and altitude, determines what lives where and when. Glaciation repeatedly extended into now-temperate middle latitudes of the Northern Hemisphere and just as often gave way to interglacial climates as warm as or warmer than today's generally clement conditions. A similar succession of glacial and interglacial stages is reflected in the more elevated Southern Hemisphere mountains. Pleistocene vegetation and contemporaneous animals migrated south and north, down and up, ahead of glacial fronts. Tundra and muskoxen that occupied Nebraska and central Europe during earlier glacial times, for example, now thrive far to the north.

Sea level fell and rose, in response to the growth and decline of glaciation, through a total range of about 140 meters (460 feet). These fluctuations were only partly balanced by the extension of coastal plains where other habitable space gave way to glaciation. The resulting emergence, then flooding, of land bridges controlled Pleistocene migration routes between some large landmasses.

Human evolution was the centerpiece of Pleistocene history. Already in Pliocene time a slender race of australopithecines had taken the first big step toward humanity with the adoption of a bipedal gait and a nearly or quite upright posture. That freed hands to do other things, leading to opposable thumbs, fully upright posture, and increases in body and brain size. Ancestral *Homo erectus* is first known from 1.64-MY-old deposits of an equatorial East African rift valley. It crossed from Africa to Eurasia when connection was established and eventually traveled all the way to Java, leaving the largest-known number of skeletal remains in eastern China.

Homo sapiens, as the subspecies often called Neanderthal Man, first appeared in Europe or maybe China, perhaps 300,000 or more years ago, but the first records of anatomically modern humans are European. A justly renowned level of material culture and craftsmanship was achieved by cave-frequenting Cro-Magnon tribes of that subspecies (*H. sapiens sapiens*) in southern France and northern Spain about 28 to 18 millennia ago. A pinnacle had been reached. Art, music, and tribal ritual took shape.

Descendants of those people eventually made their way to Australia, Oceania, and the Americas, where they have been held to blame for continuing extinctions of other animals on a tragic level. Climatic change following the last Pleistocene glaciation 12 to 10 millennia ago, however, probably deserves a fair share of blame in some regions.

Despite the successes achieved by modern technology in banishing or ameliorating natural hazards and obstacles, problems in that field remain. While gaining the unprecedented convenience and frequency of contemporary travel, modern humans have lost much of the flexibility of their nomadic forerunners in moving from the paths of danger, escaping congestion, or resettling on more favorable terrain. Many of the hazards and limitations could be lessened if populations were fewer and less congested. More, and more extensive natural reserves are needed to protect or preserve other forms of life and distinctive habitats—to maintain or reestablish the balance best assured by diversity.

Good management of our oasis in space and the human estate calls for appropriate conservation of resources, restoration of degraded environments, and avoidance or amelioration of natural hazards. Stabilization, then reduction, of populations is needed, followed eventually by the achievement of a balance among population size, density, location, resources, and needs consistent with a globally high quality of life.

EPILOGUE

Nearly 5 billion years of Earth history have here passed in review. Before that, allow another 10 to 15 billion years of cosmic history back to the Big Bang. One cannot reflect on such a history and the probability that another 100 billion years may pass before the next big bang without wondering how long our planet may last. Will it last until the Sun dies, perhaps 5 billion years from now? Will it freeze or burn? Will we have descendants on this planet when the crunch comes? Will kinfolk live on other planets, or will surviving Earthlings be alone in the galaxy and perhaps the universe? How will such kinfolk (if any) look, think, and act? How will they communicate? Will their numbers be tolerable or oppressive? Will they have survived by brains or brawn, by compassion or greed? Will their extinction be a tragedy or an act of grace?

We can only think about such questions. But because we can think, and can forsee the consequences of our actions and inactions, we owe it to our descendants to reflect deeply on such matters and to make a start at whatever providential moves are now feasible to assure lives of high quality for all while life endures. I have written my share about such matters elsewhere. I invite the reader to her or his own informed reflections and decisions.

If you ask, as thinking young people sometimes do, "What is the purpose of life?" I would propose a goal within reach of all: to live it with as much grace and integrity as possible, to enjoy and improve it while you have it, and to leave the world no worse for your having been there. You alone can choose the course.

APPENDIX

SUPPLEMENTAL

READING

CHAPTER ONE

Beatty, J. K., O'Leary, V., and Chaikin, A., eds. 1981. *The New Solar System*. Cambridge, Mass.: Sky Publishing. 224 pp.

Glass, B. P. 1982. *Introduction to Planetary Geology*. London / New York: Cambridge University Press. 469 pp.

Harrison, E. R. 1981. *Cosmology*. London / New York: Cambridge University Press. 430 pp.

Kaufmann, W. J., III. 1985. *Universe*. New York: W. H. Freeman. 594 pp.

Wasson, J. T. 1985. *Meteorites: Their Record of Early Solar-System History*. New York: W. H. Freeman. 267 pp.

Weinberg, Steven. 1977. *The First Three Minutes*. New York: Basic Books. 177 pp.

CHAPTER TWO

Brandt, J. C., ed. 1981. *Comets: Readings from Scientific American*. New York: W. H. Freeman. 92 pp.

Holland, H. D. 1984. *The Chemical Evolution of the Atmosphere and Oceans*. Princeton, N.J.: Princeton University Press. 582 pp.

Ringwood, A. E. 1979. *Origin of the Earth and Moon*. New York / Berlin: Springer Verlag. 295 pp.

Walker, J. C. G. 1977. *Evolution of the Atmosphere*. New York: Macmillan; London: Collier Macmillan. 318 pp.

Wetherill, G. W. 1972. "The Beginning of Continental Evolution." In *The Upper Mantle*, ed. A. R. Ritsema, 31–46. Amsterdam / New York: Elsevier. 637 pp.

CHAPTER THREE

Albritton, C. C., Jr. 1980. *The Abyss of Time: Changing Conceptions of Earth's Antiquity after the Sixteenth Century*. San Francisco: Freeman Cooper. 251 pp.

Berry, W. B. N. 1968. *Growth of a Prehistoric Time Scale*. New York: W. H. Freeman. 158 pp.

Hallam, Anthony. 1983. *Great Geological Controversies*. London / New York: Oxford University Press. 182 pp.

Reinecke, H. E., and Singh, I. B. 1980. *Depositional Sedimentary Environments*. New York / Berlin: Springer-Verlag. 549 pp.

Rupke, N. A. 1983. *The Great Chain of History: William Buckland and the English School of Geology*. London / New York: Oxford University Press. 322 pp.

Steno, Nicholas. [1669] 1968. *The Prodromus*. Trans. J. G. Winter, with Introduction by G. W. White. New York: Macmillan. 283 pp.

APPENDIX

CHAPTER FOUR

Dalrymple, G. B. 1984. "How Old Is the Earth: A Reply to 'Scientific creationism.' " In *Evolutionists Confront Creationists*, eds. Frank Awbrey and W. M. Thwaites. American Association for the Advancement of Science, Pacific Division, Proceedings of the 63d Annual Meeting, v. 1, pt. 3, 54–118.

Faure, Gunter. 1977. *Principles of Isotope Geology.* New York: Wiley. 464 pp.

Harper, C. T., ed. 1973. *Geochronology: Radiometric Dating of Rocks and Minerals.* Stroudsburg, Pa.: Dowden, Hutchinson and Ross. 469 pp.

Landes, D. S. 1983. *Revolution in Time: Clocks and the Making of the Modern World.* Cambridge, Mass.: Harvard University Press. 482 pp.

Whitrow, G. J. 1980. *The Natural Philosophy of Time.* 2d ed. Oxford: Clarendon Press. 399 pp.

CHAPTER FIVE

Ager, D. V. 1981. *The Nature of the Stratigraphical Record.* 2d ed. New York: Halsted Press. 122 pp.

Cox, Allan, Dalrymple, G. B., and Doell, R. R. 1967. "Reversals of the Earth's Magnetic Field." *Scientific American*, 216(2):44–54.

Harland, W. B., et al. 1982. *A Geologic Time Scale.* London / New York: Cambridge University Press. 131 pp.

Kauffman, E. G., and Hazel, J. E., eds. 1977. *Concepts and Methods in Biostratigraphy.* Stroudsburg, Pa.: Dowden, Hutchinson and Ross. 658 pp.

Simpson, G. G. 1983. *Fossils and the History of Life.* New York: W. H. Freeman. 239 pp.

CHAPTER SIX

McCall, G. J. H., ed. 1977. *The Archean: Search for the Beginning.* Stroudsburg, Pa.: Dowden, Hutchinson and Ross. 505 pp. (See chapters by Wetherill, Green, Sutton and Watson, Windley and Bridgwater, Watson, and Windley.)

Moorbath, S. 1977. "The Oldest Rocks and the Growth of Continents." *Scientific American*, 236(3):92–104.

Morey, G. B., and Hanson, G. N., eds. 1980. *Selected Studies of Archean Gneisses and Lower Proterozoic Rocks of the Southern Canadian Shield.* Geological Society of America, Special Paper 182, 175 pp.

Ritsema, A. R., ed. 1972. *The Upper Mantle.* Amsterdam / New York: Elsevier. 644 pp.

Windley, B. J. 1984. *The Evolving Continents.* 2d ed. New York: Wiley. 399 pp.

CHAPTER SEVEN

Condie, K. C. 1981. *Archean Greenstone Belts.* Amsterdam / New York: Elsevier. 434 pp.

Glover, J. E., ed. 1971. *Symposium on Archean Rocks.* Geological Society of Australia, Inc., Special Publication 3, 569 pp.

Hunter, D. R., ed. 1981. *Precambrian of the Southern Hemisphere.* Amsterdam / New York: Elsevier. 882 pp. (See chapters by Hallberg and Glickson and by Anhaeusser and Wilson.)

Kröner, A., Hanson, G. N., and Goodwin, A. N., eds. 1984. *Archean Geochemistry.* New York / Berlin: Springer-Verlag. 286 pp.

Tankard, A. J., et al. 1982. *Crustal Evolution of Southern Africa: 3.8 Billion Years of Earth History.* New York / Berlin: Springer-Verlag. 524 pp.

Windley, B. F. 1976. *The Early History of the Earth.* New York: Wiley-Interscience. 619 pp.

CHAPTER EIGHT

Button, Andrew, et al. 1981. "The Cratonic Environment." In *Precambrian of the Southern Hemisphere*, ed. D. R. Hunter, 501–639. Amsterdam / New York: Elsevier. 882 pp.

O'Nions, R. K., Hamilton, P. J., and Evensen, N. M. 1980. "The Chemical Evolution of the Earth's Mantle." *Scientific American*, 242(5):91–101.

Pretorius, D. A. 1975. "The Depositional Environment of the Witwatersrand Goldfields." *Minerals Science and Engineering*, 7:18–47.

Taylor, S. R., and McLennan, S. M. 1985. *The Continental Crust: Its Composition and Evolution.* Oxford: Blackwell Scientific Publications. 312 pp.

Veizer, Ján. 1983. "Geologic Evolution of the Archean–Early Proterozoic Earth." In *Earth's Earliest Biosphere*, ed. J. W. Schopf, 240–258. Princeton, N.J.: Princeton University Press. 543 pp.

CHAPTER NINE

Hamilton, Warren. 1977. "Plate Tectonics and Man." *U.S. Geological Survey Annual Report* FY 1976, 39–53.

Jones, D. L., Cox, Alan, Coney, P. J., and Beck, M. 1982. "Growth of Western North America." *Scientific American*, 247(5):70–84.

Kröner, A., ed. 1981. *Precambrian Plate Tectonics.* Amsterdam / New York: Elsevier. 781 pp.

Marvin, U. B. 1973. *Continental Drift: The Evolution of a Concept.* Washington, D.C.: Smithsonian. 239 pp.

Wegener, A. L. [1929] 1966. *The Origin of Continents and Oceans.* Trans. John Biram. New York: Dover. 246 pp.

Wyllie, P. J. 1976. *The Way the Earth Works.* New York: Wiley. 296 pp.

CHAPTER TEN

Campbell, F. H. A., ed. 1981. *Proterozoic Basins of Canada.* Geological Survey of Canada, Paper 81–10, 444 pp.

Holland, H. D. 1984. *The Chemical Evolution of the Atmosphere and Oceans.* Princeton, N.J.: Princeton University Press. 582 pp.

Holland, H. D., and Schidlowski, M., eds. 1982. *Mineral Deposits and the Evolution of the Biosphere.* New York / Berlin: Springer-Verlag. 332 pp.

Ponnamperuma, Cyril, ed. 1983. *Cosmochemistry and the Origins of Life.* Dordrecht: Reidel. 386 pp.

Schopf, J. W., ed. 1983. *Earth's Earliest Biosphere: Its Origin and Evolution.* Princeton, N.J.: Princeton University Press. 543 pp.

Trendall, A. F., and Morris, R. C., eds. 1983. *Iron-formation: Facts and Problems.* Amsterdam / New York: Elsevier. 558 pp.

CHAPTER ELEVEN

Campbell, F. H. A., ed. 1981. *Proterozoic Basins of Canada.* Geological Survey of Canada, Paper 81–10, 444 pp.

Margulis, Lynn. 1981. *Symbiosis in Cell Evolution.* New York: W. H. Freeman. 419 pp.

Medaris, L. G., Jr., Byers, C. W., Mickelson, D. M., and Shanks, W. C., eds. 1983. *Proterozoic Geology: Selected Papers from an International Proterozoic Symposium.* Geological Society of America, Memoir 161, 315 pp.

Punukollu, S. N., and Andrews-Speed, C. P., eds. 1984. *Proterozoic—Evolution, Mineralisation, and Orogenesis. Precambrian Research*, Special Issue, 25(1–3). 348 pp.

Trompette, R., and Young, G. M., eds. 1981. *Upper Precambrian Correlations. Precambrian Research*, Special Issue, 15(3–4):185–422.

Walter, M. R., ed. 1976. *Stromatolites.* Amsterdam / New York: Elsevier. 790 pp.

APPENDIX

CHAPTER TWELVE

Chumakov, N. M. 1981. "Upper Proterozoic Glaciogenic Rocks and Their Stratigraphic Significance." *Precambrian Research*, 15:373–395.

Crowell, J. C. 1983. "Ice Ages Recorded on Gondwanan Continents." *Transactions of the Geological Society of South Africa*, 86:237–262.

Frakes, L. A. 1979. *Climates Throughout Geologic Time*. Amsterdam / New York: Elsevier. 310 pp.

Hambrey, M. J., and Harland, W. B., eds. 1981. *Earth's Pre-Pleistocene Glacial Record*. London / New York: Cambridge University Press. 1,004 pp.

John, Brian. 1979. *The World of Ice*. London: Orbis Publishing. 120 pp.

Spencer, A. M. 1971. *Late pre-Cambrian Glaciation in Scotland*. Memoirs of the Geological Society of London 6, 98 pp.

CHAPTER THIRTEEN

Cloud, Preston. 1968. "Pre-Metazoan Evolution and the Origins of the Metazoa." In *Evolution and Environment*, ed. E. T. Drake, 1–72. New Haven, Conn.: Yale University Press. 470 pp.

Cloud, Preston, and Glaessner, M. F. 1982. "The Ediacarian Period and System." *Science*, 217(4652):783–792.

Conway-Morris, S., and Whittington, H. B. 1979. "The Animals of the Burgess Shale." *Scientific American*, 241(1):122–133.

Encyclopaedia Britannica. Current Edition. Articles on the Paleozoic Systems.

Gray, Jane, Boucot, A. J., and Berry, W. B. N., eds. 1981. *Communities of the Past*. New York: Van Nostrand Reinhold. 640 pp.

Taylor, T. N. 1981. *Paleobotany*. New York: McGraw-Hill. 589 pp.

Ziegler, Bernard. [1980] 1983. *Introduction to Paleobiology: General Paleontology*. Trans. R. Muir. Chichester, Eng.: Ellis Horwood. 225 pp.

CHAPTER FOURTEEN

Colbert, E. H. 1976. "When Reptiles Ruled." In *Our Continent*, ed. E. H. Colbert, 73–97. Washington, D.C.: National Geographic. 398 pp.

Colbert, E. H. 1981. *Evolution of the Vertebrates*. 3d ed. New York: Wiley-Interscience. 510 pp.

Encyclopaedia Britannica. Current Edition. Articles on Mesozoic, and on the Triassic, Jurassic, and Cretaceous.

Moullade, M., and Nairn, A. E. M., eds. 1978, 1983. *The Phanerozoic Geology of the World II: The Mesozoic, A and B*. Amsterdam / New York: Elsevier. 529 pp. and 450 pp.

Ostrom, J. H. 1979. "Bird Flight: How Did It Begin?" *American Scientist*, 67(1):46–56.

Thomas, R. D. K., and Olson, E. C., eds. 1980. *A Cold Look at the Warm Blooded Dinosaurs*. Boulder, Colo.: Westview Press. 514 pp.

CHAPTER FIFTEEN

Darlington, P. J. 1980. *Evolution for Naturalists: The Simple Principles and Complex Realities*. New York: Wiley-Interscience. 262 pp.

Endler, J. A. 1986. *Natural Selection in the Wild*. Princeton, N.J.: Princeton University Press. 336 pp.

Gould, S. J. 1977. *Ontogeny and Phylogeny*. Cambridge, Mass.: Harvard University Press. 501 pp.

Laporte, L. F., ed. 1982. *The Fossil Record and Evolution*. New York: W. H. Freeman. 225 pp.

Miller, Jonathan, and van Loon, Borin. 1982. *Darwin for Beginners*. New York: Pantheon Books. 176 pp.

Nitecki, M. H., ed. 1984. *Extinctions*. Chicago: University of Chicago Press. 354 pp.

Wilford, John N. 1985. *The Riddle of the Dinosaur*. New York: Knopf. 304 pp.

CHAPTER SIXTEEN

Berggren, W. A., and Van Couvering, J. A. 1974. "The Late Neogene: Biostratigraphy, Geochronology, and Paleoclimatology of the Last 15 Million Years in Marine and Continental Sequences."
Palaeogeography, Palaeoclimatology, and Palaeoecology, 16(1–2). 216 pp.

Berger, W. H., and Crowell, J. C., eds. 1982. *Climate in Earth History*. Washington, D.C.: National
Academy Press. 198 pp.

Curry, Dennis, et al. 1978. "A Correlation of Tertiary Rocks in the British Isles." *Geological Society of
London, Special Report 12*. 72 pp.

Encyclopaedia Britannica. Current Edition. Articles on Cenozoic, Mammalia, and Tertiary.

Hsu, K. J. 1983. *The Mediterranean Was a Desert*. Princeton, N.J.: Princeton University Press. 197 pp.

Kennett, J. P. 1982. *Marine Geology*. Englewood Cliffs, N.J.: Prentice-Hall. 813 pp.

Ostrom, J. H. 1976. "The Triumph of the Mammals." In *Our Continent*, ed. E. H. Colbert, 117–141.
Washington, D.C.: National Geographic. 398 pp.

CHAPTER SEVENTEEN

Büdel, Julius. [1977] 1982. *Climatic Geomorphology*. Trans. Leonore Fischer and Detlev Busche.
Princeton, N.J.: Princeton University Press. 443 pp.

Colbert, E. H., ed. 1976. *Our Continent: A Natural History of North America*. Washington, D.C.:
National Geographic. 398 pp. (Especially p. 141ff. with a chapter by Björn Kurten on Pleistocene life and chapters on the natural regions of North America.)

Denton, G. H., and Hughes, J. T. 1981. *The Last Great Ice Sheets*. New York: Wiley-Interscience. 489
pp.

Hunt, C. B. 1974. *Natural Regions of the United States and Canada*. New York: W. H. Freeman. 725
pp.

Leakey, R. E., and Lewin, Roger. 1977. *Origins*. London: Macdonald and Jane's. 264 pp.

Martin, P. S., and Klein, R. G. 1984. *Quaternary Extinctions: A Prehistoric Revolution*. Tucson: University of Arizona Press. 892 pp.

Shelton, J. S. 1966. *Geology Illustrated*. New York: W. H. Freeman. 434 pp.

Wood, B., Martin, L., and Andrews, P., eds. 1986. *Major Topics in Primate and Human Evolution*.
London / New York: Cambridge University Press. 364 pp.

INDEX

INDEX

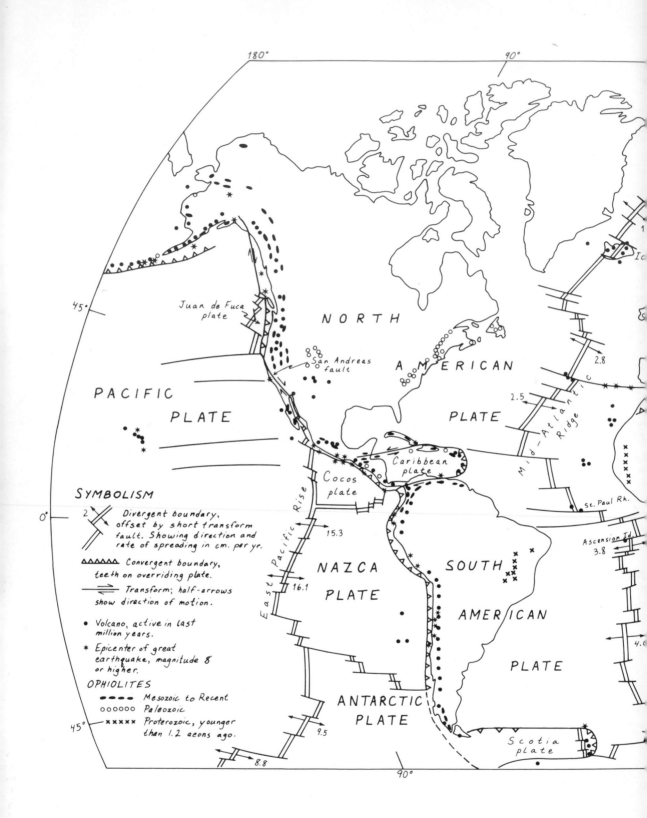

SYMBOLISM

2 ⤢ Divergent boundary, offset by short transform fault. Showing direction and rate of spreading in cm. per yr.

△△△△△△ Convergent boundary, teeth on overriding plate.

⇄ Transform; half-arrows show direction of motion.

• Volcano, active in last million years.

✳ Epicenter of great earthquake, magnitude 8 or higher.

OPHIOLITES

••••• Mesozoic to Recent

ooooooo Paleozoic

xxxxx Proterozoic, younger than 1.2 aeons ago.

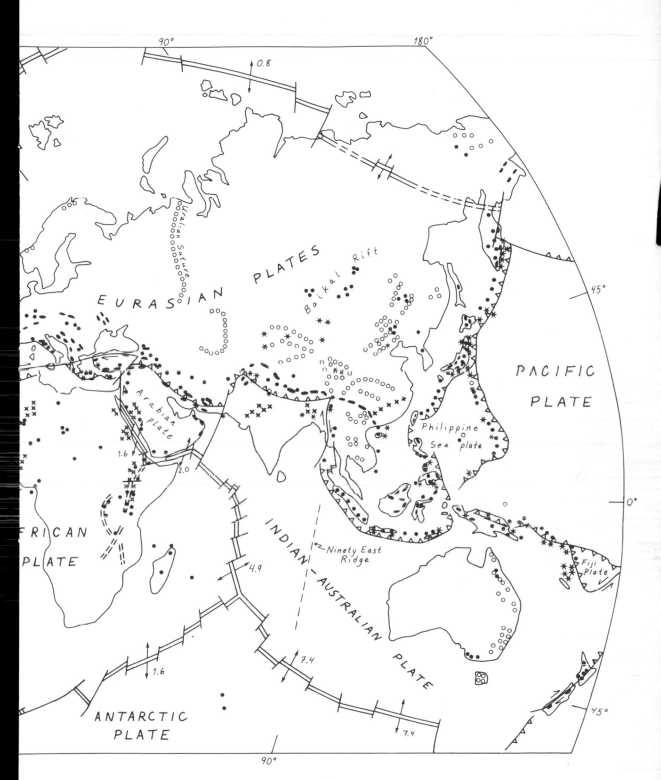